T0186529

ANTONI ZYGMUND was born in Warsaw on December 26, 1900. His early interests were in science, especially astronomy, and he planned to become an astronomer. In 1919 Zygmund entered the University of Warsaw, which had become Polish in 1915. The university was in the initial stages of development, and the teaching of astronomy was still being organized. This situation, together with the interest in mathematical problems he had developed through his independent study of astronomy, pushed him toward pure mathematics.

There had emerged in Poland a group of talented young mathematicians who were to develop what was later known as the Polish Mathematical School. Zygmund became caught up in this mathematical fervor. In addition, he was influenced by two of his older schoolmates, Stanislaw Saks and Zygmunt Zalcwasser. Sak's *Analytic Functions,* written together with Zygmund, is still considered one of the best books on the topic. Alexander Rajchman, then an instructor at the University of Warsaw, had made important contributions to the theory of formal multiplication and uniqueness of trigonometric series, and Zygmund became particularly interested in the problems of this area. One may thus say that Zygmund was primarily a disciple of Saks, Zalcwasser, and Rajchman.

Zygmund obtained his doctorate in 1923. A year eariler he had been appointed instructor at the Polytechnic School in Warsaw. In 1926 he was appointed docent at the University of Warsaw and was also awarded a Rockefeller Fellowship to spend a year in England with Hardy in Oxford and Littlewood in Cambridge. Unexpectedly, he encountered there one more mathematician who influenced him very deeply: R. E. A. C. Paley.

In 1930 Zygmund was appointed professor at the University of Wilno. There he encountered Jozef Marcinkiewicz, who became first his student and then his collaborator. At the outbreak of World War II, both Marcinkiewicz and Zygmund were mobilized as reserve officers in the Polish army. Marcinkiewicz was taken prisoner in the East and disappeared without trace. Zygmund succeeded in returning to Wilno to his wife and son. Through the help of friends, he managed to leave Poland and find his way to the United States early in 1940, where he recieved an appointment at Mount Holyoke College in Massachusetts.

In 1945 he moved to the University of Pennsylvania, and two years later he was invited by Professor Marshall Stone to join the Department of Mathematics of the University of Chicago. A most impressive group of outstanding mathematicians was to gather there--A. A. Albert, S. S.Chern, S. MacLane, L. M. graves, A. Weil, I. Kaplansky, P. Halmos, I. Segal, and E. Spainer among them. Zygmund remained at the University of Chicago for the rst of his career; he retired in 1980 as the Gustave and Ann Swift Distinguished Service Professor of Mathematics. During these years he developed what has come to be known as the Chicago School of Analysis. The mathematical influence Zygmund exerted through his work, his teaching, and his students has spread across the United States and beyond--to South America and Europe.

Professor Zygmund has recieved many honors and held many important positions in the mathematical community. He holds honorary degrees from Washington University, the University of Torun in Poland, the University of Paris, and the University of Uppsala in Sweden. The learned societies of which he is a member include the National Academy of Sciences of the United States; the Polish Academy of Sciences; the National Academy of Sciences of Buenos Aires, Argentina; the Royal Academy of Sciences of Madrid; and the National Academy of Sciences of Palermo, Italy.

CONFERENCE
ON
HARMONIC ANALYSIS

THE WADSWORTH MATHEMATICS SERIES

Series Editors

Raoul H. Bott, *Harvard University*

David Eisenbud, *Brandeis University*

Hugh L. Montgomery, *University of Michigan*

Paul J. Sally, Jr., *University of Chicago*

Barry Simon, *California Institute of Technology*

Richard P. Stanley, *Massachusetts Institute of Technology*

W. Beckner, A. Calderón, R. Fefferman, P. Jones, *Conference on Harmonic Analysis in Honor of Antoni Zygmund*

M. Behzad, G. Chartrand, L. Lesniak-Foster, *Graphs and Digraphs*

J. Cochran, *Applied Mathematics: Principles, Techniques, and Applications*

A. Garsia, *Topics in Almost Everywhere Convergence*

K. Stromberg, *An Introduction to Classical Real Analysis*

CONFERENCE ON HARMONIC ANALYSIS

IN HONOR OF ANTONI ZYGMUND

VOLUME II

Edited by

William Beckner
Alberto P. Calderón
Robert Fefferman
Peter W. Jones

University of Chicago

CRC Press
Taylor & Francis Group
Boca Raton London New York

CRC Press is an imprint of the
Taylor & Francis Group, an **informa** business

CRC Press
Taylor & Francis Group
6000 Broken Sound Parkway NW, Suite 300
Boca Raton, FL 33487-2742

© 1983 by Taylor & Francis Group, LLC
CRC Press is an imprint of Taylor & Francis Group, an Informa business

No claim to original U.S. Government works

Visit the Taylor & Francis Web site at
http://www.taylorandfrancis.com

and the CRC Press Web site at
http://www.crcpress.com

These proceedings are dedicated to
Professor Antoni Zygmund
in celebration of his eightieth birthday.

Antoni Zygmund

PREFACE

Between March 23 and 28, 1981, some two hundred mathematicians
gathered at the University of Chicago to honor Professor Antoni
Zygmund on the occasion of his eightieth birthday. It was a public
tribute to the undisputed highest authority in Fourier series and
one of the foremost analysts in the world.

Ever since the publication of its first edition in 1935,
Professor Zygmund's famous *Trigonometric Series* has been one of
the most influential books in the modern mathematical literature,
making a permanent imprint on the mathematical work and outlook of
many of the most distinguished present-day analysts.

His extensive research has covered many topics in Fourier
series, real variables, and related fields. Many of his results—
for example, those in the fine theory of differentiation, interpolation of linear operations, and singular integrals—now occupy a
permanent and central place in analysis and have deeply influenced
the modern development of other important areas of analysis, such
as the theory of partial differential equations.

Professor Zygmund has also been an extraordinary teacher. He
has had a large number of students—many of whom became top-ranking
mathematicians themselves—and has had a profound influence in the
development of analysis in several countries, including the United
States, France, Argentina, Spain, and, of course, his native Poland. It is largely because of this influence that the United
States has become the leader in the mathematical areas he culti-
vates.

These proceedings reflect the impact of Professor Zygmund's
work and teaching. Most of the contributions to the conference are
on topics he has either initiated or permanently shaped.

The Editors

CONTENTS

PART ONE: INTRODUCTORY PAPER   1

E. M. Stein/*The development of square functions in the work of
     A. Zygmund*   2

PART TWO: TRIGONOMETRIC SERIES   31

W. Beckner, S. Janson, and D. S. Jerison/*Convolution inequalities
     on the circle*   32

G. Gasper/*A convolution structure and positivity of a generalized
     translation operator for the continuous q-Jacobi poly-
     nomials*   44

G. Halász and H. L. Montgomery/*Bernstein's inequality for finite
     intervals*   60

J. -P. Kahane/*Slow points of Gaussian processes*   67

S. K. Pichorides/*Notes on trigonometric polynomials*   84

M. H. Taibleson and G. Weiss/*Certain function spaces connected
     with almost everywhere convergence of Fourier series*
     95

M. Zafran/*Exponential estimates in multiplier algebras*   114

PART THREE: FOURIER ANALYSIS ON Rⁿ AND REAL ANALYSIS   **123**

A. E. Gatto, C. E. Gutiérrez, and R. L. Wheeden/*On weighted fractional integrals*   124

R. R. Gundy/*The density of the area integral*   138

R. A. Hunt and D. S. Kurtz/*The Hardy-Littlewood maximal function on L(p, 1)*   150

R. A. Hunt, D. S. Kurtz, and C. J. Neugebauer/*A note on the equivalence of* A$_p$ *and Sawyer's condition for equal weights*   156

M. Jodeit, Jr./*On the decomposition of* L$_1^1$(R) *functions into humps*   159

B. Muckenhoupt/*On inequalities of Carleson and Hunt*   179

A. Ruiz/*On the restriction of Fourier transforms to curves*   186

T. Walsh/*Minimal smoothness for a bound on the Fourier transform of a surface measure*   213

PART FOUR: SINGULAR INTEGRALS AND PSEUDODIFFERENTIAL
             OPERATORS   **257**

R. Arocena and M. Cotlar/*A generalized Herglotz-Bochner theorem and* L^2*-weighted inequalities with finite measures*   258

D. L. Burkholder/*A geometric condition that implies the existence of certain singular integrals of Banach-space-valued functions*   270

R. R. Coifman, A. McIntosh, and Y. Meyer/*Estimations* L^2 *pour les noyaux singuliers*   287

A. Córdoba/*Vector valued inequalities for multipliers*   295

M. Cotlar and C. Sadosky/*On some* L^p *versions of the Helson-Szegö theorem*   306

R. Fefferman/*Some topics in Calderón-Zygmund theory*   318

D. H. Phong and E. M. Stein/*Singular integrals with kernels of mixed homogeneities*   327 x   Contents

D. F. Shea and S. Wainger/*An application of singular integrals to a growth problem for entire functions of finite order* 340

PART FIVE: HARDY SPACES   347

L. Carleson/*On H^∞ in multiply connected domains*   349

S. -Y. A. Chang/*Two remarks about H^1 and BMO on the bidisc*   373

A. E. Gatto, J. R. Jiménez, and C. Segovia/*On the solution of the equation $\Delta^m F = f$ for $f \in H^p$*   394

J. E. Gilbert, R. A. Kunze, R. J. Stanton, and P. A. Tomas/*Higher gradients and representations of lie groups*   416

P. W. Jones/*Interpolation between Hardy spaces*   437

U. Neri/*Weighted H^1-BMO dualities*   452

D. Sarason/*The Shilov and Bishop decompositions of $H^\infty + C$*   461

J. -O. Strömberg/*A modified Franklin system and higher-order spline systems on R^n as unconditional bases for Hardy spaces*   475

A. Uchiyama/*A constructive proof of the Fefferman-Stein decomposition of BMO on simple martingales*   495

PART SIX: DIFFERENTIATION THEORY   507

S. Chanillo and R. L. Wheeden/*Relations between Peano derivatives and Marcinkiewicz integrals*   508

M. de Guzmán/*Some results and open problems in differentiation of integrals*   526

A. Nagel, W. Rudin, and J. H. Shapiro/*Tangential boundary behavior of harmonic extensions of L^p potentials*   533

G. V. Welland/*Strong differentials in L^p*   549

PART SEVEN: PARTIAL DIFFERENTIAL EQUATIONS   555

M. Cwikel, E. B. Fabes, and C. E. Kenig/*On the lack of L^∞-estimates for solutions of elliptic systems or equations with complex coefficients*   557

E. B. Fabes, D. S. Jerison, and C. Kenig/*Boundary behavior of solutions to degenerate elliptic equations*   577

C. Fefferman and D. H. Phong/*Subelliptic eigenvalue problems*   590

D. Gurarie and M. A. Kon/*Summability and bounds for multipliers of elliptic operators on R^n*   607

L. I. Hedberg/*On the Dirichlet problem for higher-order equations*   620

J. E. Lewis and C. Parenti/*L^p-estimates for a singular hyperbolic equation*   634

B. Marshall/*Mixed norm estimates for the Klein-Gordon equation*   638

PART EIGHT: OTHER TOPICS RELATED TO HARMONIC ANALYSIS   651

R. Beals and R. R. Coifman/*Scattering and inverse scattering related to evolution equations*   653

J. Burbea/*Norm inequalities for holomorphic functions of several complex variables*   661

A. P. Calderón/*On the Radon transform and some of its generalizations*   673

C. F. Dunkl/*An addition theorem for Heisenberg harmonics*   690

M. R. Herman/*Sur les difféomorphismes du cercle de nombre de rotation de type constant*   708

Y. Meyer/*Sur un problème de Michael Herman*   726

S. Janson/*Minimal and maximal methods of interpolation of Banach spaces*   732 xii   Contents

P. Koosis/*A theorem of Khrushchev and Peller on restrictions of analytic functions having finite Dirichlet integral to closed subsets of the unit circumference*   740

H. P. Lotz, N. T. Peck, and H. Porta/*Semiembeddings in linear topological spaces*   749

P. Malliavin/*Diffusion on the loops*   764

H. N. Mhaskar/*Weighted analogues of Nikolskii-type inequalities and their applications*   783

Y. Sagher/*An application of the approximation functional in interpolation theory*   802

A. L. Shields/*An analogue of the Fejér-Riesz theorem for the Dirichlet space*   810

N. T. Varopoulos/*Potential theory and diffusion on Riemannian manifolds*   821

ANTONI ZYGMUND

Antoni Zygmund was born in Warsaw, Poland, on December 26, 1900,
the son of Wincenty Zygmund and Antonina Perkowska Zygmund. After
elementary school, he entered high school (gymnasium) in Warsaw
in 1912. His early interests were in science, especially in
astronomy, and he planned to become an astronomer.

At the outbreak of the First World War, his family was evacu-
ated to Poltava, in the Ukraine, where it remained until autumn
of 1918 before returning to (by then) independent Poland. In 1919
Zygmund entered the University of Warsaw, which had become Polish
in 1915. The university was in its initial stages of development
and the teaching of astronomy was still being organized. This situation, together with the interest in mathematical problems he had
developed through independent study of astronomy, pushed Zygmund
toward pure mathematics.

At that time there had emerged in Poland a group of talented
young mathematicians who exerted a powerful attraction on university students. They were to develop what was later known as the
Polish Mathematical School. Prominent among these young mathematicians were Waclaw Sierpinski (1882-1961), Stefan Mazurkiewicz
(1888-1945), and Zygmunt Janiszewski (1888-1920). They founded a
new mathematical periodical, *Fundamenta Mathematicae*, devoted
primarily to the theory of sets and related topics, which had a
strong impact on the subsequent development of mathematics in
Poland. Zygmund was caught up in all this mathematical fervor.
In addition, he was influenced by two of his older schoolmates,
Stanislaw Saks and Zygmunt Zalcwasser. Saks was to play a very
important role in the development of analysis in Poland. His
primary interest was real variables, mainly differentiability and
integrability of functions. His *Theory of the Integral*, first
published in French in 1933, is a classic on the subject, and
his *Analytic Functions*, written together with Zygmund, is still

considered one of the best books on this topic. Alexander Rajchman, then an instructor and docent at the University of Warsaw, had made very important contributions to the theory of formal multiplication and uniqueness of trigonometric series, and Zygmund became particularly interested in the problems of this area. One may thus say that Zygmund was primarily a disciple of Saks and Rajchman.

Zygmund obtained his doctorate in 1923. A year earlier, he had been appointed instructor at the Polytechnical School in Warsaw, a position he kept until his departure to Wilno in 1930. In 1926 he was appointed docent at the University of Warsaw and also was awarded a Rockefeller Fellowship to spend a year in England with Hardy in Oxford and Littlewood in Cambridge. Unexpectedly, he encountered there one more mathematician who influenced him very deeply: R. E. A. C. Paley, a student of Littlewood, with whom Zygmund maintained a fruitful collaboration until Paley's early accidental death in 1933, a severe blow to Zygmund both personally and professionally.

In 1930, after his return from England, Zygmund was appointed professor at the University of Wilno. There he encountered Józef Marcinkiewicz, who became first his student and then his collaborator. Marcinkiewicz, whose name is familiar to everyone interested in functional analysis and Fourier series, was an extraordinary mathematician. His collaboration with Zygmund lasted almost ten years and produced a number of important results.

At the outbreak of the Second World War, in 1939, both Marcinkiewicz and Zygmund were mobilized as reserve officers in the Polish Army. Marcinkiewicz was taken prisoner in the East and disappeared without trace, most likely executed in captivity. Zygmund succeeded in returning to Wilno to his wife and son. Through the help of friends, mainly Professors Jacob Tamarkin and Jerzy Neyman, he managed to leave Poland and find his way to the United States, where he arrived early in 1940. The specter of impending mobilization and war weakened the need for university teachers and made jobs difficult to obtain. Nevertheless, Zygmund received an appointment as professor at Mount Holyoke College in Massachusetts.

In 1945 Zygmund moved to the University of Pennsylvania in Philadelphia. Two years later he was invited by Professor Marshall Stone to join the Department of Mathematics of the University of Chicago. A most impressive group of outstanding mathematicians was to gather there, A. A. Albert, S. S. Chern, S. MacLane, L. M. Graves, A. Weil, I. Kaplansky, P. Halmos, I. Segal, and E. Spanier among them. Zygmund went to the University of Chicago and remained there until he retired in 1980 as the Gustave and Ann Swift Distinguished Service Professor of Mathematics. At Chicago he had a large following and developed what has come to be known as the Chicago School of Analysis. The mathematical influence Zygmund exerted through his work, his teaching, and his students has spread across the United States and beyond, to South America and Europe.

Professor Antoni Zygmund has been awarded many honors and has held important positions in the mathematical community. He has honorary degrees from Washington University, the University of Toruń in Poland, the University of Paris, and the University of Uppsala in Sweden. The learned societies of which he is a member include the National Academy of Sciences of the United States; the Polish Academy of Sciences; the National Academy of Sciences of Buenos Aires, Argentina; the Royal Academy of Sciences of Madrid; and the National Academy of Science of Palermo, Italy.

Alberto P. Calderón

ACKNOWLEDGMENT

The Chicago Conference on Harmonic Analysis
March 23-28, 1981

was sponsored by

The National Science Foundation
grant number
MCS-8019742

and

The University of Chicago.

PART FIVE

HARDY SPACES

ON H^{∞} IN MULTIPLY CONNECTED DOMAINS

Lennart Carleson
Institut Mittag-Leffler, Djursholm, Sweden

1

In the paper [3] Forelli proved the following theorem. Let D be a
finitely connected domain, such that no boundary component is a
point. Let D^{∞} be the universal covering surface of D. Then there
is a projection E from the space $H^{\infty}(D^{\infty})$ of bounded analytic func-
tions on D^{∞} to $H^{\infty}(D)$ so that if $f \in H^{\infty}(D^{\infty})$ and $f_0 \in H^{\infty}(D)$, then

$$E(f_0 \cdot f) = f_0 E(f).$$

This immediately proves that the Corona theorem holds for D.
Namely, if F_0, $G_0 \in H^{\infty}(D)$ are such that $|F_0| + |G_0| \geq \delta > 0$, then
we have F, $G \in H^{\infty}(D^{\infty})$ so that

$$FF_0 + GG_0 \equiv 1 \text{ on } D^{\infty}.$$

If we apply E to this equation, we obtain solutions $E(F)$, $E(G)$ in
$H^{\infty}(D)$.

If one could make the projection explicit, one would have
an efficient method to study $H^{\infty}(D)$. The most obvious candidate for
a projection is the following. We use this notation: points in some
fixed copy D^* of D are denoted z^*. Equivalent points are denoted z
with notation $z \sim z^*$. Assume that $P(z) \in H^{\infty}(D^{\infty})$ exists so that

$$\sum_{z \sim z^*} P(z) \equiv 1, \ z^* \ \epsilon \ D^*,$$ (1)

and

$$\sum_{z \sim z^*} |P(z)| \le C,$$ (2)

where C is independent of $z^* \ \epsilon \ D^*$. Then, clearly,

$$f_0(z^*) = E(f) = \sum_{z \sim z^*} P(z)f(z)$$ (3)

is a projection of the desired type.

We can use the explicit form of E to study the distribution of zeros of functions in $H^\infty(D^*)$. Let $g^*(z^*; c^*)$ be the Green's function for D^* with pole c^*. g^* always exists if $H^\infty(D^*)$ is non-trivial. $g(z; c)$ is the Green's function for D^∞. Then if $c \sim c^*$,

$$g^*(z^*; c^*) \equiv \sum_{z \sim z^*} g(z; c) \equiv \sum_{c \sim c^*} g(z; c).$$ (4)

Let $a_\nu^* \ \epsilon \ D^*$ be a sequence of points that are zeros of a function $f_0(z^*) \ \epsilon \ H^\infty(D^*)$. Then by Jensen's formula,

$$\sum_{\nu=1}^{\infty} g^*(a_\nu^*) < \infty.$$ (5)

Theorem 1. *If a partition function P exists and a_ν^* satisfies (5), then a_ν^* are zeros of some function in $H^\infty(D^*)$.*

Proof. The formula (4) shows that the sequence a_ν of all points $\sim a_\nu^*$, $\nu = 1, 2, \ldots$, is a Blaschke sequence on (the disc) D^∞. Let $A(z)$ be the corresponding Blaschke product. Let $B(z)$ be the analogous Blaschke product for the sequence of points $c \sim c^*$. We note that

$$-\log|B(z)| = g^*(z^*; c^*), \ z \sim z^*.$$

It follows, since $|A(c)|$ is independent of $c \sim c^*$ and $\neq 0$, that

$$|A(z)| + |B(z)| \ge \delta > 0.$$

Let

$$FA + GB \equiv 1$$

for F, $G \in H^{\infty}(D^{\infty})$. Define

$$f(z^*) = \sum_{z \sim z^*} P(z)F(z)A(z).$$

Then clearly $f(a_j^*) = 0$ and $f(c^*) = 1$. Observe that this construction does not show that f has a_j^* as exact zero set.

I am grateful to J. Garnett for many suggestions of improvements in the presentation.

2

Let us first show that P exists for a finitely connected domain D.
It is convenient to represent D^{∞} as the unit disc U modulo a group $\widehat{G}$ of linear transformations. Let $\widehat{F}$ be the fundamental domain of $\widehat{G}$ that contains 0 and is bounded by $2s$ circles orthogonal to ∂U.
We obtain new copies of $\widehat{F}$ by reflecting in orthogonal circles.
After n successive reflections, we have $2s \cdot (2s - 1)^{n-1} < (2s)^n$ orthogonal circles Γ_n. Then the length of each circle is $\geq a^n$ and the total projection on $|z| = 1 \leq 2\pi \cdot b^n$, where a, b are constants and $b < 1$. Furthermore, if $\gamma \in \Gamma_n$, then $\text{dist}(\gamma, \Gamma_{n+1}) \geq c$ (length γ). We shall need these simple facts later.

Let $p_0 \in C^{\infty}(U)$ and assume that $p_0 \equiv 1$ in $\widehat{F}$, $p_0 \geq 0$ and $p_0 \equiv 0$ outside the $2s$ copies of $\widehat{F}$ that are adjacent to $\widehat{F}$.
Define

$$p(z) = p_0(z) / \sum_{S \in \widehat{G}} p_0(Sz), \quad z \in U,$$

so that

$$\sum_{S \in \widehat{G}} p(Sz) \equiv 1.$$

Observe that $p(z)$ has the same support as $p_0(z)$. We wish to construct $P(z)$ by solving the $\bar{\partial}$-problem for $p(z)$. Let therefore $f(z; \zeta)$, z, $\zeta \in U$, be functions in $H^{\infty}(U)$ for every $\zeta \in U$ and be uniformly bounded for $\zeta \in \text{supp}(p)$ and finally satisfy $f(\zeta, \zeta) = 1$.
Then

$$P(z) = p(z) - \frac{1}{\pi i} \iint_U \frac{f(z; \zeta)}{z - \zeta} \overline{\partial} p(\zeta) d\xi \ d\eta = p(z) - Q(z) \tag{6}$$

satisfies (1) and (2) if

$$\sum_{S \in \textcircled{G}} Q(Sz) \equiv 0, \tag{7}$$

$$\sum_{S \in \textcircled{G}} |Q(Sz)| \le C. \tag{8}$$

Denote by $K(z; \zeta) = \dfrac{f(z; \zeta)}{z - \zeta}$ and assume that $f(z; \zeta)$ is defined for $\zeta \in \textcircled{F}$. We then define $K(z; \zeta)$ for $\zeta \in U$ so that the following identity holds:

$$S'(\zeta) \cdot K(Sz; S\zeta) = K(z; \zeta). \tag{9}$$

This defines $K(z; \zeta)$ uniquely and one can notice that $K(z; \zeta)$ has a pole with residue 1 for all $\zeta \in U$.

To prove (7) we note the following identity:

$$\pi i \sum_S Q(Sz) = \sum_S \iint_U K(Sz; \zeta) \overline{\partial} p(\zeta) d\xi \ d\eta$$

$$= \sum_S \iint_U K(z; S^{-1}(\zeta)) \frac{1}{S'(S^{-1}(\zeta))} \overline{\partial} p(\zeta) d\xi \ d\eta.$$

Make here the substitution $\zeta = S(\zeta')$. The equality can be continued by

$$\sum_S \iint_U K(z; \zeta') \overline{S'(\zeta')} (\overline{\partial} p)(S\zeta') d\xi' \ d\eta'$$

$$= \iint_U K(z; \zeta') \overline{\partial} \sum_S p(S\zeta') \ d\xi' \ d\eta' = 0.$$

To prove (8) it is enough to prove

$$\sum_{S \in \textcircled{G}} |f(Sz; \zeta)| \le C,$$

for $\zeta \in \textcircled{F}$. Choosing S corresponding to the $2s$ reflections, we obtain the same result for $\zeta \in \text{supp}(p)$.

We now recall the preliminary discussion of $\widehat{G}$. Define

$$h(\theta) = (n + 1)^2 \text{ for } e^{i\theta} \epsilon R_n = \text{int}(\Gamma_n)\backslash\text{int}(\Gamma_{n+1}).$$

Then

$$\int h(\theta)d\theta \le 2\pi \cdot \sum(n + 1)^2 b^n.$$

Define

$$H(z) = \exp\left\{-\frac{1}{2\pi}\int_{-\pi}^{\pi}\frac{e^{i\theta} + z}{e^{i\theta} - z} h(\theta)d\theta\right\}.$$

Then for $z \epsilon R_n$

$$\left|H(z)\right| \le \exp\{-c_0 n^2\}$$

with $c_0 > 0$ independent of n. In $\widehat{F}$, $\left|H(z)\right| \ge c_0$. We can therefore choose

$$f(z; \zeta) = H(z)/H(\zeta), \quad \zeta \epsilon \widehat{F},$$

and have

$$\sum\left|f(Sz; \zeta)\right| \le \frac{1}{c_0}\sum_n (2s)^n e^{-c_0 n^2}.$$

3

This method can now be generalized to very general domains D. We shall here consider the case when $\partial D = E \subset \mathbf{R}$. It is necessary [1] that E has positive measure, to avoid $H^\infty(D)$ trivial. Unfortunately, I cannot prove the existence of P under this assumption alone. It is natural that the case when certain parts of E are very thin compared to Lebesgue measure are particularly difficult. We avoid this possibility by assuming that E is *homogeneous* in the following sense. There exists a constant c, only depending on E, so that for all $x_0 \epsilon E$ and *all* $t > 0$,

$$m((x_0 - t, x_0 + t) \cap E) \ge ct. \tag{10}$$

Observe that (10) implies that $\infty \notin D$. The restriction in (10) as $t \to \infty$ is, however, not serious. If E is compact satisfying (10)

$t \leq 1$, we can obtain (10) for all t by making a suitable inversion $z \rightarrow (z - x_0)^{-1}$. From this observation it follows that all Cantor sets, even with a variable number of divisions in every step, satisfy (10). Our main result is then:

Theorem 2.   A partition function P satisfying (1) *and* (2) *exists for homogeneous linear sets.*

The construction is best carried out in the given domain D and on its covering surface D^∞. We choose $p(z) \equiv 1$ in D^* and $\equiv 0$ on the rest of D^∞. $\bar{\partial} p \, d\xi \, d\eta$ will then be Lebesgue measure on the complementary intervals of E and formula (9) will simply be reversal of signs. The function $Q(z)$ is therefore

$$\int_{CE} K(z; \xi) \, d\xi, \tag{11}$$

where K has a pole with residue 1 at ξ. To deal with the singularity, we define

$$K(z; \xi) = \frac{f(z; \xi)}{z^* - \xi} B(z; \xi) \cdot F(z; \xi) \tag{12}$$

with the following understanding:

(a) $f(z; \xi) \in H^\infty(D^\infty)$ and $f(\xi, \xi) = 1$ and $|f(z; \xi)|$ only depends on $z^* \sim z$.

(b) $B(z; \xi)$ is the Blaschke product with zeros at all points $w \sim \xi$ on D^∞ except $w = \xi \in D^*$.

(c) $F(z; \xi) \in H^\infty(D^\infty)$ and

$$F(\xi, \xi) = \frac{1}{B(\xi, \xi)} .$$

$K(z; \xi)$ is then a kernel with the correct pole and (7) holds if (8) holds.

(a') To make $Q(z)$ bounded, we shall construct $f(z; \xi)$ so that

$$\int_{CE} \frac{|f(z^*; \xi)|}{|z^* - \xi|} \, d\xi \leq C. \tag{13}$$

The conditions (a) and (a') are the requirements on $f(z; \xi)$.

(b') For the Blaschke product we need

$$|B(\xi, \xi)| \geq c_0 > 0. \tag{14}$$

This, however, is an immediate consequence of our assumption on E. Namely,

$$-\log B(z; \xi) = g^*(z^*; \xi) - g(z; \xi). \tag{15}$$

The right-hand side is independent of changes of scale: $z \to Mz$, $\xi \to M\xi$. We may therefore assume $\xi = 0$, $1 \in E$, $(-1, 1) \subset D$. Then

$$\left| g(z^*; 0) - \log \frac{1}{|z^*|} \right| \leq C, \quad |z^*| = \frac{1}{2}$$

$$\left| g^*(z^*; 0) - \log \frac{1}{|z^*|} \right| \leq C, \quad |z^*| = \frac{1}{2}$$

and by the maximum principle in (15), (14) holds.

Now observing $|B(z; \xi)| \leq 1$, we can estimate (8) by

$$\sum_{z \sim z^*} \int \frac{|f(z^*; \xi)|}{|z^* - \xi|} |F(z; \xi)| \, d\xi \leq \left(\sup_{\xi} \sum_{z \sim z^*} |F(z; \xi)| \right) \cdot C$$

by (13). Considering (b'), our requirement on $F(z; \xi)$ is therefore, by the maximum principle,

(c') $F(\xi, \xi) = 1,$ $\xi \in \mathbf{R} \backslash E$

$$\sum_{z \sim z^*} |F(z; \xi)| \leq C, \quad z^* \in E.$$

To be able to construct $f(z; \xi)$ and $F(z; \xi)$, we need the estimates of what now corresponds to the Lebesgue measure in Sec. 2, that is, harmonic measure on D^∞. If $\sigma \subset \partial D^\infty$ with projection $\sigma^* \subset \partial D^*$, we obtain majorants of harmonic measure for (σ, D^∞) by considering (σ^*, D^*). Such estimates are the object of the following section.

4

We shall here study harmonic measure for the plane domain D^*. We omit in this section the $*$-notation, since all sets are schlicht. We denote the harmonic function in D whose boundary values are $= 1$ on F and 0 on $\partial D \backslash F$ by $u(z; F)$. For the method used, compare Benedicks [2].

Lemma 1. Let E be homogeneous, normalized so that $(-1, 1) \subset D$ and $1 \in E$. Then there are numbers $\theta > 0$ and $C < \infty$, only depending on the constant c in (10) so that

$$u(0; E \cap (|x| \geq R)) \leq C \cdot R^{-\theta} \tag{16}$$

for all $R \geq 1$.

Remark. P. Jones has pointed out to me that these types of inequalities are essentially Wiener criteria for the Dirichlet problem and can be found in Wiener's paper [5].

Proof. We first note that $\theta \leq \frac{1}{2}$, since this is the case if $E = (1, \infty)$. It does not seem likely that $\theta = \frac{1}{2}$ in the general case.

Let $A = c^{-1}$ and let $K(x, t)$ be the Poisson kernel for the complement of $(-\infty, -A) \cup (A, \infty)$. We assume also $R = A^n$, $n \geq 3$. Finally, denote by m_n the supremum of $u(x)$, $|x| < 1$, under the conditions made. We have for $-1 < x < 1$

$$u(x) \leq \int_{|t| \geq A} K(x; t) u(t) dt = \sum_{\nu = 1}^{n-2} \int_{|t| = A^\nu}^{A^{\nu+1}} K(x; t) u(t) dt + O(A^{-n}).$$

When $A^\nu \le |t| \le A^{\nu+1}$, $\nu \le n - 2$,

$$u(t) \le m_{n-\nu-1}.$$

Furthermore,

$$\int_{\substack{|t|=A^\nu \\ t \notin E}}^{A^{\nu+1}} K(x; t)dt \le q_\nu, \quad -1 < x < 1,$$

where

$$\Sigma q_\nu = q < 1$$

and q only depends on c, since $m(E \cap (A^\nu < |t| < A^{\nu+1})) \ge c \cdot A^\nu$. It follows that, for A large enough,

$$m_n \le \sum_{\nu=1}^{n-2} q_\nu m_{n-\nu-1} + O(2^{-n}).$$

q_ν also satisfies

$$0 < q_\nu < c \cdot 2^{-\nu}. \tag{17}$$

If we introduce

$$m(z) = \sum_0^\infty m_n z^n$$

$$q(z) = \sum_0^\infty q_\nu z^\nu,$$

the inequalities can be written

$$m(z)(1 - q(z)) << \Sigma 2^{-n} z^n.$$

Now since $q(1) = q < 1$ and q_ν satisfies (17), it follows that $(1 - q(z))^{-1}$ is holomorphic in $|z| < 1 + \delta$, $\delta = \delta(c)$. Furthermore,

$$m(z) << \left(\sum_0^\infty q^j(z)\right)(\Sigma 2^{-n} z^n)$$

and

$$m_n \le (1 - \delta)^n \le CR^{-\theta},$$

for some $\theta > 0$.

Lemma 1 takes care of the behavior of harmonic measure at large distances. We also need information on the local behavior.

Lemma 2. Let $I = (-1, 1)$ *be a complementary interval.* Let $F \subset E$ *have small measure,* $mF = \varepsilon$. *Then*

$$u(0; F) < C_0 \exp\left\{-c_0\left(\log \frac{1}{\varepsilon}\right)^{1/2}\right\}$$

where c_0, C_0 *only depend on* c.

Proof. Divide I into N equal intervals I_j of length $2/N$. N will be fixed later and will not depend on ε. Define

$$\psi_j(\varepsilon) = \sup \frac{N}{2} \int_{I_j} u(x; F)\,dx \tag{18}$$

for all configurations with c fixed. Consider some fixed configuration and some $x \in I_j$. Let $K(x; t)$ be the Poisson kernel for $(|t| \geq 1)$. Then

$$u(x; F) = \int_{|t| \geq 1} K(x; t)u(t)\,dt = \int_F + \int_S \tag{19}$$

where $S = \cup J_k$ and J_k are the complementary intervals of E. Let A be a (large) number to be defined later only depending on ε.

Denote by J_k^* the interval J_k expanded symmetrically around its midpoint in scale A. We divide $\{J_k\}$ into two groups:

I. $|J_k^* \cap F| > A\varepsilon |J_k^*|$.

II. The remaining intervals J_k.

There exists a disjoint subset of the J_k^*'s of group I so that these expanded by the factor 3 cover $\bigcup_{(I)} J_k$. We denote the expanded intervals J_k^* without introducing a new indexing. Then corresponding to case I

$$\sum_{(I)} |J_k| \leq \sum_{(I)} m(S \cap J_k^*) \leq \frac{3}{A\varepsilon} \sum_{(I)} m(J_k^* \cap F) \leq \frac{3}{A}.$$

The contribution to the right-hand side of (19) of $J_k \subset S$ so that J_k belongs to group I has the estimate

$$\sum_{(I)} \int_{J_k} K(x; t)dt \le C \cdot A^{-1/2},$$

since $K(x, t)$ has the explicit form ($|t| > 1$)

$$K(x, t) = \frac{2}{\pi} \frac{\sqrt{1 - x^2}}{|1 - x| \cdot |t + 1| + |1 + x| \cdot |t - 1|} \cdot \frac{1}{\sqrt{t^2 - 1}}. \qquad (20)$$

From this follows also that the first term in (19) is

$$O((\varepsilon(1 - |x|)^{-1})^{1/2}).$$

If J_k is in group II, then $J_k^* \cap F$ behaves relatively to J_k as F does to $(-1, 1)$ except that ε has been changed to $A^2\varepsilon$. Write

$$J_k = \bigcup_{\nu=1}^{N} J_k(\nu), \quad |J_k(\nu)| = (1/N) \cdot |J_k|.$$

By the definition of ψ_j and by Lemma 1, we have

$$\frac{1}{|J_k(\nu)|} \int_{J_k(\nu)} u(t)dt \le \psi_\nu(A^2\varepsilon) + C \cdot A^{-\theta}.$$

Write

$$a_{j\nu} = \sum_k \frac{N}{2} \int_{I_j} K(x, t_\nu(J_k)) dx \cdot |J_k|,$$

where t_ν is the endpoint of $J_k(\nu)$, closest to 0. Then

$$\psi_j(\varepsilon) \le C(N)(\varepsilon^{1/2} + A^{-1/2}) + \sum_\nu a_{j\nu}(\psi_\nu(A^2\varepsilon) + O(A^{-\theta})).$$

We shall show in a moment that if N is large (not depending on ε),

$$\sum_\nu a_{j\nu} \le q < 1, \qquad (21)$$

where q only depends on c in (10). Accepting (21) and setting $h(\varepsilon) = \sup_j \psi_j(\varepsilon)$, we have

$$h(\varepsilon) \le C(\varepsilon^{1/2} + A^{-\theta}) + qh(A^2\varepsilon).$$

We take $A = \exp\left\{c'\left(\log \frac{1}{\varepsilon}\right)^{1/2}\right\}$ and iterate $\left[\left(\log \frac{1}{\varepsilon}\right)^{1/2}\right]$ times and obtain the desired estimate.

In proving (21), we may restrict ourselves to intervals J_k located in $(-A, A)$ and assume that there are finitely many intervals. Fix $\delta = \delta(c) > 0$.

Case 1. $J_k \subset \{1 \le |t| \le 1 + \delta\}$.

By (13), there are disjoint sets $E_k \subset E$ such that

$$\sup_{E_k} |t| \le \inf_{J_k} |t|$$

and

$$|E_k| \ge \lambda(c) |J_k|.$$

Then, since $K(x; t)$ is monotone for $t > 1$ and $t < -1$,

$$\sum_{\text{Case 1}} \left(\sum_\nu \frac{|J_k|}{N} K(x; t_\nu) - \int_{J_k(\nu)} K(x; t)dt \right) \le \sum \frac{1}{\lambda N} \int_{E_k} K(x; t)dt$$

$$\le \frac{1}{\lambda N} \int_{1 \le |t| \le 1+\delta} K(x; t)dt.$$

Case 2. J_k intersects $|t| \ge 1 + \delta$.

By (10), $J_k \subset \{|t| \ge 1 + \delta'\}$ for some $\delta'(c) > 0$, and

$$\sum_{\text{Case 2}} \left(\sum_\nu \frac{|J_k|}{N} K(x; t_\nu) - \int_{J_k(\nu)} K(x; t)dt \right) \le \sum \frac{|J_k|}{N} \sup_{J_k} |K'(x; t)|. \quad (22)$$

Observe now the explicit expression (20) for K. The inequality (22) can be continued by

$$\frac{\text{Const.}}{N} \int_{|t| \ge 1+\delta'} |K'(x; t)|dt \le \frac{\text{Const.}}{N}$$

uniformly for $-1 < x < 1$. Consequently,

$$\sum_\nu a_{j\nu} \le \frac{N}{2} \int_{|t| \ge 1+\delta'} K(x; t)dt + O(N^{-1})$$

$$\le \frac{N}{2} \int_{I_j} dx \int_{|t| \ge 1+\delta'} K(x; t)dt + O(N^{-1}).$$

By (10), the first term is $\le q' < 1$ and hence (21) follows for N large enough.

Lemma 3. If in Lemma 2 we assume that $(-\infty, 1)$ is a complementary arc, the conclusion of Lemma 2 is still valid.

Proof. Let L be a large number and set $G = (-\infty, -L)$ and $E_1 = E \cup G$. Harmonic measures with respect to the outside of E are denoted u_1. Then by Lemma 1,

$$u_1(0; G) \leq C \cdot L^{-\theta},$$

and by Lemma 2,

$$u_1\left(-\frac{L}{2}; F\right) \leq C_0 \exp\left\{-c_0\left(\log \frac{L}{\varepsilon}\right)^{1/2}\right\}.$$

By Harnack's inequality, there is a constant λ so that

$$u_1(0; F) \leq L^\lambda \cdot u_1\left(-\frac{L}{2}; F\right).$$

Finally,

$$u(0; F) \leq u_1(0; F) + u_1(0; G) \leq L^\lambda C_0 \exp\left\{-c_0\left(\log \frac{L}{\varepsilon}\right)^{1/2}\right\} + C \cdot L^{-\theta}.$$

We choose $L = \exp\left\{c'\left(\log \frac{1}{\varepsilon}\right)^{1/2}\right\}$ and obtain the desired result.

Lemma 4. Let $(-1, 1) \subset D$ and $1 \in E$. Let $h_0(\sigma)$ be harmonic measure at 0 of subsets σ of E. Then h_0 is absolutely continuous and

$$\int_E h_0'(x)\exp\{c_1(\overset{+}{\log} h_0')^{1/2}\}dx \leq C_1$$

with c_1 and C_1 only depending on c.

Proof. By Lemma 3, h_0 is absolutely continuous. Let F_n be the set where $2^n \leq h_0' < 2^{n+1}$. Then $mF_n < 2^{-n}$ and by Lemma 3,

$$h_0(F_n) \leq C_0 \exp\{-c_0'n^{1/2}\}.$$

Hence

$$\{h_0' \exp\int c_1(\overset{+}{\log} h_0')^{1/2}\}dx \leq \sum \exp\{c_1'n^{1/2}\}h_0(F_n) \leq C_1$$

if c_1 is small enough.

5

We shall now start the program outlined in Sec. 3 and first construct our functions $f(z; \xi)$. The construction is essentially the same as used by P. Jones in [4].

We denote by $\varphi(x)$ the function $x^2/(1 + x^2)^2$. For $\xi \notin E$ let $\delta(\xi)$ be the distance from ξ to E. We solve the Dirichlet problem for the complement $D^* = D$ of E with boundary values, ξ fixed, $\xi \notin E$,

$$u(x; \xi) = \log \varphi\left(\frac{|x - \xi|}{\delta(\xi)}\right) - 2\log\left\{1 + \int_{\substack{\delta(t) \leq \delta(\xi) \\ t \notin E}} \frac{\varphi\left(\frac{|x - t|}{\delta(t)}\right)}{|x - t|} \, dt\right\}. \quad (23)$$

Clearly $u \leq 0$. To show that this is indeed possible, we need to prove that $u(x; \xi)$ is summable with respect to harmonic measure h. If h_ξ denotes harmonic measure at ξ, we shall show

$$\int_E u(x; \xi) \, dh_\xi(x) \geq -C, \quad (24)$$

where C only depends on the constant of homogeneity.

We first estimate

$$I_1 = \int_E \log \varphi\left(\frac{|x - \xi|}{\delta(\xi)}\right) dh_\xi(x).$$

We can normalize so that $\xi = 0$ and belongs to a complementary interval $(-A, 1)$, $A \geq 1$. Then

$$-I_1 = -\int_{x|\geq 1} \log \varphi(x) \, dh_0(x) = -\int_1^\infty h_0(|t| \leq x)\frac{\varphi'(x)}{\varphi(x)} \, dx$$

$$\leq -C\int_1^\infty x^{-\theta}\frac{\varphi'(x)}{\varphi(x)} \, dx < -C\int_1^\infty (\log \varphi(x))x^{-\theta - 1}dx$$

by Lemma 1.

For the second term, I_2, in (23) we have the estimate, since $ab \leq e^{a-1} + b \log b$,

$$|I_2| \le 2 \int_E \log\left[1 + \int_{\delta(t) \le 1}\right] \cdot (h_0'(x)\,|x|^{\theta+1})\frac{dx}{|x|^{\theta+1}} \tag{25}$$

$$\le 2 \int_E (h_0'\,|x|^{\theta+1})\log(h_0'\,|x|^{\theta+1})\frac{dx}{|x|^{\theta+1}} + 2 \cdot e^{-1}\int_E\left(1 + \int_{\delta(t) \le 1}\right)\frac{dx}{|x|^{\theta+1}}.$$

Here

$$\int_E h_0'(x)\log|x|\,dx \le C \sum 2^{-n\theta} \cdot n$$

by Lemma 1 and

$$\int_E h_0'(x)\log h_0'(x)\,dx \le C$$

by Lemma 4. For the last integral in (25), we have the estimate
for $x \ge 1$

$$\int_1^\infty \frac{dx}{x^{\theta+1}}\int_{\delta(t)\le 1}\frac{\varphi\left(\dfrac{|x-t|}{(t)}\right)}{\dfrac{|x-t|}{\delta(t)}}\frac{dt}{\delta(t)} \le C\sum_{n=0}^\infty 2^{-n-n\theta}\int_{\delta(t)\le 1}dt\int_{2^n}^{2^{n+1}}(\quad)\frac{dx}{\delta(t)} \tag{26}$$

and similarly for $x \le -1$. Here

$$\int_{2^n}^{2^{n+1}}(\quad)\frac{dx}{\delta(t)} \le \int_0^\infty \frac{\varphi(u)}{u}\,du. \tag{27}$$

Also, if $t \ge 2^{n+2}$, the left-hand side of (27) has the improved
estimate [since $\delta(t) \le 1$]

$$\int_{t/2}^\infty \frac{\varphi(u)}{u}\,du < C \cdot t^{-2}.$$

Hence (26) has the bound

$$C\sum_{n=0}^\infty 2^{-n-n\theta}\left(2^n + \int_{2^n}^\infty \frac{dt}{t^2}\right) \le C.$$

This proves (24).

Let $v(z;\xi)$ be a conjugate function of $u(z;\xi)$ on D^∞ and
define

$$f(z;\xi) = C(\xi)e^{u(z;\xi) + iv(z;\xi)}$$

where $C(\xi) \le C$ is chosen so that $f(\xi,\xi) = 1$. We have just proved
that $f(z;\xi)$ satisfies (a) of Sec. 3. It remains to prove (a'),
that is, (13).

By subharmonicity, it is enough to prove (13) for $z^* = x$ real belonging to E. Then (13) equals

$$\int_{CE} \frac{e^{u(x;\,\xi)}\,d\xi}{|x - \xi|} = \int_{CE} \frac{\varphi\left(\dfrac{|x - \xi|}{\delta(\xi)}\right)}{\left(1 + \displaystyle\int_{\substack{\delta(t) < \delta(\xi) \\ t\,\in\,CE}}\right)^2} \, d\xi \leq \int_0^\infty \frac{du}{(1 + u)^2} = 1,$$

simply by change of variables. The construction of $f(z; \xi)$ is therefore complete.

6

It remains to construct $F(z; \xi)$. Again we may normalize so that $\xi = 0$ and $\delta(0) = 1$. We now shall work on D^∞ and need some notation. A (half-)sheet of the covering surface is defined by a sequence $\Omega = (\omega_1, \omega_2, \ldots, \omega_n)$ of complementary intervals of E, $\omega_{i+1} \neq \omega_i$. Depending on if we first go up or down, Ω defines two (half-)sheets of D^∞. We are not going to introduce any separate notation to distinguish these two cases. n is the length $|\Omega|$ of Ω. There is also a natural partial ordering: $\Omega < \Omega'$, if $\Omega' = (\Omega, \Omega'')$. We are going to make the construction of $|F(x)|$, $x \in \partial D^\infty$, and then define $F(z)$ on D^∞ by solving the Dirichlet problem for $\log|F(x)|$. If h_z denotes harmonic measure on ∂D^∞ at z, the crucial estimate is

$$\int \log|F(x)|\,dh_0(x) \geq -C. \tag{28}$$

This will take care of the first condition in (c') in Sec. 3.

The length of a complementary interval ω is denoted $|\omega|$ and its midpoint $\mu(\omega)$. To avoid special conventions later, let us assume that all $|\omega|$'s are different. Sometimes the notation ω will refer to a particular location of the slit ω in the covering surface D^∞; this will be clear from the context.

Let S be a collection of intervals ω in the ground sheet. Let Γ be the part of ∂D^∞ that is reached through these intervals:

$$\Gamma = \cup \, \partial\Omega, \ \Omega \supset \omega, \ \omega \in S.$$

Then the following analogues of Lemmas 1 and 3 of Sec. 3 hold.

Lemma 1'.   If all $\omega \in S$ are located in $|x| \geq R$, then

$$h_0(\Gamma) \leq C \cdot R^{-\theta}.$$

Lemma 3'.   If $\varepsilon = \sum |\omega|, \ \omega \in S$, then

$$h_0(\Gamma) \leq C' \exp \left\{ -c' \left(\log \frac{1}{\varepsilon} \right)^{1/2} \right\}.$$

The lemmas follow immediately from Lemmas 1 and 3 if we replace E by $E \cup \bigcup_{\omega \in S} \omega$, and $h_0(\Gamma)$ by plane harmonic measure of $(\cup \, \omega)^-$.

Let t be real.   Define

$$\psi(t; \, \omega) = \frac{|\omega|^2}{(t - \mu(\omega))^2 + |\omega|^2}$$

and

$$R(t; \, \omega) = \psi(t; \, \omega) \cdot \left(1 + \sum_{|\sigma| \leq |\omega|} \psi(t; \, \sigma) \right)^{-2}.$$

To describe $|F(x)|$, we define inductively functions $\varphi(x; \, \Omega)$ for x real.

We first set $\varphi(x; \, \emptyset) \equiv 1$.   Let $\Omega = (\omega_1, \ldots, \omega_{n-1}, \omega_n) = (\Omega', \omega_n)$.   We distinguish two cases.

A.   If $(n \geq 2)$, $|\omega_n| \geq 2|\omega_{n-1}|$ and $|\mu(\omega_n) - \mu(\omega_{n-1})| \leq |\omega_n|$, we define

$$\varphi(x; \, \Omega) = R(x; \, \omega_{n-1}) \cdot \psi(\alpha_n; \, \omega_{n-1}) \varphi(\mu_{n-1}; \, \Omega')$$

where α_n is the endpoint of ω_n that is closest to μ_{n-1}.

```
                                          μ_n
                                          x
|—————————————|      |————————————————|        |———————————|
     ω_{n-1}                  α_n        ω_n
```

B.   In all other cases, also for $n = 1$,

$$\varphi(x;\ \Omega) = R(x;\ \omega_n) \cdot \varphi(\mu(\omega);\ \Omega').$$

Let $\gamma > 0$ be a constant only depending on c to be defined shortly.   On the boundary of the half-sheet Ω, we define

$$|F(x)| = \gamma^n \cdot \varphi(x^*;\ \Omega).$$

We now verify that this definition satisfies the second condition of (c') in Sec. 3 if γ is small enough.

Denote by

$$S_n(x) = \sum_{|\Omega| = n} \varphi(x;\ \Omega).$$

Every Ω of length $n + 1$ can be written $\Omega = (\Omega^0,\ \omega_{n+1})$ with Ω^0 of length n or more generally $(p \geq 1)$

$$\Omega = (\Omega^{(p)},\ \omega_{n-p+1},\ \ldots,\ \omega_{n+1}).$$

For every p we consider the collection $\mathscr{C}(p)$ of Ω's so that case B occurs for $(\omega_{n-p},\ \omega_{n-p+1})$ with p minimal, $p > 0$. $p = n$ corresponds to no case B.   We split the sum defining $S_{n+1}(x)$ according to $\mathscr{C}(p)$. If $|\Omega| = n + 1$ and $\Omega \in \mathscr{C}(p)$, we factorize $\varphi(x;\ \Omega)$ using the definitions:

$$\varphi(\mu_{n-p+1};\ \Omega^{(p)}) \cdot R(\mu_{n-p+1};\ \omega_{n-p+1}) \cdot$$

$$\cdot R(\mu_{n-p+2};\ \omega_{n-p+1}) \cdot \psi(\alpha_{n-p+2};\ \omega_{n-p+1})$$

$$\vdots$$

$$R(\mu_n;\ \omega_{n-1}) \cdot \psi(\alpha_n;\ \omega_{n-1})$$

$$R(x;\ \omega_n) \cdot \psi(\alpha_{n+1};\ \omega_n).$$

We delete $R(\mu_{n-p+1};\ \omega_{n-p+1})$ and all ψ-factors except the last one.   We then sum over the variables ω_i in the following order:

$$\Omega^{(p)},\ \omega_{n-p+1},\ \ldots,\ \omega_{n-1};\ \omega_{n+1},\ \omega_n$$

and observe the following inequalities:

$$\sum_{\Omega^{(p)}} \varphi(\mu_{n-p+1};\ \Omega^{(p)}) \leq \|S_{n-p+1}\|_\infty$$

$$\sum_{\omega_{n-j-1}} R(\mu_{n-j}; \ \omega_{n-j-1}) \leq C$$

$$\sum_{\omega_{n+1}} \psi(\alpha_{n+1}; \ \omega_n) \leq C$$

$$\sum_{\omega_n} R(x; \ \omega_n) \leq C.$$

The third inequality holds because the intervals ω_{n+1} that are in case A relatively some fixed ω_n have to grow exponentially. Using this subdivision $\mathscr{C}(p)$ we obtain

$$|S_{n+1}(x)| \leq \sum_{p=1}^{n+1} C^p \|S_{n-p+1}\|_\infty. \tag{29}$$

We now bound $\|S_n\|_\infty$ inductively. Clearly $\|S_0\| = 1$. For $n = 1$, we always have definition B and therefore $\|S_1\| \leq C$. Suppose now $\|S_\nu\|_\infty \leq (2C)^\nu$ for $\nu \leq n$. Then (29) implies

$$\|S_{n+1}\|_\infty \leq \sum_{p=1}^{n+1} C^p (2C)^{n-p+1} < (2C)^{n+1} \sum_{1}^{\omega} 2^{-\nu} = (2C)^{n+1}.$$

Hence if $\gamma < \gamma(c)$,

$$\sum \gamma^n \|S_n(x)\|_\infty \leq C,$$

which proves the second condition (c') in Sec. 3.

7

The estimate (28) depends on the exponential decay of harmonic measure. In the group realization in the unit disc, this decay is due to the fact that the fundamental domains intersect $|z| = 1$ in sets of positive measure.

Lemma 5. Let $\Gamma_n = \bigcup_{|\Omega| \geq n} \partial\Omega$. Then there exists $\rho = \rho(c) < 1$ so that

$$h_0(\Gamma_n) \leq \rho^n.$$

Proof. Let $u(z)$ be the harmonic measure. Take Ω' of length $n - 1$. $\Omega' = (\omega_1, \ldots, \omega_{n-1})$. Let $v(z^\star)$ be harmonic in the *plane* slit domain connected through ω_{n-1}, with boundary values $= 0$ on the two sides of E and $= 1$ on the two sides of ω, $\omega \neq \omega_{n-1}$. The homogeneity condition gives

$$v(z^\star) \leq \rho < 1, \ z \ \epsilon \ \omega_{n-1}.$$

Hence for $z \ \epsilon \ \omega_{n-1}$ in Ω',

$$u(z) \leq v(z^\star) \leq \rho.$$

Proceeding inductively, we find $u(z) \leq \rho^n$ in the ground sheets and have proved Lemma 5.

This lemma shows that the factor γ^n is of no importance for (28). In the sheets Ω of length n the boundary values have two components: one factor, R, depending on x and one constant, A_n, depending on the history of the sheet. Let us first analyze the x-factor and begin with case B.

$$\omega_{n-1} \qquad\qquad \omega_n$$

Let $u(z)$ have boundary values $r(x) = -\log R(x; \omega_n)$ on $\partial\Omega$ and $= 0$ otherwise. If $H(x; \xi)$ is the plane harmonic measure discussed in Sec. 4, we have for $z \ \epsilon \ \omega_n$, $z^\star = \xi$,

$$u(z) \leq \int - \log R(x; \omega_n) \, dH(x; \xi) = \int r(x) H'(x) \, dx.$$

Let us normalize so that $\delta(\xi) = 1$, $L = |\omega_n|/\delta(\xi)$. Then by the lemmas in Sec. 4,

$$u(z) \leq C \int_1^\infty \psi(x)^{-\theta/4} \frac{dx}{x^{1+\theta}} + C \int_1^\infty \sum_{|\sigma| \leq L} \psi(x; \sigma) \frac{dx}{x^{1+\theta}}.$$

To estimate the last sum we need a bound for

$$I(\sigma) = \int_1^\infty \frac{|\sigma|^2}{(x - \mu(\sigma))^2 + |\sigma|^2} \frac{dx}{x^{1+\theta}}.$$

We first observe that if $\mu(\sigma) > 2L$ and $|\sigma| \le L$, we have

$$I(\sigma) \le \text{Const.} \ \frac{|\sigma|}{\mu(\sigma)^{1+\theta}}.$$

Hence $\sum I(\sigma)$ for this case is bounded. Let us now assume $2^{\ell-1} < \mu(\sigma) \le 2^{\ell} < 2L$. Splitting the x-integral into $(1, 2^{\ell-2})$, $(2^{\ell-2}, 2^{\ell+1})$, $(2^{\ell+1}, \infty)$ we easily find

$$I(\sigma) \le \text{Const.} \ \left(\frac{|\sigma|^2}{2^{2\ell}} + \frac{|\sigma|}{2^{\ell(1+\theta)}} \right).$$

For every $\ell = 1, 2, \ldots, [\log L] + 1$, we sum over σ and obtain a fixed bound:

$$\sum_{(\ell)} I(\sigma) \le \text{Const.} \ \left(2^{-2\ell} \sum_{(\ell)} |\sigma|^2 + 2^{-\ell(1+\theta)} \sum_{(\ell)} |\sigma| \right)$$

$$\le \text{Const.} \ \left(2^{-2\ell} \ \underset{(\ell)}{\text{Max}} |\sigma| + 2^{-\ell(1+\theta)} \right) \cdot 2^{\ell} \le \text{Const.}$$

Together this implies $u(z) = O(\log L)$, that is, in our original situation

$$u(x) \le \text{Const.} \ \log \frac{|\omega_n|}{\delta(x)} = \Delta(x; \ \omega_n), \ x \ \epsilon \ \omega_n. \tag{30}$$

Suppose now that (ω_{n-1}, ω_n) is in case A.

We proceed as before and consider $u(z)$ with boundary values $r(x) = -\log \psi(\alpha_n; \omega_{n-1}) - \log R(x; \omega_{n-1})$. Normalizing again so that $|\omega_{n-1}| = 1$, we find for $x \ \epsilon \ \omega_n$

$$u(x) \le \text{Const.} \ \left[\log\left(\frac{|\omega_{n-1}|}{\delta(x)} + 1 \right) - \log \psi(\alpha_n; \omega_{n-1}) \right] \tag{31}$$

$$= \Delta^*(x; \ \omega_{n-1}, \ \omega_n).$$

Observe here that $-\log \psi(\alpha_n; \omega_{n-1}) \le - C \cdot \log(\text{dist.}(\omega_{n-1}, \ \omega_n))$.

Let us now keep ω_{n-1} fixed and consider all possible continuations $\omega_n \neq \omega_{n-1}$. For each such ω_n we have Ω with a variable boundary term $r(x) = r(x; \omega_n)$ as above. A majorant of the total contributions of these $r(x; \omega_n)$ to the value of $\log|F(z)|$ for $z \in \omega_{n-1}$ is obtained by solving the *plane* Dirichlet problem for $\mathbf{R}^2 \backslash \omega_{n-1}$ with boundary values $\Delta(x; \omega_n)$ or $\Delta^*(x; \omega_{n-1}, \omega_n)$ according to case B or A. Since everything used is scale invariant, we may assume $\delta(z) = 1$, and it follows that the total contribution is bounded on ω_{n-1}. The contribution to $\log|F(0)|$ from these variable factors in the boundary values can therefore be majorized by the harmonic measure $h_0(\Gamma)$ for $\Gamma = \bigcup_{\omega_n} \partial\Omega(\omega_1, \ldots, \omega_{n-1}, \omega_n)$, since this harmonic measure is bounded from below on ω_{n-1}. The variable factors therefore satisfy (28).

We must now finally analyze the constant $A(\Omega) = \varphi(\mu; \Omega')$, $\mu = \mu_n$ or μ_{n-1} according to cases B and A. These constants are products of factors λ

$$\lambda = \begin{cases} \varphi(\alpha_j; \omega_{j-1}) \cdot R(\mu_j; \omega_{j-1}) & \text{(AA)} \\[1mm] \varphi(\alpha_j; \omega_{j-1}) \cdot R(\mu_{j+1}; \omega_{j-1}) & \text{(AB)} \\[1mm] R(\mu_j; \omega_j) & \text{(BA)} \\[1mm] R(\mu_{j+1}; \omega_j) & \text{(BB)} \end{cases}$$

where (XY) indicates that (ω_{j-1}, ω_j) is case X and (ω_j, ω_{j+1}) is case Y. If $\Omega_0 = (\omega_1, \ldots, \omega_{j-1}, \omega_j, \omega_{j+1})$, then the factor λ is present in all constants $A(\Omega)$ with $\Omega \geq \Omega_0$. Since we are considering logarithms, this means that the contribution to (28) from each such factor is

$$(\log \lambda) \cdot h_0(\Gamma(\Omega_0)), \tag{32}$$

where

$$\Gamma(\Omega_0) = \bigcup_{\Omega \geq \Omega_0} \partial\Omega.$$

We must therefore prove that we can add (32) over all possible choices of Ω_0. The discussion is quite similar to the "variable"

case treated above.   We now base it on Lemmas 1' and 3' of Sec. 6.
We observe that if μ is the center of some ω, then

$$C^{-1} \leq R(x;\ \sigma)/R(\mu;\ \sigma) \leq C \tag{33}$$

for $|x - \mu| \leq \dfrac{|\omega|}{3}$.

Case AA

We keep ω_{j-1}, ω_j fixed and consider the harmonic function
on D^∞ with boundary values $= \log \lambda$ on $\bigcup_{\omega_{j+1}} \Gamma(\Omega_0)$ and $= 0$ otherwise.
Clearly its values are $O(\log \lambda)$ on ω_j.   Using Lemma 1, we see that
the harmonic function with such boundary values on ω_j, as ω_j varies
is uniformly bounded on ω_{j-1}.   This case therefore follows as above.

Case AB

The factor $\varphi(\alpha_j;\ \omega_{j-1})$ is treated as in case AA.   As above,
the factor $-\log R(\mu_{j+1};\ \omega_j)$ contributes an estimate as in (31):

$$\Delta^*(x;\ \omega_{j-1},\ \omega_j) \text{ on } \omega_j.$$

As there, we can now let ω_j vary and obtain a bounded result on
ω_{j-1}.

Case BA

Since $\log R(\mu_j;\ \omega_j)$ is bounded, this is a trivial case.

Case BB

$$\mu_{j+1}$$

┠────────┨   ┠─────┨   ┠─┨   ┠─────────

      ω_{j-1} ω_j ω_{j+1}

As ω_{j+1} varies, we obtain the estimate (30), $\Delta(x; \omega_j)$, on ω_j and then again a bounded estimate on ω_{j-1}.

This completes the proof.

REFERENCES

1. Ahlfors, L., and Beurling, A. Conformal invariants and function-theoretic nullsets. *Acta Math.* 83 (1950):101-129.

2. Benedicks, M. Positive harmonic functions vanishing on the boundary of certain domains in R^n. *Ark. Mat.* 18 (1980):53-72.

3. Forelli, F. Bounded holomorphic functions and projections. *Ill. J. Math.* 10 (1966):367-380.

4. Jones, P. L^∞-estimates for the $\bar{\partial}$-problem. To appear in *Acta Math.*

5. Wiener, N. The Dirichlet problem. *J. Math. Phys.* 3 (1924): 127-146.

TWO REMARKS ABOUT H¹ AND BMO ON THE BIDISC

Sun-Yung A. Chang
University of California, Los Angeles

1. INTRODUCTION

In [11], B. Maurey proved the existence of an unconditional basis
for H^1 of the unit disc; his proof is based on Banach algebra
techniques and is nonconstructive.   In [3], L. Carleson constructed explicitly an unconditional basis for Re H^1, and in [13],
P. Wojtaszczyk proved that the orthogonal Franklin system also is
an unconditional basis.   The proof in both [3] and [13] used
explicit description of the dual space of H^1 of the disc, namely
BMO.   Although such explicit description for H^1 of the bidisc is
still unknown, in Sec. 2 of this paper, we will use instead a
Carleson measure characterization of the dual of H^1 of the bidisc
(cf. [6]) to prove that the product Franklin system is an unconditional basis for Re H^1 of the bidisc.   The existence of such basis
for the bidisc case is also mentioned in Maurey.

In Sec. 3 of the paper, we show the existence of H^p-
solutions $p < \infty$ for a special case of the corona problem of the
bidisc.   The proof generalizes T. Wolff's proof of the problem on
the disc.   We also use properties of Carleson measures on the

Research supported in part by NSF Grant and Mittag-Leffler
Institute.

bidisc developed in [4] and [10]. A probabilistic approach to this result has been obtained by N. Varopoulous [12] (cf. also the related work of E. Amar [1]).

The author would like to thank Professor Carleson for many discussions about the subjects.

2. UNCONDITIONAL BASIS FOR $H^1(T^2)$

In this section, we establish the double Franklin system as uncon-ditional basis for $H^1(T^2)$.

Let $\{v\}$ denote the collection of dyadic subintervals of $[0, 1]$, that is,

$$v = \left[\frac{k}{2^n}, \frac{k-1}{2^n}\right), \ 0 \le k \le 2^n - 1, \ n = 0, 1, 2, \ldots,$$

and $\{f_v\}$ denote the Franklin system associated with $\{v\}$. For defi-nitions and detailed properties of the Franklin system, cf. [7, 8]. For our purpose here, we will only use the following specific prop-erties of the system, which are established in [7]:

(1) $\{f_v\}$ is an orthonormal set of functions which together with the constant function forms a basis for $L^2([0, 1])$.

(2)   Each fixed f_v satisfies

(a) $\int_0^1 f_v(t)\,dt = 0$, and there exists a constant C, and $q < 1$ independent of v such that

(b) $|f_v(t)| \le \dfrac{C}{|v|^{1/2}} q^{\frac{d(t,v)}{|v|}}$,

where d denotes the usual Lebesgue distance function and $|v|$ the Lebesgue measure of v.

(c) $|f_v(t_1) - f_v(t_2)| \le \dfrac{C|t_1 - t_2|}{|v|^{3/2}} q^{\frac{d([t_1, t_2], v)}{|v|}}$.

Let ψ be a C^1 function defined on the real line supported on $[-1, 1]$ with ψ even and

$$\int_{-1}^{1} \psi(t)\,dt = 0.$$

We will use the following notation: if $y > 0$, $\psi_y(t) = \frac{1}{y}\psi(t/y)$ and if $y = (y_1, y_2)$ and $t = (t_1, t_2) \in \mathbf{R}^2$, then

$$\psi_y(t) = \psi_{y_1}(t_1)\psi_{y_2}(t_2);$$

for each dyadic interval I, let I_+ denote the region

$$\left\{(t, y) : t \in I, \frac{|I|}{2} < y \leq |I|\right\};$$

for each dyadic rectangle $R = I \times J$, R_+ will denote the region $I_+ \times J_+$.

Although we do not know a specific description of the dual space of $H^1(T^2)$ [while on the unit circle T, Fefferman [9] characterized BMO (functions of bounded mean oscillation) as the dual of $H^1(T)$], one has [6] the following "Carleson measure" type of description of the space:

Lemma 1 ([6], Sec. 2). *Suppose $f \in L^1(T^2)$ and satisfies*

$$\int f(x_1, x_2)\,dx_1 = \int f(x_1, x_2)\,dx_2 = 0$$

for all $(x_1, x_2) \in T^2$. Then f is in the dual of $H^1(T^2)$ if and only if there exists some constant C such that

$$\sum_{\substack{R \subset \Omega \\ R \text{ dyadic Rectangle}}} \iint_{R_+} |f \star \psi_y(t)|^2 \frac{dt\,dy}{y_1 y_2} \leq C|\Omega| \qquad (*)$$

for all open set $\Omega \subset T^2$.

Remark. One can think of the above result as a statement about the dual space of $H^1([0, 1] \times [0, 1])$ after the usual identification of T^2 as $[0, 1] \times [0, 1]$ with functions extending periodically at the endpoint $0, 1$; that is, $f(0, x) = f(1, x)$ for $0 \leq x \leq 1$. The

Franklin basis [after deleting the function $f(x) = x$] all satisfy the periodic property.

The following result, which is a generalization of Lemmas 2, 3, and 4 in Carleson's paper [3] to the case of the product domain, is the main result of this section.

Theorem 1.   If $f \in L^2(T^2)$ satisfies

$$\int f(x_1, x_2) dx_1 = \int f(x_1, x_2) dx_2 = 0,$$

and with expansion

$$f(x_1, x_2) = \Sigma\, C_{\mu\nu} f_\mu(x_1) f_\nu(x_2),$$

then condition (*) in Lemma 1 is equivalent to

$$\sum_{\mu \times \nu \subset \Omega} |C_{\mu\nu}|^2 \le C|\Omega| \qquad\qquad (\text{**})$$

for all open set Ω contained in T^2.

Following the same logic as in [3], one obtains immediately:

Corollary 1.   The double Franklin system $\left\{ f_\mu \times f_\nu \right\}_{\substack{\text{all } \mu,\ \nu \text{ dyadic} \\ \text{intervals}}}$ forms an unconditional basis of $H^1(T^2)$.

With similar reasoning and proof as in [5], we obtain the following "unconditional" characterization of H^1.

Corollary 2. $H^1(T^2) = \{ f \in L^1(T^2) : \Sigma\, C_{\mu\nu} f_\mu f_\nu$ converges unconditional in L^1, where

$$C_{\mu\nu} = \iint_{T^2} \{ f(x_1, x_2) f_\mu(x_1) f_\nu(x_2) dx_1\ dx_2 \}.$$

Carleson's [3] basis gives geometric insight about the structure of H^1. It remains open whether the product form of his basis also forms unconditional basis for $H^1(T^2)$. (The proof given below for Theorem 2 does not work for Carleson's basis.)   On the other hand, it looks highly likely that recent work of P. Sjolin and J. O. Strömberg about characterization of $H^p(T)$, $0 < p < 1$, using Franklin basis and its higher splines, can be generalized to the product setting.

The following lemma lists elementary estimates that we will use in proving Theorem 1.

Lemma 2.   Suppose μ, I *are dyadic intervals;* $\tilde{I}$ *the interval with same center as* I *and three times the length; then*

(1)   *If* $|\mu| \leq |I|$, $\mu \subset \tilde{\tilde{I}}$, *then* $S_I^2(f_\mu) \leq C \dfrac{|\mu|^3}{|I|^3}$,

(2)   *If* $|\mu| \leq |I|$, $\mu \cap \tilde{\tilde{I}} = \phi$, *then* $S_I^2(f_\mu) \leq C \dfrac{|\mu|}{|I|} q^{\frac{2d(\tilde{I}, \mu)}{|\mu|}}$,

(3)   *If* $|\mu| > |I|$, *then* $S_I^2(f_\mu) \leq C \dfrac{|I|^3}{|\mu|^3} q^{\frac{2d(\tilde{I}, \mu)}{|\mu|}}$,

where C *is a constant depending only on* $\| \psi \|_\infty$, $\| \psi' \|_\infty$, *and*

$$S_I^2(f_\mu) = \iint\limits_{I^+} |f_\mu * \psi_y(t)|^2 \frac{dt\ dy}{y}.$$

We will hence call a dyadic interval μ of type (i) with respect to I if condition $(*)$ in Lemma 1 is satisfied $1 \leq i \leq 3$. For two dyadic rectangles $T = \mu \times \nu$ and $R = I \times J$, we say T is of type (i, j) with respect to R if μ is of type (i) w.r.t. I and ν is of type (j) w.r.t. J.

Proof of Lemma 2.

(1)   If $|\mu| \leq |I|$ and $\mu \cap \tilde{\tilde{I}}$, then for $(t, y) \in I_+$,

$$|(f_\mu * \psi_y)(t)| = \left| \int f_\mu(x) \psi_y(t - x)\, dx \right|$$

$$= \left| \int f_\mu(x) (\psi_y(t - x) - \psi_y(t - C_\mu))\, dx \right|$$

$$\leq \frac{\| \psi' \|_\infty}{|I|^2} \frac{1}{|\mu|^{1/2}} \int |x - C_\mu| q^{\frac{d(x, \mu)}{|\mu|}}\, dx$$

$$\leq \frac{\| \psi' \|_\infty}{|I|^2} \frac{1}{|\mu|^{1/2}} \sum_{n=0}^{\infty} n |\mu|^2 q^{(n-1)} \leq C \frac{|\mu|^{3/2}}{|I|^2},$$

where C_μ is the center of the interval μ. In the second step above we have used property (a), in the third step property (b) of the Franklin function f_μ. The required estimate of $S_I^2(f_\mu)$ in the lemma follows directly from the estimate on $|f_\mu * \psi_y|$.

(2) Suppose $|\mu| \leq |I|$, $\mu \cap \tilde{\tilde{I}} = \phi$; notice that for each $(t, y) \in I_+$, $\psi_y(t - x)$ is supported on $\tilde{I}$ for each x; so if we divide $\tilde{I}$ into dyadic intervals I_j with $|I_j| = |\mu|$, then

$$|(f_\mu * \psi_y)(t)| = \left| \int f_\mu(x) \psi_y(t - x)\, dx \right|$$

$$\leq \sum_j \int_{I_j} |f_\mu(x)| \frac{\|\psi\|_\infty}{|I|}\, dx$$

$$\leq \frac{\|\psi\|_\infty}{|I|} \sum_{j=1}^N \frac{|I_j|}{|\mu|^{1/2}} q^{\frac{d(\tilde{I}, \mu)}{|\mu|} + j} \quad \text{where } N = \frac{|\tilde{I}|}{|\mu|}$$

$$\leq C \frac{1}{|I|} |\mu|^{1/2} q^{\frac{d(\tilde{I}, \mu)}{|\mu|}}.$$

The second to last step follows from property (b) of the Franklin basis f_μ.

(3) If $|\mu| > |I|$, then by property (c) of the Franklin function f_μ, we have, for $(t, y) \in I_+$,

$$|f_\mu * \psi_y(t)| = \left| \int_{\tilde{I}} (f_\mu(x) - f_\mu(C_I)) \psi_y(t - x)\, dx \right|$$

$$\leq \frac{C}{|\mu|^{3/2}} \int_{\tilde{I}} |x - c_1| \frac{\|\psi\|_\infty}{|I|} q^{\frac{d([x, C_I], \mu)}{|\mu|}}\, dx$$

$$\leq C \frac{1}{|\mu|^{3/2}} |I| q^{\frac{d(\tilde{I}, \mu)}{|\mu|}},$$

where C_I is the center of the interval I.

Notation. For fixed dyadic rectangles $\mu \times \nu$, and $R = I \times J$. Let

$$S_R^2(f) = \iint_{R_+} |(f * \psi_y)(t)|^2 \frac{dt\, dy}{y_1 y_2}.$$

Notice

$$S_R(f_\mu f_\nu) = S_I(f_\mu) S_J(f_\nu).$$

Proof of Theorem 1. $(*) \Rightarrow (**)$ Assume

$$\sum_{R \subset \Omega} S_R^2(f) \le A|\Omega|$$

for all open set Ω. We will begin to estimate $\sum_{\mu \times \nu \subset \Omega} |C_{\mu\nu}|^2$, where

$$C_{\mu\nu} = \int_0^1 \int_0^1 f(x_1,\ x_2) f_\mu(x_1) f_\nu(x_2)\, dx_1\, dx_2.$$

For each fixed μ, ν

$$|C_{\mu\nu}| \le \sum_R S_R(f) S_R(f_\mu f_\nu)$$

$$|C_{\mu\nu}|^2 \le C \left(\sum_R S_R^2(f) S_R(f_\mu f_\nu) \right) \left(\sum_R S_R(f_\mu f_\nu) \right)$$

$$\le C \sum_R S_R^2(f) S_R(f_\mu f_\nu)$$

The last step follows from the observation that

$$\sum_R S_R(f_\mu f_\nu) < \infty.$$

To see this, it suffices to verify the corresponding one-variable statement

$$\sum_I S_I(f_\mu) < \infty,$$

The latter follows easily from the estimates in Lemma 2 after we divide the intervals I into the three categories with the fixed interval μ of types (1), (2), (3) w.r.t. I accordingly. For example, let μ_N be the Nth interval adjacent to μ and of the same length $|\mu_N| = |\mu|$ ($\mu_0 = \mu$; $N > 0$ counting from the left, $N < 0$ to the right), then

$$\cdots \quad \mu_{-2} \quad \mu_{-1} \quad \mu \quad \mu_1 \quad \mu_2 \quad \mu_3 \quad \cdots$$
$$\overset{\shortparallel}{\mu_0}$$

$$\sum_{\substack{I \\ \mu \text{ of type (2)} \\ \text{w.r.t } I}} S_I(f_\mu) \leq \sum_{N=-\infty}^{\infty} \sum_{I \subset \mu_N} \left|\frac{I}{\mu}\right|^{3/2} q^{\frac{d(\tilde{I}, \mu)}{|\mu|}}$$

$$\leq 2 \sum_{N=0}^{\infty} \left(\sum_{n=1}^{\infty} \left(\frac{1}{2^n}\right)^{3/2} \cdot 2^n q^{(|N|-1)} \right) < \infty.$$

Types (1), (3) intervals could be handled similarly.   Thus

$$\sum_{\mu \times \nu \subset \Omega} |C_{\mu\nu}|^2 \leq \sum_{\mu \times \nu \subset \Omega_0} \sum_R S_R^2(f) S_R(f_\mu f_\nu) = \sum_R S_R^2(f) \sum_{\mu \times \nu \subset \Omega_0} S_R(f_\mu f_\nu)$$

$$= \sum_{i,j=1}^{3} A_{ij},$$

where

$$A_{ij} = \sum_R S_R^2(f) \sum_{\substack{\mu \times \nu \subset \Omega \\ \mu \times \nu \text{ of type } (i,j) \text{ w.r.t. } R}} S_R(f_\mu f_\nu).$$

We will now estimate each term A_{ij} separately and indicate that each of them has a bound $\leq cA|\Omega|$. It is clear that the terms A_{12}, A_{21} can be handled the same way, and so can A_{13}, A_{31} and A_{23}, A_{32}. We will illustrate below ways to estimate terms A_{12}, A_{22}; the other terms A_{11}, A_{13}, A_{23}, A_{33} can be handled by similar methods.

Estimate of A_{12}
_____

$$A_{12} = \sum_{R=I \times J} S_R^2(f) \sum_{\substack{\mu \times \nu \subset \Omega \\ |\mu| \leq |I|, \mu \subset \tilde{\tilde{I}} \\ |\nu| \leq |J|, \nu \cap \tilde{\tilde{J}} = \phi}} S_I(f_\mu) S_J(f_\nu).$$

For each integer N, let J_N be interval Nth adjacent to J as in the figure.   Let

$$R_N^k = \left\{ I \times J : \frac{1}{2^k}|I \times J| < |\tilde{\tilde{I}} \times \tilde{\tilde{J}}_N \cap \Omega| \leq \frac{1}{2^{k-1}}|I \times J| \right\};$$

then

$$A_{12} \leq \sum_{R=I \times J} S_R^2(f) \sum_{\substack{|\mu| \leq |I|, \mu \subset \tilde{\tilde{I}}}} S_I(f_\mu) \sum_{N=2}^{\infty} \sum_{\substack{\nu \subset J_N \\ \mu \times \nu \subset \Omega}} S_J(f_\nu)$$

$$\leq \sum_{N=2}^{\infty} \sum_{k=0}^{\infty} \sum_{\substack{R \in R_N^k \\ R=I \times J}} S_R^2(f) \sum_{\substack{|\mu| \leq |I| \\ \mu \subset \tilde{\tilde{I}}}} S_I(f_\mu) \sum_{\substack{\nu \subset J_N \\ \mu \times \nu \subset \Omega}} S_I(f_\nu)$$

$$= \sum_{|N|=2}^{\infty} \sum_{k=0}^{\infty} \sum_{R \in R_N^k} S_R^2(f) \Delta_N^k.$$

If $R = I \times J \in R_N^k$, $\mu \subset \tilde{\tilde{I}}$ with $|\mu| = \frac{1}{2^n}|I|$, $\nu \subset J_N$ with
$|\nu| = \frac{1}{2^m}|J|$, then for each fixed m, $n \geq 0$ maximum number of such
$\mu \times \nu \subset \Omega$ is $\leq 2^{m+n+1-k}$.   Thus

$$\Delta_N^k \leq \sum_{\ell=0}^{\infty} \left(\sum_{n+m=k+\ell} \left(\frac{1}{2^n}\right)^{3/2} \left(\frac{1}{2^m}\right)^{1/2} q^{2^m|N|} \right) 2^{\ell}$$

$$\leq C\left(\frac{1}{2^k}\right)^{3/2} q^{|N|} \left(\frac{1}{1-q^{|N|}}\right)^2.$$

And

$$A_{12} \leq C \sum_{|N|=2}^{\infty} \sum_{k=0}^{\infty} \sum_{R \in \mathscr{R}_N^k} S_R^2(f) \left(\frac{1}{2^k}\right)^{3/2} q^{|N|} \left(\frac{1}{1-q}\right)^2$$

$$\leq C \sum_{N=2}^{\infty} \sum_{k=0}^{\infty} \left(\frac{1}{2^k}\right)^{3/2} q^{|N|} A \left| \bigcup_{R \in \mathscr{R}_N^k} R \right| \quad \text{[by condition (*)]}$$

$$\leq CA \sum_{N=2}^{\infty} \sum_{k=0}^{\infty} \left(\frac{1}{2^k}\right)^{3/2} q^{|N|} Nk \, 2^k \cdot |\Omega| \leq CA|\Omega|.$$

Estimate of A_{22}

$$A_{22} = \sum_{R=I \times J} S_R^2(f) \left(\sum_{\substack{\mu \times \nu \subset \Omega \\ |\mu| \leq |I|, \mu \cap \tilde{\tilde{I}} = \phi}} S_I(f_\mu) \right) \left(\sum_{\substack{|\nu| \leq |J| \\ \nu \cap \tilde{J} = \phi}} S_J(f_\nu) \right).$$

We will handle both the terms in μ and ν as the ν-term in
the estimate of A_{12}. The only additional observation needed here
is the following: if there exists $\mu \times \nu \subset \Omega$, with

$$|\mu| = \frac{|I|}{2^n}, \quad \mu \subset I_N \, |\nu| = \frac{|I|}{2^m}, \quad \nu \subset J_M \quad (n, m \geq 0, N, M = \pm 2, \pm 3, \ldots),$$

then $I \times J$ is contained in a set of measure $\leq (|N|+1)2^n$
$(|M|+1)2^m |\mu \times \nu|$, and the collection of such $I \times J$ is contained
in a set $\Omega_{2^n|N|, 2^m|M|}$ of measure $\leq 2^n(|N|+1)2^m(|M|+1)|\Omega|$.   From
this it follows that

$$A_{22} \leq \sum_{|N|=1}^{\infty} \sum_{|M|=1}^{\infty} \sum_{n=0}^{\infty} \sum_{m=0}^{\infty} \sum_{R \subset \Omega_{2^n|N|, 2^m|M|}} (2^n)^{1/2} (2^m)^{1/2} q^{|N|2^n} q^{|M|2^m} S_R^2(f)$$

$$\leq \left(\sum_{N=1}^{\infty} \sum_{n=0}^{\infty} (2^n)^{1/2} q^{N2^n} 2^n (|N|+1) \right)^2 A|\Omega_0| \leq CA|\Omega_0|.$$

$(**) \Rightarrow (*)$ Assume there exists a constant A such that

$$\sum_{\mu \times \nu \subset \Omega} |C_{\mu\nu}|^2 \le A|\Omega|$$

for all open set Ω.   We will verify the same inequality for

$$\sum_{R \subset \Omega} S_R^2(f),$$

where $f = \sum C_{\mu\nu} f_\mu f_\nu$.   Fix an open set Ω, let $\mathcal{R}_0 = \{$dyadic rectangles $R \subset \widetilde{\widetilde{\Omega}}\}$, let $f_1 = \sum_{\mu \times \nu \in \mathcal{R}_0} C_{\mu\nu} f_\mu f_\nu$; $f_2 = f - f_1$; then

(4) $\displaystyle\sum_{R \subset \Omega} S_R^2(f_1) \le \sum_R S_R^2(f_1) \le \|f_1\|_2^2 = \sum_{\mu \times \nu \in \mathcal{R}_0} |C_{\mu\nu}|^2 \le CA|\Omega|.$

(5) $\displaystyle\sum_{R \subset \Omega} S_R^2(f_2) \le \sum_{R \subset \Omega} \left(\sum_{\mu \times \nu \notin \mathcal{R}_0} S_R(f_\mu f_\nu) |C_{\mu\nu}| \right)^2$

$$\le \sum_{R \subset \Omega} \left(\sum_{\mu \times \nu \notin \mathcal{R}_0} S_R(f_\mu f_\nu) |C_{\mu\nu}|^2 \right) \sum_{\mu \times \nu \in \mathcal{R}_0} S_R(f_\mu f_\nu)$$

$$\le C \sum_{R \subset \Omega} \sum_{\mu \times \nu \notin \mathcal{R}_0} S_R(f_\mu f_\nu) |C_{\mu\nu}|^2 = \sum_{i,j=1}^{3} S_{ij}$$

where

$$S_{ij} = \sum_{R \subset \Omega} \sum_{\substack{\mu \times \nu \notin \mathcal{R}_0 \\ \mu \times \nu \text{ of type } (ij) \text{ w.r.t. } R}} S_R(f_\mu f_\nu) |C_{\mu\nu}|^2 \qquad (S_{11} = 0).$$

The difference between the term A_{ij} and S_{ij} lies in the interchange of the role played by R and $\mu \times \nu$ in the formula.   We again will only illustrate the estimate of the terms, say, S_{23}, S_{31}, to indicate the general pattern that each $S_{ij} \le CA|\Omega|$.

Estimate of S_{23}

By Lemma 2, we have

$$S_{23} \le \sum_{\substack{I \times J \subset \Omega}} \sum_{\substack{|\mu| \le |I|, \mu \cap \widetilde{I} = \phi \\ |\nu| > |J|}} \left| \frac{\mu}{I} \right|^{1/2} q^{\frac{d(\widetilde{I}, \mu)}{|\mu|}} \left| \frac{J}{\nu} \right|^{3/2} q^{\frac{d(\widetilde{J}, \nu)}{|\nu|}} |C_{\mu\nu}|^2$$

$$= \sum_{I \times J \subset \Omega} \sum_{|N|=2}^{\infty} \sum_{M=0}^{\infty} \sum_{\mu \subset I_N} \sum_{\substack{M \le \frac{d(\widetilde{J}, \nu)}{|\nu|} \le M+1}} \left| \frac{\mu}{I} \right|^{1/2} q^{(N-1)\left|\frac{I}{\mu}\right| + \frac{d(I_N^c, \mu)}{|\mu|}} \left| \frac{J}{\nu} \right|^{3/2} q^M |C_{\mu\nu}|^2$$

$$\leq \sum_{|N|=2}^{\infty} \sum_{M=0}^{\infty} q^M \sum_{k=0}^{\infty} \sum_{\mu \times \nu \in \mathscr{R}_{N,M}^k} \Delta_{N,M}^k |C_{\mu\nu}|^2$$

where

$$\mathscr{R}_{N,M}^k = \left\{ \mu \times \nu : \frac{1}{2^k} |\tilde{\mu} \times \tilde{\nu}| \leq \left| (\tilde{\mu} \times \tilde{\nu}_{M+1}) \cap \left(\bigcup_{I \times J \subset \Omega} I_N \times J \right) \right| \right.$$

$$\left. < \frac{1}{2^{k-1}} |\tilde{\mu} \times \tilde{\nu}| \right\}$$

and for each $\mu \times \nu \in \mathscr{R}_{N,M}^k$,

$$\Delta_{N,M}^k = \sum_{\substack{I \times J \subset \Omega, \\ \mu \subseteq I_N, \ M \leq \frac{d(J,\nu)}{|\nu|} < M+1}} \left| \frac{\mu}{I} \right|^{1/2} \left| \frac{J}{\nu} \right|^{3/2} q^{(N-1)\left|\frac{I}{\mu}\right|}.$$

Observe that $\mu \subseteq I_N$, $M \leq \dfrac{d(J,\nu)}{|\nu|} < M+1$ imply that

$$|(\tilde{\mu} \times \tilde{\nu}_{M+1}) \cap (\tilde{I}_N \times \tilde{J})| = |\tilde{\mu} \times \tilde{J}|;$$

thus $\mu \times \nu \in \mathscr{R}_{N,M}^k$ implies that $|\tilde{J}| \leq \dfrac{1}{2^{k-1}} |\tilde{\nu}|$, and for each ℓ with $\left| \dfrac{J}{\nu} \right| = \dfrac{1}{2^{k+\ell}}$, the number of possible such J is $< 2^\ell$, so

$$\Delta_{N,M}^k \leq \sum_{\ell=0}^{\infty} \left(\frac{1}{2^{k+\ell}} \right)^{3/2} \cdot 2^\ell \sum_{m=0}^{\infty} \left(\frac{1}{2^m} \right)^{1/2} q^{(N-1)2^m} \leq Cq^N \left(\frac{1}{2^k} \right)^{3/2}.$$

We also have

$$\left| \bigcup_{\mu \times \nu \in \mathscr{R}_{N,M}^k} \mu \times \nu \right| \leq N(M+1)k2^k |\Omega|.$$

Thus

$$S_{23} \leq C \sum_{|N|=2}^{\infty} \sum_{M=0}^{\infty} q^M \sum_{k=0}^{\infty} q^N \left(\frac{1}{2^k} \right)^{3/2} \sum_{\mu \times \nu \in \mathscr{R}_{N,M}^k} |C_{\mu\nu}|^2$$

$$\leq C \sum_{N=2}^{\infty} \sum_{M=0}^{\infty} q^{N+M} ANM \left(\sum_{k=0}^{\infty} \left(\frac{1}{2^k} \right)^{1/2} k \right) |\Omega| \leq CA |\Omega| \quad \text{[by condition (**)]}.$$

Estimate of S_{31}

Again by Lemma 2, we have

$$S_{31} \leq \sum_{\substack{I \times J \subset \Omega \\ |\mu| > |I| \\ |\nu| \leq |J|, \ \nu \subset \tilde{J}}} \left| \frac{I}{\mu} \right|^{3/2} \left| \frac{\nu}{J} \right|^{3/2} q^{\frac{d(\tilde{I},\mu)}{|\mu|}} |C_{\mu\nu}|^2$$

$$= \sum_{\substack{I \times J \subset \Omega}} \sum_{N=0}^{\infty} q^N \sum_{\substack{N \leq \frac{d(\tilde{I},\mu)}{|\mu|} < N+1 \\ |\nu| \leq |J|, \ \nu \subset \tilde{J}}} \left| \frac{I}{\mu} \right|^{3/2} \left| \frac{\nu}{J} \right|^{3/2} |C_{\mu\nu}|^2.$$

If we consider the set $\mathscr{L}_N^k = \left\{ \mu \times \nu : \frac{1}{2^k} |\tilde{\mu} \times \tilde{\nu}| \leq |(\tilde{\mu}_{N+1} \times \nu) \cap \tilde{\tilde{\Omega}}| < \frac{1}{2^k} |\tilde{\mu} \times \hat{\nu}| \right\}$, then similar simple geometric arguments as before indicate that for each fixed $\mu \times \nu \in \mathscr{L}_N^k$, the number of $I \times J$ such that

$$N \leq \frac{d(\tilde{I}, \mu)}{|\mu|} < N + 1, \quad \nu \subset \tilde{\tilde{J}},$$

is less than 2^ℓ when $\left|\frac{I}{\mu}\right| = \frac{1}{2^{k+\ell}}$ for each $\ell \geq 0$. Thus

$$S_{31} \leq \sum_{N=0}^{\infty} q^N \sum_{k=0}^{\infty} \sum_{\mu \times \nu \in \mathscr{L}_N^k} \left(\sum_{\ell=0}^{\infty} \left(\frac{1}{2^{k+\ell}} \right)^{3/2} \cdot 2^\ell \sum_{m=0}^{\infty} \left(\frac{1}{2^m} \right)^{3/2} \cdot 2 \right) |C_{\mu\nu}|^2$$

$$\leq C \sum_{N=0}^{\infty} N q^N A |\Omega| \leq C |\Omega|.$$

Combining (4), (5), and the estimate on S_{ij}, we have established (**) from (*) and hence the proof of Theorem 2.

3. H^p-SOLUTION FOR CORONA PROBLEM ON THE BIDISC

In this section we will give another application of the characterization of the dual space of H^1 of the bidisc by a Carleson measure condition. Specifically, we will show that the Corona problem for two functions has an H^p-solution for all $p < \infty$. The proof we will give below is a direct generalization of T. Wolff's proof of the Corona problem on the disc. It does not depend on any explicit kernel formula for the solution of $\bar{\partial}$-equations. But the proof does not yield bounded solutions for the Corona problem as desired.

Theorem 2. *Suppose* f_1, f_2 *are bounded analytic functions defined on the bidisc* D^2, *and satisfy*

$$\sum_{j=1}^{2} |f_j(z)| \geq \delta$$

for all $z \in D^2$; *then there exist* g_1, $g_2 \in \bigcap_{p > \infty} H^p(D^2)$ *with*

$$\sum_{j=1}^{2} f_j(z) g_j(z) \equiv 1.$$

With the same reasoning as in T. Wolff's proof, Theorem 2 will be a consequence of Lemma 3 below. For any function g defined on D^2, we will use the notation $\partial_i g = \frac{\partial g}{\partial z_i}$, $\overline{\partial}_i g = \frac{\partial g}{\partial \overline{z}_i}$, $i = 1,\ 2$, $dA = \partial z_1 \overline{\partial} z_1 \partial z_2 \overline{\partial} z_2$. If μ is any Carleson measure on D or D^2, we will say $\mu \in C$ and denote its measure constant (cf. [2]) by $\|\mu\|_C$.

Lemma 3.   Suppose g is a smooth function defined on D^2 and satisfies

(a) $\left| (\partial_1 \partial_2 g)(z_1,\ z_2) \right| \log \dfrac{1}{|z_1|} \log \dfrac{1}{|z_2|}\ dA \in C$ *on* D^2,

(b) $\displaystyle\iint_{D^2} \left| \partial_1 g(z_1,\ z_2) \right| \left| \partial_2 h(z_1,\ z_2) \right|\ \log \dfrac{1}{|z_1|} \log \dfrac{1}{|z_2|}\ dA$

      $\leq C \| h \|_{H^1}$ *for all* $h \in H^1$,

(c)   *same condition as in (b) with* $z_1,\ z_2$ *interchanged,*

(d) $\left| g^2(z_1,\ z_2) \right| \log \dfrac{1}{|z_1|} \log \dfrac{1}{|z_2|}\ dA \in C$ *on* D^2.

Then there exists a solution v on D^2 with $\overline{\partial}_1 \overline{\partial}_2 v = g$, the radial limit of v exists on T^2, and $\| v \|_{L^\infty(T^2)}$ is bounded by a constant depending only on the Carleson measure constant in (a)-(d).

One should remark that by Carleson measure on D^2 we mean a measure that satisfies the Carleson-type geometric condition w.r.t. open set as described in [4, 10]. Condition (b) [also (c)] is a mixed Carleson measure condition for one and two variables that is satisfied, for example, when

$$\left| \overline{\partial}_1 g(z_1,\ z_2) \right| \log \frac{1}{|z_1|}\ dA_1$$

is a Carleson measure in D with $\left\| \overline{\partial}_1 g(z_1,\ z_2) \log \dfrac{1}{|z_1|} \right\|_C$ uniform in z_2.

To prove Lemma 3, we will first establish some function theoretic facts of functions defined on T^2. Let $P_{i,\ j}$ denote the projection from $L^2(T^2)$ to the closed subspace of L^2 generated by $z_1^{in} z_2^{jm}$ $n \geq 0$, $m \geq 0$, $i,\ j = \pm 1$; $H_{ij} = \{ h \in L^\infty(T^2),\ P_{ij}(h) = 0 \}$.

Notice that if we let H denote the biharmonic in D^2 of $h \in L^2$, then $h \in H_{-1,-1}$ if and only if $\bar{\partial}_1 \bar{\partial}_2 H = 0$. Furthermore, if we define BMO(T^2) as the dual of $H^1(T^2)$, and define the norm of $\varphi \in$ BMO(T^2) as

$$\| \varphi \|_* = \inf_{\varphi = \sum_{i,j=\pm 1} P_{ij}(\varphi_{ij})} \sum_{i,j=\pm 1} \| \varphi_{ij} \|_\infty ,$$

then the following lemma is immediate:

Lemma 4. $d_{L^\infty(T^2)}(f, H_{i,j}) \approx \| P_{ij}(f) \|_*$ for all $f \in L^\infty(T^2)$ for all $i, j = \pm 1$.

If we use the notation $H^1(T^2) \equiv \{f \in L^1(T^2), P_{ij}(f) \in L^1$ for all $i, j = \pm 1\}$ and let $\mathcal{H}^1$ denote the analytic function in $H^1(T^2)$, that is, $\mathcal{H}^1 = \{f \in L^1(T^2), f = P_{11}(f)\}$, then the following lemma is also immediate:

Lemma 5. For all $f \in L^\infty(T^2)$,

$$\| P_{-1,-1}(f) \|_* = \sup_{\substack{h \in H^1 \\ \|h\|_{H^1} \le 1}} \left| \int_{T^2} f P_{11}(h) \right| = \sup_{\substack{h \in \mathcal{H}^1 \\ \|h\|_{\mathcal{H}^1} \le 1}} \left| \int_{T^2} f h \right|.$$

Proof of Lemma 3. Since g is smooth, there exists some $v_0 \in C^\infty(\bar{D}^2)$ that satisfies the equation $\bar{\partial}_1 \bar{\partial}_2 v_0 = g$. To prove solution v exists with the desired bound on its norm, we need only to estimate $d_{L^\infty}(v_0, H_{-1,-1})$. Applying Lemmas 4 and 5, we have

$$d_{L^\infty}(v_0, H_{-1,-1}) \le \sup_{h \in \mathcal{H}^1} \left| \int_{T^2} v_0 h \right|.$$

If we assume w.l.o.g. that

$$\int_T v_0(t_1, t_2) dt_2 = \int_T v_0(t_1, t_2) dt_1 = 0,$$

then by Green's theorem,

$$\int_{T^2} v_0 h = C \iint_{D^2} \Delta_1 \Delta_2 (v_0 h) \log \frac{1}{|z_1|} \log \frac{1}{|z_2|} dA.$$

Observe that if we take the biharmonic extension of h in D^2, then

$$\Delta_1 \Delta_2 (v_0 h) = \Delta_1 ((\Delta_2 v_0) h + \nabla_2 v_0 \cdot \nabla_2 h)$$

$$= (\Delta_1 \Delta_2 v_0) h + \overline{\partial}_1 (\Delta_2 v_0) \cdot \partial_1 h + \overline{\partial}_2 (\Delta_1 v) \cdot \partial_2 h + \overline{\partial}_1 \overline{\partial}_2 v_0 \cdot \partial_1 \partial_2 h$$

$$= \partial_1 \partial_2 g \cdot h + \partial_1 g \cdot \partial_2 h + \partial_2 g \cdot \partial_1 h + g \partial_1 \partial_2 h.$$

We can then apply condition (a) to (b) listed to estimate each term independently (cf. [4, 10]).

To obtain Theorem 2 as a consequence of Lemma 3, we make the following observation, which is a direct consequence of Tom Wolff's theorem.

Lemma 6. *Suppose* h *is smooth and satisfies*

(a) $|h(z)|^2 \log \dfrac{1}{|z|} \, dz \, d\overline{z} \, \epsilon \, C$ *in* D

   with constant $\| \ \|_C \leq M.$

(b) $|\partial h(z)| \log \dfrac{1}{|z|} \, dz \, d\overline{z} \, \epsilon \, C$ *in* D

Then

(1)   *there exists a function* v *with* $\overline{\partial} v = h$ *on* D *and*
   $\| v |_T \|_\infty \leq CM.$

(2)   *if* v_0 *is any smooth function* $\overline{\partial} v_0 = h$ *on* $\overline{D}$, *and* V_0 *is*
   the harmonic extension into D *of* $v_0 |_T$, *then*

$$|\overline{\partial} V_0|^2 \log \frac{1}{|z|} \, dz \, d\overline{z} \, \epsilon \, C$$

   with measure constants $\leq CM.$

Proof.   (1) is Tom Wolff's result.

(2)   Choose v as in (1); then $\| v |_T \|_\infty \leq CM$ implies that

$$|\nabla V|^2 \log \frac{1}{|z|} \, dz \, d\overline{z} \, \epsilon \, C$$

with measure constant $\leq CM$ when V is the harmonic extension of v in D. If v_0 is another solution with $\overline{\partial} v = \overline{\partial} v_0 = h$, then $(v - v_0) |_T \, \epsilon \, H^\infty(T)$; hence $\overline{\partial} V_0 = \overline{\partial} V$ in D when V_0 is the harmonic extension of $v_0 |_T$.

Proof of Theorem 2.   We will follow the same pattern of proof as in
[2]. Let f_1 , f_2 be two bounded analytic functions satisfying
$|f_1(z)| + |f_2(z)| \geq 2\delta$ on D^2. Assume f_1, f_2 are smooth in $\overline{D^2}$,
choose C^∞ function $u : \mathbf{C} \to \mathbf{C}$ with $u(z) = 0$ when $|z| < \frac{1}{2}$, $u(z) = 1$
when $|z| \geq 1$.   Let

$$\psi_j(z) = \frac{1}{f_j(z)} \frac{u(f_j(z)/\delta)}{\sum_{j=1}^{2} u(f_j(z)/\delta)}, \quad j = 1, 2;$$

then $\sum_j \psi_j f_j = 1$.   To modify ψ_j to analytic solutions ϕ_j, define
$g_1(z) = \psi_1(z) + f_2(z)k(z)$, $g_2(z) = \psi_2(z) - f_1(z)k(z)$, then
$g_1 f_1 + g_2 f_2 = 1$ and solve

$$(6) \quad \begin{cases} \overline{\partial}_j g_1 = \overline{\partial}_j(\psi_1) + f_2 \overline{\partial}_j k = 0 \\ \overline{\partial}_j g_2 = \overline{\partial}_j(\psi_2) - f_1 \overline{\partial}_j k = 0 \end{cases}, \quad j = 1, 2,$$

that is, we want to solve and obtain $\| k|_{T^2} \|_* \leq$ constant depends
on δ, $\| f_j \|_\infty$, $j = 1, 2$.

$$(7) \quad \overline{\partial}_j k = -\frac{\overline{\partial}_j \psi_1}{f_2} = \frac{\overline{\partial}_j \psi_2}{f_1}, \quad j = 1, 2,$$

which is well defined, since $f_1 \psi_1 + f_2 \psi_2 = 1$.   Let h_j denote

$$-\frac{\overline{\partial}_j \psi_1}{f_2} = \frac{\overline{\partial}_j \psi_2}{f_1}.$$

To solve $\overline{\partial}_j k = h_j$, $j = 1, 2$, with the desired bound, we apply Lemmas
3 and 6 in the following manner.

First notice that h_j, $j = 1, 2$, satisfies conditions (a) and
(b) in Lemma 3 with the Carleson measure norm of

$$|h_1(z_1, z_2)|^2 \log \frac{1}{|z_1|} dA_1, \quad |\partial_1 h_1(z_1, z_2)| \log \frac{1}{|z_1|} dA_1,$$

$$|h_2(z_1, z_2)|^2 \log \frac{1}{|z_2|} dA_2, \quad |\partial_2 h_2(z_1, z_2)| \log \frac{1}{|z_2|} dA_2$$

depending only on δ and $\| f_j \|_\infty$, $j = 1, 2$, and uniform in
$(z_1, z_2) \in D^2$; we will henceforth denote such norm by M.   (The same

proof is in T. Wolff's article.)   Notice also that $\overline{\partial}_2 h_1 = \overline{\partial}_1 h_2$;
applying (2) of Lemma 6, we obtain:

(8)   There exists v_0 satisfying $\overline{\partial}_j v_0 = h_j$, $j = 1, 2$, and if v_0
       is the biharmonic extension (that is, harmonic extension
       in z_1, z_2 separately) of $v_0|_{T^2}$, then

and
$$|\overline{\partial}_1 V_0(z_1, z_2)|^2 \log \frac{1}{|z_1|} \, dA_1$$

$$|\overline{\partial}_2 V_0(z_1, z_2)|^2 \log \frac{1}{|z_2|} \, dA_2$$

are both Carleson measures with constant M uniformly for
$(z_1, z_2) \in D^2$.

Next, if we let $g = \overline{\partial}_2 h_1 = \overline{\partial}_1 h_2$, assume for a moment that g
satisfies conditions (a)-(d) in Lemma 3 with measure constant $\leq CM$.
Then Lemma 3 implies:

(9)   There exists v with $\overline{\partial}_1 \overline{\partial}_2 v = g$ and $\| v|_{T^2} \|_\infty \leq CM$ and
       $v = v_0 + F$ in D^2 for some F biharmonic in D^2 with
       $\overline{\partial}_1 \overline{\partial}_2 F = 0$.

We now take the function v in (9) and try to solve

(10) $\begin{cases} \overline{\partial}_1 k_1 = h_1 - \overline{\partial}_1 v \\ \overline{\partial}_2 k_1 = 0 \end{cases}$ also $\begin{cases} \overline{\partial}_1 k_2 = 0 \\ \overline{\partial}_2 k_2 = h_2 - \overline{\partial}_2 v. \end{cases}$

We wish to obtain k_1, k_2 satisfy (10) and with $\| k_j|_{T^2} \|_\infty \leq CM$.
(This would yield a bounded solution for the Corona problem.)   But
unfortunately, we are only able to obtain solutions with
$\| k_j|_{T^2} \|_* \leq CM$, $j = 1, 2$, instead [see Remark (1) at the end].
To see this, first observe that from (8) and (9), $\overline{\partial}_2(h_1 - \overline{\partial}_1 v) =$
$\overline{\partial}_1(h_2 - \overline{\partial}_2 v) = 0$, so both equations in (10) are well-set.   We will
now apply the one-variable technique in part (1) of Lemma 6 to
solve (10).   To be able to do so, we first check that both $h_1 - \overline{\partial}_1 v$

and $h_2 - \bar{\partial}_2 v$ satisfy the conditions (a) and (b) listed in Lemma 6, and with the measure constants uniform for all z D^2. Now if we denote V the biharmonic extension of v in D^2, then $\| v |_{T^2} \|_{L^\infty} \leq CM$ implies that

$$\| \, |\bar{\partial}_j V(z_1, \, z_2)|^2 \, \log \frac{1}{|z_j|} \, dA \, \|_C \leq CM, \, j = 1, \, 2.$$

Since $V = V_0 + F$ by (9), and

$$\| \, |\bar{\partial}_j V_0|^2 \, \log \frac{1}{|z_j|} \, dA \, \|_C \leq CM, \, j = 1, \, 2,$$

by (8), we get the same property

$$\| \, |\bar{\partial}_j F|^2 \, \log \frac{1}{|z_j|} \, dA \, \|_C \leq CM, \, j = 1, \, 2,$$

for F. Thus

$$h_j - \bar{\partial}_j v = h_j - \bar{\partial}_j (v_0 + F) = -\bar{\partial}_j F$$

satisfy (a), and

$$\partial_j (h \quad - \bar{\partial}_j v) = \partial_j h_j - \partial_j \bar{\partial}_j v_0 - \partial_j \bar{\partial}_j F = 0$$

satisfy (b) trivially for $j = 1, 2$. If we apply Wolff's Lemma 6 to the first set of equations in (10), we obtain a solution u satisfying $\bar{\partial}_1 u = h_1 - \bar{\partial}_1 v$ with $\| u |_{T^2} \|_{L^\infty} \leq CM$. Let k_1 be the orthogonal projection from $L^2(T)$ to $H^2(T)$ in z_2 of u; then k_1 satisfy both $\bar{\partial}_1 k_1 = h_1 - \bar{\partial}_1 v$ and $\bar{\partial}_2 k_1 = 0$ with $\| k_1 \|_* \leq C \| u \|_{L^\infty} \leq CM$. Similarly, we obtain k_2 satisfy the end set equations in (10).

To finish the proof of Theorem 2, we need to verify the function $g = \bar{\partial}_1 h_2 = \bar{\partial}_2 h_1$ satisfies all the conditions (a)-(d) in Lemma 3. The verifications are lengthy but relatively routine. It only depends on the following facts about H^∞ functions f, g in D^2. (Cf. [4, 10].)

(1) $\left| \dfrac{\partial f}{\partial z_j}(z_1, \, z_2) \right|^2 \, \log \dfrac{1}{|z_j|} \, dA_j \, \epsilon \, C, \, j = 1, 2,$ uniform in $z \, \epsilon \, D^2$.

(2) $\left|\dfrac{\partial f}{\partial z_j}(z_1,\ z_2)\right|^2\left|\dfrac{\partial g}{\partial z_i}(z_1,\ z_2)\right|^2\ \log\dfrac{1}{|z_j|}\ \log\dfrac{1}{|z_i|}\ dA\ \epsilon\ C,$

$$i,\ j\ =\ 1,\ 2;\ i\ \neq\ j.$$

(3) $\left|\dfrac{\partial^2 f}{\partial z_1\partial z_2}(z_1,\ z_2)\right|^2\ \log\dfrac{1}{|z_1|}\ \log\dfrac{1}{|z_2|}\ dA\ \epsilon\ C.$

Furthermore, all the Carleson measure constants of above are bounded by constants depending only on $\|f\|_\infty$, $\|g\|_\infty$. We will leave the details to the reader.

Remarks

(1) Given h satisfy $\partial_1 h = \overline{\partial}_2 h = 0$ and

$$\left\|\ |h(z_1,\ z_2)|^2\ \log\dfrac{1}{|z_1|}\ dA_1\right\|_C \leq M$$

uniformly in z_2, there may not exist a bounded solution for the equations

$$\begin{cases} \overline{\partial}_1 k\ =\ h \\[2mm] \overline{\partial}_2 k\ =\ 0. \end{cases}$$

This indicates that the BMO solution is about the best we can hope for for (10) under the given setting. I have learned the following example from Mok (private communication): Let $f = \log(1 - \overline{z}_1 z_2)$ and $h = \dfrac{\partial}{\partial \overline{z}_1}f$; then $\partial_1 h = \overline{\partial}_2 h = 0$, and

$$\left\|\ |h|^2\ \log\dfrac{1}{|z_1|}\ dA_1\right\|_C \leq \|F(z_1,\ z_2)\|_{BMO(z_1)} \leq C \cdot 2\pi$$

for all $z_2\ \epsilon\ D$. Suppose k is a solution with k bounded and $\overline{\partial}_1 k = h$, $\overline{\partial}_2 k = 0$; then $g = k - f$ is analytic in both z_1 and z_2. $g(0,\ 0) = k(0,\ 0)$ is finite, and when restricting to the subspace $z = z_1 = z_2$, and when $|z| = r$, we have $|g(z) + \log(1 - r^2)|$ bounded by $\|k\|_\infty$. Integrating over $|z| = r$, and letting $r \to 1$, yields a contradiction.

(2) Assuming only f_1, f_2, ..., f_n exists in H^p ($p > 1$) in Theorem 2 to start with, N. Varapoulous [12] proved that existence of some g_1, ..., g_n ϵ H^q for some q depends on p with

$$\sum_{j=1}^{n} f_j g_j = 1.$$

It looks like methods used in this section also generalize to that setting (using area-function characterizations of H^p instead of Carleson measure properties of H^∞ in the formulation of Lemmas 3 and 6).

REFERENCES

1. Amar, E. Généralisation d'un théorème de Wolff à la boule de **C**". Preprint.

2. Carleson, L. The Corona problem. *Proc. Fifteenth Scan. Congress 968. Lecture Notes in Math*, vol. 118, Springer-Verlag.

3. Carleson, L. An explicit unconditional basis in H^1. Institut Mittag-Leffler, Rap. No. 2 (1980).

4. Chang, S. Y. Carleson measure on the bi-disc. *Ann. Math.* 109 (1979):613-620.

5. Chang, S. Y., and Ciesielski, Z. Spline characterization of H^1. Preprint.

6. Chang, S. Y., and Fefferman, R. A continuous version of duality of H^1 and BMO on the bidisc. *Ann. Math.* (1980).

7. Ciesielski, Z. Properties of the orthonormal Franklin system. *Studia Math.* 73 (1963):141-157.

8. Ciesielski, Z. Properties of the orthonormal Franklin system II. *Studia Math.* 27 (1966):289-323.

9. Fefferman, C. Characterization of bounded mean oscillation. *Bull. Amer. Math. Soc.* 77 (1971):587-588.

10. Fefferman, R. Functions of bounded mean oscillation on the bi-disc. *Ann. Math.* (1979).

11. Maurey, B. Isomorphismes entre espaces H_1. *Acta Math.* 145 (1980):79-120.

12. Varapoulous, N. Probabilistic approach to some problems in complex analysis. Preprint.

13. Wojtaszczyk, P. The Franklin system is an unconditional basis in H^1. Preprint.

ON THE SOLUTION OF THE EQUATION $\Delta^m F = f$ For $f \in H^p$

Angel B. Gatto , Carlos Segovia
University of Buenos Aires

Julio R. Jiménez
University Central de Venezuela

DEFINITIONS AND NOTATION

Let $L^q(\text{loc})$, $1 < q < \infty$, be the space of the real-valued functions $f(x)$ on R^n that belong locally to L^q. We endow $L^q(\text{loc})$ with the topology generated by the seminorms

$$|f|_{q,B} = (|B|^{-1} \int_B |f(x)|^q \, dx)^{1/q},$$

where B is a ball in R^n and $|B|$ denotes its Lebesgue measure. For $f(x)$ in $L^q(\text{loc})$, we define a maximal function $n_{q,m}(f; x)$ as

$$n_{q,m}(f; x) = \sup_{\rho > 0} \rho^{-2m} |f|_{q, B(x, \rho)},$$

where m is a positive integer.

Let $\mathscr{P}_{2m-1}$ be the linear subspace of $L^q(\text{loc})$ formed by all the polynomials of degree at most $2m - 1$. This subspace is of finite dimension and therefore a closed subspace of $L^q(\text{loc})$. We denote by E^q_{2m-1} the quotient space of $L^q(\text{loc})$ by $\mathscr{P}_{2m-1}$. If $F \in E^q_{2m-1}$, we define the seminorm

$$\|F\|_{q,B} = \inf\{|f|_{q,B} : f \in F\}.$$

The family of all these seminorms induces on E^q_{2m-1} the quotient topology. This space E^q_{2m-1} is locally convex and complete.

Supported in part by Consejo Nacional de Investigaciones Cientificas y Técnicas, República Argentina.

For F in E^q_{2m-1}, we define a maximal function $N_{q,m}(F; x)$ as

$$N_{q,m}(F; x) = \inf\{n_{q,m}(f; x) : f \in F\}.$$

We say that an element F in E^q_{2m-1} belongs to $\mathscr{H}^p_{q,m}$, $0 < p \le 1$, if the maximal function $N_{q,m}(F; x)$ belongs to $L^p(R^n)$. The "norm" of F in $\mathscr{H}^p_{q,m}$ is defined as $\|F\|_{\mathscr{H}^p_{q,m}} = \|N_{q,m}(F; \cdot)\|_{L^p}$.

To simplify the notation, we write $\mathscr{H}^p$ and $N(F; x)$ instead of $\mathscr{H}^p_{q,m}$ and $N_{q,m}(F; x)$, respectively.

Let us denote by $\mathcal{S}$ the set of all C^∞ functions ϕ satisfying that for any pair of nonnegative integers k and h

$$P_{k,h}(\phi) = \max_{|\alpha| \le h} \sup_{x \in R^n} |\partial^\alpha \phi(x)| (1 + |x|)^k$$

is finite. The vector space $\mathcal{S}$ endowed with the family of norms $\{P_{k,h}\}$ becomes a locally convex and metrizable topological vector space. As usual, the dual space $\mathcal{S}'$ of $\mathcal{S}$ is called the space of tempered distributions. If $\phi(x)$ belongs to $\mathcal{S}$, let $\phi_t(x) = t^{-n}\phi(x/t)$.

We recall the definition of the Hardy spaces H^p on R^n. For $0 < p \le 1$, H^p is the space of all tempered distributions f such that

$$f^*(x) = \sup_{|x-y| < t} |(f * \phi_t)(y)|$$

belongs to L^p, for some $\phi \in \mathcal{S}$ with $\int \phi(x)\,dx \ne 0$. The norm $\|f\|_{H^p}$ is defined as the norm of $f^*(x)$ in L^p. A function $b(x)$ on R^n is called a p-atom if there is a ball B containing the support of $b(x)$ such that $\|b\|_\infty \le |B|^{-1/p}$ and $\int b(x)x^\alpha dx = 0$ for all multi-indexes α, $|\alpha| \le N$. The number N is assumed to be as large as necessary. It is well known that given f in H^p, there exist a numerical sequence $\{\lambda_i\}$ and a sequence $\{b_i(x)\}$ of p-atoms such that

$$f = \Sigma \lambda_i b_i$$

and $\Sigma |\lambda_i|^p \approx \|f^*\|^p_{L^p}$. For more details, see [3].

We shall say that a class A E^q_{2m-1} is a p-atom in $\mathscr{H}^p_{q,m}$ if there exist a representative $a(x)$ of A and a ball B such that

$$\operatorname{supp}(a) \subset B$$

and

$$N(A; x) \leq |B|^{-1/p}.$$

Let $F \in E_{2m-1}^q$ and $f \in F$. Since f belongs to $L^q(\text{loc})$, $\Delta^m f$ is defined in the sense of distributions. On the other hand, since any two representatives of F differ in a polynomial of degree smaller than $2m$, we get that $\Delta^m f$ is independent of the representative $f \in F$ chosen. Therefore, for $F \in E_{2m-1}^q$, we shall define $\Delta^m F$ as the distribution $\Delta^m f$, where f is any representative of F.

STATEMENT OF THE RESULTS

The main result of this paper is contained in the following.

Theorem 1. Let $n(2m + n/q)^{-1} < p \leq 1$. The operator Δ^m defines a one-to-one application from $\mathcal{H}_{q,m}^p$ onto H^p. Moreover, there exist two positive and finite constants c and c' such that

$$c\|F\|_{\mathcal{H}_{q,m}^p} \leq \|\Delta^m F\|_{H^p} \leq c'\|F\|_{\mathcal{H}_{q,m}^p} \qquad (1)$$

hold for every $F \in \mathcal{H}_{q,m}^p$.

The case when the hypothesis $n(2m + n/q)^{-1} < p$ does not hold is trivial since:

Theorem 2. If $n(2m + n/q)^{-1} \geq p$, $p \leq 1$, then $\mathcal{H}_{q,m}^p = \{0\}$.

We also give a decomposition into atoms for elements of $\mathcal{H}_{q,m}^p$. The method used to obtain the decomposition was developed in [4].

Theorem 3. Let F belong to $\mathcal{H}_{q,m}^p$, $n(2m + n/q)^{-1} < p \leq 1$. Then, there exist a numerical sequence $\{\mu_i\}$ and a sequence of p-atoms $\{A_i\}$ in $\mathcal{H}_{q,m}^p$ such that $F = \sum \mu_i A_i$ in E_{2m-1}^q. Moreover, there exist two positive constants c_1 and c_2 such that

$$c_1 \|F\|_{\mathscr{H}_{q,m}^P}^P \leq \Sigma |\mu_i|^P \leq c_2 \|F\|_{\mathscr{H}_{q,m}^P}^P .$$

SOME PREVIOUS LEMMAS

Lemma 1.   The maximal function $n(f; x)$ associated to a function f in $L^q(\mathrm{loc})$ is lower semicontinuous.

Proof.   See Lemma 6 of [1].

Lemma 2.   Let f_1 and f_2 be two representatives of an element $F \in E_{2m-1}^q$, and $P = f_1 - f_2$. There exists a constant c_α such that

$$|(\partial^\alpha P)(y)| \leq c_\alpha (n(f_1; x_1) + n(f_2; x_2))(|x_1 - y| + |x_2 - y|)^{2m-|\alpha|}$$

holds for every x_1, x_2, and y in R^n.

Proof.   The proof is the same as Lemma 3 in [1].

Lemma 3.   Let F belong to E_{2m-1}^q with $N(F; x_0) < \infty$.   Then:

(i)   There exists a unique f in F such that $n(f; x_0) < \infty$ and, therefore, $n(f; x_0) = N(F; x_0)$.

(ii)   For any ball B, there is a finite constant c depending on x_0 and B such that if f is the unique representative of F given in (i), then

$$\|F\|_{q,B} \leq |f|_{q,B} \leq c\, n(f; x_0) = c\, N(F; x_0).$$

The constant c can be chosen independently of x_0 provided that x_0 varies in a compact set.

Proof.   Let us prove (i).   Assume that f_1 and f_2 belong to F and both $n(f_1; x_0)$ and $n(f_2; x_0)$ are finite.   Call P the polynomial $f_1 - f_2$ of degree at most $2m - 1$.   Applying Lemma 2 with

$x_1 = x_2 = x_0 = y$ and varying α, $0 \leq |\alpha| \leq 2m - 1$, we get $\partial^\alpha P(x_0) = 0$, which shows that $P(x) = 0$.

Let us consider part (ii). We can assume $N(F; x_0) < \infty$, since, otherwise, there is nothing to prove. By part (i) already proved, let f be the unique representative of F satisfying $n(f; x_0) = N(F; x_0)$. For a given ball B, let ρ_0 be the least value of ρ for which $B \subset B(x_0, \rho)$. Then

$$\|F\|_{q, B} \leq |f|_{q, B} \leq \rho_0^{2m} |B|^{-1/q} |B(x_0, \rho_0)|^{1/q} \rho_0^{-2m} |f|_{q, B(x_0, \rho_0)}$$

$$\leq c_{x_0, B} n(f; x_0) = c_{x_0, B} N(F; x_0).$$

Corollary 1. *If $\{F_k\}$ is a sequence of elements of E_{2m-1}^q converging to F in $\mathscr{H}^p$, $0 < p \leq 1$, then $\{F_k\}$ converges to F in E_{2m-1}^q.*

Proof. For any ball B, by Lemma 3, we have

$$\|F_k - F\|_{q, B}^q \leq c^p |B|^{-1} \int_B N(F_k - F, x_0)^p \, dx_0 \leq c' \|F_k - F\|_{\mathscr{H}^p}^p,$$

which proves the corollary.

Lemma 4. *Let $\{F_k\}$ be a sequence in E_{2m-1}^q such that for a given point x_0, the series $\sum_k N(F_k; x_0)$ is finite. Then:*

(i) *The series $\sum_k F_k$ converges in E_{2m-1}^q to an element F and $N(F; x_0) \leq \sum_k N(F_k; x_0)$.*

(ii) *If f_k is the unique representative of F_k satisfying $n(f_k; x_0) = N(F_k; x_0)$, then $\sum_k f_k$ converges in $L^q(\mathrm{loc})$ to a function f that is the unique representative of F satisfying $n(f; x_0) = N(F; x_0)$.*

Proof. Since $\sum N(F_k; x_0)$ is finite, by Lemma 3, there exists a representative f_k of each F_k satisfying $n(f_k; x_0) = N(F_k; x_0) < \infty$. For any ball B and by Lemma 3, we have

$$\left| \sum_{k=1}^K f_k - \sum_{k=1}^J f_k \right|_{q, B} \leq \sum_{k=J+1}^K |f_k|_{q, B} \leq c \sum_{k=J+1}^K N(F_k; x_0).$$

This shows that $\sum_k f_k$ converges in L^q (loc) to a function f. Let us denote by F the class of f in E^q_{2m-1}. Since $n(f; x_0)$ $\leq \sum_k n(f_k; x_0) = \sum_k N(F_k; x_0) < \infty$, we get that f is the unique representative of F satisfying $n(f; x_0) = N(F; x_0)$. On the other hand,

$$\left\| \sum_{k=1}^{K} F_k - F \right\|_{q, B} \leq \left| \sum_{k=1}^{K} f_k - f \right|_{q, B},$$

which shows the convergence in E^q_{2m-1} of the series $\sum_k F_k = F$.

Corollary 2. The space $\mathcal{H}^p$, $0 < p \leq 1$, is complete.

Proof. It is enough to show that if $\{F_k\}$ is a sequence such that $\sum_k \|F_k\|_{\mathcal{H}^p}^p$ is finite, then $\sum_k F_k$ converges in $\mathcal{H}^p$. Observe that for $0 < p \leq 1$,

$$\int \left(\sum_k N(F_k; x) \right)^p dx \leq \sum_k \int N(F_k; x)^p dx < \infty.$$

This implies that $\sum_k N(F_k; x)$ is finite almost everywhere. Then, by Lemma 4, $\sum_k F_k$ converges in E^q_{2m-1} to an element F. Now

$$N \left(\sum_{k=1}^{K} F_k - F; x \right) \leq \sum_{k=K+1}^{\infty} N(F_k; x).$$

Integrating the pth power of this inequality, we obtain

$$\left\| \sum_{k=1}^{K} F_k - F \right\|_{\mathcal{H}^p}^p \leq \sum_{k=K+1}^{\infty} \|F_k\|_{\mathcal{H}^p}^p,$$

which shows that the series $\sum_k F_k$ converges in $\mathcal{H}^p$.

Lemma 5. If $g(x)$ belongs to L^q(loc) and $\phi \in \mathcal{S}$, then for $k > 2m+n$,

$$\int |g(y)| |\phi(y)| dy \leq c \, p_{k, 0}(\phi) \, n(g; 0).$$

Proof. Since $|\phi(y)| \leq p_{k, 0}(\phi) (1 + |y|)^{-k}$, then

$$\int |g(y)| |\phi(y)| dy \leq p_{k, 0}(\phi) \int |g(y)| (1 + |y|)^{-k} dy.$$

Now,

$$\int_{B(0, 1)} |g(y)| (1 + |y|)^{-k} dy \leq \int_{B(0, 1)} |g(y)| dy \leq c \, n(g; 0).$$

Moreover,

$$\int_{B(0,\ 2^{j+1}) \sim B(0,\ 2^j)} |g(y)| \, (1 + |y|)^{-k} \, dy \le 2^{-jk} \int_{B(0,\ 2^{j+1})} |g(y)| \, dy \le c 2^{(2m+n-k)j} n(g; 0).$$

Since $2m + n - k < 0$, adding up these estimates, we get the lemma.

Corollary 3. If $g(x) \in L^q(\text{loc})$ and there is a point x_0 such that $n(g; x_0) < \infty$, then $g(x)$ defines a tempered distribution that satisfies

$$\left| \int g(y) \phi(y) \, dy \right| \le c \, p_{k,0}(\phi) n(g; x_0) (1 + |x_0|)^k$$

if $k > n + 2m$ and $\phi(y) \in S$.

Proof. Using Lemma 5, we have

$$\int |g(y)| \, |\phi(y)| \, dy = \int |g(x_0 + y)| \, |\phi(x_0 + y)| \, dy$$

$$\le c \, p_{k,0}(\phi(x_0 + y)) \, n(g(x_0 + y); 0).$$

Since $p_{k,0}(\phi(x_0 + y)) \le (1 + |x_0|)^k p_{k,0}(\phi)$ and $n(g(x_0 + y); 0) = n(g; x_0)$, we get the corollary.

Lemma 6. Let $g(x) \in L^q(\text{loc}) \cap S'$ and $f = \Delta^m g$ in the sense of distributions. Then if $\phi \in S$,

$$f^*(x) = \sup_{|x-y| < t} |(f^* \phi_t)(y)| \le c \, p_{k,2m}(\phi) n(g; x) \qquad (2)$$

holds for every x in R^n.

Proof. Let $|x - y| < t$. Since $f = \Delta^m g$ is a tempered distribution, we have

$$(f^* \phi_t)(y) = \langle f_z, \phi_t(y - z) \rangle = \langle \Delta_z^m g, \phi_t(y - z) \rangle$$

$$= t^{-2m} \langle g_z, (\Delta^m \phi)_t (y - z) \rangle.$$

Therefore,

$$(f^* \phi_t)(y) = t^{-2m} \int g(z) (\Delta^m \phi)_t (y - z) \, dz.$$

By a change of variables, we get

$$(f^* \phi_t)(y) = t^{-2m} \int g(x + tu) (\Delta^m \phi) (t^{-1}(y - x) - u) \, du.$$

Applying Lemma 5, we obtain

$$\left| (f^{*}\phi_{t})(y) \right| \leq c\, n(g;\, x)\, p_{k,0}\left((\Delta^{m}\phi)(t^{-1}(y-x)-u) \right). \tag{3}$$

Since $|x-y| < t$, we have

$$1 + |u| \leq 1 + |u - t^{-1}(y-x)| + |t^{-1}(y-x)|$$

$$\leq 2\left(1 + |u - t^{-1}(y-x)| \right),$$

then

$$p_{k,0}\left((\Delta^{m}\phi)(t^{-1}(y-x)-u) \right) \leq 2^{k}\, p_{k,2m}(\phi).$$

This and (3) show that

$$\left| (f^{*}\phi_{t})(y) \right| \leq c'\, p_{k,2m}(\phi)\, n(g;\, x),$$

for $|x-y| < t$, which implies (2).

Let $h(x)$ be the function defined as

$$h(x) = |x|^{2m-n}\, \lg|x| \quad \text{if } n \text{ is even and } 2m-n \geq 0, \text{ and}$$

$$h(x) = |x|^{2m-n} \qquad \text{otherwise.} \tag{4}$$

It is well known that if $b(x)$ is a bounded function with compact support, its potential $B(x)$, defined as

$$B(x) = \int h(x-y)\, b(y)\, dy,$$

is a locally bounded function and $\Delta^{m}B = b$ in the sense of distributions. For these potentials, we shall prove the following.

Lemma 7.   Let $a(x)$ be a p-atom with vanishing moments up to the order $2m-1$ and assume that $B(0, d)$ is the ball containing the support of $a(x)$ in the definition of p-atom.   If

$$A(x) = \int h(x-y)\, a(y)\, dy,$$

then for $|x| \geq 2d$ and every α, there exists c_{α} such that

$$\left| (\partial^{\alpha}A)(x) \right| \leq c_{\alpha}\, d^{-n/p + 2m + n}\, |x|^{-n-|\alpha|}$$

holds.

Proof. Since $a(x)$ has vanishing moments up to the order $2m - 1$ and $|x| > 2d$, we have

$$(\partial^\alpha A)(x) = \int (\partial^\alpha h)(x - y) a(y) dy$$

$$= \int \left[(\partial^\alpha h)(x - y) - \sum_{|\beta| \leq 2m - 1} (\partial^{\alpha + \beta} h)(x)(-y)^\beta / \beta! \right] a(y) dy$$

$$= \int \left[\sum_{|\beta| = 2m} (\partial^{\alpha + \beta} h)(x - \theta y)(-y)^\beta / \beta! \right] a(y) dy,$$

where $0 < \theta < 1$. Taking into account that

$$|x - \theta y| \geq |x| - |y| \geq |x|/2,$$

and since for $|\beta| = 2m$, $(\partial^{\alpha + \beta} h)(x)$ is a homogeneous function of degree $-n - |\alpha|$, we obtain

$$|(\partial^\alpha A)(x)| \leq c \int_{|y| \leq d} |x - \theta y|^{-n - |\alpha|} dy \ d^{-n/p} d^{2m} \leq c \ d^{-n/p + 2m + n} |x|^{-n - |\alpha|}.$$

Lemma 8. *If $h(x)$ is the kernel defined in (4) and $|\alpha| = 2m$, then*

(i)   *$(\partial^\alpha h)(x)$ is a homogeneous function of degree $-n$ and C^∞ on $R^n \sim \{0\}$.*

(ii) $\displaystyle \int_{|x| = 1} (\partial^\alpha h)(x) dx = 0.$

Proof. Part (i) follows immediately from the definition of $h(x)$.

To prove (ii), we define for $0 < \rho < 1$ and $|\theta| = 1$, the function $F(\rho\theta)$ as

$$F(\rho\theta) = |\Sigma_{n-1}|^{-1} \int_{|z| = 1} h(z + \rho\theta) dz. \tag{5}$$

By a change of variables, and since h is a radial function, it follows that $F(\rho\theta) = f(\rho)$. Then, interchanging the order of integration, we get

$$f(\rho) = |\Sigma_{n-1}|^{-1} \int_{|\theta| = 1} \left(|\Sigma_{n-1}|^{-1} \int_{|z| = 1} h(z + \rho\theta) dz \right) d\theta$$

$$= |\Sigma_{n-1}|^{-1} \int\limits_{|z|=1} \left(|\Sigma_{n-1}|^{-1} \int\limits_{|\theta|=1} h(z + \rho\theta) \, d\theta \right) dx.$$

Now, since $\Delta^m h(x) = 0$ for $x \neq 0$, applying Pizetti's formula (see, for instance, [2]) to the innermost integral of the last term, we obtain

$$f(\rho) = |\Sigma_{n-1}|^{-1} \int\limits_{|z|=1} \left(\sum_{k=0}^{m-1} c_k \rho^{2k} (\Delta^k h)(z) \right) dz,$$

which shows that $f(\rho)$ is a polynomial of degree less than $2m - 1$. Then, recalling that $f(\rho) = F(\rho\theta)$, if we take $2m$ derivatives with respect to ρ at $\rho = 0$ in (5), it follows that

$$0 = \int\limits_{|z|=1} D_\rho^{2m} [h(z + \rho\theta)]_{\rho=0} \, dz = \sum_{|\alpha|=2m} \int\limits_{|z|=1} (\partial^\alpha h)(z) \, dz \; \theta^\alpha (2m)!/\alpha!$$

for every θ. This implies (ii).

PROOF OF THE RESULTS

Proof of Theorem 1. Let $F \in \mathcal{H}^p$. Since $N(F; x)$ is finite a.e. on R^n, by Corollary 3, every representative of F is a function in $L^q(\text{loc}) \cap \mathcal{S}'$. Thus, if $\phi \in \mathcal{S}$ and $\int \phi(x) \, dx \neq 0$, by Lemma 6, we get

$$(\Delta^m F)^*(x) \leq c \, p_{k, 2m}(\phi) N(F; x).$$

Taking the p-power and integrating, we obtain that $\Delta^m F$ belongs to H^p and $\|\Delta^m F\|_{H^p} \leq c\|F\|_{\mathcal{H}^p}$. This proves the second inequality in (1).

The fact that Δ^m is one to one on $\mathcal{H}^p$ follows from Lemma 9 of [1].

To prove that the application is onto, let us consider a p-atom $a(x)$ satisfying the hypotheses of Lemma 7. If $A(x) = \int h(x - y) a(y) \, dy$, we define

$$R(x, z) = A(x + z) - \sum_{|\alpha| \leq 2m-1} (\partial^\alpha h)(x - y) a(y) \, dy \; z^\alpha/\alpha!.$$

We shall prove first that for $|x| > 4d$, the following esti-mates hold: if $|z| < |x|/2$, then

$$|R(x, z)| \leq c\, d^{-n/p}(d/|x|)^{2m+n}|z|^{2m},\tag{6}$$

if $|z| \geq |x|/2$ and $|x + z| > 2d$, then

$$|R(x, z)| \leq c\, d^{-n/p}d^{2m+n}|x + z|^{-n} + c\, d^{-n/p}(d/|x|)^{2m+n}|z|^{2m},\tag{7}$$

and if $|z| \geq |x|/2$ and $|x + z| \leq 2d$, then

$$|R(x, z)| \leq c\, d^{-n/p}d^{2m}(d/|x|)^{-\mu} + c\, d^{-n/p}(d/|x|)^{2m+n}|z|^{2m}.\tag{8}$$

To prove (6), we observe that for $0 < \theta < 1$,
$|x - \theta z| > |x| - |z| > |x|/2 > 2d$. Then, by the mean value
theorem and Lemma 7, we get

$$|R(x, z)| \leq \sum_{|\alpha| = 2m} |(\partial^\alpha A)(x + \theta z)||z|^{|\alpha|}/\alpha! \leq c\, d^{-n/p}(d/|x|)^{2m+n}|z|^{2m}.$$

Now, let $|z| \geq |x|/2$. We have

$$|R(x, z)| \leq |A(x + z)| + \sum_{|\alpha| \leq 2m-1} |(\partial^\alpha A)(x)||z|^{|\alpha|}/\alpha!.$$

To estimate the right-hand side of this inequality, we
shall consider first the terms under the summation sign. Since
$|x| > 4d > 2d$, by Lemma 7, and observing that $|z|/|x| > 1/2$, we
have

$$|(\partial^\alpha A)(x)||z|^{|\alpha|}/\alpha! \leq c\, d^{-n/p}d^{2m+n}|x|^{-n+|\alpha|}|z|^{|\alpha|}$$
$$\leq c\, d^{-n/p}(d/|x|)^{2m+n}|z|^{2m}.\tag{9}$$

As for the other term, $|A(x + z)|$, we distinguish the cases
$|x + z| > 2d$ and $|x + z| \leq 2d$. In the case $|x + z| > 2d$, we apply
Lemma 7 for $\alpha = 0$, obtaining

$$|A(x + z)| \leq c\, d^{-n/p}d^{2m+n}|x + z|^{-n}.$$

This estimate together with (9) give (7). For $|x + z| \leq 2d$,
we consider the cases n even and $2m - n \geq 0$, and n odd or
$2m - n < 0$. In the first case, since $|x|^{2m-n}$ is a polynomial of
degree smaller than $2m$, we have

$$A(x + z) = c \int |x + z - y|^{2m-n} \lg|x + z - y| \, a(y) dy$$

$$= c \int |x + z - y|^{2m-n} (\lg|x + z - y| - \lg|x|) a(y) dy.$$

Since $|x + z - y|/|x| \leq 3d/4d < 1$, we get

$$|A(x + z)| \leq c \, d^{-n/p} \int_{|y| \leq d} |x + z - y|^{2m-n} \lg(|x|/|x + z - y|) dy.$$

Recalling that $\lg t \leq \mu^{-1} t^\mu$ for any $0 < \mu$ and $t \geq 1$, and since $|y| \leq d$ implies $|x + z - y| \leq 3d$, we get

$$|A(x + z)| \leq c \, d^{-n/p} |x|^\mu \int_{|x+z-y| \leq 3d} |x + z - y|^{2m-n-\mu} dy$$

$$\leq c \, d^{-n/p} d^{2m} (d/|x|)^{-\mu}.$$

This estimate together with (9) give (8). For the case n odd or $2m - n < 0$, the proof is simpler since no logarithm appears, and the estimates obtained hold with $\mu = 0$.

Let us estimate

$$\rho^{-2m} \left(\rho^{-n} \int_{|z| \leq \rho} |R(x, z)|^q dz \right)^{1/q}, \quad \rho > 0.$$

According to the estimates obtained for $R(x, z)$ above, we split the domain of integration into three subsets:

$$D_1 = \{z : |z| \leq \rho, \ |z| < |x|/2\},$$

$$D_2 = \{z : |z| \leq \rho, \ |z| \geq |x|/2, \ |x + z| > 2d\},$$

and

$$D_3 = \{z : |z| \leq \rho, \ |z| \geq |x|/2, \ |x + z| \leq 2d\}.$$

By (6), we have

$$\rho^{-2m} \left(\rho^{-n} \int_{D_1} |R(x, z)|^q dz \right)^{1/q} \leq c \, d^{-n/p} (d/|x|)^{2m+n}. \tag{10}$$

Let us assume $D_2 \neq \phi$. By (7), we get

$$\rho^{-2m} \left(\rho^{-n} \int_{D_2} |R(x, z)|^q dz \right)^{1/q}$$

$$\leq c \, d^{-n/p} d^{2m+n} \rho^{-2m} \left(\rho^{-n} \int_{|x+z| > 2d} |x + z|^{-nq} dz \right)^{1/q} + c \, d^{-n/p} (d/|x|)^{2m+n}.$$

Since

$$\int_{|x+z| > 2d} |x + z|^{-nq} dz \leq c \, d^{-nq+n},$$

and observing that $D_2 \neq \phi$ implies $\rho > |x|/2$, we get

$$\rho^{-2m}\left(\rho^{-n}\int_{D_2}|R(x,\,z)|^q\,dz\right)^{1/q} \leq c\,d^{-n/p}(d/|x|)^{2m+n}. \tag{11}$$

For the case $D_3 \neq \phi$, by (8), we have

$$\rho^{-2m}\left(\rho^{-n}\int_{D_3}|R(x,\,z)|^q\,dz\right)^{1/q}$$

$$\leq c\,d^{-n/p}\,d^{2m}\,(d/|x|)^{-\mu}\,\rho^{-2m}\left(\rho^{-n}\int_{|x+z|\leq 2d}dz\right)^{1/q} + c\,d^{-n/p}\,(d/|x|)^{2m+n}.$$

The assumption $D_3 \neq \phi$ implies $\rho > |x|/2$.   Therefore,

$$\rho^{-2m}\left(\rho^{-n}\int_{D_3}|R(x,\,z)|^q\,dz\right)^{1/q}$$

$$\leq c\,d^{-n/p}(d/|x|)^{2m+n/q-\mu} + c\,d^{-n/p}(d/|x|)^{2m+n}. \tag{12}$$

The estimates (10), (11), and (12) and $|x|/d > 4$ imply

$$N(A;\,x) \leq c\,d^{-n/p}(d/|x|)^{2m+n/q-\mu}. \tag{13}$$

For $|x| \leq 4d$, we shall prove the following estimate for $R(x,\,z)$:

$$|R(x,\,z)| \leq c|z|^{2m}\left(d^{-n/p} + \sum_{|\alpha|=2m}K_\alpha^*(a)\,(x)\right), \tag{14}$$

where K_α^* is the maximal singular integral operator corresponding to the kernel $\partial^\alpha h(x)$; see Lemma 8.   We have

$$R(x,\,z) = \int\left[h(x+z-y) - \sum_{|\alpha|\leq 2m-1}(\partial^\alpha h)(x-y)z^\alpha/\alpha!\right]a(y)\,dy$$

$$= \int_{|x-y|<2|z|} + \int_{|x-y|\geq 2|z|} = I_1 + I_2.$$

We can write, at least if y does not belong to the segment $[x,\,x+z]$,

$$U = h(x+z-y) - \sum_{|\alpha|\leq 2m-1}(\partial^\alpha h)(x-y)z^\alpha/\alpha!$$

$$= (2m-1)\sum_{|\alpha|=2m-1}(z^\alpha/\alpha!)\int_0^1 \partial^\alpha h(x+tz-y)(1-t)^{2m-2}dt$$

$$+ \sum_{|\alpha|=2m-1}(\partial^\alpha h)(x-y)z^\alpha/\alpha!.$$

Since for $|\alpha| = 2m - 1$ the derivatives $\partial^\alpha h(x)$ are homogeneous functions of degree $-n + 1$, we get

$$|U| \le c\left(\int_0^1 |x + tz - y|^{-n+1}(1 - t)^{2m-2}dt + |x - y|^{-n+1}\right)|z|^{2m-1}.$$

Therefore, since $|x - y| < 2|z|$ implies $|x + tz - y| \le 3|z|$, we get the estimate

$$|I_1| \le \int_{|x-y| < 2|z|} |U|\,|a(y)|\,dy$$

$$\le c\,d^{-n/p}\left[\int_0^1 (1 - t)^{2m-2}\left(\int_{|x+tz-y| \le 3|z|} |x + tz - y|^{-n+1}dy\right)dt\right.$$

$$\left. + \int_{|x-y| < 2|z|} |x - y|^{-n+1}dy\right]|z|^{2m-1} \le c\,d^{-n/p}|z|^{2m}.$$

To estimate I_2, we write

$$U = h(x + z - y) - \sum_{|\alpha| \le 2m-1} (\partial^\alpha h)(x - y)z^\alpha/\alpha!$$

$$= \left[h(x + z - y) - \sum_{|\alpha| \le 2m} (\partial^\alpha h)(x - y)z^\alpha/\alpha!\right] + \sum_{|\alpha| = 2m} (\partial^\alpha h)(x - y)z^\alpha/\alpha!$$

$$= U_1 + U_2.$$

Let us estimate U_1. We have that for some $0 < s < 1$,

$$|U_1| \le \sum_{|\alpha| = 2m+1} |(\partial^\alpha h)(x + sz - y)|\,|z|^{2m+1}/\alpha!.$$

Since $|x - y| > 2|z|$ implies $|x + sz - y| > |x - y|/2$, and recalling that for $|\alpha| = 2m + 1$ the derivatives $\partial^\alpha h$ are homogeneous functions of degree $-n - 1$, we get $|U_1| \le c|x - y|^{-(n+1)}|z|^{2m+1}$.

Therefore,

$$|I_2| = \left|\int_{|x-y| > 2|z|} U\,a(y)\,dy\right|$$

$$\le \int_{|x-y| > 2|z|} |U_1|\,|a(y)|\,dy + \left|\int_{|x-y| > 2|z|} U_2\,a(y)\,dy\right|$$

$$\le c\left(d^{-n/p} + \sum_{|\alpha| = 2m} K_\alpha^*(a)(x)\right)|z|^{2m}.$$

This ends the proof of (14).

Using (14), we get that for $|x| \le 4d$,

$$N(A; x) \leq c\left(d^{-n/p} + \sum_{|\alpha| = 2m} K_\alpha^*(a)(x)\right).\tag{15}$$

The estimates (13) and (15) for $N(A; x)$ allow us to write

$$\int N(A; x)^p dx = \int_{|x| > 4d} + \int_{|x| \leq 4d} \leq c\, d^{-n} \int_{|x| > 4d} (d/|x|)^{(2m + n/q - \mu)p} dx$$

$$+ c\, d^{-n} \int_{|x| \leq 4d} dx + c \sum_{|\alpha| = 2m} \int_{|x| \leq 4d} K_\alpha^*(a)(x)^p dx,$$

which by Hölder inequality is bounded by

$$c\left(1 + \sum_{|\alpha| = 2m} d^{n(1 - p/2)}\left(\int K_\alpha^*(a)(x)^2 dx\right)^{p/2}\right).$$

Since K_α^* is of strong type 2-2, and since $\|a\|_2 \leq c\, d^{n(p-2)/2p}$, we get

$$\int N(A; x)^p dx \leq c.\tag{16}$$

Let ϕ belong to H^p. We know that there exist a numerical sequence $\{\lambda_i\}$ and a sequence $\{a_i(x)\}$ of p-atoms in H^p such that $\phi = \sum \lambda_i a_i$ and $\sum |\lambda_i|^p \leq c\|\phi\|_{H^p}$. If $A_i(x)$ are the potentials of the p-atoms $a_i(x)$, then from (16) we get $\sum |\lambda_i| \|A_1\|_{\mathcal{H}^p}^p < \infty$. By Corollary 2, there exists F in $\mathcal{H}^p$ such that $F = \sum \lambda_i A_i$. Recalling that we have already shown that Δ^m is a continuous operator from $\mathcal{H}^p$ into H^p, we get

$$\Delta^m F = \sum \lambda_i \Delta^m A_i = \sum \lambda_i a_i = \phi.$$

This shows that Δ^m is onto H^p. Moreover,

$$\|F\|_{\mathcal{H}^p}^p \leq \sum |\lambda_i|^p \|A_i\|_{\mathcal{H}^p}^p \leq c \sum |\lambda_i|^p \leq c'\|\phi\|_{H^p}^p = c'\|\Delta^m F\|_{H^p}^p.$$

This ends the proof of the theorem.

Proof of Theorem 2. Let $G \in \mathcal{H}_{q, m}^p$ and assume $G \neq 0$. Then there exists a representative $h(x)$ of G that is not a polynomial of degree less than or equal to $2m - 1$. It is easy to see that there exists a ball B with center at 0 and radius $r > 0$ such that for every $P \in \mathcal{P}_{2m-1}$,

$$\left(\int_B |h(y) - P(y)|^q dy\right)^{1/q} \geq a > 0.$$

Let x be a point such that $|x| > r$ and let $\rho = 2|x|$. Then $B \subset B(x, \rho)$. Let $g(x)$ be any representative of G. Then $g = h - P$ for some $P \in \mathscr{P}_{2m-1}$ and

$$\rho^{-2m}|g|_{q, B(x, \rho)} \geq |B(0, 1)|^{-1/q} 2^{-2m-n/q}|x|^{-2m-n/q}a = c|x|^{-2m-n/q},$$

showing that $N(F; x) \geq c|x|^{-2m-n/q}$, for $|x| > r$. Taking the pth power and integrating, we get a contradiction. Therefore, $G = 0$ and the theorem is proved.

To prove Theorem 3, a Calderón-Zygmund-type lemma is required. The statement of this lemma depends on a partition of unity given in the following:

Lemma 9 (a partition of unity). Let $\Omega \neq \phi$ be an open subset of R^n not coinciding with R^n. There exists a sequence $\{\phi_k\}_{k=1}^{\infty}$ of C_0^{∞} functions satisfying the following conditions

   (i) $0 \leq \phi_k(x) \leq 1$ and $\Sigma \phi_k(x) = \chi_{\Omega}(x)$.

   (ii) *For each* k, *there is a ball* $B_k = B(x_k, r_k) \subset \Omega$
         such that $\mathrm{supp}(\phi_k) \subset B_k$, *and for every*
         $z \in B_k$, $r_k \leq d(z, C\Omega) \leq c\, r_k$.

   (iii) *For every* k, *the ball* $B(x_k, 2r_k)$ *is contained in* Ω.
         Moreover, there exists an integer M *such that the*
         number of balls $B(x_j, 2r_j)$ *that intersect* $B(x_k, 2r_k)$
         does not exceed M.

   (iv) *For every multiindex* α, *we have*

         $$|(\partial^{\alpha}\phi_k)(x)| \leq c_{\alpha}r_k^{-|\alpha|},$$

         where c_{α} *does not depend on* k.

Proof. See [5].

Lemma 10 (Calderón-Zygmund-type lemma). Let F be an element of E_{2m-1}^q whose maximal function belongs to $L^p(R^n)$. Let

$n(2m + n/q)^{-1} < p \leq 1$. Given $t > 0$, let $\Omega = \Omega_t = \{x : N(F; x) > t\}$.
By Lemma 1, this is an open subset. Let $\{\phi_k\}_{k=1}^{\infty}$ be the partition
of unity associated to Ω in Lemma 9. For each k, let y_k be a point
of $C\Omega$ such that $d(B(x_k, 2r_k), C\Omega) = d(B(x_k, 2r_k), y_k)$. For a given
representative $f(x)$ of F, by Lemma 3, there is a polynomial
$P(y_k, y)$ of $\mathcal{P}_{2m-1}$ satisfying $N(F, y_k) = n(f(y) - P(y_k, y); y_k)$.
For $k = 1, 2, \ldots$, we define the functions $w_k(y) = \phi_k(y)(f(y) - P(y_k, y))$. These functions $w_k(y)$ belong to $L^q(\text{loc})$. Let us denote
by W_k the class of $w_k(x)$ in E_{2m-1}^q. Then the following conditions
are satisfied:

(i) $N(W_k; x) \leq c\, N(F; x)$ if $x \epsilon B(x_k, 2r_k)$.

(ii) $N(W_k; x) \leq ct\,(r_k/(|x - x_k| + r_k))^{2m+n/q}$ if $x \notin B(x_k, 2r_k)$.

(iii) The series $\sum_k N(W_k; x)$ is pointwise convergent for almost
      every x in R^n. Moreover,

$$\int \left(\sum_k N(W_k; x) \right)^p dx \leq \sum_k \int N(W_k; x)^p dx \leq c \int_\Omega N(F; x)^p dx.$$

(iv) The series $\sum_k W_k = W$ converges in E_{2m-1}^q, and for almost
      every x,

$$N(W; x) \leq \sum_k N(W_k; x).$$

(v) $\int N(W; x)^p dx \leq c \int_\Omega N(F; x)^p dx.$

(vi) If $G = F - W$, there exists a representative of G having
      continuous derivatives up to the order $2m - 1$, and the
      derivatives of order $2m - 1$ satisfy a Lipschitz condition
      with constant not exceeding ct. Moreover, $N(G; x) \leq ct$.

All the constants c appearing in the statement of the lemma
are finite and independent of F and $t > 0$.

Proof. Let us prove (i). Let $x \epsilon B(x_k, 2r_k)$. Assume $N(F; x) < \infty$.
Otherwise, there is nothing to prove. Let f be a representative of
F and $P(x, y)$ the polynomial of degree less than $2m$ satisfying

$n(f(y) - P(x, y); x) = N(F; x)$. We define the polynomial $Q_k(x, y)$ as

$$Q_k(x, y) = \sum_{|\alpha| \leq 2m-1} \partial_y^\alpha (\phi_k(y) P(x, y) - P(y_k, y)) \Big|_{y=x} (y - x)^\alpha / \alpha!.$$

Let us estimate $\rho^{-2m} |w_k(y) - Q_k(x, y)|_{q, B(x, \rho)}$.

By the Leibnitz formula and reordering the sum, we can write

$$Q_k(x, y) = \sum_{|\alpha| \leq 2m-1} \left[\partial_y^\alpha (P(x, y) - P(y_k, y)) \Big|_{y=x} (y - x)^\alpha / \alpha! \right.$$
$$\left. \times \left(\sum_{|\gamma| = 2m-1-|\alpha|} (\partial^\gamma \phi_k)(x) (y - x)^\gamma / \gamma! \right) \right]. \tag{17}$$

From the hypothesis we get $|x_k - y_k| \leq 8r_k$. Therefore,

$$|y - x| + |y - y_k| \leq \rho + |y - x| + |x - x_k| + |x_k - y_k|$$
$$\leq 10(\rho + r_k).$$

Then by Lemma 2 and since $t < N(F; x)$, we get

$$|\partial_y^\alpha (P(x, y) - P(y_k, y))| \leq c_\alpha N(F; x) (\rho + r_k)^{2m-|\alpha|}. \tag{18}$$

Assume first that $\rho \geq 2r_k$. Then

$$|w_k(y) - Q_k(x, y)|$$
$$\leq \phi_k(y) |f(y) - P(x, y)| + \phi_k(y) |P(x, y) - P(y_k, y)| + |Q_k(x, y)|. \tag{19}$$

For the second term of the right-hand side of this inequality, taking $\alpha = 0$ in (18) and since $0 \leq \phi_k(x) \leq 1$, we obtain

$$\phi_k(y) |P(x, y) - P(y_k, y)| \leq c N(F; x) \rho^{2m}.$$

Since $|x - y| \leq 10 r_k$, by Lemma 2, we have

$$|\partial_y^\alpha (P(x, y) - P(y_k, y))|_{y=x}| \leq c N(F; x) r_k^{2m-|\alpha|}.$$

Using this estimate, condition (iv) of Lemma 9, and (17), it follows that

$$|Q_k(x, y)| \leq \sum_{|\alpha| \leq 2m-1} c N(F; x) r_k^{2m-|\alpha|} \rho^{|\alpha|} \left(\sum_{|\gamma| \leq 2m-1-|\alpha|} c r_k^{-|\gamma|} \rho^{|\gamma|} \right).$$

Recalling that $\rho \geq 2r_k$, we obtain for the third term on the right-hand side of (19) the estimate

$$|Q_k(x, y)| \leq c \, N(F; x)\rho^{2m}.$$

Integrating (19) over $B(x, \rho)$ and using the estimates obtained, we get

$$\rho^{-2m}|w_k(y) - Q_k(x, y)|_{q, B(x, \rho)}$$

$$\leq \rho^{-2m}|f(y) - P(x, y)|_{q, B(x, \rho)} + c \, N(F; x). \qquad (20)$$

For the case when $\rho < 2r_k$, we rewrite $Q_k(x, y)$ as

$$Q_k(x, y) = \sum_{|\beta| \leq 2m - 1} \left[(\partial^\beta \phi_k)(x)(y - x)^\beta/\beta! \right.$$

$$\left. \times \left(\sum_{|\gamma| \leq 2m - 1 - |\beta|} \partial_y^\gamma (P(x, y) - P(y_k, y)) \Big|_{y = x} (y - x)^\gamma/\gamma! \right) \right].$$

Adding and subtracting the expression

$$\phi_k(y)P(x, y) + \sum_{|\beta| \leq 2m - 1} (\partial^\beta \phi_k)(x)(P(x, y) - P(y_k, y))(y - x)^\beta/\beta!$$

to $Q_k(x, y)$, we obtain

$$|w_k(y) - Q_k(x, y)| \leq \phi_k(y)|f(y) - P(x, y)|$$

$$+ \left| \left[\phi_k(y) - \sum_{|\beta| \leq 2m - 1} (\partial^\beta \phi_k)(x)(y - x)^\beta/\beta! \right](P(x, y) - P(y_k, y)) \right|$$

$$+ \left| \sum_{|\beta| \leq 2m - 1} \left[(\partial^\beta \phi_k)(x)(y - x)^\beta/\beta! \right]\left[P(x, y) - P(yk, y) \right.\right.$$

$$\left.\left. - \sum_{|\gamma| \leq 2m - 1 - |\beta|} \partial_y^\gamma (P(x, y) - P(y_k, y)) \Big|_{y = x} (y - x)^\gamma/\gamma! \right] \right|$$

$$\leq |f(y) - P(x, y)| + A_1 + A_2. \qquad (21)$$

If $y \in B(x, \rho)$ and considering condition (iv) of Lemma 9 and (18) we get

$$A_1 = \left| \sum_{|\beta| = 2m} \partial^\beta \phi_k(\xi)(y - x)^\beta/\beta! \right| \left| P(x, y) - P(y_k, y) \right|$$

$$\leq c \, r_k^{-2m} \rho^{2m} N(F; x) r_k^{2m}$$

$$= c \, N(F; x)\rho^{2m}.$$

As for A_2, similar arguments show that

$$A_2 = \left| \sum_{|\beta| \le 2m-1} \left[(\partial^\beta \phi_k)(x) (y-x)^\beta/\beta! \right. \right.$$

$$\left. \left. \times \left(\sum_{|\gamma| = 2m-|\beta|} \partial_y^\gamma (P(x,y) - P(y_k, y)) \right) \right|_{y=\xi} (y-x)^\gamma/\gamma! \right) \right] \right|$$

$$\le c \sum_{|\beta| \le 2m-1} r_k^{-|\beta|} \rho^{|\beta|} N(F; x) (\rho + r_k)^{|\beta|} \rho^{2m-|\beta|} \le c N(F; x) \rho^{2m}.$$

Integrating (21) and taking into account the estimates obtained for A_1 and A_2, we get that (20) holds also for $\rho < 2r$. This shows condition (i).

Let us prove condition (ii). We shall estimate

$$\rho^{-2m} |w(y)|_{q, B(x, \rho)}.$$

This expression is equal to zero unless $B(x, \rho) \cap B(x_k, r_k) \ne \phi$. Since $x \notin B(x_k, 2r_k)$, we have $2r_k \le |x - x_k| < \rho + r_k$. Then, $r_k < \rho$ and $|x - x_k| < 2\rho$. Thus, $|x - x_k| + r_k < 3\rho$. On the other hand, by condition (ii) of Lemma 9, $B(x_k, r_k) \subset B(y_k, 7r_k)$. Then

$$\int_{B(x, \rho)} |w_k(y)|^q \, dy \le \int_{B(x_k, r_k)} |w_k(y)|^q \, dy$$

$$\le \int_{B(y_k, 7r_k)} |f(y) - P(y_k, y)|^q \, dy.$$

Since $N(F; y_k) < t$, we get

$$\rho^{-2m} |w_k|_{q, B(x, \rho)} \le c\, N(F; y_k) (r_k/\rho)^{2m+n/q}$$

$$\le ct(r_k/(|x - x_k| + r_k))^{2m+n/q}, \qquad (22)$$

which implies condition (ii).

Let us prove condition (iii). Since $0 < p < 1$ and taking into account the estimates (i) and (ii) already obtained, we have

$$\int (\Sigma_k N(W_k; x))^p \, dx \le \Sigma_k \left\{ \int_{B(x_k, 2r_k)} N(W_k; x)^p \, dx + \int_{CB(x_k, 2r_k)} N(W_k; x)^p \, dx \right\}$$

$$\leq c \Sigma_k \int_{B(x_k, \, 2r_k)} N(F; \, x)^p \, dx + ct^p \int_{CB(x_k, \, 2r_k)} (r_k / (|x - x_k| + r_k))^{(2m+n/q)p} \, dx$$

$$\leq c \int_\Omega N(F; \, x)^p \, dx + ct^p \Sigma_k r_k^n \leq c \int_\Omega N(F; \, x)^p \, dx + ct^p |\Omega|$$

$$\leq c \int_\Omega N(F; \, x)^p \, dx.$$

Condition (iv) is a consequence of condition (iii) and Lemma 4. As for condition (v), it follows from conditions (iii) and (iv), already proved.

Let us prove condition (vi). From condition (iii) we get a point $x_0 \notin \Omega$ such that $\Sigma_k N(W_k; \, x_0) < \infty$. Since $x_0 \notin B(x_k, \, 2r_k)$ for every $k = 1, 2, \ldots,$ from (22), we have that $w_k(x)$ is the only representative of W_k satisfying $n(w_k; \, x_0) = N(W_k; \, x_0)$. Then by Lemma 4 the series $w = \Sigma_k w_k$ converges in $L^q(\text{loc})$ and supplies the representative of $W = \Sigma_k W_k$ satisfying $n(w; \, x_0) = N(W; \, x_0)$. Then the function $g(x) = f(x) - w(x)$ is a representative of $G = F - W$. Since $g(x)$ is equal to $f(x)$ on $C\Omega$ and $g(x) = \Sigma_k \phi_k(y) P(y_k, \, y)$ on Ω, this function $g(x)$ coincides with the function $F_2(y)$ defined by A. P. Calderón in the statement of Theorem 5 in [1]. Therefore, $N(G, \, x) \leq ct$ and $g(x)$ coincides almost everywhere with a function having continuous derivatives up to the order $2m - 1$. Moreover, the derivatives of order $2m - 1$ satisfy a Lipschitz condition with constant ct.

Proof of Theorem 3. The method we shall use to obtain the atomic decomposition is that developed in [4]. Following that method, as a consequence of Lemma 10, we can show that if H is an element of E_{2m-1}^q satisfying $N_{q, \, m}(H; \, x) \leq 1$ and $\int N(H; \, x)^r \, dx < \infty$, for some $n(2m + n/q)^{-1} < r < p \leq 1$, then there exist a numerical sequence $\{\lambda_j\}$ and a sequence of p-atoms $\{A_j\}$ of E_{2m-1}^q such that $H = \Sigma_j \lambda_j A_j$ in $\mathcal{H}_{q, \, m}^p$. Moreover, $\Sigma |\lambda_j|^p \leq c \int N_{q, \, m}(H; \, x)^r \, dx$. The proof of this result is essentially that of Lemma 4.2 of [4]. After this result is obtained, the proof of Theorem 3 can be obtained following the same lines of the proof of Theorem 4.3 of [4].

REFERENCES

1. Calderón, A. P.   Estimates for singular integral operators in terms of maximal functions.   *Studia Math.* 44 (1972):563-582.

2. Courant, R., and Hilbert, D.   *Methods of Mathematical Physics*, vol. 2.   New York: Interscience Publ., 1953.

3. Latter, R. H.   A characterization of $H^p(R^n)$ in terms of atoms. *Studia Math*. 42 (1978):93-101.

4. Macías, R. A., and Segovia, C.   A decomposition into atoms of distributions on spaces of homogeneous type.   *Advances in Math*. 33 (1977):271-309.

5. Stein, E. M.   *Singular Integrals and Differentiability Properties of Functions*.   Princeton, N.J.: Princeton Univ. Press, 1970.

HIGHER GRADIENTS AND
REPRESENTATIONS OF LIE GROUPS

J.E. Gilbert, P.A. Tomas
University of Texas, Austin

R.A. Kunze
University of California, Irvine

R.J. Stanton
Ohio State University

INTRODUCTION

To construct an H^p theory for $\mathbf{R}^n$, Stein and Weiss [13] showed that
a solution ω of the system

$$\text{div } \omega = 0, \quad \text{curl } \omega = 0$$

automatically is harmonic and $|\omega|^p$ subharmonic when $p \geq (n-2)/n-1$.
The existence of boundary values of H^p functions, $p \geq (n-2)/n-1$,
can then be established, permitting the development of an H^p theory
for solutions of the div-curl system [13]. Unlike the dimension 2
case, however, where boundary value results may also be based on
factorization techniques, this H^p theory is available as yet only
for such p. So-called "higher-gradient" systems of differential
equations were introduced to provide an H^p theory in n dimensions
for a full range of p. Stein and Weiss [13] in three dimensions
and then later in general Calderón and Zygmund [3] and Stein and
Weiss [12] established subharmonicity results for solutions to mth-
order higher-gradient systems for $p \geq (n-2)/m + n - 2$. Calderón

All authors were supported in part by NSF grants.

and Stein later observed that if d is any constant coefficient, first-order differential system whose solutions $df = 0$ are necessarily harmonic, then there exists some $p_0 < 1$ so that $|f|^p$ is subharmonic when $df = 0$ and $p \geq p_0$. This allows great freedom in the construction of H^1 theories via systems of differential operators and singular integrals.

The purpose of this note is to develop for noncompact Riemannian symmetric spaces an analogue of classical H^p theory.

In the theory envisaged, by contrast with the Euclidean theory, we shall see here that the only nontrivial H^2 spaces are those constructed through higher gradients (more precisely, operators with the same principal symbol). Riesz transforms provide solutions to Euclidean higher-gradient systems; we shall give group-theoretic interpretations and generalizations of this result. We shall also exhibit the analogy of higher-gradient systems with the Schmid $\mathcal{D}$ operator [10]. The latter was used by Schmid to realize discrete series representations of semisimple Lie groups. We shall use the higher-gradient systems to realize another important class of irreducible unitary representations of semisimple Lie groups. Connections between the representation theory of a noncompact semisimple group G and the H^2 theory of the symmetric space G/K will be used frequently.

To orient the reader, we shall sketch the constituents of the H^p theory proposed:

(1)   The theory should be based on elliptic systems of first-order differential operators that commute with the action of G. The kernel of the system should be contained in the kernel of a second-order elliptic differential operator.

(2)   A theory of boundary values and of Poisson kernels should exhibit isomorphisms of spaces of functions on G/K with corresponding function spaces on the boundary.

(3)   There should be a Riesz transform characterization of boundary values. This characterization should give a complete account of the overdetermined nature of the elliptic system.

(4)   There should be a molecular decomposition of H^p functions.

(5)   The differential equations, Riesz transforms, Poisson kernels, and boundary value operators should all commute with the action of G. In particular, H^2 should be invariant and one should obtain irreducible unitary representations from the action of G on H^2.

The construction of a full H^p theory in general is a formidable task, which we shall not accomplish in this note. Here we shall confine our discussion to the construction of nontrivial H^2 spaces for the Lorentz groups $SO(2k, 1)$. In the last section of the paper we shall discuss extensions of these results.

We wish to thank Professors B. Blank and A. W. Knapp for many valuable conversations on these topics.

NOTATION

In general, we shall follow the conventions of Knapp and Wallach [8].

Let G be a noncompact connected semisimple Lie group with finite center, Lie algebra $\mathscr{g}$ and Cartan decomposition $\mathscr{g} = \mathscr{k} \oplus \mathscr{p}$. Let K be the analytic subgroup corresponding to $\mathscr{k}$, and we assume rank G = rank K. Thus if $\mathscr{t}$ is a Cartan subalgebra, we may take $\mathscr{t}$ compact and $\mathscr{t} \subseteq \mathscr{k}$. Let Δ_k be the compact roots, ρ_k and ρ_n are half the sum of the positive compact and noncompact roots.

If $\mathscr{g} = \mathscr{k} \oplus \alpha \oplus \mathscr{n}$ is an Iwasawa decomposition, let A, N, and $V = \theta N$ be the corresponding analytic subgroups. Then

$g = \exp H(g)N(g)\kappa(g)$. Let ρ_+ be half the sum of the positive restricted roots, with multiplicites. Let M be the centralizer of α in K, M' the normalizer, and W the Weyl group M'/M.

If τ is an irreducible unitary representation of K on V, we let $C^\infty(G, \tau) = C^\infty(G, V) = \{f : G \to V | f(kg) = \tau(k)f(g) \text{ and } f \in C^\infty\}$.

EUCLIDEAN ANALOGUES

There is a simple Euclidean analogue of G/K, which clarifies the connection between Euclidean and non-Euclidean H^p theories.

We define affine motion groups through the semidirect product on $\mathbf{R}^n \times SO(n)$ defined by

$$(x, \theta) \cdot (y, \Psi) = (x + R_\theta y, \theta \Psi),$$

where R_θ is the rotation corresponding to θ.

The symmetric space $\mathbf{R}^n \times SO(n)/SO(n)$ is then isomorphic to $\mathbf{R}^n$, but the full motion group acts on the symmetric space by translations and by rotations. Differential operators invariant under the action of the group must therefore be constant coefficient and commute with rotations. It is not difficult to see that such first-order operators cannot be scalar valued, but must take values in a vector space on which rotations can be made to act. The classical systems of differential operators used to define H^p theory—div-curl, higher gradients, or the d, d^* system—all can be seen to possess rotational invariance, and therefore may be viewed as operators on a symmetric space.

The construction of such systems of operators was unified and generalized by Stein and Weiss [12]. They start with an irreducible unitary representation τ of $SO(n)$ on a vector space V. If $f : \bar{\mathbf{R}}^n \to V$ is a vector value function and we choose coordinates $\bar{x} = \Sigma x_i \bar{e}_i$, we may form the invariant operator

$$\nabla f = \sum \frac{\partial}{\partial x_i} f \otimes \bar{e}_i ;$$

∇ takes C^∞, V-valued functions to C^∞, $V \otimes \mathbf{C}^n$-valued functions. The group $SO(n)$ acts on $V \otimes \mathbf{C}^n$, by $\rho \cdot V \otimes z = \tau(\rho)V \otimes \rho \cdot z$, where $\rho \cdot z$ is the standard action of rotations on $\mathbf{C}^n$. There is a canonical subspace V_0 of $V \otimes \mathbf{C}^n$ on which $SO(n)$ acts irreducibly; it is called the Cartan composition of V and $\mathbf{C}^n$. Let P_0 denote the projection of $V \otimes \mathbf{C}^n$ onto $V_0^\perp$, and define the operator $\overline{\partial}$ on f by $\overline{\partial}f = (P_0\nabla)f$. Stein and Weiss [12] defined the $\overline{\partial}$ operator, showed that it reduced, for special choices of τ to div-curl, higher gradients, and so on, and showed that if $\overline{\partial}f = 0$, the components of f are harmonic. They also showed that solutions to the system $\overline{\partial}f = 0$ enjoy subharmonicity properties, so that one may construct an H^p theory for $\mathbf{R}^n+$ and obtain boundary values for $1 \leq p < \infty$. We shall sketch such a theory, for it indicates what we may expect in the non-Euclidean case.

We first define a boundary and a normal direction in a canonical way, by selecting a highest-weight vector $\overline{v}_0 + i\widetilde{v}_1$ for the standard representation of $SO(n)$ on $\mathbf{R}^n$. Then we define $\mathbf{R}^n+ = \{\overline{v}|\overline{v} \cdot \overline{v}_0 > 0\}$, and we define M to be the subgroup of $SO(n)$ that fixes v_0, and M_0 the subgroup that fixes $\overline{v}_0$ and $\overline{v}_1$. If $\xi \in \mathbf{R}^{n-1}$, let δ_ξ be the rotation that maps $\overline{v}_1$ to $\xi/|\xi|$ on the unit sphere. It is possible to construct a space $W \subset V$ on which $\tau|_{M_0}$ acts, and which characterizes the solutions to the τ system as follows.

If H^p is defined in the obvious way, then f is in H^p if and only if there is an $F : \mathbf{R}^{n-1} \to W$, $F \in L^p$, such that

$$\int_{\mathbf{R}^{n-1}} f(x + y)e^{ix\xi}\,dx = e^{-y|\xi|}\tau(\delta_\xi)\hat{F}(\xi).$$

The homogeneous multiplier $\tau(\delta_\xi)$ thus plays the role of the Riesz transforms. We have not been able to give a complete characterization of W, but if ϕ_τ is the highest-weight vector for (τ, V), it is known that $\tau(M_0)\phi_\tau \subset W$. In all known examples actual equality holds.

We shall need a further understanding of this theory to the special case that τ is the representation of $SO(n)$ on symmetric m-tensors. Then $f : \mathbf{R}^n \to V$ satisfies the system $\overline{\partial} f = 0$ if and only if ∇f is a symmetric, traceless $m + 1$-tensor. This occurs if and only if f can be written as $f = \nabla^m u$ for a harmonic function u. If f is in H^p, then f has boundary values and $f\big|_{\mathbf{R}^{n-1}} = (-\Delta)^{-m/2} (\nabla_T)^m F$, where ∇_T is the boundary gradient operator and F is a complex-valued function. In this case, then, W is one-dimensional and the Riesz transforms may be given by iterates of the classical Riesz transforms, as in Calderón and Zygmund [3].

DIFFERENTIAL OPERATORS

To construct invariant differential operators on the non-Euclidean symmetric spaces G/K, we shall generalize the procedure of Stein and Weiss.

The maximal compact subgroup K plays the role of $SO(n)$, and the space $\exp \not{p}$ that of $\mathbf{R}^n$. But $\not{p}$ and K cannot be split apart, as $[\not{p}, \not{p}] \subset \not{k}$, and matters cannot be reduced to vector-valued functions on G/K. Moreover, solutions to any of the natural analogues of the Euclidean equation are eigenfunctions of the Casimir operator, which is not an elliptic operator. Both of these difficulties can be overcome in a standard way; our operators will act on function $f : G \to V$ with invariance properties. Technically speaking, our operators will act on sections of a homogeneous vector bundle over G/K, rather than on actual functions, but for the most part we shall not need this characterization.

We fix an order Δ_k^+ on the positive compact roots and choose an irreducible unitary representation τ of K on a space V, with highest-weight vector λ. The representation Ad of K acts on $\not{p}_{\mathbf{c}}$; we choose any order Δ_n^+ on the noncompact roots so that the highest-weight vector of Ad on $\not{p}_{\mathbf{c}}$ is a positive noncompact root. We let

V_0 be the complement of the irreducible subspace that contains the highest-weight vector of $\tau \otimes \mathrm{Ad}$ on $V \otimes \mathfrak{h}_{\mathbb{C}}$ and let P_0 be projection onto V_0.

Let

$$C^{\infty}(G, V) = \{f : G \to V \mid f(kg) = \tau(k)f(g), f \epsilon C^{\infty}\}.$$

The gradient operator may be defined from $C^{\infty}(G, V)$ into $C^{\infty}(G, V \otimes \mathfrak{h}_{\mathbb{C}})$ by the formula $\nabla f = \Sigma X_i f \otimes \overline{X}_i$, where $\{X_i\}$ is any orthonormal basis of $\mathfrak{h}_{\mathbb{C}}$ and conjugation is with respect to $\mathscr{g}$ in $\mathscr{g}_{\mathbb{C}}$. We define $\mathscr{I}$ from $C^{\infty}(G, V)$ by

$$\mathscr{I}f = (P_0\nabla)f.$$

The operator $\mathscr{I}$ commutes with the right action of G and is a generalization of the $\overline{\partial}$ system of Stein and Weiss. We have avoided the $\overline{\partial}$ notation because, in the case G/K is a Hermitian symmetric space, $\mathscr{I}$ is not the $\overline{\partial}$ operator on the Dolbeault complex.

The operator $\mathscr{I}$ is similar to the operator D of Schmid [10]. Note that

$$V \otimes \mathfrak{h}_{\mathbb{C}} = \sum_{\beta \epsilon \Delta_n} m_\beta V_{\lambda + \beta};$$

define

$$V^- = \sum_{-\beta \epsilon \Delta_n} m_\beta V_{\lambda + \beta}.$$

Let $P_- : V \otimes \mathfrak{h}_{\mathbb{C}} \to V_-$ be orthogonal projection, and $Df = P_- \circ \nabla f$.

Note $\ker \mathscr{I} \subset \ker D$. Knapp and Wallach [8] show that if $Df = 0$, then f is an eigenfunction of the Casimir, with eigenvalue $|\lambda + \rho - \rho_n|^2 - |\rho|^2$. Moreover, Schmid [10] shows that the principal symbol of D is $P_-(v \otimes \xi)$; that of $\mathscr{I}$ is $P_0(v \otimes \xi)$. Since $V_- \subset V_0$, $\ker P_0 \subset \ker P_-$ and the ellipticity of D implies that of $\mathscr{I}$. The Schmid operator D is elliptic if $< \lambda - 2\rho, \alpha > \geq 0$ for $\alpha \epsilon \Delta_k^+$. In general, D will not be elliptic, but the $\mathscr{I}$ operator will be. This will require separate proof.

In the case of motion groups, it is possible to show that $\mathscr{I}$ and the $\overline{\partial}$ system of Stein and Weiss are equivalent. We define

$T : C^\infty(\mathbf{R}^n \times \mathrm{SO}(n, V)) \to C^\infty(\mathbf{R}^n) \otimes V$ by $(Tf)(x) = f(x, e)$. One readily computes that $T\mathscr{Y}f T^{-1} = 2\bar{\partial}Tf$. The proof makes use of the fact that for motion groups $[\not{p}, \not{p}] = 0$. The $\mathscr{Y}$ system is therefore equivalent to a system of first-order differential equations acting on vector-valued functions on $\mathbf{R}^n$. In the case of semisimple groups, $[\not{p}, \not{p}] \subset \not{k}$ and the $\mathscr{Y}$ system does not reduce to a system of equations acting on valued functions on G/K, but must be viewed as an operator on G, acting on functions with transformation properties.

EXISTENCE RESULTS

We shall state our results for $\mathrm{SO}(4, 1)$, with the understanding that they are valid for $\mathrm{SO}(2k, 1)$. If J is the matrix defining the indefinite form

$$\sum_{J=1}^{4} x_j^2 - x_5^2,$$

then we define

$$\mathrm{SO}(4, 1) = \{g \in GL(5) \,|\, gJg^T = J,\ \det g = 1,\ g_{55} \geq 1\}.$$

We choose a Cartan involution $\theta X = -X^T$. If E_{ij} is the matrix with a 1 in the (i, j) position and zeros elsewhere, we set $Y_j = E_{j5} + E_{5j}$. Then the root vectors of $\mathrm{Ad}K$ on $\not{p}_{\mathbf{c}}$ are

$$E_{\alpha_1} = Y_1 + iY_2, \quad E_{\alpha_2} = Y_3 + iY_4;$$

and

$$\Delta_n = \{\pm\alpha_1,\ \pm\alpha_2\}, \quad \Delta_k = \{\pm(\alpha_1 + \alpha_2),\ \pm(\alpha_1 - \alpha_2)\}.$$

We begin our construction of differential operators with a choice of positive roots for Δ_k; we shall choose $\Delta_k^+ = \{\alpha_1 + \alpha_2,\ \alpha_1 - \alpha_2\}$. We also choose orderings on Δ_n; if E_{α_1} is the highest weight for Ad on $\not{p}_{\mathbf{c}}$, we have two compatible orderings, $\Delta_1 = \{\alpha_1,\ \alpha_2\}$ and $\Delta_2 = \{\alpha_1,\ -\alpha_2\}$. If λ is a given dominant integral weight of K, there corresponds an irreducible unitary representation $\tau = \tau_\lambda$ of K. We may then construct the $\mathscr{Y}$ operator, which is independent of Δ_n order, and we may construct, for each order Δ_i, a Schmid operator D_i.

Generally, ker $\mathscr{T}$ is contained in ker D_i but is not equal. Therefore, if λ is such that ker D_i leads to a discrete series or limit of discrete series in either order, then any invariant H^2 space must be a proper subspace of an irreducible representation and therefore trivial. In the Euclidean theory, Stein and Weiss [12] showed that any representation τ of K gives rise to an H^p theory on $\mathbf{R}^n_+$. In the semisimple case, only certain τ will lead to nontrivial H^2 theories. To be precise, we must have $\Lambda = \lambda + \delta_k - \delta_n$ not dominant with respect to Δ_i for either i, and if Λ is singular, we cannot have $\langle \Lambda, \alpha_2 \rangle = 0$. A simple computation shows that this is valid for $\lambda = m\alpha_1$, where $m \in \mathbf{Z}^+$. The representations of $K = SO(4)$ that have highest weight $m\alpha_1$ are the representations of the rotation group on symmetric m-tensors. Therefore, the systems of differential equations that we obtain are algebraically the same as the higher-gradient systems, that is, they have the same principal symbol, and are easily seen to be elliptic.

For the remainder of the paper, we shall assume $\tau = \tau_\lambda$ for $\tau = m\alpha_1$, $m \in \mathbf{Z}^+$. The relationship with the Schmid operators may then be analyzed more precisely:

$$D_1 f = 0 \text{ iff } \nabla f \in \sum_{\alpha \in \Delta_1} m_\lambda \tau_{\lambda + \alpha} = (m + 1)\alpha_1 \oplus (m\alpha_1 + \alpha_2),$$

$$D_2 f = 0 \text{ iff } \nabla f \in \sum_{\alpha \in \Delta_2} m_\lambda \tau_{\lambda + \alpha} = (m + 1)\alpha_1 \oplus (m\alpha_1 - \alpha_2),$$

$$\mathscr{T} f = 0 \text{ iff } \nabla f \in (m + 1)\alpha_1.$$

Lemma 1. ker $\mathscr{T}$ = ker $D_1 \cap$ ker D_2. Moreover, $\mathscr{T}$ is elliptic and the kernel of $\mathscr{T}$ is contained in the eigenspace of the Casimir with eigenvalue $m^2 + m - 2$.

This identifies the candidates for nontrivial H^2 spaces; the main result of this note is to prove that they are in fact nontrivial.

Theorem 1. *There is a positive definite inner product* $\langle \ , \ \rangle$ *on*
$C^{\infty}(G, V)$ *such that the closure of* $C^{\infty} \cap \ker \mathscr{T}$, *called* H^2, *is a non-*
trivial Hilbert space. *Moreover, the right regular representation*
$\pi(h)f(g) = f(gh)$ *forms an irreducible unitary representation of* G
on H^2.

Remarks. The sketch of the proof will occupy most of the remainder
of the paper. We shall therefore discuss the central ideas and
difficulties.

      Knapp and Wallach [8] constructed a Szegö map $\mathcal{S}$ for dis-
crete series, which maps certain nonunitary principal series into
the kernel of the Schmid operator. We verify that the parameters
are independent of Δ_n order and therefore map into the kernel of $\mathscr{T}$.
The Szegö map is nontrivial but has a large kernel; moreover, the
Casimir has a highly degenerate principal symbol near the boundary,
and in fact smooth solutions of the $\mathscr{T}$ system have zero boundary
values. To identify precisely the kernel and range of the Szegö
map, we use multiplicity formulas of Hotta and Parthasarathy [6],
originally developed for discrete series.

      To complete the proof, we need to find a positive definite
inner product. In the case of discrete series, the inner product
is that of $L^2(G, V)$, and the proof that the L^2 kernel of D is
nontrivial requires extra work. Briefly, Atiyah and Schmid [1]
observe the existence of a cocompact discrete subgroup Γ for which
the Γ-index of D in $L^2(G, V)$ is related to the index of D on
$L^2(\Gamma \backslash G/K, V)$. This latter may be computed by the L^2-index theorem;
the computation relates the index of D to certain Chern classes.

      It should be observed that these techniques are particularly
adapted to L^2 theory and cannot be applied to nondiscrete series
representations. The representations we realize are specifically
nondiscrete series. A substitute is the theory of limits of complementary series, constructed by Knapp and Stein [7]. This requires
us to relate the kernel of the intertwining operators to the kernel

of the Szegö map. To accomplish this, we follow Blank [2] in
defining a nonzero boundary value map, whose kernel can be
analyzed precisely. We then obtain an inner product from the
theory of complementary series.

The next section of the paper constructs Riesz transforms.
We cannot readily use harmonic function theory to characterize the
overdeterminedness of the $\mathcal{J}$ system; the Casimir vanishes near the
boundary. Instead we use an abstract embedding result of Langlands
on the classification of irreducible unitary representations. The
resulting Riesz transforms are like ∇^m but do not have the homogeneity of singular integrals; H^2 is not isomorphic to L^2.

In the final section of the paper we briefly discuss molecular decompositions. The existence of molecular decomposition for
H^2 follows from an abstract result of Kunze [9]; we give a more
concrete realization that relates the decomposition to the molecular decomposition for Bergman spaces given by Coifman and Rochberg
[4].

SZEGÖ MAPPING

Since ker $\mathcal{J}$ = ker $D_1 \cap$ ker D_2, we may use discrete series techniques
to provide a mapping of boundary values into ker $\mathcal{J}$. As in the
Euclidean example we discussed earlier, the normal direction must
be carefully chosen. We let $\boldsymbol{a} = \mathbf{R}(E_{\alpha_2} + E_{-\alpha_2})$, A the corresponding
analytic subgroup, $G = ANK$ the Iwasawa decomposition, ρ_+ half the
sum of the positive restricted roots counted with multiplicity, and
M the centralizer of α in K. Now $\pm\alpha_2$ is a fundamental sequence of
Knapp and Wallach (since it is simple) and we may construct the corresponding Szegö map. Let σ_λ be the representation of M on W with
highest weight $m\alpha_1$, and let $\nu = (1 - z)\rho_+$. The Szegö map
$S_z : C^\infty(K, \sigma_\lambda) \to C^\infty(G, \tau_\lambda)$ is defined by

$$S_z F(g) = \int_K e^{\nu H(kg^{-1})} \tau_\lambda^{-1}(k(kg^{-1})) F(k) dk.$$

Proposition 1. *If* $z = 1/3$, S_z *is a nontrivial map of* $C^\infty(K, \sigma_\lambda)$ *into* $C^\infty(G, \tau_\lambda) \cap \ker \mathscr{I}$.

Proof. Knapp and Wallach identify parameters ν_i, σ_i for which S_{ν_i} maps $C^\infty(K, \sigma_\lambda)$ into $C^\infty \cap \ker D_i$. One easily checks that the parameters are independent of order Δ_i, and are $\nu = 2/3\rho_+$, $\sigma_\lambda \sim m\alpha_1$. To show S_z is nontrivial, Knapp and Wallach let $P_{\lambda\sigma}$ be the projection of V to W, ϕ_λ the highest vector of τ_λ, and $F_\lambda(k) = P_{\lambda\sigma}\tau_\lambda(k)\phi_\lambda$. Then

$$(S_\nu F_\lambda(e), \phi_\lambda) = \int_K |F|^2;$$

the latter is positive, since F is continuous and $F(e) = \phi_\lambda$.

Remark. Knapp and Wallach reformulate this result as follows. Let $\nu' = 2\rho_+ - \nu$; the nonunitary principal series $U(\sigma_\lambda, \nu')$ acts on the space

$$\{F : G \to W | F(manx) = e^{\nu' H(a)\sigma_\lambda}(m) F(x)\}$$

by the right regular representation. If F in $C^\infty(K, \sigma_\lambda)$ is appropriately extended, the Szegö map may be computed as

$$S_z F(x) = \int_K \tau_\lambda(k^{-1}) F(kx) dk,$$

from which it is clear that S_z is equivariant.

Theorem 2. $S_{1/3}$ *is a map onto* $C^\infty(G, \tau_\lambda) \cap \ker \mathscr{I}$. *Moreover,* $\ker \mathscr{I}$ *is irreducible.*

Proof. We proceed through an identification of $\ker \mathscr{I}$ and again make use of discrete series techniques; in this case we compute precisely the multiplicities of the K-isotypic subspaces that occur in the decomposition of the right regular representation of G on $\ker D_i$. The relevant result is Theorem 1 of Hotta and Parthasarathy [6]. The result requires $\langle \lambda + \rho_k - \langle \phi \rangle, \alpha \rangle \geq 0$ for every

$\alpha \in \Delta_k^+$, $\phi \subset \Delta_i$, where $\langle \phi \rangle = \sum_{\alpha \in \phi} \alpha$. It is easily checked that the result applies if $m \geq 1$.

Let $m_\lambda(\mu)$ be the multiplicity of μ in the right regular representation of G in ker $D_1 \cap$ ker D_2. Then $m_\lambda(\mu) \leq b_\lambda(\mu)$, where

$$b_\lambda(\mu) = \sum_{s \in W_k} \det(s) \phi [s(\mu + \rho_k) - (\lambda + \rho_k)],$$

where

$$\phi(\mu) = \sum_{k=0}^\infty \phi_k(\mu);$$

$\phi_0(0) = 1$, and $\phi_k(\mu)$ = number of distinct ways μ can be written as a sum of exactly k elements of Δ_i, = 1, 2. Moreover, equality $m_\lambda(\mu) = b_\lambda(\mu)$ holds if λ is far from the walls, which means, in the case $\lambda = m\alpha_1$, $m \geq 2$.

Since $\Delta_i = \{\alpha_1, \pm\alpha_2\}$, there can be no cancellation and $\phi_k(p\alpha_1 + q\alpha_2) = \delta_{k,p+q}$.

Lemma 2. Let $\lambda = m\alpha_1$, $m \geq 1$. If $\mu = a\alpha_1 + b\alpha_2$, then $m_\lambda(\mu) = 0$ if $b \neq 0$ or $a < m$; if $b = 0$, $a \geq m$, then $m_\lambda(\mu) \leq 1$. If $a \geq 2$, $m_\lambda(\mu) = 1$.

The lemma may be proved by straightforward computations. It is important to note that we need to use both the Δ_1 and Δ_2 orders to compute $b_\lambda(\mu)$, to show $m_\lambda(\mu) = 0$ if $b \neq 0$.

To complete the proof of the theorem, we make three observations:

(a) $m_\lambda(\lambda) = 1$.

(b)   Let E_m be the sum of the K-isotypic subspaces of $C^\infty(G, \tau_\lambda)$ with highest weights $a\alpha_1$, $a \geq m$. Then E_m is irreducible under $\mathcal{G}_\mathbf{C}$.

(c)   The K-finite subspaces of ker $\mathcal{I}$ and Im S are irreducible under the actions of $\mathcal{G}_\mathbf{C}$. (Here $S \equiv S_{1/3}$.)

From (a), (b), and invariance, ker $\mathscr{J}$ is E_m. From the invariance of Im S and its nontriviality, it follows that the K-finite image of S is E_m. Hence S is onto, as was to be shown. To prove (a), note that $m_\lambda(\lambda) \leq 1$, so that we need only show $m_\lambda(\lambda) \neq 0$. Let $\chi_\lambda = \text{trace } \tau_\lambda$; it suffices to show $\chi_\lambda * \text{Im } S \neq 0$. But S commutes with the action of K, so that

$$\chi_\lambda * S(F_\lambda) = S(\chi_\lambda * F_\lambda) = S(F_\lambda) \neq 0.$$

The proof of (b) is immediate from the simple structure of E_m; the proof of (c) comes from the invariance of $\mathscr{J}$ and S. This completes the proof of the theorem.

BOUNDARY VALUES

To complete our analysis, we shall need a knowledge of the C^∞ kernel of S. We shall obtain such information from boundary values. Following Blank [2], we note that if $g = a_t \in A$ and w_0 is chosen in the Weyl group with $w_0^{-1} a_t w_0 = a_{-t}$, then

$$S_a F(a_t) = \int e^{\nu H(ka_{-t})} \tau_\lambda^{-1}\big(\kappa(ka_{-t})\big) F(k)\, dk$$

$$= \int e^{\nu H(ka_t w_0)} \sigma_\lambda^{-1}\big(\kappa(ka_t w_0)\big) F(kw_0)\, dk$$

$$= \int e^{\nu H(ka_t)} \tau_\lambda^{-1}\big(\kappa(ka_t)w_0\big) F(kw_0)\, dk$$

$$= \iint_{M \times V} e^{\nu H(m\kappa(x)a_t)} e^{2\rho_+ H(x)} \tau_\lambda^{-1}\big(\kappa(m\kappa(x)a_t)w_0\big) F(m\kappa(x)w_0)\, dx\, dm$$

but

$$H\big(m\kappa(x)a_t\big) = H\big(\kappa(x)a_t\big) = -H(x) + H(xa_t)$$

and

$$\kappa\big(m\kappa(x)a_t\big) = m\kappa\big(\kappa(x)a_t\big) = m\kappa(xa_t)$$

so that

$$S_z F(a_t) = \iint e^{-\nu H(x)} e^{2\rho H(x)} e^{\nu H(xa_t)} \tau_\lambda^{-1}\big(m\kappa(xa_t)w_0\big) F(m\kappa(x)w_0)\, dx\, dm$$

$$= \int_V e^{(2\rho_+ - \nu)H(x)} \, e^{\nu H(xa_t)} \tau_\lambda^{-1}\big(\kappa(xa_t)w_0\big) F\big(\kappa(x)w_0\big) dx.$$

Now letting $\nu' = 2\rho_+ - \nu$ and viewing F as an element of $U(\sigma_\lambda, \nu')$, with transformation properties under MAN, we obtain

$$S_z F(a_t) = \int e^{\nu H(xa_t)} \, \tau^{-1}\big((a_t^{-1}xa_t)w_0\big) F(xw_0) dx.$$

But

$$e^{\nu H(a_t^{-1}xa_t)} e^{\nu H(a_t)} = e^{\nu H(\delta_t x)} e^{\nu H(a_t)}$$

so that $S_z F(a_t)$ is

$$e^{\nu H(a_t)} \int_V e^{\nu H(\delta_t x)} \tau_\lambda^{-1}\big(\kappa(\delta_t x)w_0\big) F(xw_0) dx,$$

and

$$\lim_{t \to \infty} e^{\nu H(a_t)} S_z F(a_t) = \lim_{t \to \infty} \int e^{2\nu H(a_t)} e^{\nu H(\delta_t x)} \tau_\lambda^{-1}\big(\kappa(\delta_t x)w_0\big) F(xw_0) dx.$$

But $\kappa(\delta_t x)w_0$ converges almost everywhere to $m(xw_0)$, and $\tau^{-1}\big(\kappa(\delta_t x)w_0\big)$ boundedly to $\sigma_\lambda^{-1}\big(m(xw_0)\big)$. Moreover,

$$e^{2\nu H(a_t)} e^{\nu H(\delta_t x)} F(xw_0)$$

converges dominatedly to $e^{\nu H(x)} F(wx_0)$ when $\nu = (1 - z)\rho_+$ satisfies Re $z > 0$. A comparison with Knapp and Stein [7] shows that

$$\lim_{t \to \infty} e^{\nu H(a_t)} S_z F(a_t) = A(w_0, \sigma_\lambda, z) F(e),$$

where A is the intertwining operator from $U(\sigma_\lambda, \nu')$ into $U(w_0\sigma_\lambda, w_0\nu')$, and $\nu = (1 - z)\rho_+$. If we define a boundary operator B from Im S_z to $U(w_0\sigma_\lambda, w_0\nu')$ by

$$BS_z F(g) = U(w_\sigma\sigma_\lambda, w_0\nu')(g) \lim_{t \to \infty} e^{\nu H(a_t)} S_z F(a_t),$$

it follows that:

Theorem 3. If Re $z > 0$, then $BS_z = A(w_0, \sigma_\lambda, z)$, and the map B is *equivariant.*

We note in particular that the above proof does not use special properties of S at $z = 1/3$, but is valid for $z > 0$.

CONSTRUCTION OF H²

To construct H^2, we shall first construct a Hilbert space on a quotient of a nonunitary principal series and then use S to lift the structure to G/K.

Theorem 4 (Knapp and Stein, [7]). The map $A(w_0, \sigma_\lambda, z)$ defines a positive definite norm on $U(\sigma_\lambda, (1 - z)\rho_+)$ if $0 < z < 1/3$. If $z = 1/3$, the norm is positive semidefinite, and there is a G-unitary Hilbert space structure on $U(\sigma_\lambda, 1/3\rho_+)/\ker A(w_0, \sigma_\lambda, 1/3)$.

Remark. The result of Knapp and Stein was used to construct complementary series. If $z = 1/3$, then $\nu = (1 - z)\rho_+ = 2/3\rho_+$. This is the nonunitary principal series from which $S_{1/3}$ maps $U(\sigma_\lambda, 1/3\rho_+)$ onto ker $\mathscr{I}$.

Corollary 1. $U(\sigma_\lambda, 1/3\rho_+)/\ker A(w_0, \sigma_\lambda, 1/3)$ has a G-unitary structure, $\| [F] \|_\sigma$.

Theorem 5. ker $S_{1/3}$ = ker $A(w_0, \sigma_\lambda, 1/3)$.

Proof. We shall establish this result for K-isotypic subspaces. Since the nonunitary principal series are admissible, all the isotypic subspaces appear with finite multiplicity, and the results below follow from finite dimensional vector space theory.

Note first that $B \circ S = A(w_0, \sigma_\lambda, z)$ for Re $z > 0$. In particular at $z = 1/3$, ker $S \subset$ ker A. Therefore, coker $A \subset$ coker S. But coker $S \cong$ Im S and at $z = 1/3$, Im S = ker $\mathscr{I}$ is irreducible. Since A commutes with G, ker A and coker A are invariant. Since coker A is not trivial, coker A = coker S and ker A = ker S.

We may now define the H^2 norm. Take $f \in C^\infty(G, \tau_\lambda) \cap$ ker $\mathscr{I}$. There is a unique class $[F]$ in $U(\sigma_\lambda, 1/3\rho_+)/\ker A(w_0, \sigma_\lambda, 1/3)$ with $S[F] = f$. Define

$$\| f \|_\tau = \| [F] \|_\sigma .$$

This is a unitary structure on ker $\mathcal{J}$:

$$\|\pi(h)f(g)\|_\tau = \|f(gh)\|_\tau = \|S[F](gh)\|_\tau$$

$$= \|S(U(\sigma_\lambda, 1/3\rho_+))(h)[F]\|_\tau$$

$$\equiv \|U(\sigma_\lambda, 1/3\rho_+)(h)[F]\|_\sigma = \|[F]\|_\sigma$$

$$\equiv \|f\|_\tau.$$

Theorem 6. Let H^2 = closure $C^\infty(G, \tau_\lambda) \cap$ ker $\mathcal{J}$. Then H^2 is a non-trivial Hilbert space. The right regular representation $\pi(h)f(g) = f(gh)$ forms an irreducible unitary representation of G on H^2.

Remark. This construction is analogous to defining $H^2(\mathbf{R}^n)$ as $L^2(\mathbf{R}^{n-1})$. It is a reasonable substitute for the index theorem, but clearly we cannot expect to do H^p theory. We discuss alternate characterizations in the last section of the paper.

RIESZ TRANSFORMS

In the Euclidean case of H^2 defined by higher gradients, every solution is determined by a single harmonic function. Since the degree of freedom of the $\mathcal{J}$ system is determined by the space $W = \text{Span } \tau_\lambda(M)\phi_\lambda$, one degree of freedom suggests an embedding of (π, H^2) into a nonunitary principal series $U(\sigma_0, \nu_0)$, where σ_0 is the trivial representation. Riesz transforms in the classical case take F into $(F, R_1F, \ldots, R_nF, \ldots, R_{i_1} \ldots R_{i_m}F)$, or, invariantly, $F \to (-\Delta)^{-m/2}\nabla^m F$. In the non-Euclidean case, the Riesz transforms will be given by the intertwining operator from $U(\sigma_0, \nu_0)$ to $U(\sigma_\lambda, 1/3\rho_+)$.

The action of the Weyl group on the infinitesimal character of one embedding yields all potential embeddings. We have obtained a quotient embedding at $\sigma_\lambda \sim m\alpha_1$, $\nu' = 1/3\rho_+$; the corresponding infinitesimal character is $(m + 1/2)\alpha_1 + \alpha_2$. After the action of the Weyl group, we obtain embeddings at $[\sigma_\lambda, i\mu]$, where $i\mu = \nu' - \rho_+$, at the following nonunitary principal series:

$$[m\alpha_1, 1/3\rho_+], \quad [m\alpha_1, -1/3\rho_+],$$

and

$$[0\alpha_1, (m + \tfrac{1}{2})2/3\rho_+], \quad [0\alpha_1, -(m + \tfrac{1}{2})2/3\rho_+].$$

Each of these pairs potentially gives (π, H^2) as a quotient and a subrepresentation of the nonunitary principal series. By the uniqueness in Langlands' classification, there is at most one quotient on any side of the unitary axis; by existence there is at least one. Since we are interested in obtaining (π, H^2) as the image of Riesz transforms, we wish to find $U(\sigma_0, \nu_0)$ for which (π, H^2) is embedded as a quotient. This occurs at

$$[m\alpha_1, 1/3\rho_+] \quad \text{and} \quad [0\alpha_1, -2/3(m + 1)\rho_+].$$

To obtain the Riesz transforms, we let $J'(\sigma_\lambda, i\mu)$ denote the irreducible quotient or subrepresentation realization of (π, H^2) at $U(\sigma_\lambda, \rho_+ + i\mu)$. We now know abstractly that $J'(0, -2/3(m + 1)\rho_+)$ is equivalent to $J'(m\alpha_1, 1/3\rho_+)$. To implement this isomorphism concretely, we let A_0 denote the intertwining map $A(w_0, m\alpha_1, 1/3)$ restricted to the quotient $J'(m\alpha_1, 1/3\rho_+)$; then A_0 is invertible, and A^{-1} gives the equivalence of $J'(m\alpha_1, -1/3\rho_+)$ with $J'(m\alpha_1, 1/3\rho_+)$. A_0 is a convolution operator on V, with kernel

$$J(\exp X) = \Omega\left(\frac{x}{||x||}\right) \Big/ ||x||^{3(1-\kappa)} = \Omega/ ||x||^2.$$

Thus A_0^{-1} is formally a fractional integral operator composed with Euclidean Riesz transforms. To finish the characterization, we need an explicit equivalence of $J'(0\alpha_1, -2/3(m + 1)\rho_+)$ and $J'(m\alpha_1, -1/3\rho_+)$. From Knapp and Stein [7], it is known that this intertwining operator is given by convolution with a distribution at the origin.

If ∇ is the invariant gradient for V, one easily checks that ∇^m performs this intertwining. The Riesz transforms are then given by $A_0^{-1}\nabla^m$, where A_0 is a smoothing operator; this composition then plays the role of $(-\nabla)^{-m/2}\nabla^m$ in Euclidean H^2 theory, but it is not a singular integral operator on V, instead being the composition of a singular integral and a fractional integral.

MOLECULAR DECOMPOSITION

The representation (π, H^2) possesses a cyclic invariant subspace that transforms by τ, since the τ-isotypic subspace certainly occurs, and it must be cyclic by irreducibility. The results of Kunze [9] now imply that H^2 is a reproducing kernel space. There is an operator-valued kernel $R(g)$ on G, $R : V \to V$, satisfying

(i) $R(e) = I$.

(ii) $R(k_1 g k_2) = \tau(k_1)R(g)\tau(k_2)$.

(iii) R is positive definite.

(iv) (π, H^2) is unitarily equivalent to the right regular representation of G on the closure of

$$\left\{\sum R(x_i^{-1}x)v_i \mid x_i \in G, \; v_i \in V\right\}$$

under the H^2 norm.

To give alternate characterizations of H^2, the precise form of R is important.

Lemma 3. $R(g)v = S_{1/3}\left(P_{\lambda_\sigma}\tau(k^{-1})v\right)(g)$.

Proof.   (i) and (ii) are immediate. The unitarity of (π, H^2) implies (iii), and the invariance and irreducibility of the relevant spaces implies (iv).

Remarks.

(1) The characterization of H^2 as a reproducing kernel space gives a molecular decomposition for H^2.

(2) In the special case $G = SL(2, \mathbf{R})$, the $\mathscr{D}$ operator is the Schmid D operator, and in appropriate local coordinates, the discrete series are realized in Bergman spaces:

$$\left\{ f : \mathbf{R}_+^2 \to C \mid \overline{\partial} f = 0, \ \iint |f(x + iy)|^2 y^{n-2} dx \ dy < \infty \right\}.$$

A molecular decomposition for such spaces was then given by Coifman and Rochberg [4]. The Coifman-Rochberg decomposition is more precise in that it relies on a fixed lattice of points $x_i \in G$. The decomposition given here may be regarded as a vector-valued analogue of classical molecular decompositions.

(3) The construction of reproducing kernels for unitary representations should lead to alternate characterizations of H^2, similar to the explicit descriptions of Bergman spaces given above. See Gross and Kunze [5] for a discussion. Such a characterization is essential to provide an H^p theory for $p \neq 2$. We shall take up this topic in a later paper.

REFERENCES

1. Atiyah, M., and Schmid, W. A geometric construction of the discrete series for semi-simple Lie groups. *Inv. Math.* 42 (1977):1-62.

2. Blank, B. E. Knapp-Wallach Szegö integrals and P-induced continuous series representations: The parabolic rank one case. Thesis, Cornell University, 1980.

3. Calderón, A., and Zygmund, A. On higher gradients of harmonic functions. *Studia Math.* 24 (1964):211-226.

4. Coifman, R. R., and Rochberg, R.   Reparesntation theorems for holomorphic and harmonic functions in L^p. *Astérisque* 77 (1980).

5. Gross, K., and Kunze, R. A.   Fourier-Bessel transforms and holomorphic discrete series. *Springer Lecture Notes in Math.* 266 (1972).

6. Hotta, R., and Parthasarathy, R.   Multiplicity formulae for discrete series. *Inv. Math.* 26 (1974):133-178.

7. Knapp, A. W., and Stein, E. M.   Intertwining operators for semi-simple groups. *Ann. Math.* 93 (1974):489-578.

8. Knapp, A. W., and Wallach, N.   Szegö kernels associated with discrete series. *Inv. Math.* 34 (1976):163-200.

9. Kunze, R. A.   Positive definite operator-valued kernels and unitary representations. *Proc. Conf. on Func. Anal.*, Irvine, Calif., 1966.

10. Schmid, W.   On the realization of the discrete series of a semi-simple Lie group. *Rice Univ. Studies* 56 (1970):99-108.

11. Stein, E. M.   Conjugate harmonic functions in several variables. *Proc. International Congress Math.* (1962):414-419.

12. Stein, E. M., and Weiss, G.   Generalizations of the Cauchy-Riemann equations and representations of the rotation group. *Amer. J. Math.* 90 (1968):163-196.

13. Stein, E. M., and Weiss, G.   On the theory of harmonic functions of several variables I. *Acta Math.* 103 (1960):26-62.

INTERPOLATION BETWEEN HARDY SPACES

Peter W. Jones
University of Chicago

1. INTRODUCTION

This paper is a survey of the results on interpolation between Hardy spaces. For $0 < p < \infty$ we denote by $H^p = H^p(\mathbf{R}^n)$ the real variables Hardy space consisting of functions f harmonic in $\mathbf{R}^{n+1}_+$ and satisfying $f^*(x) \in L^p(\mathbf{R}^n)$. Here f^* is the nontangential maximal function defined by

$$f^*(x) = \sup_{|x-y| < t} |f(y, t)|.$$

We always identify H^p functions with their distributional boundary values. When $n = 1$ we also consider for $0 < p < \infty$ the analytic Hardy spaces $H^p_a = H^p_a(\mathbf{R})$ consisting of functions f analytic on $\mathbf{R}^2_+$ and satisfying $f^* \in L^p$. The norm of a function $f \in H^p$ is given by $\|f\|_{H^p} = \|f^*\|_{L^p}$ with a similar definition for H^p_a. When $p = \infty$ there are several possible spaces that are natural "endpoint spaces" of the classes H^p and H^p_a. The first and most obvious is L^∞. When considering the classes H^p_a, we should consider as an endpoint space the class H^∞_a, the ring of bounded analytic functions on $\mathbf{R}^2_+$ endowed with the supremum norm. A third possibility is to consider BMO = BMO$(\mathbf{R}^n)$, the space of functions of bounded mean oscillation. We also consider the space BMO$_a$ = BMO$_a(\mathbf{R})$ with the obvious definition.

At first sight there seems to be no reason to distinguish between the spaces $H_a^p(\mathbf{R})$ and $H^p(\mathbf{R})$. When $0 < p < \infty$, the spaces $H_a^p(\mathbf{R})$ and $H^p(\mathbf{R})$ are isomorphic. (This follows from the Burkholder-Gundy-Silverstein theorem [2].) Since the Hilbert transform is bounded on $BMO(\mathbf{R})$, it also follows that the spaces $BMO_a(\mathbf{R})$ and $BMO(\mathbf{R})$ are isomorphic. The real story is that $\mathrm{Re}\ (H_a^\infty) \neq \mathrm{Re}\ (L^\infty(\mathbf{R}))$ (the Hilbert transform is not bounded on L^∞); $H_a^\infty \subsetneq L^\infty(\mathbf{R})$ and it is a much smaller space. Consequently, it must be harder to interpolate between H_a^p and H_a^∞ than it is to interpolate between $H^p(\mathbf{R})$ and $L^\infty(\mathbf{R})$.

We will consider two methods of interpolation, namely the complex and real methods. The reader is referred to [1] for a full account of these methods. (The reader should be aware, however, that the results for interpolation between Hardy spaces that are stated in that book are not stated correctly!) Before proceeding to the statements of the known results, we point out that a good majority of these results are due to Professor Zygmund and his mathematical descendants.

2. THE COMPLEX METHOD

In this section we discuss the complex method of interpolation, due to Calderón [4]. In this method the spaces between X_0 and X_1 are denoted by $(X_0, X_1)_\theta$, where $0 < \theta < 1$. The complex method of interpolation can be viewed as a generalization of the celebrated Riesz-Thorin theorem. In his 1939 paper [25], Thorin proved the following result (modulo the modern notation).

Theorem 1 (Riesz-Thorin). If $1 \leq p_0 < p_1 \leq \infty$, then $(L^{p_0}, L^{p_1})_\theta = L^p$,
$$\frac{1}{p} = \frac{1 - \theta}{p_0} + \frac{\theta}{p_1}.$$

If we combine Thorin's result with the Marcel Riesz theorem on conjugate functions [21], proved in 1927, we obtain our first result:

$$(H_a^{p_0}, H_a^{p_1})_\theta = H_a^p, \ 1 < p_0 < p_1 < \infty, \ \frac{1}{p} = \frac{1 - \theta}{p_0} + \frac{\theta}{p_1}. \qquad (1)$$

The next result in this direction is due to Salem and Zygmund [23]. In that paper, the authors were looking for a simpler proof of a theorem due to Littlewood and Paley [19]:

$$\int_0^{2\pi} \frac{|S_{n(\theta)}f(\theta)|^p}{\log[n(\theta) + 2]} \, d\theta \leq C_p \int_0^{2\pi} |f(e^{i\theta})|^p \, d\theta, \ 1 < p \leq 2.$$

Here

$$S_{n(\theta)}f(\theta) = \sum_{-n(\theta)}^{n(\theta)} a_k e^{ik\theta}$$

is a partial sum of

$$f(e^{i\theta}) = \sum_{-\infty}^{\infty} a_k e^{ik\theta}.$$

The Littlewood-Paley result is not terribly difficult for $p = 2$, or for $p = 1$ with f in H_a^1 of the circle. Salem and Zygmund had the idea to prove the L^p result, $1 < p < 2$, by interpolating H_a^1 and H_a^2 and then invoking the Marcel Riesz theorem on conjugate functions. [Notice that I am ignoring the difference between $H_a^p(\mathbf{R})$ and $H_a^p(\mathbf{T})$ — almost any theorem for one space has an immediate analogue for the other.] Here is the result Salem and Zygmund proved to deduce the Littlewood-Paley theorem of above:

Theorem 2. Let

$$f(x) = \sum_0^\infty x_v z^v$$

be regular for $|z| < 1$ and belong to the class H^1/α ($\alpha > 0$). Let $M(\alpha, \beta)$ denote the maximum of the finite complex bilinear form

$$\sum_{j=0}^m \sum_{h=0}^n a_{jh} x_j y_h$$

when the complex variables x_j, y_h are subject to the conditions

$$\sum_{0}^{n} |y_k|^{1/\beta} \leq 1 \qquad (\beta \geq 0)$$

and

$$\int_{0}^{2\pi} |f(re^{i\theta})|^{1/\alpha} d\theta \leq 1 \qquad (\alpha > 0),$$

the second condition holding for every $f(z)$ belonging to $H^{1/\alpha}$ and having $x_0, x_1, \ldots, x_m$ as its $m + 1$ first expansion coefficients, and for every nonnegative r inferior to 1. Then if the point (α, β) lies on the segment joining the points (α_1, β_1) and (α_2, β_2), that is to say, if $\alpha = t\alpha_1 + (1 - t)\alpha_2$ and $\beta = t\beta_1 + (1 - t)\beta_2$ $(0 < t < 1)$, we have

$$M(\alpha, \beta) \leq C(\alpha_1, \alpha_2) M^t(\alpha_1, \beta_1) M^{1-t}(\alpha_2, \beta_2),$$

the constant $C(\alpha_1, \alpha_2)$ depending on α_1 and α_2 only.

In 1950 Calderón and Zygmund [6] extended this result to (essentially) obtain

$$(H_a^{p_0}, H_a^{p_1})_\theta = H_a^p, \quad 0 < p_0 < p_1 < \infty, \quad \frac{1}{p} = \frac{1 - \theta}{p_0} + \frac{\theta}{p_1}. \qquad (2)$$

Actually, they only showed that $(H_a^{p_0}, H_a^{p_1}) \supset H_a^p$, that being the only useful result. The opposite inclusion is true, but nobody bothered to prove it until much later (see [5]). The idea behind the proof of (2) goes back to Thorin's 1948 thesis [24]. One first selects an integer n such that $np_0 > 1$. For $f \in H_a^p$, one writes $f = BG = B(G^{1/n})^n$, where B is a Blaschke product and G is zero-free. By this trick and a clever argument (due as far as I can tell to Calderón and Zygmund), one reduces oneself to proving $(H_a^{np_0}, H_a^{np_1}) = H_a^{np}$, which follows from (1) since $1 < np_0 < np_1 < \infty$. Notice that in results (1) and (2), the endpoint $p = \infty$ is excluded. This is because the above proofs use the Marcel Riesz theorem, which fails for $p = \infty$.

After the paper of Calderón and Zygmund, there was a hiatus of some twenty years; it was not until 1972 that C. Fefferman and E. M. Stein [10] gave a new result. In their *Acta* paper they showed

$$(H^1, L^{P_1})_\theta = L^P, \ 1 < p_1 < \infty, \ \frac{1}{p} = 1 - \theta + \frac{\theta}{p_1} \tag{3}$$

and

$$(L^{P_0}, \text{BMO})_\theta = L^P, \ 1 < p_0 < \infty, \ \frac{1}{p} = \frac{1 - \theta}{p_0}. \tag{4}$$

Fefferman and Stein deduced their results from properties of their "sharp function," which is defined by

$$f^\#(x) = \sup_Q \frac{1}{|Q|} \int_Q \left| f(t) - \frac{1}{|Q|} \int_Q f(y)\,dy \right| dt,$$

where the above supremum is taken over all cubes Q containing x. Clearly $f^\# \in L^P$ whenever $f \in L^P$ and $1 < p \le \infty$ and $f^\# \in L^\infty$ whenever $f \in \text{BMO}$. Fefferman and Stein's theorem on the sharp function states that if $1 \le p_0 \le p < \infty$ and $f \in L^{P_0}$, then $\|f\|_{L^P} \le C \|f^\#\|_{L^P}$. From this result it is not hard to deduce (4). Then (3) follows from (4) and H^1, BMO duality. As an exercise, the reader can verify that this approach does not allow one to compute the spaces (H^1, L^∞) or $(H^1, \text{BMO})_\theta$. It should be pointed out, however, that results (3) and (4) applied to L^2 are probably the most important results in the entire area. They give, for example, very traceable methods for proving L^P boundedness of convolution operators.

In 1977 Calderón and Torchinsky [5] extended the results of Fefferman and Stein to show

$$(H^{P_0}, H^{P_1})_\theta = H^P, \ 0 < p_0 < p_1 < \infty, \ \frac{1}{p} = \frac{1 - \theta}{p_0} + \frac{\theta}{p_1} \tag{5}$$

and

$$(H^{P_0}, L^\infty)_\theta \subset (H^{P_0}, \text{BMO})_\theta \subset H^P, \ 0 < p_0 < \infty, \ \frac{1}{p} = \frac{1 - \theta}{p_0}. \tag{6}$$

We will outline a proof of (5) later in this section—it is slightly different from the original proof of Calderon and Torchinsky. The result (6) was obtained by a careful examination of square functions.

This brings us to the most recent developments, which handle the case where one of the endpoints corresponds to $p = \infty$. In 1980 Jones [16] proved the following results:

$$(H_a^{p_0}, H_a^\infty)_\theta = H_a^p, \ 0 < p_0 < \infty, \ \frac{1}{p} = \frac{1 - \theta}{p_0} \tag{7}$$

and

$$(H_a^{p_0}, \text{BMO}_a)_\theta = H_a^p, \ 0 < p_0 < \infty, \ \frac{1}{p} = \frac{1 - \theta}{p_0}. \tag{8}$$

Notice that (8) follows immediately from (6) and (7) because $(H_a^{p_0}, H_a^\infty)_\theta \subset (H_a^{p_0}, \text{BMO}_a)_\theta$. As a direct corollary of (7), (8), and H^1, BMO duality, one obtains

$$(X_0, X_1)_\theta = L^p; \ \frac{1}{p} = 1 - \theta, \tag{9}$$

where X_0 equals either $H^1(\mathbf{R})$ or $L^1(\mathbf{R})$ and where X_1 equals either $H_a^\infty + \overline{H_a^\infty}$, $L^\infty(\mathbf{R})$, or BMO$(\mathbf{R})$. Here is a rough idea of how one would go about proving (7) when $p_0 = 1$, $p = 2$, and $\theta = 1/2$. Take $f \in H_a^2$ with norm 1 and tile $\mathbf{R}_+^2$ into regions $\mathcal{R}_j$. The regions $\mathcal{R}_j$ should satisfy two properties. Firstly, there should be an integer $m(j)$ such that $|f(z)| \approx 2^{m(j)}$ on a large portion of $\mathcal{R}_j$. Secondly, $\sigma \equiv \Sigma\sigma_j$ should be a Carleson measure, where σ_j is arclength measure on $(\partial\mathcal{R}_j) \cap R_+^2$. Now put $f_j = f\chi_{\mathcal{R}_j}$ and notice that $f = \Sigma f_j$. We would like to solve the $\overline{\partial}$ problem $\overline{\partial}h_j = \overline{\partial}f_j$ with good L^∞ estimates on $\Sigma|h_j|$. To this end we invoke a magic formula. Let $\mu = \sigma/\|\sigma\|_c$, where $\|\sigma\|_c$ is the norm of the Carleson measure σ, and put

$$K_0(\mu, z, \zeta) = \frac{1}{z - \zeta} \exp\left\{\iint_{\text{Im } w \leq \text{Im } \zeta} \left(\frac{-i}{z - \overline{w}} + \frac{i}{\zeta - \overline{w}}\right) d\mu(w)\right\},$$

$$K(\mu, z, \zeta) = \frac{2i}{\pi}\left(\frac{\text{Im } \zeta}{z - \zeta}\right)K_0(\mu, z, \zeta).$$

Notice that $K(\mu, z, \zeta)$ is holomorphic (in z) on $R_+^2 \backslash \zeta$ and has a simple pole at $z = \zeta$. Now put

$$h_j(z) = \iint_{R_+^2} K(\phi, z, \zeta)\overline{\partial}f_j(\zeta).$$

Then $\overline{\partial}h_j = \overline{\partial}f_j$ and by linearity, $\Sigma h_j \equiv 0$. Let $g_j = f_j - h_j$ and let $\alpha(\zeta) = 1 - \zeta$. Then $G_\zeta = \Sigma 2^{m(j)\alpha(\zeta)}g_\zeta$ is a holomorphic Banach space-valued function in $\{0 < \text{Re } \zeta < 1\}$, and using the properties of $K(\mu, z, \zeta)$, one can show

$$\|G_{it}\|_{H_a^1} \leq C, \ t \in R$$

and

$$\| G_{1+it} \|_{H_a^{\infty}} \leq C, \ t \ \epsilon \ \mathbf{R}.$$

This does the job because

$$G_{\theta} = G_{1/2} = \Sigma(f_j - h_j) = \Sigma f_j = f.$$

The idea of using Carleson measures and solutions of $\bar{\partial}$ to attack problems like this has a long history. We say no more than the idea is a descendant of ideas coming from Carleson's solution of the Corona problem [7] and Hörmander's later modifications [12]. The reader is referred to [16] for a further discussion.

In 1980, after (7) and (8) had been obtained, Janson and Jones [15] completed the classification of the interpolation spaces in the complex method by showing

$$(H^{p_0}, L^{\infty})_{\theta} = H^p, \ 0 < p_0 < \infty, \ \frac{1}{p} = \frac{1 - \theta}{p_0} \tag{10}$$

and

$$(H^{p_0}, \mathrm{BMO})_{\theta} = H^p, \ 0 < p_0 < \infty, \ \frac{1}{p} = \frac{1 - \theta}{p_0}. \tag{11}$$

As in the one-dimensional case, (11) follows from (6) and (10), and one can show

$$(X_0, X_1)_{\theta} = L^p, \ \frac{1}{p} = 1 - \theta, \tag{12}$$

where X_0 equals either H^1 or L^1 and where X_1 equals either L^{∞} or BMO. The proof of (10) heavily uses the theory of H^p atoms. (As far as I know, the first person to use atoms explicitly in proving interpolation theorems was Peetre in his 1979 paper [20]. The idea was, however, lurking in the papers [8, 9, and 22].) A function a on $\mathbf{R}^n$ is an H^p atom if a is supported on a cube Q,

$$\| a \|_{L^{\infty}} \leq |Q|^{-1/p}$$

and

$$\int x^{\alpha} a \ dx = 0$$

for all monomials $x^{\alpha} = x_1^{\alpha_1} \ldots x_n^{\alpha_n}$ satisfying $|\alpha| \leq \alpha(p)$, where $\alpha(p)$ is a certain nonincreasing function of p. The fundamental result in this area, due to Coifman [8] when $n = 1$ and Latter [17] when $n > 1$, is:

Theorem 3 (on the atomic decomposition). Suppose $f \in H^p$, $0 < p < \infty$, and suppose $0 < p_0 < p$. Then there are sequences $\{a_j\}$ of H^p atoms and $\{\lambda_j\}$ of complex numbers such that

$$\int x^{\alpha} a_j \; dx = 0, \quad |\alpha| \le \alpha(p_0)$$

and

$$\sum |\lambda_j|^p \le C(p, \; p_0) \|f\|_{H^p}^p.$$

Other proofs of this theorem can be found in [4, 18, and 26]. (The reader should be aware, however, that none of these papers state their results as done above.) A combinatorial argument presented in [15] shows that one may refine the atomic decomposition to obtain the following result.

Theorem 4. Suppose $0 < p_0 < p < \infty$ and suppose $\|f\|_{H^p} = 1$. Then there are dyadic cubes Q_j and functions a_j supported on the triples of Q_j such that $f = \sum a_j$. Furthermore, there are integers $m(j)$ such that

(α) $\|a_j\|_{L^{\infty}} \le C 2^{m(j)}$,

(β) $\sum 2^{pm(j)} |Q_j| \le C$,

(γ)   if $Q_k \subset Q_j$ and $k \ne j$ then

$$m(k) > m(j) \text{ and } Q_k \subsetneq Q_j,$$

(δ) $\displaystyle\sum_{Q_k \subset Q_j} |Q_k| \le 2|Q_j|$,

(ε) $\int x^{\alpha} a_j \; dx = 0, \quad |\alpha| \le \alpha(p_0)$,

(ζ)   if A is any finite set of indices, then for every x there exists $j(x) \in A$ such that

$$\sum_{j \in A} |a_j(x)| \le C 2^{m(j(x))} \chi_{3Q_{j(x)}}(x).$$

Now let $\alpha(\zeta) = \dfrac{p}{p_0}(1 - \zeta) - 1$ and let $G_{\zeta} = \sum 2^{m(j)\alpha(\zeta)} a_j$. Then the conclusions of the above theorem show $\|G_{it}\|_{H^{p_0}} \le C$, $t \in \mathbf{R}$, and if $0 < \varphi < 1$, $\dfrac{1}{r} = \dfrac{1 - \varphi}{p_0}$, then $\|G_{\varphi + it}\|_{H^r} \le C_r$, $t \in \mathbf{R}$. This

yields the result of Calderón and Torchinsky (5). (In fact, the
proof of (5) is much simpler once one has the theorem on the atomic
decomposition. From the special *form* of the functions $\lambda_j a_j$ in [4],
[17], or [18], one funds scalars β_j and a holomorphic function $a(\zeta)$
such that $G_\zeta = \sum \beta_j^{a(\zeta)} \lambda_j a_j$ is an appropriate holomorphic function.)
The point of the refinement of the atomic decomposition is that by
(δ), a Carleson packing condition, the functions a_j are "almost dis-
joint." One can find functions $\tilde{a}_j$ such that $\| f - \sum \tilde{a}_j \|_{H^p} < 1/2$
and almost every point x is in the support of at most C functions
$\tilde{a}_j$. Now let $\alpha(\zeta) = \frac{p}{p_0}(1 - \zeta) - 1$ and put $G_\zeta = \sum 2^{m(j)\alpha(\zeta)} \tilde{a}_j$. Then

$$\| G_{1+it} \|_{L^\infty} \leq C, \ t \in \mathbf{R}.$$

The functions $\tilde{a}_j$ are also constructed so that

$$\| G_{it} \|_{H^{p_0}} \leq C, \ t \in \mathbf{R}.$$

Since

$$\| f - G_\theta \|_{H^p} = \| f - \sum \tilde{a}_j \|_{H^p} < 1/2,$$

an iteration argument now shows that $H^p \subset (H^{p_0}, L^\infty)_\theta$.

Very recently, Tom Wolff [27] has shown that when $p_0 = 1$,
(10) follows from (3) and

Theorem 5 (Wolff's theorem). Let A_j be Banach spaces, $1 \leq j \leq 4$,
and suppose $A_1 \cap A_4 \subset A_2 \cap A_3$ and $A_1 \cap A_4$ is dense in A_2 and A_3.
Suppose also there exist $0 < \psi_2, \psi_3 < 1$ such that $(A_1, A_3)_{\psi_2} = A_2$
and $(A_2, A_4)_{\psi_3} = A_3$. Then there exist $0 < \theta_j < 1$ such that
$(A_1, A_4)_{\theta_j} = A_j, \ j = 2, 3.$

The result $(H^1, L^\infty)_\theta = L^p$ now follows from $(H^1, L^3)_\theta = L^p$,
$(L^2, L^\infty)_\theta = L^p$, and Wolff's theorem. Notice, however, that this
method *cannot* be used to give a quick proof of (7) because if (7)
holds for one value of $p_0 < \infty$, then it follows immediately for all
$0 < p_0 < \infty$ by factorization.

The results for the complex method are summarized in
Table 1.

TABLE 1

X_0	X_1	$(X_0, X_1)_\theta$	Restrictions
$H_a^{p_0}$	$H_a^{p_1}$	H_a^p	$0 < p_0 < p_1 \leq \infty, \quad \dfrac{1}{p} = \dfrac{1-\theta}{p_0} + \dfrac{\theta}{p_1}$
$H_a^{p_0}$	BMO_a	H_a^p	$0 < p_0 < \infty, \quad \dfrac{1}{p} = \dfrac{1-\theta}{p_0}$
$H^1(\mathbf{R})$ or $L^1(\mathbf{R})$	$H_a^\infty + \overline{H_a^\infty}$	$L^p(\mathbf{R})$	$\dfrac{1}{p} = 1 - \theta$
H^{p_0}	H^{p_1}	H^p	$0 < p_0 < p_1 < \infty, \quad \dfrac{1}{p} = \dfrac{1-\theta}{p_0} + \dfrac{\theta}{p_1}$
H^{p_0}	L^∞ or BMO	H^p	$0 < p_0 < \infty, \quad \dfrac{1}{p} = \dfrac{1-\theta}{p_0}$
L^{p_0}	BMO	L^p	$1 \leq p_0 < \infty, \quad \dfrac{1}{p} = \dfrac{1-\theta}{p_0}$

3. THE REAL METHOD

In this section we discuss the real method of interpolation. The
reader is referred to [1] for basic properties and definitions.
In the real method the spaces between X_0 and X_1 are denoted by
$(X_0, X_1)_{\theta, q}$, where $0 < \theta < 1$ and $0 < q \leq \infty$. The first nontrivial
result in this area came in 1963 and is due to Igari [13 and 14].
Unfortunately, his work seems to have been largely unnoticed. I,
myself, was unaware of his work until the conference in honor of
Professor Zygmund, at which time Guido Weiss kindly pointed it out
to me. In [14], Igari shows that if $f \in L^p$ and $\alpha > 0$, then
$f = g + h$ where $\| h \|_{L^\infty} \leq C\alpha$ and

$$\| g \|_{H^1} \leq C \int_{\{f^* > \alpha\}} f^*(x)\, dx$$

Igari proved this by using a Calderón-Zygmund decomposition. From this result, Igari deduced the following for sublinear operators T:

If $T : H^{p_j} \to L^{q_j}$, $1 \leq p_j \leq q_j < \infty$, then $T : H^p \to L^q$, where   (13)

$$\frac{1}{p} = \frac{1 - \theta}{p_0} + \frac{\theta}{p_1}, \frac{1}{q} = \frac{1 - \theta}{q_0} + \frac{\theta}{q_1}, \text{ and } 0 < \theta < 1,$$

and

If $T : H_a^{p_j} \to L^{q_j}$, $1 \leq p_j \leq q_j < \infty$, then $T : H_a^p \to L^q$, where   (14)

$$\frac{1}{p} = \frac{1 - \theta}{p_0} + \frac{\theta}{p_1}, \frac{1}{q} = \frac{1 - \theta}{q_0} + \frac{\theta}{q_1}, \text{ and } 0 < \theta < 1.$$

Notice that Igari did not state any results about interpolation between H^1 and L^∞. On the other hand, with a little bit of functional analysis plus Igari's result on decompositions of L_p functions, one obtains

$$(H^{p_0}, L^\infty)_{\theta, q} = H^{p, q}, \quad 1 \leq p_0 < \infty, \quad \frac{1}{p} = \frac{1 - \theta}{p_0}.$$ (15)

Here $H^{p, q}$ denotes the class of harmonic functions f having the property that f^* is in the Lorentz space $L^{p, q}$. Result (15) was rediscovered by Riviere and Sagher in their 1973 paper [22]. In 1974, C. Fefferman, Riviere, and Sagher [9] extended result (15) to show

$$(H^{p_0}, H^{p_1})_{\theta, q} = H^{p, q}, \quad 0 < p_0 < p_1 < \infty, \quad \frac{1}{p} = \frac{1 - \theta}{p_0} + \frac{\theta}{p_1}$$ (16)

and

$$(H^{p_0}, L^\infty)_{\theta, q} = H^{p, q}, \quad 0 < p_0 < \infty, \quad \frac{1}{p} = \frac{1 - \theta}{p_0}.$$ (17)

The methods used by these three authors are closely related to those used to prove the atomic decomposition theorem (see Sec. 2). Indeed, Latter's proof in [17] of the atomic decomposition used some functions first appearing in [9] as a key ingredient. In turn, the form of the atomic decomposition as presented in [4, 17, 18, and 26] can be used to deduce (16) and (17). Suppose, for example, that $f \epsilon H^p$, $0 < p < \infty$, $\|f\|_{H^p} = 1$ and let

$\Omega_k = \{f^* > 2^k\}$. Then for each $k \in \mathbf{Z}$ there are functions $a_{k, j}$ supported on Ω_k such that

$$\| a_{k, j} \|_{L^\infty} \leq C 2^k, \int x^\alpha a_{k, j} \, dx = 0$$

for $0 \leq |\alpha| \leq \alpha(p_0)$, and such that every point is in the support of at most C functions $a_{k, j}$. Furthermore, $\sum_{k, j} a_{k, j} = f$. If $\alpha > 0$, let

$$f_1^\alpha = \sum_{2^k > \alpha} \sum_j a_{k, j}$$

and let $f_2^\alpha = f - f_1^\alpha$. Using this splitting, one can rather easily verify that (17) holds for $0 < p_0 \leq 1$ (for $1 < p_0 < \infty$, it is classical). Result (16) now follows from the reiteration theorem (see, for example, [1]).

The results of C. Fefferman, Rivière, and Sagher were extended by Hanks [11] in 1977 to include L^p and BMO as endpoints. His results are

$$(L^{p_0}, \text{BMO})_{\theta, q} = L^{p, q}, \quad 0 < p_0 < \infty, \quad \frac{1}{p} = \frac{1 - \theta}{p_0} \tag{18}$$

and

$$(H^{p_0}, \text{BMO})_{\theta, q} = H^{p, q}, \quad 0 < p_0 < \infty, \quad \frac{1}{p} = \frac{1 - \theta}{p_0}. \tag{19}$$

Hanks deduces (19) from (18). The idea is that (19) follows from (18) when $p > 1$. (There is a small slip in Hank's proof at this point—he asserts that $H^{p_0} \subset L^{p_0}$ when $p_0 < 1$, which is false. Hank's argument can easily be fixed by noticing that $p_\varepsilon * H^{p_0} \subset L^{p_0}$ and $p_\varepsilon * \text{BMO} \subset \text{BMO}$, where p_ε is the Poisson kernel of height ε.) The full result in (19) now follows from the reiteration theorem.

The classification of the intermediate spaces in the real method was completed by Jones [16] in 1980. The result is

$$\left(H_a^{p_0}, H_a^\infty\right)_{\theta, q} = \left(H_a^{p_0}, \text{BMO}_a\right)_{\theta, q} = H_a^{p, q}, \quad 0 < p_0 < \infty, \quad \frac{1}{p} = \frac{1 - \theta}{p_0}. \tag{20}$$

This result follows from Theorem 6 below and the reiteration theorem.

Theorem 6. *Suppose* $0 < p_0 < p < \infty$, $\alpha > 0$, *and suppose* $f \in H_a^p$. *Then* $f = F_\alpha + f_\alpha$, *where* $F_\alpha \in H_a^{p_0}$, $f_\alpha \in H_a^\infty$, *and*

$$\| F_\alpha \|_{H_a^{p_0}}^{p_0} \leq C_{p_0} \int_{\{f^* > a\}} |f^*(x)|^{p_0}\, dx,$$

$$\| f_\alpha \|_{H_a^\infty} \leq C\alpha.$$

We illustrate the proof of the theorem when $p_0 = 1$ and $p = 2$. Let $O_\alpha = \{x : f^\alpha(x) > \alpha\}$ and let T_α denote the union of the tents lying over O_α, that is, the complement of the usual saw tooth region. Let $H_\alpha(z) = f(z)\chi_{T_\alpha}(z)$. Then $\overline{\partial}H_\alpha$ is a Carleson measure supported on the boundary of T_α and $\| \overline{\partial}H_\alpha \|_C \leq \sqrt{2}\alpha$. Let K be the kernel constructed in Sec. 2 and let

$$g_\alpha(z) = \iint_{\mathbf{R}_+^2} K(\overline{\partial}H_\alpha / \| \overline{\partial}H_\alpha \|_C,\ z,\ \zeta)\overline{\partial}H_\alpha(\zeta).$$

Then

$$f = H_\alpha - g_\alpha + (f - H_\alpha + g_\alpha) = F_\alpha + f_\alpha \in H_a^1 + H_a^\infty$$

and

$$\| f_\alpha \|_{H_a^\infty} \leq C\alpha, \quad \| F_\alpha \|_{H_a^1} \leq C \int_{O_\alpha} f^*(x)\, dx.$$

The results for the real method are summarized in Table 2. The index q can take any value $0 < q \leq \infty$.

TABLE 2

X_0	X_1	$(X_0, X_1)_{\theta, q}$	Restrictions		
$H_a^{p_0}$	$H_a^{p_1}$	$H_a^{p, q}$	$0 < p_0 < p_1 \leq \infty$,	$\dfrac{1}{p} = \dfrac{1 - \theta}{p_0} + \dfrac{\theta}{p_1}$	
$H_a^{p_0}$	BMO$_a$	$H_a^{p, q}$	$0 < p_0 < \infty$,	$\dfrac{1}{p} = \dfrac{1 - \theta}{p_0}$	
H^{p_0}	H^{p_1}	$H^{p, q}$	$0 < p_0 < p_1 < \infty$,	$\dfrac{1}{p} = \dfrac{1 - \theta}{p_0} + \dfrac{\theta}{p_1}$	
H^{p_0}	L^∞ or BMO	$H^{p, q}$	$0 < p_0 < \infty$,	$\dfrac{1}{p} = \dfrac{1 - \theta}{p_0}$	
L^{p_0}	BMO	$L^{p, q}$	$0 < p_0 < \infty$,	$\dfrac{1}{p} = \dfrac{1 - \theta}{p_0}$	

REFERENCES

1. Bergh, J., and Löfström, J. *Interpolation Spaces*. New York: Springer-Verlag, 1976.

2. Burkholder, D. L.; Gundy, R. F.; and Silverstein, M. L. A maximal function characterization of the class H^p. *Trans. Amer. Math. Soc.* 157 (1971):137-153.

3. Calderón, A. P. An atomic decomposition of distributions in parabolic H^p spaces. *Advances in Math.* 25 (1977):216-225.

4. Calderón, A. P. Intermediate spaces and interpolation, the complex method. *Studia Math.* 24 (1964):113-190.

5. Calderón, A. P., and Torchinsky, A. Parabolic maximal functions associated with a distribution, II. *Advances in Math.* 24 (1977):101-171.

6. Calderón, A. P., and Zygmund, A. On the theorem of Hausdorff-Young and its extensions. *Ann. Math.* study no. 25:166-188. Princeton University Press, 1950.

7. Carleson, L. Interpolation by bounded analytic functions and the Corona theorem. *Ann. Math.* 76 (1962):547-549.

8. Coifman, R. R. A real variable characterization of H^p. *Studia Math.* 51 (1974):269-274.

9. Fefferman, C.; Rivière, N. M.; and Sagher, Y. Interpolation between H^p spaces: The real method. *Trans. Amer. Math. Soc.* 191 (1974):75-81.

10. Fefferman, C., and Stein, E. M. H^p spaces of several variables. *Acta Math.* 129 (1972):137-193.

11. Hanks, R. Interpolation by the real method between BMO, L^α ($0 < \alpha < \infty$) and H^α ($0 < \alpha < \infty$). *Indiana Math. J.* 26 (1977):679-690.

12. Hörmander, L. Generators for some rings of analytic functions. *Bull. Amer. Math. Soc.* 73 (1967):943-949.

13. Igari, S. An extension of the interpolation theorem of Marcinkiewicz. *Proc. Japan Acad.* 38 (1962):731-734.

14. Igari, S. An extension of the interpolation theorem of Marcinkiewicz II. *Tôhoku Math. J.* 15 (1963):343-358.

15. Janson, S., and Jones, P. W. Interpolation between H^p spaces: The complex method. To appear in *J. Func. Anal.*

16. Jones, P. W. L^∞ estimates for the $\overline{\partial}$ problem in a half-plane. To appear in *Acta Math.*

17. Latter, R. H. A characterization of $H^p(\mathbf{R}^n)$ in terms of atoms. *Studia Math.* 62 (1978):93-101.

18. Latter, R. H., and Uchiyama, A. The atomic decomposition for parabolic H^p space. *Trans. Amer. Math. Soc.* 253 (1979):391-398.

19. Littlewood and Paley. Theorems on Fourier series and power series (III). *Proc. London Math. Soc.* 43 (1937):105-126.

20. Peetre, J. Two observations on a theorem of Coifman. *Studia Math.* 64 (1979):191-194.

21. Riesz, M. Sur les fonctions conjugées. *Math. Zeit.* 27 (1927):218-244.

22. Rivière, N. M., and Sagher, Y. Interpolation between L^∞ and H^1, the real method. *J. Func. Anal.* 14 (1973):401-409.

23. Salem, R., and Zygmund, A. A convexity theorem. *Proc. Nat. Acad. Sci.* 34 (1948):443-447.

24. Thorin, G. O. Convexity theorems. *Medd. Lunds Mat. Sem.* 9 (1948):1-57.

25. Thorin, G. O. An extension of a convexity theorem due to M. Riesz. *Medd. Lunds Univ. Mat. Sem.* 4 (1939).

26. Wilson, J. M. A simple proof of the atomic decomposition for $H^p(\mathbf{R}^n)$, $0 < p \le 1$. To appear in *Studia Math.*

27. Wolff, T. H. A note on interpolation spaces. To appear in the proceedings of the 1981 conference in honor of N. M. Rivière, held at the University of Minnesota.

WEIGHTED H¹-BMO DUALITIES

Umberto Neri
University of Maryland

1. ATOMIC PREDUALS

The simplest examples of the dualities to be considered here occur between *atomic* h^1 and *weighted BMO* over spaces of homogeneous type. Thus, let $X = (X, d)$ be a space of homogeneous type ([2], Sec. 2) with respect to a pair of positive, mutually absolutely continuous Borel measures μ and ν. On X, we define the function space $BMO_\mu(d\nu)$ to be the set of all $f \in L^1(d\nu)$ such that

$$\mu(B)^{-1} \int_B |f - f_{\nu(B)}| \, d\nu \leq C \tag{1}$$

uniformly for all balls $B \subset X$, where $f_{\nu(B)} = \nu(B)^{-1} \int_B f \, d\nu$.

If $\nu = \mu$, this space is the usual BMO. Otherwise, we shall call it *weighted BMO*, or BMV (for *bounded mixed variance*) as in [15].

To find the predual of BMV = $BMO_\mu(d\nu)$, we look for ν-measurable functions a on X such that

$$|\langle a, f \rangle| = \left| \int_X af \, d\nu \right| \leq \mu(B)^{-1} \int_B |f - f_{\nu(B)}| \, d\nu \tag{2}$$

The author was partially supported by the University of Minnesota and by a grant of the C.N.R. (National Research Council of Italy).

for each $f \in$ BMV.   Clearly, any function a with

$$\int a \, d\nu = 0, \quad \text{supp}(a) \subset B, \quad \text{and} \quad \| a \|_\infty \leq \mu(B)^{-1}$$

for some ball B in X will fulfill (2).   We call these a *atoms of type* (μ, ν) and we denote by $h^1_\mu \, (d\nu)$ the space of all (infinite) linear combinations

$$\Sigma \lambda_j a_j, \quad \text{with } \Sigma |\lambda_j| < \infty$$

and a_j being such atoms.   Note that for $\mu = \nu$, our atoms are just $(1, \infty)$ atoms in the terminology of [2].

Adapting some Coifman-Weiss arguments in [2], we have that if μ and ν are related by appropriate A_p conditions, then

$$[h^1_\mu(d\nu)]^* = \text{BMO}_\mu (d\nu) \tag{3}$$

with action $\langle a, f \rangle$ as above on the atoms $a \in h^1_\mu(d\nu)$ (see [15]).

2. UNWEIGHTED H'-BMO THEORY IN NONSMOOTH DOMAINS

Let us consider bounded, simply connected domains D in $\mathbf{R}^{n+1}$, $n \geq 2$, say, where the following general principle of harmonic functions (and other solutions of "good" P.D.E.'s) holds: "nontangential boundedness" implies the existence (a.e.) of "nontangential limits" on ∂D.   Thus, in decreasing order of generality, we may consider

(A)   N.T.A. domains (in the sense of Jerison and Kenig [12]),

(B)   Lipschitz domains (for example, Hunt-Wheeden [11], Dahlberg [3]),

(C) C^1 domains (as in [6, 7, 8] and so forth).

On $X = \partial D$, we may have (Lebesgue) surface measure $d\sigma$ and/or harmonic measure $d\omega$, evaluated at some fixed $P_0 \in D$.   The general problem is as follows: to find H^1-spaces of harmonic functions in D identifiable (via nontangential limits) with atomic h^1-spaces on ∂D, and hence yielding new preduals for BMO, or weighted BMO, on ∂D.

In case (A), if B is any "surface ball" (that is, intersection of a Euclidean ball, centered on ∂D, with ∂D itself), then $\sigma(B)$ may be infinite. So we must define BMO(∂D) as BMO$_\omega(d\omega)$, taking $\mu = \nu = \omega$ in formula (1) of Sec. 1. The simplest way to introduce a Hardy H^1 space of harmonic functions is to solve a Dirichlet problem for the Laplacean with appropriate L^1 data. In short, introduce the class

$$H^1_{\mathscr{D}}(D, \ d\omega) = \{u : \Delta u = 0 \text{ in } D, \ u(P_0) = 0, \ (Nu) \in L^1(d\omega) \text{ on } \partial D\}$$

normed with $\| u \| \ = \ \| Nu \|_{L^1(d\omega)}$, where Nu denotes a nontangential maximal function of u (see [12]). Then Jerison and Kenig proved that this class of harmonic functions can be identified with the space $H^1_{\mathscr{D}}(\partial D, \ d\omega)$ of nontangential limits of such functions. Moreover, we have a duality

$$[H^1_{\mathscr{D}}(\partial D, \ d\omega)]^* = \text{BMO}_\omega(d\omega) \tag{4}$$

and an atomic representation

$$H^1_{\mathscr{D}}(\partial D, \ d\omega) = h^1_\omega(d\omega) \tag{4'}$$

in terms of atoms of type (ω, ω).

In the easier setting of C^1 domains, (4) and (4′) were first proved by Fabes, Kenig, and Neri in [8]. In the setting of N.T.A. domains, no other H^1-BMO dualities are known.

Before passing to cases (B) and (C), we must recall some definitions. A positive function w on ∂D is in $A_p(d\sigma)$, respectively $B_p(d\sigma)$, $1 < p < \infty$ if

$$(A_p) \qquad \left\{ \sigma(B)^{-1} \int_B w \ d\sigma \right\} \left\{ \sigma(B)^{-1} \int_B (1/w)^{1/(p-1)} d\sigma \right\}^{p-1} \leq C_p,$$

respectively

$$(B_p) \qquad \left\{ \sigma(B)^{-1} \int_B w^p \, d\sigma \right\}^{1/p} \le C_p \left\{ \sigma(B)^{-1} \int_B w \, d\sigma \right\},$$

for some constants $C_p > 0$ independent of the surface balls B. For brevity, let

$$A_\infty = \bigcup_{p>1} A_p, \qquad B_\infty = \bigcup_{p>1} B_p,$$

$$A_* = \bigcap_{p>1} A_p, \qquad B_* = \bigcap_{p>1} B_p.$$

As is well known, these conditions play a key role for the validity of weighted L^p estimates of maximal operators, singular integrals, and so forth. Now, a basic difference between C^1 and Litschitz domains is as follows:

If D is Lipschitz, then $K = (d\omega/d\sigma) \in \{A_\infty(d\sigma) \cap B_2(d\sigma)\}$ (see [3, 5]);

If D is C^1, then K and K^{-1} are in $A_*(d\sigma) \cap B_*(d\sigma)$ (cf. [8, Sec. 1]).

This difference reflects the fact that on the boundary of a C^1 domain we can use the L^p theory of *Cauchy-Calderón singular integrals* [1, 6], while this theory is still unavailable for Lipschitz domains. In 1981, Coifman, McIntosh, and Meyer (in press) proved the boundedness of these operators also in this case. As a result, the H^1-BMO theory of harmonic functions has been developed more fully for C^1 domains.

From now on, D shall always denote a C^1 domain in $\mathbf{R}^{n+1}$. Here, Fabes and Kenig [7] solved the Neumann problem

$$\Delta U = 0 \text{ in } D, \quad \partial U/\partial N_Q = g \in h_\sigma^1(d\sigma) \text{ on } \partial D,$$

by means of single-layer potentials and found solutions U with gradients satisfying $N(|\nabla U|) \in L^1(d\sigma)$. In other words, the atomic space $h_\sigma^1(d\sigma)$ is contained in a class of Neumann data:

$$H_{\mathcal{N}}^1(\partial D, \, d\sigma) = \left\{ f \in L^1(d\sigma), \, \int f \, d\sigma = 0, \, f = \partial U/\partial N_Q \right.$$

$$\left. \text{for some harmonic } U \text{ in } D \text{ with } N(|\nabla U|) \in L^1(d\sigma) \right\}.$$

Moreover, they proved the duality

$$[H_{\mathcal{N}}^1(\partial D,\ d\sigma)]^* = \mathrm{BMO}_\sigma(d\sigma) \qquad\qquad (5)$$

along with the atomic representation

$$H_{\mathcal{N}}^1(\partial D,\ d\sigma) = h_\sigma^1(d\sigma). \qquad\qquad (5')$$

Since $K \in A_\infty(d\sigma)$ implies that $\mathrm{BMO}_\omega(d\omega) = \mathrm{BMO}_\sigma(d\sigma)$, comparing (4) and (5) we observe the existence of two distinct preduals of $\mathrm{BMO}(\partial D)$ for C^1 domains: the first being associated to a Dirichlet problem, the second to a Neumann problem.

3. WEIGHTED H¹-BMO DUALITIES

With $\mu = \sigma$ and $d\nu = d\omega$ on ∂D, we define $\mathrm{BMO}_\sigma(d\omega)$ as in Sec. 1 and let $h_\sigma^1(d\omega)$ denote the corresponding atomic predual. Thus, for any atom $a \in h_\sigma^1(d\omega)$, we have

$$\int a\ d\omega = 0, \quad \mathrm{supp}(a) \subset B \quad \text{and} \quad \|a\|_\infty \leq \sigma(B)^{-1}.$$

Let us solve the Dirichlet problem

$$\Delta u = 0 \text{ in } D, \qquad \text{nontang. } \lim u = a \text{ on } \partial D$$

for an atom a in $h_\sigma^1(d\omega)$, by means of the Poisson integral $u = \mathscr{P}a$. That is, for each $X \in D$ and general f on ∂D,

$$[\mathscr{P}f](X) = \int_{\partial D} f\ d\omega^X = \int_{\partial D} f(y)k(X,\ y)\,d\omega(y),$$

where $k(X,\ y)$ is the *kernel function* for D normalized at $X = P_0$.

Now, letting

$$H_{\mathscr{D}}^1(D,\ d\sigma) = \{u : \Delta u = 0 \text{ in } D,\ u(P_0) = 0,\ \text{and}\ (Nu) \in L^1(d\sigma)\},$$

we have the duality

$$[H_{\mathscr{D}}^1(D,\ d\sigma)]^* = \mathrm{BMO}_\sigma(d\omega) \qquad\qquad (6)$$

and the atomic representation

$$H^1_{\mathscr{D}}(\partial D, \ d\sigma) = h^1_\sigma(d\omega) \tag{6'}$$

(see [10]). We note that (6) was announced earlier by Dahlberg [4], but his proofs have been unavailable to us.

An important step in the proof of (6) is given by the following:

Lemma. There exists a geometric constant $C > 0$ such that

$$\| N(\mathscr{P}a) \|_{L^1(d\sigma)} \le C, \ \text{for all atoms } a \ \epsilon \ h^1_\sigma(d\omega). \tag{*}$$

Proof. Let $a \ \epsilon \ h^1_\sigma(d\omega)$ be an atom and fix a surface ball $B = B(x_0, \ r) \supset \text{supp}(a)$. By a result of Jerison and Kenig [12, Theorem 7.1], there exists a constant $M > 1$ such that for all $j \ \epsilon \ \mathcal{N} = \{1, \ 2, \ 3, \ \ldots\}$, all x_0 and x_1 in ∂D and all $X \ \epsilon \ D$, if $|X - x_0| > M^j|x_1 - x_0|$, then

$$|k(X, \ x_1) - k(X, \ x_0)| \le M(1 - M^{-1})^j k(X, \ x_0). \tag{7}$$

Consider the M-adic dilations of B, that is, the surface balls $B_j = B(x_0, \ M^j r)$, $j \ \epsilon \ \mathcal{N}$. On B_1, letting $d\sigma = K^{-1} \ d\omega$ and using Schwarz's inequality, we see that with $u = \mathscr{P}a$,

$$\int_{B_1} |Nu| \ d\sigma \le \left\{ \int_{B_1} (Nu)^2 \ d\omega \right\}^{1/2} \left\{ \int_{B_1} K^{-2} \ d\omega \right\}^{1/2}$$

$$\le C_2 \| Nu \|_{L^2(d\omega)} \sigma(B_1)^{1/2} [\sigma(B_1)/\omega(B_1)]^{1/2},$$

since $K \ \epsilon \ A_2(d\sigma)$. Hence, by the $L^2(d\omega)$ boundedness of $a \to Nu$ and the definition of a, it follows that

$$\int_{B_1} |Nu| \ d\sigma \le C'_2 \left\{ [\omega(B)/\sigma(B)] \ [\sigma(B_1)/\omega(B_1)] \right\}^{1/2} \le C.$$

On $\partial D \backslash B_1$ we apply (7), noting that $k(X, \ x_0) \int a \ d\omega = 0$. Thus, for any $x \ \epsilon \ \partial D \backslash B_j$ and any $X \ \epsilon \ \Gamma_x$ (truncated inner cone with vertex at x), we have

$$|u(X)| \le \int_B |k(X, \ y) - k(X, \ x_0)| \, |a(y)| \, d\omega(y)$$

(continued)

$$\leq M(1-M^{-1})^j \int_B k(X, x_0)\sigma(B)^{-1} d\omega(y)$$

$$\leq M(1-M^{-1})^j \sigma(B)^{-1}\omega(B)\omega(B_j)^{-1}.$$

Hence, with $\varepsilon = M^{-1}$, we see that for all $x \in \partial D\backslash B_j$,

$$[Nu](x) \leq \varepsilon^{-1}(1 - \varepsilon)^j \sigma(B)^{-1}\omega(B)/\omega(B_j). \qquad (8)$$

Since D is a C^1 domain, $(d\sigma/d\omega) \in A_p(d\omega)$ for all $1 < p < \infty$. So a simple argument in [8, Lemma 2.6] shows that for any $\eta > 0$ there exists a constant $C_\eta > 0$ such that

$$[\omega(B)/\omega(B_j)][\sigma(B_j)/\sigma(B)] \leq C_\eta M^{\eta j}.$$

Therefore,

$$\int_{\partial D\backslash B_1} |Nu| d\sigma \leq \varepsilon^{-1}C_\eta \sum_{j\geq 2} [(1 - \varepsilon)M^\eta]^j < \infty$$

if we fix $\eta > 0$ so small that $(1 - \varepsilon)M^\eta < 1$. The proof is complete.

Alternatively, looking at the atomic space $h_\omega^1(d\sigma)$, we can also solve the Neumann problem by means of single-layer potentials in analogy to [7]. This approach leads to a duality of the form

$$\text{BMO}_\omega(d\sigma) = [H_N^1(\partial D, d\omega)]^* \qquad (9)$$

and a corresponding atomic representation [9].

Let me close by mentioning an interesting open problem. The new results of Jerison and Kenig [13] suggest the possibility of solving the Neumann problem with L^p data, $1 < p < 2$, for a *Lipschitz* domain D. By [13], since $h_\sigma^1(d\sigma) \subset L^2(d\sigma)$, we know that a solution U of

$$\Delta U = 0 \text{ in } D, \qquad \partial U/\partial N_Q = a \in h_\sigma^1(d\sigma),$$

always exists. If an H^1-BMO duality of this type still holds, one would like to prove that, for some absolute constant $C > 0$,

$$\| N(|\nabla U|) \|_{L^1(d\sigma)} \leq C$$

whenever a is an atom of type (σ, σ). At present, all this is unknown.

REFERENCES

1. Calderón, A. P. Cauchy integrals on Lipschitz curves and related operators. *Proc. Nat. Acad. Sci.* 74 (1977):1324-1327.

2. Coifman, R. R., and Weiss, G. Extensions of Hardy spaces and their use in analysis. *Bull. Amer. Math. Soc.* 83 (1977):569-645.

3. Dahlberg, B. E. J. Estimates of harmonic measure. *Archiv. for Rat. Mech. and Anal.* 65 (1977):272-288.

4. Dahlberg, B. E. J. A note on H^1 and BMO, "A tribute to Ake Pleijel." Uppsala University, 1980.

5. Dahlberg, B. E. J. On the Poisson integral for Lipschitz and C^1 domains. *Studia Math.* 66 (1979):13-24.

6. Fabes, E.; Jodeit, M.; and Rivière, N. Potential techniques for boundary value problems on C^1 domains. *Acta Math.* 141 (1978):165-186.

7. Fabes, E., and Kenig, C. On the Hardy space H^1 of a C^1 domain. *Arkiv. för Mat.* To appear.

8. Fabes, E.; Kenig, C.; and Neri, U. Carleson measures, H^1 duality and weighted BMO in nonsmooth domains. *Indiana Univ. Math. J.* 30 (1981):547-581.

9. Fabes, E.; Kenig, C.; and Neri, U. Neumann problem in weighted H^1 and preduals of BMV. Preprint, Univ. of Maryland, Feb. 1982.

10. Fabes, E.; Kenig, C.; and Neri, U. Weighted H_D^1-BMO dualities in C^1 domains. Preprint, Univ. of Maryland, July 1981.

11. Hunt, R. A., and Wheeden, R. L. On the boundary values of harmonic functions. *Trans. Amer. Math. Soc.* 132 (1968):307-322.

12. Jerison, D. S., and Kenig, C. E. Boundary behavior of harmonic functions in nontangentially accessible domains. *Advances in Math.* To appear.

13. Jerison, D. S., and Kenig, C. E. The Neumann problem on Lipschitz domains. *Bull. Amer. Math. Soc.* (N.S.) 4 (1981): 203-207.

14. Muckenhoupt, B., and Wheeden, R. On the dual of weighted H^1 of the half-space. *Studia Math.* 63 (1978):57-79.

15. Neri, U. On functions with bounded mixed variance. *Studia Math.* To appear in vol. 75 (1982).

THE SHILOV AND BISHOP DECOMPOSITIONS OF $H^{\infty} + C$

Donald Sarason
University of California, Berkeley

1. INTRODUCTION

A compact Hausdorff space X admits several natural decompositions relative to any closed unital subalgebra A of $C(X)$. Two of these decompositions reflect, in some sense, the extent to which A is self-adjoint. The sets in the coarser of the two decompositions, the Shilov decomposition [12], are the maximal level sets of the largest C^*-subalgebra of A. Those in the finer of the decompositions, the Bishop decomposition [2, 5], are the maximal sets of antisymmetry of A. (A subset S of X is a set of antisymmetry of A if any function in A that is real valued on S must be constant on S.)

      This paper is concerned with the case where $X = M(L^{\infty})$ and $A = H^{\infty} + C$. By L^{∞} is meant the C^*-algebra of essentially bounded, complex-valued, measurable functions with respect to Lebesgue measure on ∂D, the unit circle in the complex plane. This algebra will be identified, in the usual manner, with the algebra of continuous functions on $M(L^{\infty})$, its Gelfand space (space of multiplicative linear functionals). By H^{∞} is meant the space of boundary functions for bounded holomorphic functions in D, the unit disc, and by C is meant $C(\partial D)$. The largest C^*-subalgebra of $H^{\infty} + C$ will be denoted, as usual, by QC, and the sets in the Shilov decomposition of $M(L^\infty)$ associated with $H^\infty + C$ will be called QC-level sets. The subsets of $M(L^\infty)$ that are sets of antisymmetry for $H^\infty + C$ will be called simply sets of antisymmetry.

The following results will be established.

1. The Bishop decomposition of $M(L^\infty)$ associated with $H^\infty + C$ is strictly finer than the Shilov decomposition.

2. If f and g are functions in L^∞ and if, for each set of antisymmetry S, either $f|S$ or $g|S$ is in $H^\infty|S$, then the same conclusion holds for each QC-level set.

The first result will be proved in Sec. 2, where a function in H^∞ will be constructed whose imaginary part is continuous at the point $z = 1$ but whose restriction to $M_1(L^\infty)$, the fiber of $M(L^\infty)$ above the point 1, is not the restriction of any function in QC. Such a function is constant on each set of antisymmetry contained in $M_1(L^\infty)$, but it cannot be constant on every QC-level set contained in $M_1(L^\infty)$ (by the Stone-Weierstrass theorem).

The second result seems to say that even though the Shilov and Bishop decompositions of $M(L^\infty)$ due to $H^\infty + C$ are different, they are nevertheless somehow closely related. It would be nice to have a geometric picture of $M(H^\infty)$ that illuminates this relation, but that awaits further study. An example in Sec. 6 shows that the second result is not merely an instance of a general fact about uniform algebras.

The second result is closely related to and leads to a refinement of results on Toeplitz operators from [1]. Its proof depends on some of the latter results and on a theorem of T. Wolff. In Sec. 3 a preliminary lemma involving a duality relation is established. Sec. 4 contains what amounts to a concrete version of the second result, a theorem on the separation of the zeros of two

Blaschke products satisfying a certain disjointness condition.   The
second result will be derived from the separation theorem in Sec. 5.

T. Wolff [14] has found a more direct proof of the separation theorem using methods of J. Garnett and P. Jones.   As will be
explained in Sec. 5, this provides a new proof of the main result
of [1].   That the separation theorem might be true was first suggested by J. Garnett.

Actually, the second result will be proved in a slightly
stronger form than is stated above.   It is known that each functional in $M(H^\infty + C)$ $[= M(H^\infty) - D]$ has a unique representing measure
on $M(L^\infty)$ [6, p. 182].   The support of any such representing measure
will be called a support set.   Support sets are easily seen to be
sets of antisymmetry (although the precise relation between support
sets and sets of antisymmetry for $H^\infty + C$ remains mysterious).   The
version of the second result to be established involves support
sets instead of sets of antisymmetry.

Although the questions addressed in this paper are motivated in part by the general theory of function algebras, their
resolution is largely a matter of classical analysis.   It is hoped
that the paper is therefore not out of place in a volume dedicated
to Professor Zygmund.

The author is grateful to Sheldon Axler, Donald Marshall,
and Thomas Wolff for their help.

In what follows, functions in L^∞ will be routinely identified with their harmonic extensions into D as well as with their
Gelfand transforms.

2. PROOF OF RESULT 1

For c a point of ∂D, we let $M_c(L^\infty)$ denote the fiber of $M(L^\infty)$ above
c.   For f in L^∞, we let $\text{dist}_c(f, QC)$ denote the distance of

$f|M_c(L^\infty)$ from $QC|M_c(L^\infty)$. As explained in Sec. 1, to establish the first result, it will be enough to construct a function f in H^∞ such that Im f is continuous at 1 but $\text{dist}_1(f, QC) > 0$. Since $\text{dist}_c(f, QC)$ is an upper semicontinuous function of c, we can guarantee the last condition by making sure $\text{dist}_c(f, QC)$ is bounded away from 0 along a sequence of points c tending to 1. The construction that follows is a substantial simplification due to D. Marshall of the author's original one (which involved Blaschke products).

Choose a sequence (c_n) of distinct points on ∂D converging to 1 and a sequence (r_n) of positive numbers tending to 0. Let u be any real-valued function on ∂D that is continuous except at the points of the sequence (c_n) and that has one-sided limits at c_n equal to r_n and $-r_n$. We can, for example, take u to be linear between successive points of the sequence (c_n). In particular, u is to be continuous at 1; it is automatic that $u(1) = 0$.

Let
$$h = e^{u+i\tilde{u}} + e^{-(u+i\tilde{u})},$$

where $\tilde{u}$ denotes the conjugate function of u. Then h is in H^∞, and the continuity of Im $h = (e^u - e^{-u})\sin \tilde{u}$ at 1 is clear. It only remains to show that $\text{dist}_{c_n}(h, QC)$ is bounded away from 0.

To simplify the notation, let c denote any one of the points c_n and r the corresponding r_n. As $z \to c$ radially, $u(z)$ tends to 0, and $\tilde{u}(z)$ tends either to $+\infty$ or to $-\infty$ [16, pp. 98, 108]. Hence $h(z) \to 0$ along some sequence tending to c. Let x be a cluster point in $M_c(H^\infty)$ of such a sequence, and let m be the representing measure of x on $M(L^\infty)$. The values of h on $M_c(L^\infty)$, and hence on supp m, lie in the ellipse E that is the image of the circle $|z| = e^r$ under the map $z \to z + 1/z$. If $h(\text{supp } m)$ were not all of E, there would be a polynomial p having positive real part on $h(\text{supp } m)$ but vanishing at 0. This would lead to the contradiction

$$0 = \text{Re } p(h(x)) = \int \text{Re } p(h)\, dm > 0,$$

so we can conclude that $h(\text{supp } m) = E$. The diameter of E exceeds 4, so, as every QC function is constant on supp m, we must have $\text{dist}_a(h, QC) \geq 2$. This completes the proof.

3. A LEMMA

By BMO and VMO we mean the spaces of functions of bounded mean oscillation and vanishing mean oscillation on ∂D [8, 11]. Their relevance here comes through the equality $QC = \text{VMO} \cap L^\infty$. The sub-spaces of BMO and VMO consisting of the functions they contain whose harmonic extensions are holomorphic will be denoted by BMOA and VMOA.

It is a consequence of Fefferman's theorem that BMOA is the dual of the Hardy space H^1 under the pairing

$$\langle f, g \rangle = \frac{1}{2\pi} \int_{-\pi}^{\pi} f\overline{g} \ d\theta, \qquad f \in H^1, \qquad g \in \text{BMOA},$$

the integral on the right existing, for example, if one of the factors in the integrand is bounded (but not in general). Similarly, H^1 is the dual of VMOA under the analogous pairing. (For definiteness, we give H^1 its usual norm and BMOA the dual norm.)

Lemma. If b is a Blaschke product, the natural map of the quotient space VMOA/(bVMOA $\cap$ VMO) into the quotient space BMOA/(bBMOA $\cap$ BMO) is an isometry.

The following observation is needed: if the Blaschke product b divides the VMOA function f, then f/b is in VMOA. This follows from the characterization of VMOA as the class of functions f in H^2 that satisfy

$$\lim_{|a| \to 1} \left[\frac{1}{2\pi} \int_{-\pi}^{\pi} |f|^2 P_a \ d\theta - |f(a)|^2 \right] = 0$$

(P_a = Poisson kernel at a). It implies that the map in the above lemma is one-to-one. Similar reasoning establishes the analogous property of BMOA.

We can establish the lemma by showing that BMOA/(bBMOA $\cap$ BMO) is the second dual of VMOA/(bVMOA $\cap$ VMO) and that the natural map of the second space into the first one is the canonical embedding. Since the dual of VMOA/(bVMOA $\cap$ VMO) is the annihilator of bVMOA $\cap$ VMO in H^1, and BMOA/(bBMOA $\cap$ BMO) is the dual of the annihilator of bBMOA $\cap$ BMO in H^1, it will suffice to show that bVMOA $\cap$ VMO and bBMOA $\cap$ BMO have the same annihilator in H^1. We shall show that the annihilator of each is $(bH^\infty)^\perp$, the annihilator of bH^∞ in H^1.

Let $QA = QC \cap H^\infty$. The following result of T. Wolff [15] is needed.

Theorem 1. *Any function in L^∞ can be multiplied into QC by an outer function in QA.*

To see that the annihilator of bVMOA $\cap$ VMO in H^1 is contained in $(bH^\infty)^\perp$, consider any function g in H^1 that annihilates bVMOA $\cap$ VMO and any function f in H^∞. By Theorem 1, there is an outer function h in QA such that hbf is in QA. Then hf is in QA (by the observation following the statement of the lemma), so g annihilates hbf. Similarly, g annihilates $phbf$ for any polynomial p, and it follows by Beurling's theorem [8, p. 110] that g annihilates bf, as desired.

To complete the proof of the lemma we need only to show that $(bH^\infty)^\perp$ annihilates bBMOA $\cap$ BMO. For this, let b_n denote the nth partial product of b. Then $(b_n H^\infty)^\perp$ consists of bounded functions. As $(bH^\infty)^\perp$ is the norm closure of $\cup (b_n H^\infty)^\perp$ (by a standard duality argument), it is spanned by bounded functions. As any bounded function in $(bH^\infty)^\perp$ obviously annihilates bBMOA, the desired conclusion follows.

4. A SEPARATION THEOREM

The relevant facts about Toeplitz operators will now be stated;
general background can be found in [10]. For f in L^∞, the
Toeplitz operator on H^2 with symbol f will be denoted by T_f, and
the closed subalgebra of L^∞ generated by H^∞ and f will be denoted
by $H^\infty[f]$.

*Theorem 2. For f and g in L^∞, the following conditions are
equivalent:*

    (i)   *$T_{\bar{f}} T_g - T_{\bar{f}g}$ is compact;*

   (ii)   *$H^\infty[f] \cap H^\infty[g] \subset H^\infty + C$;*

  (iii)   *For every support set S, either $f|S$ or $g|S$ is in $H^\infty|S$.*

*Theorem 3. Let φ and ψ be Blaschke products with disjoint zero
sequences (z_n) and (w_n). The following conditions are equivalent:*

    (i)   *$T_\varphi T_{\bar{\psi}} - T_{\varphi\bar{\psi}}$ is compact;*

   (ii)   *$\lim_{|z| \to 1} \max(|\varphi(z)|, |\psi(z)|) = 1$;*

  (iii)   *For any positive number ε, there is a function h in H^∞
         satisfying $\|h\|_\infty < 1 + \varepsilon$, $h(z_n) = 1$ except for finitely
         many n, and $h(w_n) = 0$ except for finitely many n.*

     The implication from condition (i) of Theorem 2 to condi-
tion (ii) is due to A. L. Volberg [13]; the other implications can
be found in [1].

     The separation theorem states that the zeros of two Blaschke
products satisfying the conditions of Theorem 3 can be separated by
a function in QC.

*Theorem 4. If φ and ψ satisfy the conditions of Theorem 3, there is
a function u in QC such that $0 \leq u \leq 1$, $u(z_n) \to 1$, and $u(w_n) \to 0$.*

The proof will be accomplished in three steps. To simplify
slightly the explanation that follows, it will be assumed that φ
and ψ have simple zeros; only a trivial rephrasing is needed to
handle the general case.

Step 1. For any positive number δ, there is a function f
in BMOA such that $f(z_n) = 1$ for all n, $f(w_n) = 0$ for all n, and
dist$(f, \text{VMOA}) < \delta$.

To establish this we use condition (iii) of Theorem 3.
Given $\varepsilon > 0$, there is, by that condition, a function h in H^∞ and a
positive integer N such that $\|h\|_\infty \leq 1$, $h(w_n) = 0$ for $n > N$, and
$h(z_n) = 1 - \varepsilon$ for $n > N$. Let $h_1 = \log(1 - h)/\log \varepsilon$. Then
$h_1(z_n) = 1$ for $n > N$, $h_1(w_n) = 0$ for $n > N$, and h_1 is in BMOA,
since its imaginary part is bounded. In fact, since

$$\| \text{Im } h_1 \|_\infty \leq \pi / \left(2 \log \tfrac{1}{\varepsilon}\right),$$

we have

$$\| h_1 - h_1(0) \|_{\text{BMO}} \leq \text{const.}/\log \tfrac{1}{\varepsilon}$$

for some absolute constant (proportional to the norm of the conjugation operator on BMO). Thus, if we choose ε sufficiently small,
we get $\| h_1 - h_1(0) \|_{\text{BMO}} < \delta$. By Theorem 1, the Blaschke product
with zero set $\{z_n, w_n : n > N\}$ can be multiplied into QA by an outer
function. That yields a function h_2 in QA that vanishes at z_n and
w_n for $n > N$ but is nonzero at z_n and w_n for $n \leq N$. Multiplying
h_2 by a suitable polynomial, we can make it satisfy $h_2(z_n) =$
$1 - h_1(z_n)$ and $h_2(w_n) = -h_1(w_n)$ for $n \leq N$. The function $f = h_1 + h_2$
then has the required properties.

Step 2. There is a function g in VMOA such that $g(z_n) = 1$
for all n and $g(w_n) = 0$ for all n.

To see this, we take any f in BMOA such that $f(z_n) = 1$ for
all n and $f(w_n) = 0$ for all n; step 1 guarantees that there is such

an f. Let $b = \varphi\psi$. By step 1, the coset of f in BMOA/(bBMOA $\cap$ BMO)
has distance 0 from the canonical image of VMOA/(bVMOA $\cap$ VMO). By
the lemma of Sec. 3, that canonical image is closed, so the coset
of f contains a function in VMO, which is the desired conclusion.

$Step$ 3.   The function u of the theorem exists.

It now follows by standard reasoning that if g is the function of step 2, then the function

$$u = \max\bigl(\min(\operatorname{Re} g, 1), 0\bigr)$$

has all of the required properties.

5. PROOF OF RESULT 2

As mentioned in Sec. 1, the result to be established is slightly
stronger than the one stated there.

$Theorem$ 5.   Let f and g be functions in L^{∞} such that, for each sup-
port set S, either $f|S$ or $g|S$ is in $H^{\infty}|S$. Then, for each QC-level
set S, either $f|S$ or $g|S$ is in $H^{\infty}|S$.

We consider first a special case. Suppose φ and ψ are
interpolating Blaschke products such that, for each support set S,
either $\overline{\varphi}|S$ or $\overline{\psi}|S$ is in $H^{\infty}|S$. Then, by virtue of the unimodularity
of φ and ψ on $M(L^{\infty})$, one sees easily that $\max(|\varphi|, |\psi|)$ is identi-
cally 1 on $M(H^{\infty} + C)$, which implies condition (ii) of Theorem 3.
The separation theorem therefore applies to φ and ψ. Because φ and
ψ are interpolating Blaschke products, the zero set of each in
$M(H^{\infty})$ is the closure of its zero set in D. Thus, the separation
theorem implies that no QC-level set in $M(H^{\infty} + C)$ contains a zero
of both φ and ψ. Suppose S is a QC-level set in $M(L^{\infty})$ and S' is
the corresponding one in $M(H^{\infty} + C)$. Standard reasoning shows that

$S' = M(H^\infty|S)$, so at least one of $\varphi|S$ and $\psi|S$ is invertible in $H^\infty|S$; in other words, either $\overline{\varphi}|S$ or $\overline{\psi}|S$ is in $H^\infty|S$. This establishes the theorem for the special case under consideration. (Incidentally, for this case, one of $\varphi|S$ and $\psi|S$ is constant for any QC-level set S [3, Prop. 4].)

Now let f and g be any functions satisfying the hypotheses of the theorem. If φ and ψ are interpolating Blaschke products that are invertible in $H^\infty[f]$ and $H^\infty[g]$, respectively, then the preceding argument applies to φ and ψ, so that for each QC-level set S, either $\overline{\varphi}|S$ or $\overline{\psi}|S$ is in $H^\infty|S$. Hence, for each QC-level set S, either $\overline{\varphi}|S$ is in $H^\infty|S$ for every interpolating Blaschke product φ that is invertible in $H^\infty[f]$, or $\overline{\psi}|S$ is in $H^\infty|S$ for every interpolating Blaschke product ψ that is invertible in $H^\infty[g]$. By the Chang-Marshall theorem [9], $H^\infty[f]$ and $H^\infty[g]$ are generated, as closed subalgebras of L^∞, by H^∞ and the complex conjugates of interpolating Blaschke products. The desired conclusion about f and g is now immediate.

By virtue of what has just been proved, the following condition can be added to the equivalent conditions (i)-(iii) of Theorem 2:

(iv) *For every QC-level set S, either $f|S$ or $g|S$ is in $H^\infty|S$.*

It is quite simple to show that this condition implies the compactness of $T_{\overline{f}}T_g - T_{\overline{f}g}$; one can use either a straightforward adaptation of the argument in [11] or the localization technique of R. G. Douglas [4].

Wolff's proof of the separation theorem mentioned in Sec. 1 is a direct (albeit complicated) analytic argument that starts from condition (ii) of Theorem 3. Coupled with the argument just given, it shows that condition (iv) follows from condition (iii) of Theorem 2 without relying on the proof of the most difficult implication of that theorem, the implication from (iii) to (i). It thus

provides a replacement for the hardest part of the proof of
Theorem 2.

P. Jones [7] has proved a Corona-type theorem that has as
a corollary the implication from condition (ii) to condition (iii)
of Theorem 3. As explained in [1], this provides yet another proof
of the difficult implication in Theorem 2. There are thus three
quite distinct proofs of that implication, all, by the way, of
comparable difficulty.

The author has been unable to decide whether the theorem
proved in this section extends from two to three functions. Another
problem of interest is that of finding a concrete structural condition on the pair f, g that is equivalent to the conditions of
Theorem 2.

6. AN EXAMPLE

To show that the theorem in Sec. 5 does not generalize to arbitrary
uniform algebras, we sketch here a simple example of a uniform
algebra A on a compact Hausdorff space X and two functions,
f and g, in $C(X)$, with the following properties: (1) on each maximal
set of antisymmetry S of A, either $f|S$ or $g|S$ is in $A|S$; (2) there
is a level set S_0 of $A \cap \overline{A}$ such that neither $f|S_0$ nor $g|S_0$ belongs
to $A|S_0$. The example is based on the standard one of an algebra
whose Shilov and Bishop decompositions are distinct. We produce X
by sewing together a square and a cylinder along an edge of the
square and the axis of the cylinder. Namely, let X be the set of
points (z_1, z_2) in $\mathbf{C}^2$ such that either $0 \leq \mathrm{Re}\ z_1 \leq 1$, $0 \leq \mathrm{Im}\ z_1 \leq 1$,
$z_2 = 0$, or $0 \leq \mathrm{Re}\ z_1 \leq 1$, $\mathrm{Im}\ z_1 = 0$, $|z_2| \leq 1$. Our square is

$$Q = \{(z_1, z_2) \in X : z_2 = 0\}$$

and our cylinder is

$$S_0 = \{(z_1, z_2) \in X : \mathrm{Im}\ z_1 = 0\}.$$

For $n = 1, 2, \ldots,$ let

$$R_n = \{(z_1, 0) \in Q : 0 < \mathrm{Re}\ z_1 < 1,\ 2^{-2n} < \mathrm{Im}\ z_1 < 2^{-2n+1}\},$$

and for $0 \leq t \leq 1,$ let

$$D_t = \{(t, z_2) \in S_0 : |z_2| < 1\}.$$

Let A be the algebra of functions in $C(X)$ that are holomorphic with respect to z_1 on each R_n and with respect to z_2 on each D_t.   (The algebra $A|Q$ is the standard example referred to above.)

The level sets of $A \cap \bar{A}$ are $\bar{R}_n$ ($n = 1, 2, \ldots$), S_0, and the remaining singletons.   The maximal sets of antisymmetry of A are $\bar{R}_n$ ($n = 1, 2, \ldots$), $\bar{D}_t$ ($0 \leq t \leq 1$), and the remaining singletons.

Let the functions f and g in $C(X)$ be defined by

$$f(z_1, z_2) = \left(\frac{1}{2} - \mathrm{Re}\ z_1\right)\bar{z}_2,\ 0 \leq \mathrm{Re}\ z_1 \leq \frac{1}{2},\ \mathrm{Im}\ z_1 = 0$$

$$= 0,\ \text{otherwise,}$$

$$g(z_1, z_2) = \left(\mathrm{Re}\ z_1 - \frac{1}{2}\right)\bar{z}_2,\ \frac{1}{2} \leq \mathrm{Re}\ z_1 \leq 1,\ \mathrm{Im}\ z_1 = 0$$

$$= 0,\ \text{otherwise.}$$

On each set of antisymmetry of A, one of these functions coincides with a function in A, namely, the zero function.   However, neither f nor g coincides with a function in A on the level set S_0.

In this example, distinct nontrivial Gleason parts of A have disjoint closures.   It appears likely that the geometric reason behind the theorem of Sec. 5 is some sort of intertwining of distinct Gleason parts of $H^\infty + C$.

In the example just given, $X = M(A)$, but a small modification produces an example in which the underlying space is the Shilov boundary.   Namely, the Shilov boundary of A is the set Y consisting of all points of Q not belonging to any R_n and all points of S_0 not belonging to any D_t.   If $B = A|Y$, then the level sets of $B \cap \bar{B}$ are the intersections with Y of the level sets of $A \cap \bar{A}$, and the maximal sets of antisymmetry of B are similarly related to those of A.   Thus, on each set of antisymmetry of B, one of the two functions $f|Y$ and $g|Y$ coincides with a function in B, but neither $f|Y$ nor $g|Y$ coincides with a function in B on the level set $S_0 \cap Y$.

REFERENCES

1. Axler, S.; Chang, S.-Y. A.; and Sarason, D.   Products of Toeplitz operators.   *Integral Equations and Operator Theory* 1 (1978):285-309.

2. Bishop, E.   A generalization of the Stone-Weierstrass theorem. *Pac. J. Math.* 11 (1961):777-783.

3. Clancey, K. F., and Gosselin, J. A.   The local theory of Toeplitz operators.   *Illinois J. Math.* 22 (1978):449-458.

4. Douglas, R. G.   Banach algebra techniques in the theory of Toeplitz operators.   *Regional Conference Series in Mathematics*, no. 15, Amer. Math. Soc., Providence, R.I., 1973.

5. Glicksberg, I.   Measures orthogonal to algebras and sets of antisymmetry.   *Trans. Amer. Math. Soc.* 105 (1962):415-435.

6. Hoffman, K.   *Banach Spaces of Analytic Functions.* Englewood Cliffs, N.J.: Prentice-Hall, 1962.

7. Jones, P. W.   Estimates for the Corona problem.   *J. Func. Anal.* To appear.

8. Koosis, P.   *Introduction to H_p Spaces.*   New York: Cambridge Univ. Press, 1980.

9. Marshall, D. E.   Subalgebras of L^∞ containing H^∞.   *Acta Math.* 137 (1976):91-98.

10. Sarason, D.   *Function Theory on the Unit Circle.*   Blacksburg, Va.: Virginia Polytechnic Inst. and State Univ., 1978.

11. Sarason, D.   On products of Toeplitz operators.   *Acta Sci. Math.* (Szeged) 35 (1973):7-12.

12. Shilov, G. E.   On rings of functions with uniform convergence. *Ukrain. Mat. Z.* 3 (1951):404-411.

13. Volberg, A. L.   Two remarks concerning the theorem of S. Axler, S.-Y. A. Chang and D. Sarason. *J. Operator Theory*. To appear.

14. Wolff, T.   Some theorems on vanishing mean oscillation. Doctoral Dissertation, Univ. of California, Berkeley, 1979.

15. Wolff, T.   Two algebras of bounded functions. To appear.

16. Zygmund, A.   *Trigonometric Series*, vol. I.   New York: Cambridge Univ. Press, 1959.

A MODIFIED FRANKLIN SYSTEM AND HIGHER-ORDER SPLINE SYSTEMS ON $\mathbf{R}^n$ AS UNCONDITIONAL BASES FOR HARDY SPACES

Jan-Olov Strömberg
University of Stockholm
Princeton University

1. INTRODUCTION

Let $H^p(I)$ be the subspace of distributions in the Hardy space $H^p(\mathbf{R})$ that are supported in $I = [0, 1]$. The Franklin system ($m = 0$) and higher-order spline system ($m > 0$) have been studied and used as unconditional basis for $H^p(I)$, $p > 1/(m + 2)$ in [1, 4, 7, and 9–11]. The existence of an unconditional basis for H^1 was first shown by B. Maurey [8] and an explicit construction was made by L. Carleson [3]. The spline system as unconditional basis for the bi-Hardy space $H^1(I \times I)$ has been studied by S. Y. A. Chang.

The purpose of this paper is to define m-order spline systems on $\mathbf{R}$ and especially get a modified Franklin system ($m = 0$). We will also get spline systems on $\mathbf{R}^n$, $n > 1$. All these systems can also be made periodical to become spline systems on $\mathbf{T}^n$ by a simple summation. (The spline systems on I do not extend periodically to the circle $\mathbf{T} = \mathbf{R}/\mathbf{Z}$ and cannot be used as unconditional basis for the real Hardy spaces $H^p(\mathbf{T})$, $0 < p \leq 1$, without modification.)

We are especially interested to see how these spline systems on $\mathbf{R}^n$ can be used as unconditional basis for $H^p(\mathbf{R}^n)$, $n \geq 1$.

2. CONSTRUCTION OF TWO FUNCTIONS

Let $m \geq 0$. We will construct one function on $\mathbf{R}$ from which we get the m-order spline system on $\mathbf{R}$ and another function on $\mathbf{R}$ that we will use together with the first one to get the m-order spline system on $\mathbf{R}^n$.

Let $A_0 = \mathbf{Z}_+ \cup \{0\} \cup \frac{1}{2} \mathbf{Z}_-$ and $A_1 = A_0 \cup \left\{\frac{1}{2}\right\}$. A_0 splits $\mathbf{R}$ into intervals $\{I_\sigma\}_{\sigma \epsilon A_0}$ (σ = left endpoint of I_σ). Let S_0^m be the subspace of functions f in $L^2(\mathbf{R})$ such that $f \epsilon C^m(\mathbf{R})$ and is a real polynomial of degree $\leq m + 1$ on each I_σ, $\sigma \epsilon A_0$. Let S_1^m be the corresponding subspace of $L^2(\mathbf{R})$ with the set A_0 replaced by A_1. Note that $S_1^m = \{f;\ f(\cdot + 1) \epsilon S_0^m\}$ and also that the set A_1 splits $\mathbf{R}$ into the same collection of intervals as A_0 did except for the interval $I_0 = [0,\ 1]$, which is split into two intervals $\left[0,\ \frac{1}{2}\right]$ and $\left[\frac{1}{2},\ 1\right]$. It follows that S_0^m has codimension 1 in S_1^m, since the $m + 1$ order derivative of a function f in S_1^m may have a jump at $\frac{1}{2}$ and if not, it is contained in S_0^m. Thus there is a function τ in $L^2(\mathbf{R})$ that is a uniquely defined modulo sign by

(i) $\tau \epsilon S_1^m$,

(ii) $\tau \perp S_0^m$, that is, $\int_{\mathbf{R}} \tau(x) f(x)\, dx = 0$ for all $f \epsilon S_0^m$,

(iii) $\| \tau \|_{L^2} = 1$.

Next let $\tilde{S}_0^m$ be the subspace of functions f in S_0^m that are supported in $[0,\ \infty)$ and let $\tilde{S}_1^m$ be the subspace of functions f in S_1^m that are supported in $[1,\ \infty)$. We observe that $\tilde{S}_1^m$ is a subspace of $\tilde{S}_0^m$ with codimension 1. Thus there is a function ρ in $L^2(\mathbf{R})$ that is a uniquely defined modulo sign by

(iv) $\rho \epsilon \tilde{S}_0^m$,

(v) $\rho \perp \tilde{S}_1^m$,

(vi) $\| \rho \|_{L^2} = 1$.

By definition, both τ and ρ are piecewise polynomials of degree $\leq m + 1$ and are in $C^m(\mathbf{R})$; furthermore, the $(m+1)$-order derivatives of them are piecewise constant with discontinuities in A_1 resp. $\mathbf{Z_+} \cup \{0\}$ and the following estimates hold with some constant r, $0 < r < 1$ (which may change from line to line):

$$|D^k \tau(x)| \leq Cr^{|x|}, \ k = 0, \ \ldots, \ m + 1, \tag{1}$$

$$|D^k \rho(x)| \leq \begin{array}{l} Cr^x, \ x \geq 0 \\ 0, \ \ \ \ x < 0 \end{array}, \ k = 0, \ \ldots, \ m + 1. \tag{2}$$

The function τ also satisfies the moment conditions

$$\int_{\mathbf{R}} \tau(x) x^\alpha \, dx = 0, \ \alpha = 0, \ \ldots, \ m + 1. \tag{3}$$

3. MAIN RESULTS

First we consider the one-dimensional case. Let $\nu = (j, k) \in \mathbf{Z} \times \mathbf{Z}$ and set

$$f_\nu(x) = 2^{j/2} \tau(2^j x - k).$$

Then we have:

Theorem 1

 (a) $\{f_\nu\}_{\nu \in \mathbf{Z} \times \mathbf{Z}}$ *is a complete orthonormal system in* $L^2(\mathbf{R})$.

 (b) *Let* $p > 1/(m + 2)$. *Then* f_ν *is in the dual space of* $H^p(\mathbf{H})$, $\nu \in \mathbf{Z} \times \mathbf{Z}$, *and if* $c_\nu = (f_\nu, f)$, *then the sum* $\sum_{\nu \in \mathbf{Z} \times \mathbf{Z}} c_\nu f_\nu$ *converges unconditionally to* f *in the* $H^p(\mathbf{R})$ *norm. Furthermore, the coefficients in a converging sum are uniquely defined; that is, if* $\sum c_\nu f_\nu$ *converges to a function* $f \in H^p(\mathbf{R})$ *in the norm, then* $c_\nu = (f_\nu, f)$.

 (c) *Let* J_ν *be the interval* $[2^{-j}k, \ 2^{-j}(k + 1)]$ *and let* χ_ν *be its characteristic function. Then for any collections of coefficients* $\{c_\nu\}_\nu$ *such that only finitely many* c_ν *are nonvanishing, we have*

$$\left\|\sum_\nu c_\nu f_\nu \right\|_{H^p(\mathbf{R})} \le C \left\| \left\{\sum_\nu |c_\nu|^2 2^j \chi_\nu \right\}^{1/2} \right\|_{L^p(\mathbf{R})}, \quad p > 1/(m + 5/2). \quad (4)$$

On the other hand, if $f \in H^p(\mathbf{R})$, $p > 1/(m + 2)$ and $c_\nu = (f_\nu, f)$, then

$$\left\| \left\{\sum_\nu |c_\nu|^2 2^j \chi_\nu \right\}^{1/2} \right\|_{L^p(\mathbf{R})} \le C \|f\|_{H^p(\mathbf{R})}. \quad (5)$$

In higher dimensions we will use both the functions τ and ρ. We want to define a system of functions on $\mathbf{R}^n$, $n > 1$. Let Ω be the set of n-tuples of functions $\omega = (\omega_1, \ldots, \omega_n)$, where ω_i is either τ or ρ but $\omega_i = \tau$ for at least one i, $1 \le i \le n$. Thus Ω consists of $2^n - 1$ n-tuples. Now we define the function τ^ω on $\mathbf{R}^n$ by

$$\tau^\omega(x) = \tau^\omega(x_1, \ldots, x_n) = \prod_{i=1}^n \omega_i(x_i), \quad x \in \mathbf{R}^n, \omega \in \Omega.$$

For example, when $n = 2$, we get the three functions $\tau(x_1)\tau(x_2)$, $\tau(x_1)\rho(x_2)$, and $\rho(x_1)\tau(x_2)$. Let $\nu = (j, k_1, \ldots, k_n) \in \mathbf{Z} \times \mathbf{Z}^n$ and define the function f by

$$f_\nu^\omega(x) = 2^{nj/2}\tau^\omega(2^j x - (k_1, \ldots, k_n)), \quad x \in \mathbf{R}^n.$$

Then:

Theorem 2

(a) $\{f_\nu^\omega\}_{\Omega \times \mathbf{Z}^{n+1}}$ *is a complete orthonormal system in* $L^2(\mathbf{R}^n)$.

(b) $\{f_\nu^\omega\}_{\Omega \times \mathbf{Z}^{n+1}}$ *is an unconditional basis for* $H^p(\mathbf{R}^n)$ *when*
     $p > n/(n + m + 1)$; *that is, the corresponding statements*
     of Theorem 1(b) *hold with the sum taken over both* ν *and* ω.

(c) *The* n-*dimensional version of* (4) *holds when* $p > p(n, m)$,
     and the n-*dimensional version of* (5) *holds when*
     $p > n/(n + m + 1)$. (*The sum is taken over both* ν *and* ω *and*
     $\mathbf{R}$ *is replaced by* $\mathbf{R}^n$ *and* J_ν *is a cube.*)

Remark 1. With the methods that we give in detail when $n = 1$, one can get $p(n, m) = 1/(1/2 + (m + 2)/n)$.

Remark 2. $2^{(n)j} \chi_\nu$ may as well be replaced by $|f_\nu^{(\omega)}|^2$ in (4) and (5).

Remark 3.   If we set $f_\nu^{\omega,T}(x) = \sum_{t \epsilon Z^n} f_\nu^\omega (x - t)$ when $\nu = (j, k_1, \dots, k_n)$, $j \geq 0$, we get functions on T^n, which together with the constant function 1 will be a complete orthonormal system. This system is an unconditional basis for the real Hardy space $H^p(T^n)$ (defined with one dilation) when $p > n/(n + m + 1)$.

4. EXPLICIT DESCRIPTION OF τ AND ρ, m = 0

Since τ and ρ are linear on the intervals between the points $\sigma \epsilon A_1$, it is enough to find their values at these points. We will in this section use the notation $\tau_\sigma = \tau(\sigma)$ and $\rho_\sigma = \rho(\sigma)$, $\sigma \epsilon A_1$.

Definition.   The *simple tent* Λ_σ in S_0^0, $\sigma \epsilon A_0$, is the function in S_0^0 that is 1 at $x = \sigma$ but is 0 at all $\sigma_1 \epsilon A_0$, $\sigma_1 \neq \sigma$.

Since τ is orthogonal to all simple tents in S_0^0 and ρ is orthogonal to all simple tents in S_0^0 with support in $[1, \infty)$, we will get the following equations after computing the integrals:

$$\tau_{\sigma-1} + 4\tau_\sigma + \tau_{\sigma+1} = 0, \qquad = 2, 3, 4, \dots,$$
$$\tau_{\sigma-1/2} + 4\tau_\sigma + \tau_{\sigma+1/2} = 0, \qquad = -1/2, -1, -3/2, \dots,$$
$$\tau_0 + 6\tau_{1/2} + 13\tau_1 + 4\tau_2 = 0,$$
$$2\tau_{-1/2} + 9\tau_0 + 6\tau_{1/2} + \tau_1 = 0,$$

and

$$\rho_{\sigma-1} + 4\rho_\sigma + \rho_{\sigma+1} = 0.$$

The characteristic equation $r^2 + 4r + 1 = 0$ has the two roots $r_1 = \sqrt{3} - 2$ and $r_2 = r_1^{-1} = -(2 + \sqrt{3})$. Since the functions must go to zero at infinity, we get

$$\tau_\sigma = \tau_1 (\sqrt{3} - 2)^{\sigma-1}, \quad \sigma = 1, 2, 3, \dots,$$
$$\tau_\sigma = \tau_0 (-\sqrt{3} - 2)^{2\sigma}, \quad \sigma = 0, -1/2, -1, \dots,$$

$$\rho_\sigma = \rho_1(\sqrt{3} - 2)^{\sigma-1}, \qquad = 1, 2, 3, \ldots,$$

and since ρ is linear on $[0, 1]$ and vanishes on the negative axis, we have $\rho_{1/2} = \frac{1}{2}\rho_1$ and $\rho_\sigma = 0$ for all $= 0, -1/2, -1, \ldots$. From the third and fourth rows of the equations above, we now get $\tau_0 = 2(\sqrt{3} - 1)\tau_1$ and $\tau_{1/2} = -(\sqrt{3} + 1/2)\tau_1$. Finally, by the normalization (iii) and (vi), one can get the values of $|\tau_1|$ and $|\rho_1|$. We leave the computation of these values to the interested readers.

It is now easy to check the estimates (1) and (2) when $k = 0, 1$ and $m = 0$. To see that the inequality (3) holds $(m = 0)$, we can truncate the functions 1 and x outside a large compact set so that the truncated functions are in S_0^0. The corresponding part of the integrals will then be zero by orthogonality and the remainder of the integral can be made arbitrarily small if the compact set is chosen large enough, since τ is rapidly decreasing at infinity.

5. IDEAS FOR (1)–(3) WHEN m > 0

The derivatives $D^{m+2}\tau$ and $D^{m+2}\rho$ (taken in the sense of distributions) are measures supported in A_1, and in this section we use the notations τ_σ and ρ_σ, $\sigma \in A_1$ for the coefficients given by

$$D^{m+2}\tau = \sum_{\sigma \in A_1} \tau_\sigma \delta_\sigma,$$

$$D^{m+2}\rho = \sum_{\sigma \in A_1} \rho_\sigma \delta_\sigma$$

(δ_σ is the delta function at σ).

Let h be the function $h(x) = \Delta^{2m+4}(x_+^{2m+3})$, where $x_+^i = \max(x^i, 0)$ and Δ^i is defined by $\Delta^1 f(x) = f(x) - f(x - 1)$, $\Delta^i = \Delta^1 \Delta^{i-1}$, $i = 2, 3, \ldots$, and set $h_k(x) = h(x - k)$, $k = 1, 2, \ldots$. Then $D^{m+2} h_k$ are functions in S_0^m, $k = 1, 2, \ldots$ (and also in $\widetilde{S}_1^m$) and by orthogonality we have

$$\int D^{m+2}h_k\tau \; dx = 0,$$

which after integration by parts, gives us the equations

$$\sum_{\sigma \in A_1} h_k(\sigma)\tau_\sigma = 0, \; k = 1, \; 2, \; \ldots \; .$$

Observe that $h_k(\sigma)$ is zero for all $\sigma \in A_1$ except $\sigma = k + 1, \; \ldots,$ $k + 2m + 3$. We get the characteristic equation

$$\sum_j h(j)r^{j-1} \equiv Q_{2m+2}(r),$$

where Q_{2m+2} is a polynomial of degree $2m + 2$. By differentiation of geometric series, we can write

$$Q_j(r) = \frac{c_j(1-r)^{j+2}}{r}\left(r\frac{d}{dr}\right)^{j+1}\frac{1}{1-r}, \; j = 0, \; 1, \; \ldots \; .$$

Using a simple change of variable ($t = 1/r$), we see that if r_1 is a root to the equation $Q_j(r) = 0$, then $1/r_1$ is a root to the same equation. By this and an elementary mean value theorem, one can conclude by induction over j that the characteristic equation $Q_{2m+2}(r) = 0$ has $2m + 2$ distinct roots of which $m + 1$ roots $r_1, \; \ldots, \; r_{m+1}$ are in the open interval $(0, -1)$ and the remaining $m + 1$ roots are in the interval $(-\infty, -1)$. We conclude that

$$\tau_\sigma = C_1 r_1^\sigma + \cdots + C_{m+1}r_{m+1}^\sigma + C_{m+2}r_1^{-\sigma} + \cdots + C_{2m+2}r_{m+1}^{-\sigma}, \; \sigma = 2, \; 3, \; \ldots,$$

for some constants $C_1, \; \ldots, \; C_{2m+2}$, and since τ is in $L^2(\mathbf{R})$, $C_{m+2}, \; \ldots, \; C_{2m+2}$ are all equal to zero. We now get the estimate (1) for $x \geq 0$ by repeated integration. In the same way we also get (1) for $x < 0$ and (2). With the same truncation argument as we used in the case $m = 0$, we will get (3) also when $m > 0$.

6. ORTHOGONALITY

First we will see that $\{f_\nu\}_{\nu \in \mathbf{Z}^2}$ is an orthogonal system in $L^2(\mathbf{R})$. Let $A_\nu = \{x; \; 2^j x - k \in A_0\}$. The basic observation is that the collection of sets $\{A_\nu\}_{\mathbb{Z}^2}$ is totally ordered by inclusion; that is, if $\nu \neq \nu'$, then either $A_\nu \not\subseteq A_{\nu'}$ (we write then $\nu \triangleleft \nu'$) or $A_{\nu'} \subsetneq A_\nu$ (we write then $\nu' \triangleleft \nu$). Thus if $\nu = (j, k)$ and $\nu' = (j', k')$, then

$$\nu \triangleleft \nu' \quad \text{is equivalent to} \quad \begin{cases} j < j' \\ j = j', \ k < k'. \end{cases}$$

Let $\nu \triangleleft \nu'$ and consider the functions f_ν and $f_{\nu'}$. By a linear transformation, $f_{\nu'}$ can be transformed to (a multiple of) τ and, by the same linear transformation, f_ν will be transformed to a function in S_0^m, and since $\tau \perp S_0^m$, we conclude that $f_{\nu'} \perp f_\nu$. This proves that $\{f_\nu\}_{\mathbb{Z}^2}$ is an orthogonal system in $L^2(\mathbb{R})$.

Next we want to see that $\{f_\nu^\omega\}_{\Omega \times \mathbb{Z}^{n+1}}$ is an orthogonal system in $L^2(\mathbb{R}^n)$, $n > 1$. Let $\omega = (\omega_1, \ldots, \omega_n)$ and $\nu = (j, k_1, \ldots, k_n)$. Set

$$\omega_{i;j,k_i}(x_i) = 2^{j/2} \omega_i(2^j x_i - k_i), \quad i = 1, \ldots, n.$$

Then

$$f_\nu^\omega(x) = \prod_{i=1}^n \omega_{i;j,k_i}(x_i),$$

and $\omega_{i;j,k_i}$ is of the form $2^{j/2}\tau(2^j x_i - k_i) = f_{j,k_i}(x_i)$ or $2^{j/2}\rho(2^j x_i - k_i)$. Let $\omega' = (\omega_1', \ldots, \omega_n')$ and $\nu' = (j', k_1', \ldots, k_n')$. Then we have

$$\int_{\mathbb{R}^n} f_\nu^\omega(x) f_{\nu'}^{\omega'}(x)\, dx = \prod_{i=1}^n \int_{\mathbb{R}} \omega_{i;j,k_i}(x_i) \omega_{i;j',k_i'}'(x_i)\, dx_i.$$

If $(\omega, \nu) \neq (\omega', \nu')$, we will see that there is an $i = i_0$ such that the integral in the x_i-direction vanishes in the product on the right-hand side. There are the following four cases:

Case 1. $j < j'$ (or similarly $j' < j$). Then we use that $\omega_i' = \tau$ for some $i = i_0$.

Case 2. $j = j'$, $\omega \neq \omega'$; then $\omega_i \neq \omega_i'$ for some $i = i_0$, say, $\omega_{i_0}' = \tau$, $\omega_{i_0} = \rho$ (or similarly, if $\omega_{i_0} = \tau$, $\omega_{i_0}' = \rho$).

Case 3. $j = j'$, $\omega = \omega'$, $(k_1, \ldots, k_n) \neq (k_1', \ldots, k_n')$; that is, $k_i \neq k_i'$ for some $i = i_0$, and $\omega_{i_0}' = \omega_{i_0} = \tau$, say, $k_{i_0}' > k_{i_0}$ (or similarly $k_{i_0} > k_{i_0'}$).

Case 4. $j = j'$, $\omega = \omega'$, $(k_1, \ldots, k_n) \neq (k_1', \ldots, k_n')$; that is, $k_i \neq k_i'$ for some $i = i_0$ and $\omega_{i_0}' = \omega_{i_0} = \rho$, say, $k_{i_0}' < k_{i_0}$ (or similarly $k_{i_0} < k_{i_0'}$).

By a linear transformation we can now transform the function $\omega_{i_0;\,j',\,k_{i_0}'}'$ to get the function τ (in cases 1-3) or the function ρ (in case 4), and by the same linear transformation $\omega_{i_0;\,j,\,k_{i_0}}$ will be transformed to a function in S_0^m (cases 1-3) resp. $\tilde{S}_1^m$ (case 4). Since τ is orthogonal to S_0^m and ρ is orthogonal to $\tilde{S}_1^m$, we conclude that the integral in the x_{i_0}-direction vanishes. This proves the orthogonality of the system $\left\{f_\nu^\omega\right\}_{\Omega \times \mathbf{Z}^{n+1}}$.

7. COMPLETENESS IN L^2

We will first show that $\left\{f_\nu\right\}_{\mathbf{Z}^2}$ is a complete system in $L^2(\mathbf{R})$.

Let $S_\nu = \{f;\ f(x) = g(2^j x - k)$ for some $g \in S_0^m\}$, $\nu = (j, k) \in \mathbf{Z}^2$, and let P_ν be the projection operator $P_\nu : L^2 \to S_\nu$. If $\nu' < \nu$, then $S_{\nu'} \not\subseteq S_\nu$ and $P_\nu P_{\nu'} f = P_{\nu'} P_\nu f = P_\nu f$, $f \in L^2(\mathbf{R})$.

Let $f \in L^2(\mathbf{R})$ and let $c_\nu = \int_{\mathbf{R}} f\, f_\nu\, dx$; then

$$c_\nu f_\nu = P_{j,\,k+1} f - P_{j,\,k} f,$$

$$\nu = (j, k),$$

and

$$f - \sum_{|j| \le j_0} \sum_{|k| \le k_0} c_\nu f_\nu = f - \sum_{|j| \le j_0} (P_{j,\,k_0+1} f - P_{j,\,-k_0} f)$$

$$= f - P_{j_0,\,k_0+1} f - \sum_{j=-j_0}^{j_0-1} (P_{j+1,\,-k_0} f - P_{j,\,k_0+1} f) + P_{-j_0,\,-k_0} f.$$

The completeness of the system $\left\{f_\nu\right\}_{\mathbf{Z}^2}$ in $L^2(\mathbf{R})$ now follows easily if we can show that if $f \in L^2(\mathbf{R})$, then

(i) $\| P_{j,k} f - f \|_{L^2} \to 0$ as $j \to +\infty$ independent of k,

(ii) $\| P_{j,k} f \|_{L^2} \to 0$ as $j \to -\infty$ independent of k,

(iii) $\| P_{j+1,-k} f - P_{j,k+1} f \|_{L^2} \to 0$ as $k \to +\infty$ for each fixed j.

(i) follows by approximation in $L^2(\mathbf{R})$ by spline functions, which we leave to the reader.

For (ii) we use that $P_{j,k} f$ is piecewise a polynomial of degree $\leq m + 1$ to get

$$\| P_{j,k} f \|_\infty \leq C2^{j/2} \| P_{j,k} f \|_{L^2} \leq C2^{j/2} \| f \|_{L^2} .$$

Thus $f - P_{j,k} f$ converges uniformly to f as j tends to $-\infty$. We conclude that

$$\| f \|_{L^2} \geq \| f - P_{j,k} f \|_{L^2} \to \| f \|_{L^2}$$

and

$$\| P_{j,k} f \|_{L^2}^2 = \| f \|_{L^2}^2 - \| f - P_{j,k} f \|_{L^2}^2 \to 0 \text{ as } j \to -\infty.$$

To show (iii), we may assume that $f \in S_{j+1,0}$ and has compact support. Then

$$P_{j+1,-k} f = f + P_{j+1,-k} f - P_{j+1,0} f = f - \sum_{i=-k}^{-1} c_{j+1,i} f_{j+1,i}$$

and

$$\sum_{i=-k}^{-1} c_{j+1,i} f_{j+1,i}(x) = \int \left\{ \sum_{i=-k}^{-1} f_{j+1,i}(y) f_{j+1,i}(x) \right\} f(y) \, dy$$

and by (1) we get

$$\left| \sum_{i=-k}^{-1} f_{j+1,i}(y) f_{j+1,i}(x) \right| \leq C2^{j/2} r^{2^{j+1}|x-y|}.$$

We conclude from this that $| P_{j+1,-k} f(x) | \leq C_f r^{2^j|x|}$ with the constant C_f independent of k. For any $\varepsilon > 0$, we can now find a large compact set K_ε independent of k such that we can write $P_{j+1,-k} f = g_k + g_k'$ with $g_k, g_k' \in S_{j+1,-k}$, $\| g_k \|_{L^2} < \varepsilon$, and g_k' supported in K_ε. We observe that $g_k' \in S_{j,k+1}$ if k is chosen large enough. Thus $P_{j,k+1} g_k' = g_k'$ and we get

$$P_{j+1,-k}f - P_{j,k+1}f = P_{j+1,-k}f - P_{j,k+1}P_{j+1,-k}f$$

$$= g_k + g_k' - P_{j,k+1}g_k - P_{j,k+1}g_k'$$

$$= g_k - P_{j,k+1}g_k$$

and

$$\| P_{j+1,-k}f - P_{j,k+1}f \|_{L^2} = \| g_k - P_{j,k+1}g_k \|_{L^2} \leq \| g_k \|_{L^2} \leq \varepsilon,$$

provided k is chosen large enough. This proves (iii) and finishes the proof of the completeness of the system $\{f_\nu\}_{\mathbf{Z}^2}$ in $L^2(\mathbf{R})$.

Next we will see that $\{f_\nu^\omega\}_{\Omega \times \mathbf{Z}^{n+1}}$ is a complete system in $L^2(\mathbf{R}^n)$. For simplicity we will only look at the case when $n = 2$. From the one-dimensional case, we conclude that the collection of products

$$\left\{ f_{j_1,k_1}(x_1) f_{j_2,k_2}(x_2) \right\}_{\mathbf{Z}^2 \times \mathbf{Z}^2}$$

is a complete orthonormal system in $L^2(\mathbf{R}^2)$. However, it involves two independent dilations and that is the reason we use the system $\{f_\nu^\omega\}_{\Omega \times \mathbf{Z}^2}$, which involves only one dilation. We split the products into three cases:

Case 1. $j_1 = j_2$. These products are still in our system as the functions f_ν^ω with $\omega = (\tau, \tau)$.

Case 2. $j_1 > j_2$. We will see that these products are in the closure of the span of the functions f_ν^ω with $\omega = (\tau, \rho)$.

Case 3. $j_1 < j_2$. These products are in the closure of the span of the functions f_ν^ω with $\omega = (\rho, \tau)$.

We will only look at case 2, as case 3 can be treated in the same way. We claim that the product $f_{j_1,k_1}(x_1) f_{j_2,k_2}(x_2)$ is in the closure of the span of functions

$$\left\{ f_{j_1,k_1}(x_1) 2^{j/2} \rho(2^{j_1} x_2 - k) \right\}_{k \in \mathbf{Z}}$$

provided $j_1 > j_2$. This will again be a one-dimensional problem. By dilation argument we may assume that $j_1 = 0$ and $j_2 < 0$. It is

enough to show that the function f_{j_2, k_2} that is in S_{j_2, k_2+1} can be
approximated in $L^2(\mathbf{R})$ by linear combinations of the functions
$\left\{\rho(\cdot - k)\right\}_{k \in \mathbf{Z}}$.

First we approximate f_{j_2, k_2} by a function f in S_{j_2, k_2+1} with
compact support. Let k_0 be so large that the interval $[-k_0, k_0]$
contains the support of f. Set $\widetilde{S}_k = \{f : f(\cdot + k) \in \widetilde{S}_0^m\}$ and let
$\widetilde{P}_k$ be the projection operator $\widetilde{P}_k : L^2(\mathbf{R}) \to \widetilde{S}_k$. Then $\widetilde{P}_{-k_0} f = f$ and
$\widetilde{P}_{k_0} f = 0$, since $f \in \widetilde{S}_{-k_0}$ and since

$$\| f \|_{L^2}^2 - \| \widetilde{P}_{k_0} f \|_{L^2}^2 = \| f - \widetilde{P}_{k_0} f \|_{L^2}^2 = \| f \|_{L^2}^2 + \| \widetilde{P}_{k_0} f \|_{L^2}^2.$$

The last inequality holds, since f and $\widetilde{P}_{k_0} f$ have disjoint supports.
Thus

$$f = \widetilde{P}_{-k_0} f - \widetilde{P}_{k_0} f = \sum_{k=-k_0}^{k_0-1} \widetilde{P}_k f - \widetilde{P}_{k+1} f,$$

and $\widetilde{P}_k f - \widetilde{P}_{k+1} f$ is a multiple of $\rho(\cdot - k)$. This shows that f can
be written as a linear combination of the functions $\left\{\rho(\cdot - k)\right\}_{k=-k_0}^{k_0-1}$
and hence f_{j_2, k_2} is in the closure of the span of the functions
$\left\{\rho(\cdot - k)\right\}_{k \in \mathbf{Z}}$. This completes the proof of the completeness of the
system $\left\{f_\nu^\omega\right\}_{\Omega \times \mathbf{Z}^3}$ in $L^2(\mathbf{R}^2)$.

8. PROOF OF THEOREM 1(a) AND (b)

First we observe that f is in $\text{Lip}(m + 1)$, that is, $D^m f$ is in $\text{Lip } 1$
and hence is in the dual space of $H^p(\mathbf{R})$ when $1/p \le 2 + m$. We will
first prove Theorem 1(c) by an area integral approach. The
inequalities (4) and (5) will then be the basic ingredients in the
proof of Theorem 1(b). At the end of this paper, we will show how
(b) can be proved directly with atoms.

Proof of (4) and (5). It is enough to show (5) for f in a dense
subspace of $H^p(\mathbf{R})$, since if f_j converges to f, so will the corresponding coefficients $c_{j;\nu}$ converge to c_ν by the duality, and hence

(5) will hold for a general f if the sum over ν is restricted to be finite. The estimate for an infinite sum then follows by monotone convergence. Thus we may assume that the Fourier transform of f is C^∞, compactly supported and vanishing near the origin.

Let $\tilde{\nu} = (2^{-j}k,\ 2^{-j}) \in \mathbf{R}^2_+$, $\nu = (j,\ k) \in \mathbf{Z}^2$, and let μ be the measure on $\mathbf{R}^2_+$ defined by $\mu = \Sigma_\nu \delta_{\tilde{\nu}}$, where $\delta_{\tilde{\nu}}$ is the Dirac measure at $\tilde{\nu}$. Set $\tau_t(x) = t^{-1}\tau(x/t)$, $\tilde{\tau}_t(x) = \tau_t(-x)$, and define the function F on $\mathbf{R}^2_+$ by

$$F(x,\ t) = \tilde{\tau}_t * f(x),$$

where f is the finite sum $\Sigma_\nu c_\nu f_\nu$ [in showing (4)] or the function given in (5). Observe that in both cases we have $F(\tilde{\nu}) = 2^{j/2}c_\nu$, $\nu \in \mathbf{Z}^2$. We define the functions

$$AF(z) = \left\{ \iint_{0 \le z - x \le t} |F(x,\ t)|^2 d\mu(x,\ t) \right\}^{1/2}$$

and

$$g^*_\lambda F(z) = \left\{ \iint_{\mathbf{R}^2_+} (1 + |x - z|/t)^{-2\lambda} |F(x,\ t)|^2 d_\mu(x,\ t) \right\}^{1/2},\ \lambda > 0.$$

Then AF is identical to the expression

$$\left\{ \sum_\nu |c_\nu|^2 2^j \chi_\nu \right\}^{1/2}.$$

Now we let ψ be a C^∞ function supported in $[-1,\ 1]$ such that

$$\int \psi(x)x^\alpha\ dx = 0,\ \alpha = 0,\ 1,\ \ldots,\ m + 1 \tag{6}$$

and

$$\int_0^\infty |\hat{\psi}(s\xi)|^2\ \frac{ds}{s} = 1 \text{ for } \xi \ne 0. \tag{7}$$

Let $\psi_s(x) = s^{-1}\psi(x/s)$ and define the function G on $\mathbf{R}^2_+$ by $G(y,\ s) = \psi_s * f(y)$ and set

$$AG(z) = \left\{ \iint_{|y-z| < s} |G(y,\ s)|^2\ \frac{dy\ ds}{s^2} \right\}^{1/2}$$

and

$$g^*_\lambda G(z) = \left\{ \iint_{\mathbf{R}^2_+} (1 + |y - z|/s)^{-2\lambda} |G(y,\ x)|^2\ \frac{dy\ ds}{s^2} \right\}^{1/2},\ \lambda > 0.$$

By A. P. Calderón and A. Torchinsky [2, Theorem 6.9, p. 56, and
Theorem 3.5, p. 20], we get

$$\| AG \|_{L^p} \sim \| f \|_{H^p}, \quad 0 < p < \infty$$

and

$$\| g_\lambda^*(\cdot) \|_{L^p} \leq C \| A(\cdot) \|_{L^p}, \quad 1/\lambda < \min(p, 2).$$

To obtain the inequalities (4) and (5), we have only to show the
pointwise estimates

$$AG(z) \leq C g_\lambda^* F(z), \quad \lambda < m + 5/2, \tag{8}$$

$$AF(z) \leq C g_\lambda^* G(z), \quad \lambda < m + 2. \tag{9}$$

By (5) we have the identity

$$F(x, t) = \iint_{\mathbf{R}_+^2} \tau_t * \psi_s (y - x) G(y, s) \frac{dy\,ds}{s},$$

and since $F(\tilde{\nu}) = c_\nu 2^{j/2}$, we also have the identity

$$G(y, s) = \iint \tau_t * \psi_s (y - x) F(x, t) t\, d\mu(x, t).$$

We have to make estimates on the convolution $\tau_t * \psi_s$. Let $D^{-1}g$
denote the primitive function of g, that is,

$$D^{-1}g(x) = \int_0^x g(y)\,dy.$$

By (3), $D^k\tau$ will satisfy (1) even for $k = -1, \ldots, -m - 2$; and
similarly, by (6), $D^k\psi$ will be supported in $[-1, 1]$ even for
$k = -1, \ldots, -m - 2$. By repeated integration by parts we get

$$|\tau_t * \psi_s(y)| = |D^{m+1}\tau_t * D^{-m-1}\psi_s(y)| \leq C(s/t)^{m+1}(1 + |y|/t)^{-N} t^{-1},$$

for any $N > 0$ when $s \leq t$, and

$$|\tau_t * \psi_s(y)| = |D^{-m-2}\tau_t * D^{m+2}\psi_s(y)| \leq C(t/s)^{m+2}(1 + |y|/s)^{-N} s^{-1},$$

for any $N > 0$ when $s \geq t$. If we use that $D^{m+2}\tau$ is a sum of point-
measures, rapidly decreasing at infinity, we can also get the
estimate

$$\left\| (1 + |\cdot|/t)^N (\tau_t * \psi_s) \right\|_{L^2} \le C(s/t)^{m+3/2} t^{-1/2},$$

for any $N > 0$ when $s \le t$.

To estimate $AG(z)$, we let $|y - z| < s$ and get from these inequalities

$$|\tau_t * \psi_s(y - x)| \le C(t/s)^\varepsilon (1 + |x - z|/t)^{-3-m+\varepsilon} t^{-1}, \text{ when } t \le s,$$

and

$$|\tau_t * \psi_s(y - x)| \le C(s/t)^\varepsilon (1 + |x - z|/t)^{-N} t^{-1}, \text{ when } t \ge s,$$

for any $N > 0$ and $\varepsilon > 0$. Plugging these estimates into the identity for $G(y, s)$ above, we get, after using the Cauchy-Schwarz inequality,

$$|G(y, s)|^2 \le C \iint_{\mathbf{R}_+^2} (\min(s/t, t/s))^\varepsilon$$

$$\times (1 + |x - z|/t)^{-5-2m+3\varepsilon} |F(x, t)|^2 d\mu(x, t).$$

Integrating over the cone $|y - z| < s$, we get, after changing the order of integration, the inequality (8) with $\lambda = m + 5/2 - 3\varepsilon/2$.

To estimate $AF(z)$, we let $|x - z| < t$ and get from the estimates of $\tau_t * \psi_s$

$$|\tau_t * \psi_s(y - x)| \le C(t/s)^\varepsilon (1 + |y - z|/s)^{-N} s^{-1}, \text{ when } s \ge t$$

and

$$\left\| (1 + |(\cdot - z)|/s)^{m+2-\varepsilon} (\tau_t * \psi_s(\cdot - x)) \right\|_{L^2} \le C(s/t)^\varepsilon s^{-1/2}, \text{ when } s \le t,$$

for any $N > 0$ and $\varepsilon > 0$. Using these estimates in the identity for $F(x, t)$ together with the Cauchy-Schwarz inequality, we get

$$|F(x, t)|^2 \le C \iint_{\mathbf{R}_+^2} (\min(s/t, t/s))^\varepsilon$$

$$\times (1 + |y - z|/s)^{-4-2m+2\varepsilon} |G(y, s)|^2 \frac{dy\ ds}{s^2}.$$

Integrating over the cone $|x - z| < t$, we get, after changing the order of integration, the inequality (9) with $\lambda = m + 2 - \varepsilon$. This finishes the proof of Theorem 1(c).

Proof of Theorem 1(b). As a consequence of Theorem 1(c), the partial sums $\sum_\nu (f_\nu, f)f_\nu$ are uniformly bounded in the H^p norm by the H^p norm of f, $f \in H^p(\mathbf{R})$, $p > 1/(m + 2)$. We will also get the following lemma.

Lemma. *The set of finite linear combinations of the functions $\{f_\nu\}$ are dense $H^p(\mathbf{R})$, $p > 1/(m + 2)$.*

Proof of the Lemma. Define $S_N f$ by

$$S_N f = \sum_{\substack{|j| < N \\ |k| < N}} (f_\nu, f)f_\nu, \; (\nu = (j, k)), \; f \in H^p(\mathbf{R}).$$

Since $L^2(\mathbf{R}) \cap H^p(\mathbf{R})$ is dense in $H^p(\mathbf{R})$, it is enough to show that

$$\| f - S_N f \|_{H^p} \to 0 \text{ as } N \to 0, \; f \in L^2(\mathbf{R}) \cap H^p(\mathbf{R}).$$

Using dominated convergence, we see that the left side of (5) goes to zero as N tends to infinity if the sum is only taken over $\nu = (j, k)$ with $|j|$ or $|k|$ larger than N. It follows that for each $\varepsilon > 0$ there is an N_ε such that $\| S_N f - S_{N_\varepsilon} f \|_{H^p}^p < \varepsilon$ for all $N \geq N_\varepsilon$. Let $M(\;)$ be the nontangential maximal function. Then we have

$$\| f - S_N f \|_{H^p}^p \leq \int_{|x| < R} |M(f - S_N f)|^p \, dx + C\int_{|x| > R} |M(f - S_{N_\varepsilon} f)|^p \, dx$$

$$+ \; C\int_{|x| > R} |M(S_N f - S_{N_\varepsilon} f)|^p \, dx.$$

The third integral is less than $C\varepsilon$ independent of R, and since $f - S_{N_\varepsilon}$ is in $H^p(\mathbf{R})$, the second integral will be less than ε if R is large enough; and finally, using Hölder's inequality, the first integral is bounded by $C_R\| M(f - S_N f) \|_{L^2}^p \leq C_R\| f - S_N f \|_{L^2}^p$, which goes to zero as N tends to infinity by the completeness of $\{f_\nu\}$ in $L^2(\mathbf{R})$. This finishes the proof of the lemma. (We have assumed that $p < 2$, which is the only case of interest.)

Now let $\{\nu_i\}_{i=1}^\infty$ be any enumeration of $\{\nu\}_{\mathbf{Z}^2}$ and let $S_K^p f$ denote the partial sum

$$\sum_{i=1}^{K} (f_{\nu_i}, f) f_{\nu_i}, \quad f \in H^p(\mathbf{R}).$$

Let $f \in H^p(\mathbf{R})$. For each $\varepsilon > 0$ we can find a finite linear combination f_ε of the functions $\{f_\nu\}$ such that $\| f - f_\varepsilon \|_{H^p} < \varepsilon$. Then $S_K^p f_\varepsilon = f_\varepsilon$ if K is large enough. Since the partial sums are uniformly bounded in the H^p norm, we get

$$\| f - S_K^p f \|_{H^p} \le C \| f - f_\varepsilon \|_{H^p} + C \| S_K^p (f - f_\varepsilon) \|_{H^p} \le C \| f - f \|_{H^p} < C\varepsilon,$$

for any $\varepsilon > 0$ provided K is large enough. This proves the unconditional convergence. The uniqueness of the coefficients is now easy. If $\sum_\nu c_\nu f_\nu$ converges to f in the H^p norm, we apply the partial sums to f_{ν_0}, which is in the dual space. On the one hand, we get the limit c_{ν_0}; on the other hand, this limit must be (f_{ν_0}, f), which shows that $c_\nu = (f_\nu, f)$. With this we have finished the proof of Theorem 1.

9. PROOFS OF THEOREM 2(b) AND (c)

The area integral approach will work in higher dimension in exactly the same way as in the one-dimensional case. It involves a summation over ω that is harmless, since Ω is a finite set. To estimate the convolutions $\tau_t^\omega * \psi_s$, we can no longer use integration by parts, but instead we can, for instance, make estimates on the Fourier transform side. With this method, we get (4) in Theorem 2(c) when $p > 1/(1/2 + (m + 3/2)/n)$ and (5), and consequently also Theorem 2(b) when $p > 1/(1/2 + (m + 2)/n)$.

Using atoms we obtain inequality (5) and Theorem 2(b) when $p > n/(n + m + 1)$. Since atoms have been used on Franklin systems on [0, 1] (see, for example, [5 and 10]), we will only sketch the proof briefly.

Let f be an atom supported in a cube Q with side s and center x_0 such that

$$\int f x^\alpha \, dx = 0, \quad |\alpha| \le m, \quad \| f \|_\infty \le s^{-n/p}$$

and let x_ν denote the point $(2^{-j} k_1, \ldots, 2^{-j} k_n) \in \mathbf{R}^n$ and $|\nu| = 2^{-j}$ for $\nu = (j, k_1, \ldots, k_n) \in \mathbf{Z}^{n+1}$. Then

$$|(f_\nu^\omega, f)| \le \| f_\nu^\omega \|_{L^1(Q)} \| f \|_\infty$$

$$\le C s^{-n/p} |\nu|^{n/2} \min\{1, \, r^{(|x_0 - x_\nu| - ns)/|\nu|}\}, \quad \text{when } |\nu| \le s,$$

and using the mth-order Taylor expansion of f_ν^ω at x_0 and the moment conditions of f, we get

$$|(f_\nu^\omega, f)| \le C s^{1+m+n-n/p} |\nu|^{-n/2-m-1} r^{|x_\nu - x_0|} |\nu|, \quad \text{when } |\nu| \ge s.$$

Let $M(\ )$ be the nontangential maximal function and let Sf be *any* partial sum

$$\sum_{\nu, \omega} (f_\nu^\omega, f) f_\nu^\omega.$$

Looking at the moment conditions of τ^ω, we find that $\left| M(\tau^\omega)(x) \right| \le C(1 + |x|)^{-m-n-2}$ and consequently

$$\left| M(f_\nu^\omega)(x) \right| \le C |\nu|^{-n/2} (1 + |x - x_\nu|/|\nu|)^{-m-n-2}.$$

By summation we now get for $\varepsilon > 0$ and $|x - x_0| \ge 2ns$

$$\left| M(Sf)(x) \right| \le \sum_{\nu, \omega} |(f_\nu^\omega, f)| \left| M(f_\nu^\omega)(x) \right| \le C(|x - x_0|/s)^{-m-n-1+\varepsilon} s^{-n/p}$$

and

$$\left\{ \sum_{\nu, \omega} |(f_\nu^\omega, f)|^2 |\nu|^{-n} \chi_\nu(x) \right\}^{1/2} \le C(|x - x_0|/s)^{-m-n-1+\varepsilon} s^{-n/p}.$$

By these estimates and L^2 estimates when $|x - x_0| < 2ns$, we get

$$\| Sf \|_{H^p} \le \| M(Sf) \|_{L^p} \le C, \quad p > n/(n + m + 1),$$

and

$$\left\| \left\{ \sum_{\nu, \omega} |(f_\nu^\omega, f)|^2 |\nu|^n \chi_\nu \right\}^{1/2} \right\|_{L^p} \le C, \quad p > n/(n + m + 1).$$

Taking limits with finite sums of atoms, we get inequality (5) and $\| Sf \|_{H^p} \le C \| f \|_{H^p}$ for all $f \in H^p(\mathbf{R}^n)$, $p > n/(n + m + 1)$.

To prove that finite linear combinations of $\{f_\nu^\omega\}$ are dense in $H^p(\mathbf{R}^n)$, $p > n/(n + m + 1)$, is now easy if we use the

L^2-completeness and that

$$\left| M(Sf - f)(x) \right| \leq C_f |x|^{-n-m-1+\epsilon} \text{ for } |x| > c_f$$

uniformly for all partial sums Sf when f is a finite sum of atoms. The rest of the proof of Theorem 2(b) is the same as in the one-dimensional case.

An open problem is to find a better range of p for the n-dimensional version of inequality (4) when $n \geq 2$. (Added in proof:) A better range of p for the n-dimensional version of (4) has been found by Peter Sjögren, and the author of this paper is now aware of methods that will give the sharp range.

REFERENCES

1. Bockariev, S. V. Existence of basis in the space of analytic functions in the disc and some properties of the Franklin system. *Mat. Sbornik* 95 (1974):3-18 (in Russian).

2. Calderón, A. P., and Torchinsky, A. Parabolic maximal functions associated with a distribution. *Adv. Math.* 16 (1975): 1-64.

3. Carleson, L. An explicit unconditional basis in H^1. Institut Mittag-Leffler, Report No. 2, 1980.

4. Ciesielski, Z. Equivalence, unconditionality and convergence a.e. of the spline bases in L_p spaces. Approximation theory, Banach Center Publications, vol. 4, pp. 55-68.

5. Ciesielski, Z. The Franklin orthogonal system as unconditional basis in Re H^1 and VMO. Preprint, 1980.

6. Ciesielski, Z., and Domsta, J. Estimates for the spline orthonormal functions and for their derivatives. *Studia Math.* 44 (1972):315-320.

7. Ciesielski, Z.; Simon, P.; and Sjölin, P. Equivalence of Haar and Franklin bases in L^p spaces. *Studia Math.* 60 (1976):195-211.

8. Maurey, B. Isomorphismes entre espace H^1. *Acta Math.* 145 (1980):79-120.

9. Schipp, F., and Simon, P.　Investigation of Haar and Franklin series in the Hardy spaces.　Preprint, 1980.

10. Sjölin, P., and Strömberg, J.-O.　Basis properties of Hardy spaces.　University of Stockholm, Report No. 19, 1981.

11. Wojtaszczyk, P.　The Franklin system is an unconditional basis in H^1.　To appear in *Arkiv för Matematik*.

A CONSTRUCTIVE PROOF OF THE FEFFERMAN-STEIN DECOMPOSITION OF BMO ON SIMPLE MARTINGALES

Akihito Uchiyama
University of Chicago

Let

$$\Omega = (0, 1].$$

Let $\mathfrak{J}$ be the σ-field of all Borel measurable sets in Ω. Then $(\Omega, \mathfrak{J}, dx)$ is a probability space. Let $d \geq 3$ be an integer. For each integer $n \geq 0$, let $\mathfrak{J}_n$ be the sub-σ-field of $\mathfrak{J}$ generated by

$$((k - 1)d^{-n}, kd^{-n}], \quad k = 1, \ldots, d^n.$$

Set

$$I_{k_1, k_2, \ldots, k_n} = ((k_1 - 1)d^{-1} + (k_2 - 1)d^{-2} + \cdots$$
$$+ (k_n - 1)d^{-n}, (k_1 - 1)d^{-1} + (k_2 - 1)d^{-2} + \cdots + k_n d^{-n}]$$

for each $k_1, \ldots, k_n \in \{1, \ldots, d\}$. For $f \in L^1(\Omega)$, set

$$f_n = E[f \mid \mathfrak{J}_n]$$

and

$$\Delta f_n = f_n - f_{n-1}.$$

Let

$$V = \left\{ \mathfrak{x} = \{x_k\}_{k=1}^d \in \mathbf{C}^d : \sum_{k=1}^d x_k = 0 \right\},$$

and let A be a linear operator from V to V. For $f \in L^1(\Omega)$ and

This work was supported in part by Science Research Foundation of Japan (General Research © 1980).

$k_1, \ldots, k_{n-1} \in \{1, \ldots, d\}$, set

$$\xi = \left\{ \Delta f_n \left(I_{k_1, k_2, \ldots, k_{n-1}, k} \right) \right\}_{k=1}^{d} \in V.$$

Define

$$\left\{ \Delta g_n \left(I_{k_1, \ldots, k_{n-1}, k} \right) \right\}_{k=1}^{d} = A\xi.$$

Set

$$g = \sum_{n=1}^{\infty} \Delta g_n. \tag{1}$$

It can be easily shown that if $f \in L^2(\Omega)$, then the right-hand side of (1) converges in $L^2(\Omega)$ and

$$\| g \|_{L^2} \leq C \| f \|_{L^2}.$$

Set

$$Tf = g. \tag{2}$$

Definition. For $f \in L^2(\Omega)$, let

$$\| f \|_{\text{BMO}} = \sup_n \left\| E \left[\sum_{k=n+1}^{\infty} |\Delta f_k|^2 \,\Big|\, \mathfrak{I}_n \right] \right\|_{\infty}^{1/2}.$$

If $\| f \|_{\text{BMO}} < \infty$, we say f belongs to BMO.

Assume that $A_1, \ldots, A_m$ are one or more linear transformations on V and let $T_1, \ldots, T_m$ be the corresponding operators defined by (2). Let $A_0 = I$ and T_0 be the identity operator. S. Janson [4] showed

Theorem

$$\text{BMO} = \sum_{j=0}^{m} T_j L^{\infty}$$

if and only if

(i) $A_1^*, \ldots, A_m^*$ *do not have a common real eigenvector, where* A_j^* *is the adjoint operator of A_j from V to V.*

This result corresponds to the Fefferman-Stein decomposition of BMO($\mathbf{R}^m$). In [4], Janson showed only the existence of $g_0, \ldots, g_m \in L^{\infty}$ such that for $f \in$ BMO,

$$f = \sum_{j=0}^{m} T_j g_j . \tag{3}$$

In this note, we give a construction of $g_0, \ldots, g_m$.

Notation. $\mathbf{a}$, $\mathbf{X}$, $\mathbf{Y}$, $\ldots$ denote $(m + 1)$-dimensional row vectors. $\mathfrak{x}$, $\mathfrak{y}$, $\mathfrak{z}$, $\ldots$ denote d-dimensional column vectors. For

$$\mathfrak{x} = \begin{pmatrix} x_1 \\ \vdots \\ x_d \end{pmatrix} \quad \text{and} \quad \mathfrak{y} = \begin{pmatrix} y_1 \\ \vdots \\ y_d \end{pmatrix},$$

$\langle \mathfrak{x}, \mathfrak{y} \rangle$ denotes $\sum_{k=1}^{d} x_k \overline{y}_k$, $\quad |\mathfrak{x}|$ denotes $\left(\sum_{k=1}^{d} |x_k|^2 \right)^{1/2}$

and similarly for $|\mathbf{a}|$. $\mathbf{X}_k$ and $\mathfrak{x}_j$ denote the kth row vector and the $(j + 1)$th column vector of the same $d \times (m + 1)$ matrix, respectively. $\mathbf{C}$ denotes the set of all complex numbers. The letter c denotes various positive constants.

Lemma 1. The condition (i) *is equivalent to the following:*

(ii) *For any*

$$(a_0, \ldots, a_m) \in \mathbf{C}^{m+1}$$

and any

$$\mathfrak{y} \in V,$$

there exist

$$\mathfrak{x}_0, \ldots, \mathfrak{x}_m \in V$$

such that

$$\sum_{j=0}^{m} A_j \mathfrak{x}_j = \mathfrak{y}, \tag{4}$$

$$\sum_{j=0}^{m} \operatorname{Re} a_j \cdot \operatorname{Re} \mathfrak{x}_j + \operatorname{Im} a_j \cdot \operatorname{Im} \mathfrak{x}_j = 0, \tag{5}$$

$$\sum_{j=0}^{m} |\mathfrak{x}_j| \leq C |\mathfrak{y}|. \tag{6}$$

where C depends on A_1, ..., A_m but is independent of $\mathfrak{y}$ and
$(a_0, ..., a_m)$.

Remark 1.   (5) means that each row vector $\mathbf{X}_1$, ..., $\mathbf{X}_d$ of the
$d \times (m + 1)$ matrix

$$(\mathfrak{x}_0, ..., \mathfrak{x}_m)$$

is orthogonal to $(a_0, ..., a_m)$ in $\mathbf{R}^{2(m+1)}$.   Similarly, for $\mathfrak{x}, \mathfrak{y} \in C^d$,

$$\mathrm{Re} \langle \mathfrak{x}, \mathfrak{y} \rangle$$

is equal to the inner product in $\mathbf{R}^{2d}$.

Proof.   (ii) → (i).   Suppose $\mathfrak{y}_0$ is a common real eigenvector of
A_1^*, ..., A_m^*.   Let a_0, ..., a_m be its eigenvalues.   If

$$\sum_{j=0}^{m} A_j \mathfrak{x}_j = \mathfrak{y}_0$$

and (5) holds, then

$$|\mathfrak{y}_0|^2 = \mathrm{Re} \left\langle \sum_{j=0}^{m} A_j \mathfrak{x}_j, \mathfrak{y}_0 \right\rangle = \mathrm{Re} \sum_{j=0}^{m} \left\langle \mathfrak{x}_j, A_j^* \mathfrak{y}_0 \right\rangle$$

$$= \mathrm{Re} \sum_{j=0}^{m} \left\langle \mathfrak{x}_j, a_j \mathfrak{y}_0 \right\rangle = \left\langle \mathrm{Re} \sum_{j=0}^{m} \overline{a}_j \mathfrak{x}_j, \mathfrak{y}_0 \right\rangle = 0,$$

and (ii) is false.

(i) → (ii).   It suffices to show

$$\left\{ \sum_{j=0}^{m} A_j \mathfrak{x}_j : \mathfrak{x}_j \in V \text{ and (5) holds} \right\} = V \qquad (7)$$

for any $(a_0, ..., a_m) \in C^{m+1}$.   Then (6) follows from a compactness
argument.   If $a_0 = 0$, then (7) is clear.

The left-hand side of (7) is a subspace of V over $\mathbf{R}$.   So,
to show (7), it suffices to show that for any $\mathfrak{y} \in V$, there exists
$\mathfrak{x}_0$, ..., $\mathfrak{x}_m \in V$ such that (5) and

$$\mathrm{Re} \left\langle \sum_{j=0}^{m} A_j \mathfrak{x}_j, \mathfrak{y} \right\rangle \neq 0 \qquad (8)$$

hold.   (Remember Remark 1.)

Case 1. $\mathfrak{y}$ is not of the form $\lambda \mathfrak{y}_0$, where $\lambda \in \mathbf{C}$ and $\mathfrak{y}_0$ is a real vector, and $a_0 \neq 0$. Set

$$\mathfrak{y} = \begin{pmatrix} y_1 \\ \vdots \\ y_d \end{pmatrix}.$$

Then, there exist two components, say, y_1 and y_2, that satisfy

$$\mathrm{Re}(ia_0 \cdot \overline{y}_1) > 0$$

and

$$\mathrm{Re}(ia_0 \cdot \overline{y}_2) < 0,$$

where $i = (-1)^{1/2}$. So, set

$$\mathfrak{r}_0 = \begin{pmatrix} ia_0 \\ -ia_0 \\ 0 \\ \vdots \\ 0 \end{pmatrix}, \quad \mathfrak{r}_1 = \cdots = \mathfrak{r}_m = 0.$$

Then (5) and (8) hold.

Case 2. $\mathfrak{y}$ is a constant multiple of a real vector and $a_0 \neq 0$. By (i), $\mathfrak{y}$ is not a common eigenvector of $A_1^*, \ldots, A_m^*$. So at least one of $\{A_j^* \mathfrak{y}\}_{j=1,\ldots,m}$, say, $A_1^* \mathfrak{y}$, is independent of $\mathfrak{y}$ over $\mathbf{C}$. Take $\tilde{\mathfrak{r}} \in V$ such that

$$\langle \tilde{\mathfrak{r}}, A_1^* \mathfrak{y} \rangle = a_0,$$
$$\langle \tilde{\mathfrak{r}}, \mathfrak{y} \rangle = 0.$$

Set

$$\mathfrak{r}_0 = \overline{a}_1 \tilde{\mathfrak{r}},$$
$$\mathfrak{r}_1 = -\overline{a}_0 \tilde{\mathfrak{r}},$$
$$\mathfrak{r}_j = 0 \text{ for other } j.$$

Then

$$\mathrm{Re}\left\langle \sum_{j=0}^m A_j \mathfrak{r}_j, \mathfrak{y} \right\rangle = \mathrm{Re}\langle A_1 \mathfrak{r}_1, \mathfrak{y} \rangle = -|a_0|^2.$$

So (8) holds. Since

$$\sum_{j=0}^m a_j \tilde{\mathfrak{r}}_j = a_0 \tilde{\mathfrak{r}}_0 + a_1 \tilde{\mathfrak{r}}_1 = 0,$$

(5) holds.

Main Lemma. Let $A_1, \ldots, A_m$ satisfy (i). Let $R > 0$ be a sufficiently large number depending only on $A_1, \ldots, A_m$. Let

$$\mathfrak{y} \in V,$$

$$\mathbf{a} = (a_0, \ldots, a_m) \in \mathbf{C}^{m+1},$$

$$b_1, \ldots, b_d \geq 0$$

be such that

$$\min(R - |\mathbf{a}|, 1) \geq \left\{\left(|\mathfrak{y}|^2 + \sum_{k=1}^{d} b_k^2\right)\Big/ d\right\}^{1/2}. \tag{9}$$

Then there exist

$$\mathfrak{r}_0, \ldots, \mathfrak{r}_m \in V \tag{10}$$

such that

$$\sum_{j=0}^{m} A_j \mathfrak{r}_j = \mathfrak{y} \tag{11}$$

and

$$R - |\mathbf{a} + \mathbf{X}_k| \geq b_k, \quad 1 \leq k \leq d, \tag{12}$$

hold.

Proof. First we show this for three special cases and then for the
general case.

   Case 1. $R - |\mathbf{a}| \geq 2d^{1/2}$.
   In this case, set

$$\mathfrak{r}_0 = \mathfrak{y}, \quad \mathfrak{r}_1 = \cdots = \mathfrak{r}_m = 0.$$

Then (11) is trivial and

$$R - |\mathbf{a} + \mathbf{X}_k| \geq R - |\mathbf{a}| - |\mathbf{X}_k| \geq 2d^{1/2} - |\mathbf{X}_k| \geq d^{1/2} \geq b_k,$$

since

$$\left(|\mathfrak{y}|^2 + \sum_{k=1}^{d} b_k^2\right)\Big/ d \leq 1.$$

   Case 2.

$$2d^{1/2} \geq R - |\mathbf{a}| \geq 1, \tag{13}$$

$$\left(|\mathfrak{y}|^2 + \sum_{k=1}^{d} b_k^2\right)\Big/ d = 1, \tag{14}$$

$$\max_{1 \le k \le d} b_k < 1/2. \tag{15}$$

By Lemma 1, there exist

$$\mathfrak{k}_0, \ \mathfrak{k}_1, \ \ldots, \ \mathfrak{k}_m \ \epsilon \ V$$

such that (4)-(6) hold. (11) is clear from (4). By (5) and (6), $\mathbf{X}_1, \ \ldots, \ \mathbf{X}_d$ are orthogonal to $\mathbf{a}$ in $\mathbf{R}^{2(m+1)}$ (see Remark 1), and

$$|\mathbf{X}_k| \le c, \ k = 1, \ \ldots, \ d.$$

So,

$$|\mathbf{a} + \mathbf{X}_k| = (|\mathbf{a}|^2 + |\mathbf{X}_k|^2)^{1/2} \le |\mathbf{a}|(1 + c|\mathbf{X}_k|^2/|\mathbf{a}|^2)$$

$$\le R - 1 + c/R \le R - b_k,$$

by (13) and (15). Thus, (11) and (12) hold.

Case 3. (13), (14), *and*

$$\max_{1 \le k \le d} b_k \ge 1/2. \tag{16}$$

We may assume

$$b_1 \ge b_2 \ge \cdots \ge b_d.$$

Then by (16) and (14),

$$\sum_{k=1}^{d} b_k/d \le \left(\sum_{k=1}^{d} b_k^2/d\right)^{1/2} - c(b_1 - b_d)^2 \le 1 - c(|\mathfrak{y}|^2 \pm (b_1 - b_d)^2).$$

So we can take

$$\mathfrak{z} = \begin{pmatrix} z_1 \\ \vdots \\ z_d \end{pmatrix} \ \epsilon \ V \cap \mathbf{R}^d$$

such that

$$|\mathfrak{z}| \le c(b_1 - b_d), \tag{17}$$

$$1 - z_k \ge b_k + c(|\mathfrak{y}|^2 + (b_1 - b_d)^2), \ 1 \le k \le d. \tag{18}$$

Set

$$\mathfrak{k}'_j = a_j \mathfrak{z}/|\mathbf{a}|, \ j = 0, \ \ldots, \ m,$$

$$\mathfrak{y}' = \mathfrak{y} - \sum_{j=0}^{m} A_j \mathfrak{k}'_j. \tag{19}$$

Then,

$$|\mathfrak{y}'| \le |\mathfrak{y}| + c(b_1 - b_d).$$

By Lemma 1, there exist

$$\mathfrak{x}_j'' \in V, \ j = 0, \ \ldots, \ m,$$

such that

$$\sum_{j=0}^{m} A_j \mathfrak{x}_j'' = \mathfrak{y}', \tag{20}$$

$$\sum_{j=0}^{m} \operatorname{Re} a_j \cdot \operatorname{Re} \mathfrak{x}_j'' + \operatorname{Im} a_j \cdot \operatorname{Im} \mathfrak{x}_j'' = 0, \tag{21}$$

$$\sum_{j=0}^{m} |\mathfrak{x}_j''| \leq c(|\mathfrak{y}| + (b_1 - b_d)). \tag{22}$$

Set

$$\mathfrak{x}_j = \mathfrak{x}_j' + \mathfrak{x}_j''.$$

By (19) and (20),

$$\sum_{j=0}^{m} A_j \mathfrak{x}_j = \mathfrak{y}.$$

So (11) holds. Since $\mathbf{a} + \mathbf{X}_k'$ and $\mathbf{X}_k''$ are orthogonal in $\mathbf{R}^{2(m+1)}$ by (21), we get

$$|\mathbf{a} + \mathbf{X}_k| = |\mathbf{a} + \mathbf{X}_k' + \mathbf{X}_k''| \leq |\mathbf{a} + \mathbf{X}_k'|(1 + c|\mathbf{X}_k''|^2 / |\mathbf{a} + \mathbf{X}_k'|^2)$$

$$\leq |\mathbf{a} + \mathbf{X}_k'| + c(|\mathfrak{y}|^2 + (b_1 - b_d)^2)/R$$

by (22). Since

$$|\mathbf{a} + \mathbf{X}_k'| = |\mathbf{a}| + z_k \leq R - 1 + z_k \leq R - b_k - c(|\mathfrak{y}|^2 + (b_1 - b_d)^2)$$

by (18), we get

$$|\mathbf{a} + \mathbf{X}_k| \leq R - b_k - c(|\mathfrak{y}|^2 + (b_1 - b_d)^2) + c(|\mathfrak{y}|^2 + (b_1 - b_d)^2)/R.$$

So (12) holds if R is large enough.

General Case. Note that by the above three cases, the case

$$R - |\mathbf{a}| \geq \left\{ \left(|\mathfrak{y}|^2 + \sum_{k=1}^{d} b_k^2 \right) \Big/ d \right\}^{1/2} = 1 \tag{9'}$$

was completely shown. In the general case, set

$$\alpha = \left(\left(|\mathfrak{y}|^2 + \sum_{k=1}^{d} b_k^2 \right) \Big/ d \right)^{1/2}.$$

By (9), $\alpha \leq 1$ and

$$R/\alpha - |\mathbf{a}/\alpha| \geq \left\{ \left(|\mathfrak{y}/\alpha|^2 + \sum_{k=1}^{d} (b_k/\alpha)^2 \right) \Big/ d \right\}^{1/2} = 1. \qquad (9'')$$

So

$$R/\alpha, \ \mathbf{a}/\alpha, \ \mathfrak{y}/\alpha, \ b_1/\alpha, \ \ldots, \ b_d/\alpha$$

satisfy (9'). So we get

$$\tilde{\mathfrak{k}}_j \, \epsilon \, V, \ j = 0, \ \ldots, \ m,$$

such that

$$\sum_{j=0}^{m} A_j \, \tilde{\mathfrak{k}}_j = \mathfrak{y}/\alpha \qquad (11')$$

and

$$R/\alpha - |\mathbf{a}/\alpha + \tilde{\mathbf{X}}_k| \geq b_k/\alpha \qquad (12')$$

hold. Set

$$\mathfrak{k}_j = \alpha \tilde{\mathfrak{k}}_j.$$

Then (11) and (12) follow from (11') and (12').

The construction of $g_0, \ \ldots, \ g_m$ *of* (3). Let

$$\|f\|_{\text{BMO}} \leq 1, \ E[f] = 0.$$

Let

$$\mathbf{g}_0 \equiv (0, \ \ldots, \ 0) \, \epsilon \, \mathbf{C}^{m+1}.$$

Let $R > 0$ be as in the main lemma. Of course,

$$R - |\mathbf{g}_0| > 1 \geq \left(E\left[\sum_{n=1}^{\infty} |\Delta f_n|^2 \, \Big| \, \mathfrak{J}_0 \right] \right)^{1/2}.$$

Assume that we have constructed the $\mathfrak{J}_n$-measurable $\mathbf{C}^{m+1}$-valued function

$$\mathbf{g}_n = (g_{0n}, \ \ldots, g_{mn})$$

such that

$$R - |\mathbf{g}_n| \geq \left(E\left[\sum_{h=n+1}^{\infty} |\Delta f_h|^2 \, \Big| \, \mathfrak{J}_n \right] \right)^{1/2}, \qquad (23)$$

$$\sum_{j=0}^{m} T_j \, g_{jn} = f_n. \qquad (24)$$

Take any $I = I_{k_1, \ldots, k_n} \, \epsilon \, \mathfrak{J}_n$. Set

$$\mathfrak{h} = \begin{pmatrix} \Delta f_{n+1}(I(1)) \\ \vdots \\ \Delta f_{n+1}(I(d)) \end{pmatrix} \in V, \tag{25}$$

$$\mathbf{a} = \mathbf{g}_n(I) \in \mathbf{C}^{m+1},$$

$$b_k = \left\{ E\left[\sum_{h=n+2}^{\infty} |\Delta f_h|^2 \Big| \mathfrak{F}_{n+1} \right](I(k)) \right\}^{1/2}, \tag{26}$$

where

$$I(k) = I_{k_1,\ldots,k_n k}.$$

Then

$$E\left[\sum_{h=n+1}^{\infty} |\Delta f_h|^2 \Big| \mathfrak{F}_n \right](I) = \left(|\mathfrak{h}|^2 + \sum_{k=1}^{d} b_k^2 \right) \Big/ d.$$

So by (23) and the main lemma, there exist $\mathbf{X}_1, \ldots, \mathbf{X}_d \in \mathbf{C}^{m+1}$ such that (10)-(12) hold. Set

$$\mathbf{g}_{n+1}(I(k)) = \mathbf{a} + \mathbf{X}_k, \quad k = 1, \ldots, d.$$

In this way, we can define $\mathfrak{F}_{n+1}$-measurable $\mathbf{g}_{n+1}$ on Ω.

By (10),

$$E[\mathbf{g}_{n+1} | \mathfrak{F}_n] = \mathbf{g}_n.$$

By (12) and (26),

$$R - |\mathbf{g}_{n+1}| \geq \left(E\left[\sum_{h=n+2}^{\infty} |\Delta f_h|^2 \Big| \mathfrak{F}_{n+1} \right] \right)^{1/2}.$$

By (24), (25), and (11),

$$\left\{ \sum_{j=0}^{m} T_j g_{j,\,n+1}(I(k)) \right\}_{k=1}^{d} = \left\{ \sum_{j=0}^{m} T_j g_{j,\,n}(I) \right\} + \sum_{j=0}^{m} A_j \mathfrak{r}_j$$

$$= \left\{ f_n(I) + \Delta f_{n+1}(I(k)) \right\}_{k=1}^{d}$$

$$= \left\{ f_{n+1}(I(k)) \right\}_{k=1}^{d}.$$

So we can define $\{\mathbf{g}_n\}_{n=1}^{\infty}$ with (23) and (24). By (23),

$$\|\mathbf{g}_n\|_{\infty} \leq R.$$

Let

$$\mathbf{g} = \lim_{n \to \infty} \mathbf{g}_n \text{ in } L^2(\Omega).$$

Then this is the desired function.

REFERENCES

1. Chao, J.-A., and Taibleson, M. H. A sub-regularity inequality for conjugate systems on local fields. *Studia Math*. 46 (1973): 249-257.

2. Fefferman, C., and Stein, E. M. H^p spaces of several variables. *Acta Math*. 129 (1972):137-193.

3. Gundy, R. F. Inegalites pour martingales un et deux indices: L'espace H^p. *Springer Lecture Notes in Math*. 774 (1980):251-334.

4. Janson, S. Characterization of H^1 by singular integral transforms on martingales and $\mathbf{R}^n$. *Math. Scand*. 41 (1977):140-152.

5. Jones, P. W. Carleson measures and the Fefferman-Stein decomposition of BMO($\mathbf{R}$). *Ann. of Math*. 111 (1980):197-208.

6. Jones, P. W. L^∞ estimates for the $\bar{\partial}$ problem in a half-plane. To appear in *Acta Math*.

7. Taibleson, M. H. *Fourier Analysis on Local Fields*. Princeton, N.J.: Princeton Univ. Press, 1975.

PART SIX

DIFFERENTIATION THEORY

RELATIONS BETWEEN PEANO
DERIVATIVES AND MARCINKIEWICZ INTEGRALS

Sagun Chanillo
Richard L. Wheeden
Rutgers University

The purpose of this paper is to describe the results obtained in our
papers [5 and 6]. To make the presentation as short as possible, we
do not prove any of the facts stated below; complete proofs can be
found in [5 and 6].

In [5], the main results are distribution function inequalities relating Marcinkiewicz integrals and Peano derivatives. These
inequalities imply as corollaries many local and global results,
including some new facts, and thus provide a fairly unified approach
to the study of Marcinkiewicz integrals. They have the advantage of
also leading easily to weighted norm inequalities for A_∞ weight
functions. The distribution function inequalities involve several
other operators, such as nontangential maximal functions and area
integrals related to differentiation. The purpose of [6] is to
study pointwise estimates that connect the Peano maximal function
of a given order with an appropriate nontangential maximal function.
This helps in understanding the hierarchy among the operators
related to Marcinkiewicz integrals and simplifies some of the
corollaries of the distribution function inequalities.

The local results mentioned above are theorems of
Marcinkiewicz [11] and Stein and Zygmund [14, 15, and 16] giving

Research partly supported by NSF Grant No. MCS80-03098.

necessary and sufficient conditions for the finiteness a.e. in a set of a Marcinkiewicz integral. The global results involve L^p and weak L^p relations between Peano maximal functions, nontangential maximal functions, and Marcinkiewicz integrals. Such results are ultimately connected to the question of when a function is the potential of an element of L^p or H^p. Global results have been proved by many authors: see, for example, [11, 20, 10, 18, 13, 8, 9, 19, and 7].

The most familiar of the Marcinkiewicz integrals is

$$\mathcal{M}_1(F)(x) = \left(\int_{\mathbf{R}^n} \frac{|F(x+u) + F(x-u) - 2F(x)|^2}{|u|^{n+2}}\, du \right)^{1/2}.$$

Frequently, in local considerations, one wishes to study the truncated integral

$$\mathcal{M}_1(F)(x,\ \delta) = \left(\int_{|u|<\delta} \frac{|F(x+u) + F(x-u) - 2F(x)|^2}{|u|^{n+2}}\, du \right)^{1/2},\ \delta > 0.$$

More generally, for $0 < \alpha < 2$, we consider

$$\mathcal{M}_\alpha(F)(x) = \left(\int_{\mathbf{R}^n} \frac{|F(x+u) + F(x-u) - 2F(x)|^2}{|u|^{n+2\alpha}}\, du \right)^{1/2}$$

and the truncated version $\mathcal{M}_\alpha(F)(x,\ \delta)$ in which the integration is only extended over $|u| < \delta$. For technical reasons, we also define

$$\mathcal{M}_\alpha^*(F)(x) = \sup_{s>0} \left(\int_{\mathbf{R}^n} \frac{|F(x+u,\ s) + F(x-u,\ s) - 2F(x,\ s)|^2}{|u|^{n+2\alpha}}\, du \right)^{1/2},$$

where $0 < \alpha < 2$ and $F(x,\ y)$ is an appropriate function in $\mathbf{R}_+^{n+1}$. We will always assume that $F(x,\ y)$ is harmonic, which in particular includes the case when $F(x,\ y)$ is the Poisson integral of a function on the boundary. Another convenient class of functions is the solutions of the heat equation, which would include Gauss-Weierstrass extensions of functions on the boundary. It can be proved that

$\mathcal{M}_\alpha^*(F)(x)$ and $\mathcal{M}_\alpha(F)(x)$ are pointwise equivalent if $F(x, y)$ is the Poisson integral of $F(x)$.

We also consider the unsymmetric Marcinkiewicz integrals

$$
\mathcal{D}_\alpha(F)(x) = \begin{cases} \left(\displaystyle\int_{\mathbf{R}^n} \frac{|F(x + u) - F(x)|^2}{|u|^{n+2\alpha}} \, du \right)^{1/2}, & 0 < \alpha < 1, \\[4mm] \left(\displaystyle\int_{\mathbf{R}^n} \frac{|F(x + u) - F(x) - u \cdot A(x)|^2}{|u|^{n+2\alpha}} \, du \right)^{1/2}, & 1 < \alpha < 2, \end{cases}
$$

where $A(x)$ is a vector with n components. We define $\mathcal{D}_1(F) = \mathcal{M}_1(F)$. Similarly, for harmonic $F(x, y)$,

$$
\mathcal{D}_\alpha^*(F)(x) = \sup_{s > 0} \begin{cases} \left(\displaystyle\int_{\mathbf{R}^n} \frac{|F(x + u, s) - F(x, s)|^2}{|u|^{n+2\alpha}} \, du \right)^{1/2}, & 0 < \alpha < 1, \\[4mm] \left(\displaystyle\int_{\mathbf{R}^n} \frac{|F(x + u, s) - F(x, s) - u \cdot \nabla_1 F(x, s)|^2}{|u|^{n+2\alpha}} \, du \right)^{1/2}, \\[3mm] \hspace{6cm} 1 < \alpha < 2, \end{cases}
$$

and $\mathcal{D}_1^*(F) = \mathcal{M}_1^*(F)$. Here ∇_1 denotes the gradient in the x-variables,

$$
\nabla_1 = \left\langle \frac{\partial}{\partial x_1}, \ \cdots, \ \frac{\partial}{\partial x_n} \right\rangle,
$$

as distinguished from the full gradient

$$
\nabla = \left\langle \nabla_1, \frac{\partial}{\partial y} \right\rangle.
$$

The truncations obtained by only integrating over $|u| < \delta$ are denoted $\mathcal{D}_\alpha(F)(x, \delta)$ and $\mathcal{D}_\alpha^*(F)(x, \delta)$. We then clearly have $\mathcal{M}_\alpha(F)(x) \le 2\mathcal{D}_\alpha(F)(x)$ and $\mathcal{M}_\alpha(F)(x, \delta) \le 2\mathcal{D}_\alpha(F)(x, \delta)$, as well as the corresponding inequalities for $\mathcal{M}_\alpha^*$ and $\mathcal{D}_\alpha^*$. If $F(x, y)$ is the Poisson integral of $F(x)$, it can be shown that $\mathcal{D}_\alpha^*(F)(x)$ and $\mathcal{D}_\alpha(F)(x)$ are pointwise equivalent, assuming when $1 < \alpha < 2$ that $\nabla_1 F(x, s)$ has a limit as $s \to 0$ and defining $A(x)$ to be this limit.

We now define the Peano maximal functions $\mathcal{P}_\alpha(F)(x)$. These are inherent in [4] and were explicitly studied in [3]. Let

$$\mathcal{P}_\alpha(F)\,(x) \;=\; \sup_{h>0} \begin{cases} h^{-\alpha}\left(h^{-n}\displaystyle\int_{|u|<h}\big|F(x+u)-F(x)\big|^2\,du\right)^{1/2}, \; 0 < \alpha \le 1, \\[2ex] h^{-\alpha}\left(h^{-n}\displaystyle\int_{|u|<h}\big|F(x+u)-F(x)-u\cdot A(x)\big|^2\,du\right)^{1/2}, \\[1ex] \hspace{6cm} 1 < \alpha \le 2, \end{cases}$$

where $A(x)$ is a vector with n-components. To eliminate problems at the boundary, we shall also consider for harmonic $F(x,\,y)$, the function

$$\mathcal{P}_\alpha^*(F)\,(x) \;=\; \sup_{s,\,h>0} \begin{cases} h^{-\alpha}\left(h^{-n}\displaystyle\int_{|u|<h}\big|F(x+u,\,s)-F(x,\,s)\big|^2\,du\right)^{1/2}, \\[1ex] \hspace{6cm} 0 < \alpha \le 1, \\[2ex] h^{-\alpha}\Big(h^{-n}\displaystyle\int_{|u|<h}\big|F(x+u,\,s)-F(x,\,s) \\[1ex] \hspace{3cm} -\,u\cdot\nabla_1 F(x,\,s)\big|^2\,du\Big)^{1/2}, \; 1 < \alpha \le 2. \end{cases}$$

As usual, if $F(x,\,y)$ is the Poisson integral of its boundary values $F(x)$ and $\nabla_1 F(x,\,s)$ has a limit as $s \to 0$ and we take $A(x)$ in the definition of $\mathcal{P}_\alpha(F)\,(x)$ to be this limit, then $\mathcal{P}_\alpha(F)\,(x)$ and $\mathcal{P}_\alpha^*(F)\,(x)$ are pointwise equivalent.

The function $\mathcal{P}_\alpha^*(F)$ is closely related to a function $Z_\alpha^*(F)$ whose study is somewhat easier. We set

$$Z_\alpha^*(F)\,(x) \;=\; \sup_{s,\,h>0} \begin{cases} h^{-\alpha}\left(h^{-n}\displaystyle\int_{|u|<h}\big|F(x+u,\,|u|+s)-F(x+u,\,s)\big|^2\,du\right)^{1/2}, \\[1ex] \hspace{6cm} 0 < \alpha \le 1, \\[2ex] h^{-\alpha}\Big(h^{-n}\displaystyle\int_{|u|<h}\big|F(x+u,\,|u|+s)-F(x+u,\,s) \\[1ex] \hspace{2.5cm} -\,|u|\frac{\partial F}{\partial s}(x+u,\,|u|+s)\big|^2\,du\Big)^{1/2}, \; 1 < \alpha \le 2. \end{cases}$$

$Z_\alpha^*(F)$ is easier to localize than $\mathcal{P}_\alpha^*(F)$, and it is possible to make pointwise estimates on $Z_\alpha^*(F)$ that yield global results for $\mathcal{P}_\alpha^*(F)$. To write down the relation between $\mathcal{P}_\alpha^*(F)$ and $Z_\alpha^*(F)$, we let

$$N^{(\alpha)}(F)\,(x) \;=\; \sup_{\Gamma(x)} \begin{cases} y^{1-\alpha}\left|\dfrac{\partial F}{\partial y}(u,\,y)\right|, \; 0 < \alpha \le 1, \\[2ex] y^{2-\alpha}\left|\dfrac{\partial^2 F}{\partial y^2}(u,\,y)\right|, \; 1 < \alpha \le 2, \end{cases}$$

where $\Gamma(x) = \{(u,\,y) : |x-u| < ay\}$, a nontangential cone with suitably large aperture a. We then have

$$\mathscr{P}_\alpha^*(F)(x) \leq c[Z_\alpha^*(F)(x) + N^{(\alpha)}(F)(x)], \tag{1}$$

except that when $\alpha = 1$, $N^{(\alpha)}(F)(x)$ must be replaced by $N(|\nabla F|)(x)$, where N denotes the ordinary nontangential maximal function. Moreover, the roles of $\mathscr{P}_\alpha^*(F)(x)$ and $Z_\alpha^*(F)(x)$ may be interchanged in (1). The proof of (1) is essentially an application of the mean value theorem.

It is also possible to show that for an arbitrary harmonic function $F(x, y)$,

$$Z_\alpha^*(F)(x) \leq c \sup_{h>0} \left(\frac{1}{h^n} \int_{|u|<h} [N^{(\alpha)}(F)(x+u)]^{p_0} \, du \right)^{1/p_0}, \tag{2}$$

where $p_0 = 2n/(n + 2\alpha)$. This inequality together with its weighted version is the main result of [6] referred to earlier. The idea of its proof is to use the mean value property for harmonic functions to estimate $Z_\alpha^*(F)(x)$ in terms of the product of two "box" maximal functions

$$T_{\lambda, r}\left(y^{k-\alpha} \frac{\partial^k F}{\partial y^k} \right)(x), \quad k - 1 < \alpha \leq k, \ k = 1, 2, \ldots,$$

for two choices of each of the parameters λ and r; here

$$T_{\lambda, r}(F)(x) = \sup_{h>0} \left(h^{-\lambda n} \iint_{\substack{|x-u|<h \\ 0<y<h}} y^{\lambda n - n - 1} |F(u, y)|^r \, du \, dy \right)^{1/r}.$$

From this estimate, (2) follows by [1] or [17].

Inequality (2) in conjunction with (1) easily implies global norm inequalities for $\mathscr{P}_\alpha^*(F)$. For example, letting

$$\|f\|_p = \left(\int_{\mathbf{R}^n} |f(x)|^p \, dx \right)^{1/p}, \quad \|f\|_{p,\infty} = \sup_{t>0} t |\{x : |f(x)| > t\}|^{1/p},$$

$0 < p < \infty$, we have:

Theorem 1. *Let* $0 < \alpha < 2$, $F(x, y)$ *be harmonic in* $\mathbf{R}_+^{n+1}$, *and assume that* $\nabla_1 F(x, y) \to 0$ *as* $y \to \infty$ *if* $\alpha = 1$. *Then*

$$\|\mathscr{P}_\alpha^*(F)\|_p \le c\|N^{(\alpha)}(F)\|_p, \quad 2n/(n+2\alpha) < p < \infty,$$

$$\|\mathscr{P}_\alpha^*(F)\|_{p_0,\infty} \le c\|N^{(\alpha)}(F)\|_{p_0}, \quad p_0 = 2n/(n+2\alpha).$$

The assumption that $\nabla_1 F(x, y) \to 0$ as $y \to \infty$ if $\alpha = 1$ can be dropped if we replace $N^{(1)}(F)$ on the right by $N(|\nabla F|)$. For another estimate of the weak norm of $\mathscr{P}_\alpha^*(F)$, see Theorem 12.

For $\alpha > 0$, we define the αth (harmonic) derivative $F^{(\alpha)}(x, y)$ of $F(x, y)$ by

$$F^{(\alpha)}(x, y) = \frac{e^{-i\beta\pi}}{\Gamma(\beta)} \int_0^\infty \frac{\partial^m F}{\partial y^m}(x, y+s) s^{\beta-1}\,ds,$$

where $m = [\alpha] + 1$ and $m = \alpha + \beta$. In this and all that follows, we assume that $F(x, y)$ satisfies a decay condition that we shall refer to as the "cone condition." This may be stated as the requirement that there exist $x_0 \in \mathbf{R}^n$ and $\eta > 0$ such that

$$\sup_{\Gamma(x_0),\, y>1} y^\eta |F(u, y)| < \infty.$$

The aperture a of the cone $\Gamma(x_0)$ is taken to be fixed and sufficiently large. The cone condition ensures, for example, that $F^{(\alpha)}(x, y)$ exists and is harmonic, and that $F^{(k)}(x, y) = \dfrac{\partial^k}{\partial y^k} F(x, y)$ for positive integers k and

$$\left(F^{(\alpha_1)}\right)^{(\alpha_2)}(x, y) = F^{(\alpha_1+\alpha_2)}(x, y).$$

We also need to introduce $N(F^{(\alpha)})(x)$, the ordinary nontangential maximal function of $F^{(\alpha)}$, as well as both the Lusin area integral

$$S(F^{(\alpha)})(x) = \left(\iint_{\Gamma(x)} |F^{(\alpha+1)}(u, y)|^2 \frac{du\,dy}{y^{n-1}}\right)^{1/2}$$

and the Littlewood-Paley-Zygmund function

$$g_\lambda^*(F^{(\alpha)})(x) = \left(\iint_{\mathbf{R}_+^{n+1}} \left(\frac{y}{y+|x-u|}\right)^{\lambda n} |F^{(\alpha+1)}(u, y)|^2 \frac{du\,dy}{y^{n-1}}\right)^{1/2}.$$

It is not difficult to show that $N^{(\alpha)}(F)(x) \le cN(F^{(\alpha)})(x)$; if α is an integer, $N^{(\alpha)}(F)(x)$ and $N(F^{(\alpha)})(x)$ of course coincide.

SUFFICIENT CONDITIONS FOR THE FINITENESS OF MARCINKIEWICZ INTEGRALS AND RELATED GLOBAL RESULTS

Our aim now is to see how the Peano maximal function $\mathscr{P}_\alpha^*(F)$ controls $\mathscr{D}_\alpha^*(F)$. As a first step in doing this, one proves that for $F(x, y)$ satisfying the cone condition,

$$\mathscr{D}_\alpha^*(F)(x) \leq c[\mathscr{A}_\alpha(F)(x) + S(F^{(\alpha)})(x)],$$

$$\mathscr{A}_\alpha(F)(x) = \sup_{s > 0}\left[\int_{\mathbf{R}^n}\left(\int_0^{|u|} yF_{yy}(x + u, y + s)\,dy\right)^2 \frac{du}{|u|^{n+2\alpha}}\right]^{1/2}.$$

$$(3)$$

Truncated versions of (3) are also valid, and these are helpful for local results. The term $\mathscr{A}_\alpha(F)(x)$ is a tangential component that can be estimated pointwise in terms of $g_\lambda^*(F^{(\alpha)})(x)$ for $\lambda < 1 + (2\alpha/n)$ (see [20, 10, and 13]. However, the restriction that λ be strictly less than $1 + (2\alpha/n)$ causes a slight loss of precision in theorems obtained by this method. Sharper results can be gotten by estimating $\mathscr{A}_\alpha(F)$ in terms of $Z_\alpha^*(F)$ through a distribution function inequality in the spirit of Burkholder and Gundy [2]. For technical reasons, we consider an approximation to $\mathscr{A}_\alpha(F)$ defined for $0 < \varepsilon, R < \infty$, by

$$\mathscr{A}_\alpha(F)(x, R, \varepsilon) = \sup_{s > \varepsilon}\left(\int_{|v| < R}\left|\int_0^{|x-v|} yF_{yy}(v, y + s)\,dy\right|^2 \frac{dv}{|x - v|^{n+2\alpha}}\right)^{1/2}.$$

$\mathscr{A}_\alpha(F)(x, R)$ is defined similarly with $\varepsilon = 0$. Note that $\mathscr{A}_\alpha(F)(x, R, \varepsilon) \uparrow \mathscr{A}_\alpha(F)(x)$ as $\varepsilon \downarrow 0$ and $R \uparrow \infty$. The first distribution function estimate is as follows.

Theorem 2. Let $F(x, y)$ be a harmonic function in $\mathbf{R}_+^{n+1}$ that satisfies the cone condition, and let $\beta \gg 1$ and $0 < \gamma < 1$. Then for $0 < \alpha < 2$ and $t > 0$,

$$|\{x : \mathscr{A}_\alpha(F)(x, R, \varepsilon) > \beta t, \ Z_\alpha^*(F)(x) + S(F^{(\alpha)})(x) + N^{(\alpha)}(F)(x) \leq \gamma t\}|$$
$$\leq c(\gamma/\beta)^2|\{x : \mathscr{A}_\alpha(F)(x, R, \varepsilon) > t\}|,$$

where c depends only on α and n.

Now let us indicate the local and global results that follow from this.

Theorem 3. Let $F(x, y)$ be a harmonic function in $\mathbf{R}_+^{n+1}$ that satisfies the cone condition and for which $N(F)(x) \in L^2_{loc}(\mathbf{R}^n)$. If $E \subset \mathbf{R}^n$ and $0 < \delta < \infty$, then

(a) $\mathscr{D}^*_\alpha(F)(x, \delta)$ is finite a.e. in E if both $\mathscr{P}^*_\alpha(F)(x)$ and $N(F^{(\alpha)})(x)$ are finite in E, $0 < \alpha < 2$.

(b) If, in addition, F satisfies the semigroup property $F(\cdot, y_1 + y_2) = F(\cdot, y_1) * P(\cdot, y_2)$, $y_1, y_2 > 0$, where P denotes the Poisson kernel, then $\mathscr{M}^*_1(F)(x, \delta)$ is finite a.e. in E if $\mathscr{P}^*_1(F)(x)$ is finite in E.

This result yields the following basic corollary.

Theorem 4. Let $F(x) \in L^2_{loc}(\mathbf{R}^n)$, E be a subset of $\mathbf{R}^n$, and $0 < \delta < \infty$.

(a) If $\alpha = 1$ and $\mathscr{P}_1(F)(x, \delta)$ is finite in E, then $\mathscr{M}_1(F)(x, \delta)$ is finite a.e. in E.

(b) Let $0 < \alpha < 2$, $\alpha \neq 1$, and $m = [\alpha] + 1$. Assume that $F \in L^p(\mathbf{R}^n)$ for some p satisfying $1 \leq p < n/(m - \alpha)$, and let $I_{m-\alpha}(F)(x)$ be the fractional integral of F of order $m - \alpha$. If both $\mathscr{P}_\alpha(F)(x, \delta)$ and $\mathscr{P}_m(I_{m-\alpha}F)(x, \delta)$ are finite in E, then $\mathscr{D}_\alpha(F)(x, \delta)$ is finite a.e. in E.

The assumption that $F \in L^p(\mathbf{R}^n)$ is made to guarantee the existence of $I_{m-\alpha}F$. Part (a) is due to Stein and Zygmund [15 and 14]; part (b) for $n = 1$ was also proved by Stein and Zygmund [16].

To give a rough idea of how Theorem 3 is deduced from Theorem 2, let $\beta \to \infty$ in Theorem 2 to get for any $k > 0$,

$$\left| \{ x : \mathscr{A}_\alpha(F)(x, R, \varepsilon) = \infty, \; Z^*_\alpha(F)(x) + S(F^{(\alpha)})(x) + N^{(\alpha)}(F)(x) \leq k \} \right| = 0.$$

This means essentially by letting $\varepsilon \to 0$ that the finiteness of $Z^*_\alpha(F)$,

$S(F^{(\alpha)})$, and $N^{(\alpha)}(F)$ in a set E implies that $\mathscr{A}_\alpha(F)(x, R)$ is finite a.e. in E, and by the truncated form of (3), that $\mathscr{D}_\alpha^*(F)(x, \delta)$ is finite a.e. in E. Now assume only that $\mathscr{P}_\alpha^*(F)$ and $N(F^{(\alpha)})$ are finite in E. Using (1) with $\mathscr{P}_\alpha^*$ and Z_α^* interchanged, and theorems of Calderón and Stein (see [14]) relating the finiteness of non-tangential maximal functions and area integrals, it then follows that $Z_\alpha^*(F)$ and $S(F^{(\alpha)})$ are finite a.e. in E. Hence, the finiteness of $N(F^{(\alpha)})$ and $\mathscr{P}_\alpha^*(F)$ in E ensures that of $Z_\alpha^*(F)$, $S(F^{(\alpha)})$, and $N^{(\alpha)}(F)$ a.e. in E. The finiteness of $\mathscr{D}_\alpha^*(F)(x, \delta)$ a.e. in E then follows. For part (b) of Theorem 3, the semigroup condition allows us to show that $N(|\nabla_1 F|)(x) \leq c\mathscr{P}_1^*(F)(x)$. Hence, by using the results of Calderón and Stein, we can conclude that if $\mathscr{P}_1^*(F)$ is finite in E, then $Z_1^*(F)$, $N^{(1)}(F)$, and $S(F^{(1)})$ are finite a.e. in E, from which part (b) of Theorem 3 follows. Theorem 4 is now almost immediate if we first note that since its conclusions are local, we may assume $F \in L^2(\mathbf{R}^n)$ and then use the pointwise equivalences of $\mathscr{P}_\alpha^*$ and $\mathscr{P}_\alpha$ and of $\mathscr{D}_\alpha^*$ and $\mathscr{D}_\alpha$.

It should be remarked that the results of Stein and Zygmund for $\alpha = 1$ were originally formulated as follows: if $F(x)$ has a first Peano derivative in E, that is, if there is a vector $A(x)$ such that

$$\left(h^{-n}\int_{|u| < h}|F(x + u) - F(x) - u \cdot A(x)|^2\, du\right)^{1/2} = o(h) \text{ as } h \to 0 \quad (4)$$

for $x \in E$, then $\mathscr{M}_1(F)(x, \delta)$ is finite a.e. in E. In fact, using the Whitney extension theorem, Calderón and Zygmund show in [4] that (4) is equivalent a.e. to the analogous condition with o replaced by O. Thus (4) is actually equivalent a.e. to

$$\left(h^{-n}\int_{|u| < h}|F(x + u) - F(x)|^2\, du\right)^{1/2} = O(h), \ h \to 0, \quad (5)$$

which is the same as the condition that $\mathscr{P}_1(F)(x, \delta) < \infty$. We would also like to point out that the proofs of Theorem 2 and the rest of the results below do not use either the Whitney extension theorem or the harmonic derivative splitting theorem given in [14]. The role of the extension theorems thus appears to be extraneous to the mechanism by which $\mathcal{M}_\alpha^*(F)$ and $\mathcal{P}_\alpha^*(F)$ are related, but is a means of passing from 0 conditions to o conditions.

By a rather standard argument, one can deduce the following global result from Theorem 2.

Theorem 5.   Let $F(x, y)$ be harmonic and satisfy the cone condition.
Then for $0 < \alpha < 2$,

(a) $\|\mathcal{D}_\alpha^*(F)\|_p \leq c\|N(F^{(\alpha)})\|_p$, $2n/(n + 2\alpha) < p < \infty$,

(b) $\|\mathcal{D}_\alpha^*(F)\|_{p,\infty} \leq c(\|N(F^{(\alpha)})\|_{p,\infty} + \|\mathcal{P}_\alpha^*(F)\|_{p,\infty})$,
     $2n/(n + 2\alpha) \leq p < \infty$.

Part (a) was essentially already known (see [14]) and can be deduced from the technique based on g_λ^* that was mentioned earlier. Note that the right side of (a) is the H^p norm of $F^{(\alpha)}$. To obtain the theorem from Theorem 2 is simple; (2) lets us omit the norm of $Z_\alpha^*(F)$ in the strong-type result and [2] lets us omit the norm of $S(F^{(\alpha)})$.

In case the harmonic function $F(x, y)$ satisfies the semi-group property, it is possible to show that for $\alpha = 1$ the norms of $N(F^{(1)})$ are majorized by those of $\mathcal{P}_1^*(F)$. Similarly, under suitable conditions, if

$$I_\beta F(x, y) = \int_0^\infty F(x, y + s)s^{\beta-1}ds, \quad \beta > 0, \tag{6}$$

then for $0 < \alpha < 2$, $\alpha \neq 1$, the norms of $N(F^{(\alpha)})$ can be majorized by those of $\mathcal{P}_m^*(I_{m-\alpha}F)$, $m = [\alpha] + 1$. Thus, we obtain the following two results from the last theorem.

Theorem 6.   Let $F(x, y)$ be a harmonic function in $\mathbf{R}_+^{n+1}$ that satis-
fies the cone condition and the semigroup property.   Then

(a) $\|\mathcal{M}_1^*(F)\|_p \leq c\|\mathcal{P}_1^*(F)\|_p$, $2n/(n + 2) < p < \infty$,

(b) $\|\mathcal{M}_1^*(F)\|_{p,\infty} \leq c\|\mathcal{P}_1^*(F)\|_{p,\infty}$, $2n/(n+2) \leq p < \infty$.

Theorem 7. *Let* $0 < \alpha < 2$, $\alpha \neq 1$, *and* $m = [\alpha] + 1$. *If* $F(x, y)$ *is harmonic in* $\mathbf{R}_+^{n+1}$ *and satisfies the cone condition for some* $\eta > m - \alpha$ *and* $I_{m-\alpha}F(x, y)$ *satisfies the semigroup property, then*

(a) $\|\mathcal{D}_\alpha^*(F)\|_p \leq c\|\mathcal{P}_m^*(I_{m-\alpha}F)\|_p$, $2n/(n+2\alpha) < p < \infty$,

(b) $\|\mathcal{D}_\alpha^*(F)\|_{p,\infty} \leq c(\|\mathcal{P}_m^*(I_{m-\alpha}F)\|_{p,\infty} + \|\mathcal{P}_\alpha^*(F)\|_{p,\infty})$,

$2n/(n+2\alpha) \leq p < \infty$.

NECESSARY CONDITIONS FOR THE FINITENESS OF MARCINKIEWICZ INTEGRALS AND RELATED GLOBAL RESULTS

In this section, the object is to use Marcinkiewicz integrals to obtain information about Peano maximal functions. To do so, we need to study in a quantitative form the "desymmetrization principle" of Stein and Zygmund [15]. The Marcinkiewicz integral in effect controls the function $\mathcal{P}_\alpha^*(F)(x)$ in terms of two distribution function inequalities. To precisely write down these results, we first define the symmetric maximal function $m_\alpha^*(F)(x)$ by

$$m_\alpha^*(F)(x) = \sup_{h,s>0} h^{-\alpha}\left(h^{-n}\int_{|u|<h} \left|F(x+u, s) + F(x-u, s) - 2F(x, s)\right|^2 du\right)^{1/2}, \quad 0 < \alpha < 2.$$

Clearly, $m_\alpha^*(F)(x) \leq \mathcal{M}_\alpha^*(F)(x)$. Moreover, if $F(x, y)$ is the Poisson integral of $F(x)$, then $m_\alpha^*(F)$ and $m_\alpha(F)$ are pointwise equivalent.

The first distribution function inequality shows that $\mathcal{M}_\alpha^*(F)$ controls $N(F^{(\alpha)})$. Actually, to be more precise, $\mathcal{M}_\alpha^*(F)$ controls $g_\lambda^*(F^{(\alpha)})$ for $\lambda > 1 + (2\alpha/n)$. For technical reasons, we study an approximation to $g_\lambda^*(F^{(\alpha)})$, namely, for $0 < \varepsilon < R < \infty$, the function

$$g_{k-\alpha,\lambda}^*(F^{(\alpha)})(x, R, \varepsilon) = \left(\iint_{\substack{|u|<R \\ \varepsilon<y<R}} \frac{y^{\lambda n + 2(k-\alpha-1)}}{(y+|x-u|)^{\lambda n}} |F^{(k)}(u, y)|^2 \frac{du\,dy}{y^{n-1}}\right)^{1/2}.$$

One can show (see [12]) that if we define

$$g^*_{k-\alpha,\,\lambda}(F^{(\alpha)})(x) = \lim_{R\uparrow\infty,\,\varepsilon\downarrow 0} g^*_{k-\alpha,\,\lambda}(F^{(\alpha)})(x,\,R,\,\varepsilon),$$

then $g^*_\lambda(F^{(\alpha)})$ and $g^*_{k-\alpha,\,\lambda}(F^{(\alpha)})$ are pointwise equivalent.

Theorem 8. *Let $F(x,\,y)$ be harmonic in $\mathbf{R}^{n+1}_+$ and satisfy*

$$\frac{\partial^k F}{\partial y^k}(x,\,y+s) = F(\cdot,\,s) * \frac{\partial^k P}{\partial y^k}(\cdot,\,y),\quad y,\,s > 0$$

for some $k \geq 2$. Given $0 < \alpha < 2$, let

$$M_t = \{x : \mathscr{M}^*_\alpha(F)(x) > t\},\quad t > 0,$$

and let $\chi^*_{M_t}$ denote the Hardy-Littlewood maximal function of χ_{M_t}. Then for $\lambda > 1 + (2\alpha/n)$, β and N suitably large, and $0 < \gamma < 1$,

$$\left|\left\{x : g^*_{k-\alpha,\,\lambda}(F^{(\alpha)})(x,\,R,\,\varepsilon) > \beta t,\ \chi^*_{M_{\gamma t}}(x) \leq \frac{1}{N}\right\}\right|$$

$$\leq c\left(\frac{\gamma}{\beta}\right)^2 |\{x : g_{k-\alpha,\,\lambda}(F^{(\alpha)})(x,\,R,\,\varepsilon) > t\}|,$$

where c depends only on k, α, λ, and n.

The second result describes how $Z^*_\alpha(F)$ and $m^*_\alpha(F)$ are related. To state it, we define a truncated version of $Z^*_\alpha(F)$ by

$$Z^*_\alpha(F)(x,\,R,\,\varepsilon) = \sup_{\substack{h>0 \\ \varepsilon < s < R}} \begin{cases} h^{-\alpha}\left(h^{-n}\displaystyle\int_{\substack{|v|<R \\ |v-x|<h}} \Big| F(v,\,|v-x|+s) \\ \qquad\qquad - F(v,\,s)\Big|^2\,dv\right)^{1/2},\ 0 < \alpha \leq 1, \\[4ex] h^{-\alpha}\left(h^{-n}\displaystyle\int_{\substack{|v|<R \\ |v-x|<h}} \Big| F(v,\,|v-x|+s) - F(v,\,x) \\ \qquad - |v-x|F_y(v,\,|v-x|+s)\Big|^2\,dv\right)^{1/2}, \\ \qquad\qquad\qquad\qquad\qquad\qquad 1 < \alpha \leq 2. \end{cases}$$

Note that $Z^*_\alpha(F)(x,\,R,\,\varepsilon) \uparrow Z^*_\alpha(F)(x)$ as $R \uparrow \infty$ and $\varepsilon \downarrow 0$.

Theorem 9. *Let $F(x,\,y)$ be harmonic in $\mathbf{R}^{n+1}_+$ and satisfy the cone condition (even with $\eta = 0$). Given $0 < \alpha < 2$, let*

$$H_t = \{x : m^*_\alpha(F)(x) + N^{(\alpha)}(F)(x) > t\},\quad t > 0,$$

except that when $\alpha = 1$, $N^{(\alpha)}(F)$ *is replaced by* $N(|\nabla F|)$. *Let* $\chi^*_{H_t}$ *denote the Hardy-Littlewood maximal function of* χ_{H_t}. *Then for suitably large* β *and* N *and* $0 < \gamma < 1$,

$$\left| \left\{ x : Z^*_\alpha(F)(x, R, \varepsilon) > \beta t, \ \chi^*_{H_{\gamma t}}(x) \leq \frac{1}{N} \right\} \right|$$

$$\leq c \left(\frac{\gamma}{\beta} \right)^2 |\{ x : Z^*_\alpha(F)(x, R, \varepsilon) > t \}|,$$

where c *depends only on* α *and* n.

In essence, Theorem 8 implies that the finiteness of $\mathcal{M}_\alpha(F)$ in a set E implies the finiteness of $N(F^{(\alpha)})$ a.e. in E, while Theorem 9 shows us that if $N^{(\alpha)}(F)$ and $m_\alpha(F)(x)$ are finite in E, then $Z_\alpha(F)$ is finite a.e. in E. From (1), we can then conclude that $\mathcal{P}_\alpha(F)(x)$ is finite a.e. in E. This strategy enables us to prove the following result.

Theorem 10. *Let* $F(x) \in L^2_{loc}(\mathbf{R}^n)$, $E \subset \mathbf{R}^n$, *and* $0 < \delta < \infty$.

   (a)   *If* $\alpha = 1$ *and* $\mathcal{M}_1(F)(x, \delta)$ *is finite in* E, *then* $\mathcal{P}_1(F)(x, \delta)$ *is finite a.e. in* E.

   (b)   *If* $0 < \alpha < 2$, $\alpha \neq 1$, *let* $m = [\alpha] + 1$. *Assume* $F \in L^p(\mathbf{R}^n)$ *for some* p *satisfying* $1 \leq p < n/(m - \alpha)$ *and let* $I_{m-\alpha}(F)(x)$ *be the fractional integral of* F *of order* $m - \alpha$. *If* $\mathcal{M}_\alpha(F)(x, \delta)$ *is finite in* E, *then both* $\mathcal{P}_\alpha(F)(x, \delta)$ *and* $\mathcal{P}_m(I_{m-\alpha}F)(x, \delta)$ *are finite a.e. in* E.

Part (a) was proved for $n = 1$ in [15] and for $n > 1$ in [14]. For $n = 1$, under the stronger assumption that $\mathcal{D}_\alpha(F)(x, \delta)$ is finite in E, part (b) was proved in [16]. It follows by combining Theorems 4 and 10 that for $F \in L^2_{loc}$, the finiteness of $\mathcal{M}_\alpha(F)(x, \delta)$ is equivalent a.e. to the finiteness of $\mathcal{D}_\alpha(F)(x, \delta)$.

We now consider the global conclusions to be drawn from the two distribution function estimates in this section. From the first one, we get the following result.

Theorem 11. *Let $F(x, y)$ be harmonic in $\mathbf{R}_+^{n+1}$ and satisfy the cone condition and $F^{(k)}(\cdot, y + s) = F(\cdot, y) * \frac{\partial^k P}{\partial s^k}(\cdot, s)$, $y, s > 0$ for some $k \geq 2$. If $0 < \alpha < 2$ and $\lambda > 1 + (2\alpha/n)$, then*

(a) $\|g_\lambda^*(F^{(\alpha)})\|_p \leq c\|\mathscr{M}_\alpha^*(F)\|_p$, $2/\lambda < p < \infty$,

(b) $\|g_\lambda^*(F^{(\alpha)})\|_{p, \infty} \leq c\|\mathscr{M}_\alpha^*(F)\|_{p, \infty}$, $2/\lambda \leq p < \infty$.

Part (a) was essentially already known by combining norm inequalities for the Littlewood-Paley functions g and g_λ^*.

   Theorem 9 yields the following result.

Theorem 12. *Let $F(x, y)$ be harmonic in $\mathbf{R}_+^{n+1}$ and satisfy the cone condition (even for $\eta = 0$). Then for $0 < \alpha < 2$,*

$$\|\mathscr{P}_\alpha^*(F)\|_{p, \infty} \leq c(\|m_\alpha^*(F)\|_{p, \infty} + \|N^{(\alpha)}(F)\|_{p, \infty}), \quad 2n/(n + 2\alpha) \leq p < \infty.$$

The corresponding strong-type statement for $2n/(n + 2\alpha) < p < \infty$ is omitted, since it gives no information not already included in Theorem 1.

   We may combine the last two results with the fact that $m_\alpha^*(F)(x) \leq \mathscr{M}_\alpha^*(F)(x)$ to get:

Theorem 13. *Let $F(x, y)$ be harmonic in $\mathbf{R}_+^{n+1}$ and satisfy the cone condition and $F^{(k)}(\cdot, y + s) = F(\cdot, y) * \frac{\partial^k P}{\partial s^k}(\cdot, s)$, $y, s > 0$, for some $k \geq 2$. If $0 < \alpha < 2$, then*

(a) $\|N(F^{(\alpha)})\|_p + \|\mathscr{P}_\alpha^*(F)\|_p \leq c\|\mathscr{M}_\alpha^*(F)\|_p$, $2n/(n + 2\alpha) < p < \infty$,

(b) $\|N(F^{(\alpha)})\|_{p, \infty} + \|\mathscr{P}_\alpha^*(F)\|_{p, \infty} \leq c\|\mathscr{M}_\alpha^*(F)\|_{p, \infty}$,
      $2n/(n + 2\alpha) \leq p < \infty$.

Part (a) actually contains no information not already given in Theorems 11 and 1 and is only included for completeness. Also, if one omits the norms of $\mathscr{P}_\alpha^*(F)$, the statements are valid for $0 < p < \infty$.

We summarize the main inequalities listed in Theorems 5 and 13 in the following corollary.

Corollary 14. Let $F(x, y)$ *be harmonic in* $\mathbf{R}_+^{n+1}$ *and satisfy the cone condition and the semigroup property* $F(\cdot, y + s) = F(\cdot, y) * P(\cdot, s)$, y, $s > 0$. *Then for* $0 < \alpha < 2$,

(a) $\left\| \mathscr{D}_\alpha^*(F) \right\|_p \approx \left\| \mathscr{M}_\alpha^*(F) \right\|_p \approx \left\| N(F^{(\alpha)}) \right\|_p$, $2n/(n + 2\alpha) < p < \infty$,

   $\left\| \mathscr{M}_1^*(F) \right\|_p \approx \left\| \mathscr{P}_1^*(F) \right\|_p$, $2n/(n + 2) < p < \infty$,

(b) $\left\| \mathscr{D}_\alpha^*(F) \right\|_{p,\infty} \approx \left\| \mathscr{M}_\alpha^*(F) \right\|_{p,\infty} \approx \left\| N(F^{(\alpha)}) \right\|_{p,\infty} + \left\| \mathscr{P}_\alpha^*(F) \right\|_{p,\infty}$, $2n/(n + 2\alpha) \leq p < \infty$,

   $\left\| \mathscr{M}_1^*(F) \right\|_{p,\infty} \approx \left\| \mathscr{P}_1^*(F) \right\|_{p,\infty}$, $2n/(n + 2) \leq p < \infty$.

If $\alpha \neq 1$, $F(x, y)$ satisfies the cone condition for some $\eta > m - \alpha$, where $m = [\alpha] + 1$, and $I_{m-\alpha}F$ satisfies the semigroup property, then by using Theorem 1 and the facts

$$\mathscr{P}_m^*(I_{m-\alpha}F) \leq c[\mathscr{P}_\alpha^*(F) + N(\left| \nabla^m I_{m-\alpha}F \right|)]$$

and

$$N\left(\frac{\partial^m}{\partial y^m} I_{m-\alpha}F \right) = cN(F^{(\alpha)}),$$

the term $N(F^{(\alpha)})$ may be replaced by $\mathscr{P}_m^*(I_{m-\alpha}F)$ in both (a) and (b) above.

We point out that if $F(x, y)$ is the Poisson integral of $F(x)$, then by the pointwise equivalences mentioned earlier, one can drop the *'s in all the preceding global results. In particular, if $F = I_\alpha f$ in the sense of (6) for $f \in H^p$, $p < n/\alpha$, then $F^{(\alpha)} = f$ and we get

$$\left\| \mathscr{D}_\alpha(F) \right\|_p \approx \left\| \mathscr{M}_\alpha(F) \right\|_p \approx \left\| f \right\|_{H^p}, \quad 2n/(n + 2\alpha) < p < n/\alpha,$$

and

$$\left\| \mathscr{D}_\alpha(F) \right\|_{p_0,\infty} \approx \left\| \mathscr{M}_\alpha(F) \right\|_{p_0,\infty} \leq c\left\| f \right\|_{H^{p_0}}, \quad p_0 = 2n/(n + 2\alpha).$$

For the last statement, we have also used Theorem 1. See also [19].

L^q-ANALOGUES AND WEIGHTS

It is not necessary to restrict attention to the L^2 metric in the results above. For $0 < \alpha < 2$ and $0 < q < \infty$, define

$$\mathscr{M}_{\alpha,q}^*(F)(x) = \sup_{s>0} \left(\int_{\mathbf{R}^n} \frac{|F(x+u,s) + F(x-u,s) - 2F(x,s)|^q}{|u|^{n+q\alpha}} \, du \right)^{1/q} .$$

Similarly, for $0 < \alpha \leq 1$, let

$$\mathscr{P}_{\alpha,q}^*(F)(x) = \sup_{s,h>0} h^{-\alpha} \left(h^{-n} \int_{|u| < h} |F(x+u,s) - F(x,s)|^q \, du \right)^{1/q},$$

$$Z_{\alpha,q}^*(F)(x) = \sup_{s,h>0} h^{-\alpha} \left(h^{-n} \int_{|u| < h} |F(x+u, |u|+s) \right.$$
$$\left. - F(x+u,s)|^q \, du \right)^{1/q} .$$

Analogous definitions can be given for $1 < \alpha \leq 2$. It is then possible to derive versions of Theorem 2 for $2 < q < \infty$ and of Theorem 9 for $0 < q < \infty$. Also, inequality (2) holds for $0 < q < \infty$ if we replace $2n/(n + 2\alpha)$ by $qn/(n + q\alpha)$. The estimate (1) holds for $0 < q < \infty$ too. Local and global results can be obtained from these facts. For example, we have:

Theorem 15. Let $F(x, y)$ be harmonic in $\mathbf{R}_+^{n+1}$ and satisfy the cone condition and the semigroup property. If $0 < \alpha < 2$ and $2 < q < \infty$, then

(a) $\left\| \mathscr{D}_{\alpha,q}^*(F) \right\|_p + \left\| \mathscr{D}_{\alpha}^*(F) \right\|_p \approx \left\| \mathscr{M}_{\alpha,q}^*(F) \right\|_p + \left\| \mathscr{M}_{\alpha}^*(F) \right\|_p \approx \left\| N(F^{(\alpha)}) \right\|_p$,
    $qn/(n + q\alpha) < p < \infty$,

(b) $\left\| \mathscr{D}_{\alpha,q}^*(F) \right\|_{p,\infty} + \left\| \mathscr{D}_{\alpha}^*(F) \right\|_{p,\infty} \approx \left\| \mathscr{M}_{\alpha,q}^*(F) \right\|_{p,\infty} + \left\| \mathscr{M}_{\alpha}^*(F) \right\|_{p,\infty}$
    $\approx \left\| N(F^{(\alpha)}) \right\|_{p,\infty} + \left\| \mathscr{P}_{\alpha,q}^*(F) \right\|_{p,\infty}$,
    $qn/(n + q\alpha) \leq p < \infty$.

The global results have weighted versions too if we replace $\| \cdot \|_p$ and $\| \cdot \|_{p,\infty}$ by their respective weighted analogues, namely,

$$\|f\|_{L^p_w} = \left(\int_{\mathbf{R}^n} |f(x)| \, p \, w(x) \, dx \right)^{1/p}$$

and

$$\|f\|_{L^p_w, \infty} = \sup_{0 < t < \infty} t^{-1} \left(\int_{\{x : |f(x)| > t\}} w(x) \, dx \right)^{1/p}.$$

The weight w is assumed to be in A_∞ and, for example in (a) of the preceding theorem, the restriction $p_0 < p < \infty$ [where $p_0 = qn/(n + q\alpha)$] is replaced by $\mu p_0 < q$ and $\mu p_0 < p$, where μ is the doubling order of w: that is, where

$$\int_{B_1} w(x) \, dx \leq c \left(\frac{|B_1|}{|B_2|} \right)^\mu \int_{B_2} w(x) \, dx$$

for all balls B_1 and B_2 with $B_2 \subset B_1$. In statement (b), which still has a term involving $\mathscr{P}^*_{\alpha, q}(F)$, the restriction $p_0 \leq p < \infty$ is replaced simply by

$$\sup_{0 < t < 1} t^p \int_{|x| < t^{-p_0/n}} w(x) \, dx < \infty.$$

REFERENCES

1. Barker, S. R. An inequality for measures on a half-space. *Math. Scand.* 44 (1979):92–102.

2. Burkholder, D. L., and Gundy, R. F. Distribution function inequalities for the area integral. *Studia Math.* 44 (1972): 527–544.

3. Calderón, A. P. Estimates for singular integral operators in terms of maximal functions. *Studia Math.* 44 (1972):167–186.

4. Calderón, A. P., and Zygmund, A. Local properties of solutions of elliptic partial differential equations. *Studia Math.* 20 (1961):171–225.

5. Chanillo, S., and Wheeden, R. L. Distribution function estimates for Marcinkiewicz integrals and differentiability. To appear in *Duke Math J.*

6. Chanillo, S., and Wheeden, R. L.   Inequalities for Peano maximal functions and Marcinkiewicz integrals.   To appear.

7. DeVore, R., and Sharpley, R.   Maximal operators.   To appear.

8. Fefferman, C. L.   Inequalities for strongly singular convolution operators.   *Acta. Math.* 124 (1970):9-36.

9. Gatto, A. B. E.; Jiménez, J. R.; and Segovia, C.   On the solutions of the equation $\Delta^m F = f$ for $f \in H^p$.   This volume, p. 394.

10. Hirschman, I. I., Jr.   Fractional integration.   *Amer. J. of Math.* 75 (1953):531-546.

11. Marcinkiewicz, J.   Sur quelques integrals du type de Dini.   *Ann. Soc. Polonaise Math.* 17 (1938):42-50.

12. Segovia, C., and Wheeden, R. L.   On certain fractional area integrals.   *J. Math. Mech.* (now *Indiana Univ. Math. J.*) 19 (1969):247-262.

13. Stein, E. M.   The characterization of functions arising as potentials I.   *Bull. Amer. Math. Soc.* 67 (1961):102-104.

14. Stein, E. M.   *Singular Integrals and Differentiability Properties of Functions*. Princeton Math. Series No. 30.   Princeton, N.J.: Princeton Univ. Press, 1970.

15. Stein, E. M., and Zygmund, A.   On the differentiability of functions.   *Studia Math.* 23 (1964):247-283.

16. Stein, E. M., and Zygmund, A.   On the fractional derivatives of functions.   *Proc. London Math. Soc.* 14A (1965):249-264.

17. Torchinsky, A.   Weighted norm inequalities for the Littlewood-Paley function g_λ^*.   *Proc. Symp. Pure Math.* 35, part 1 (1979): 125-131.

18. Waterman, D.   On an integral of Marcinkiewicz.   *Proc. Internat. Congress of Mathematicians* 2 (1954):185-186.

19. Weiss, G.   Lecture notes.

20. Zygmund, A.   On certain integrals.   *Trans. Amer. Math. Soc.* 55 (1944):170-204.

SOME RESULTS AND OPEN
PROBLEMS IN DIFFERENTIATION OF INTEGRALS

Miguel de Guzmán
Universidad Complutense de Madrid

The purpose of this brief note is to present in an expository way some developments in the theory of differentiation of integrals in which one can strongly feel the influence of A. Zygmund, in their origin, their style, and their motivation. For more technical details the reader is referred to the recent work of the author in 1981.

The theory of differentiation, born from the fundamental theorem of Lebesgue in 1910, has expanded into two main branches, one more abstract and global, in the direction of the Lebesgue-Radon-Nikodym theorem, and the other more local and concrete. It has been in the latter direction where the influence of Zygmund has been more strong and decisive. Through the problems he has raised, some solved by himself, some by others, he has indicated one direction of research that has proved extremely useful, both for the development of the theory in itself and for the links that have been established with other branches of the modern real analysis and in particular with the real variable developments of the last thirty years in Fourier analysis.

1

When one considers the classical theorem of Lebesgue about the differentiation of the integral of a function f in $L^1(\mathbf{R}^n)$ through cubic
intervals or through spheres containing the point at which one
differentiates, several questions arise in a natural way. Restrict-
ing ourselves to $\mathbf{R}^2$ for the sake of clarity, one can ask, for
example: (1) What happens if instead of square intervals one considers rectangles?; (2) How about rectangles in one fixed direction for each point x of the plane, such directions varying from
point to point?

These two simple questions, their surprising answers, in a
good part given by Zygmund, and the subsequent questions and developments they have stimulated have strongly marked the ulterior
research in the field.

2

In 1927, Nikodym constructed a set N of measure 1 in the unit
square Q of $\mathbf{R}^2$ so that for each point $x \in N$ there is a straight
line $\ell(x)$ such that $\ell(x) \cap N = \{x\}$. From this fact Zygmund, as
Nikodym points out at the end of his paper, deduced in an easy way
that the Lebesgue differentiation theorem is false when one takes,
instead of square intervals, rectangles containing the point at
which one differentiates, and this is so even if one imposes that
$f \in L^\infty(\mathbf{R}^2)$; that is, the Lebesgue density theorem does not hold for
arbitrary rectangles. Moreover, one observes from the same fact of
the existence of the Nikodym set that even if one fixes for each
point x of $\mathbf{R}^2$ a single direction $d(x)$ and takes the means of the
integral $\int f$ for $f \in L^\infty(\mathbf{R}^2)$ over the rectangles containing x in that
direction $d(x)$, the differentiation property may completely fail.

3

Some years after this result came the strong density theorm of Saks
in 1933 and the more complete results of Zygmund [18] and Jessen,
Marcinkiewicz, and Zygmund [4] telling that rectangular parallele-
pipeds in $\mathbf{R}^n$ are a good basis for the differentiation of $\int f$, $F \in L \times$
$(\log^+ L)^{n-1}(\mathbf{R}^n)$, provided one fixes the direction of their axes.

The general philosophy underlying many of the questions proposed by Zygmund in differentiation has been inspired by these
results. A differentiation basis in $\mathbf{R}^n$ can have too many sets to
present good differentiation properties, not differentiating the
integrals of a desirable class of functions, like, for instance, L^p.
A restriction imposed on the geometry of the basis or a restriction
on the space of functions whose integrals we would like to differentiate, or both, can give an acceptable theorem of differentiation.

4

An instance of the foregoing situation is the following. According
to the Jessen-Marcinkiewicz-Zygmund theorem, if in $\mathbf{R}^n$ one considers
the basis of intervals with *no restriction* on their sides, one
obtains the differentiation of $\int f$, for $f \in L(\log^+ L)^{n-1}(\mathbf{R}^n)$. What
will happen if we consider the basis of intervals in $\mathbf{R}^n$ with s
($s \leq n$) of their sides of the same length, the others being arbitrary? This was the question raised and answered by Zygmund in
1967. This basis differentiates $L(\log^+ L)^{n-s}(\mathbf{R}^n)$.

5

In this same spirit was raised by Zygmund the question about the
role that a space of functions in $\mathbf{R}^2$ like $L(\log^+ L)^{1/2}(\mathbf{R}^2)$ can play

in differentiation. Perhaps can one say that a basis of intervals of side lengths d, D, with $1 > D > d > D^2$, differentiates $L(\log^+ L)^{1/2}(\mathbf{R}^2)$? This question was answered by R. Moriyón in 1975. Such a basis does not differentiate any space worse in the appropriate sense than $L(\log^+ L)(\mathbf{R}^2)$.

However, the question about the possible role in differentiation of a space like $L(\log^+ L)^{1/2}(\mathbf{R}^2)$ remains still open. Can one construct a differentiation basis such that it differentiates $L(\log^+ L)^{1/2}(\mathbf{R}^2)$ and no worse space?

6

Another problem with the same flavor suggested by Zygmund about bases of parallelepipeds was the following: Let $\phi(s_1, s_2, \ldots, s_{n-1})$ be a function from $[0, \infty) \times [0, \infty) \times \cdots \times [0, \infty)$ to $[0, \infty)$, which is separately increasing in each variable with $\phi(0, 0, \ldots, 0) = 0$. Consider the basis of all intervals in $\mathbf{R}^n$ with side lengths s_1, $s_2, \ldots, s_{n-1}, \phi(s_1, \ldots, s_{n-1})$. What are the differentiation properties of this basis? One good guess, looking at the Jessen-Marcinkiewicz-Zygmund theorem, could be that it differentiates $L(\log^+ L)^{n-2}(\mathbf{R}^n)$. That for $n = 3$ it is so has been proved by A. Córdoba in 1978. The question for $n > 3$ is still open.

7

Another interesting question proposed by Zygmund was connected with the following result of Saks [13]: there exists a function f in $L^1(\mathbf{R}^2)$ such that the basis of intervals of $\mathbf{R}^2$ differentiates $\int f$ at almost no point x of $\mathbf{R}^2$ (the upper derivative of $\int f$ at almost each x is $+\infty$). The question, both natural and practical, proposed by Zygmund is now the following: given $g \in L^1(\mathbf{R}^2)$, is it possible to

choose a direction so that the basis of all rectangles in such
direction differentiates $\int g$ [that is, the derivative of $\int g$ at almost
each $x \in \mathbf{R}^2$ is $g(x)$]? If the answer were positive, for a given g,
a change of axes would ease many considerations. However, the
answer, found by Marstrand in 1977, is negative. The situation is
similar in some other more general cases studied by El Helou [2]
and L. Melero [7].

8

The main method by which one has traditionally attacked problems in
the local theory of differentiation of integrals has been to obtain
a covering lemma in some form similar to the classical one of
Vitali [16] by which Lebesgue obtained his theorem. Its connection
with the Hardy-Littlewood maximal theorem and the further manipula-
tion of this latter tool by an adequate process of iteration have
given rise to most of the positive results in the field. Zygmund
has been very deeply conscious of this strong dependence of the
theory on geometric facts and sought himself and stimulated others
to seek methods in differentiation not involving covering lemmas.
Negative results such as the one of Nikodym quoted in Sec. 2, the
one of Saks in Sec. 3, or that of Marstrand in Sec. 7 did not
involve covering methods, but these tools proved rather ineffective
to produce positive results and to attack some of the many interesting open problems in differentiation. For this reason, one can say
that the recent methods involving the application of powerful techniques of Fourier analysis to solve problems that still seem far
from being successfully tackled by other more classical geometric
considerations come to fulfill Zygmund's expectations. Such
methods were first introduced by Nagel, Rivière, and Wainger [9,
10, and 11] and were later perfected by Stein and Wainger [15].

It is now justifiably expected that such methods may give a
complete answer to the question formulated in Sec. 1 (2): Let be
given for each point x of $\mathbf{R}^2$ a direction $d(x)$. For each x consider
the family $B(x)$ of all rectangles centered at x, one of whose sides
has direction $d(x)$. What are the differentiation properties of this
basis? The basis can be very bad, as the Nikodym set shows, but if
the field of directions varies in a sufficiently smooth way, the
basis is rather good. This problem and many others connected with
it have been successfully explored by Stein and Wainger by means of
their useful tools.

REFERENCES

1. Córdoba, A. $s \times t \times \phi(s, t)$. Mittag-Leffer Institute, Report
    No. 9 (1978).

2. El Helou, J. Recouvrement du tore T^q par des ouverts
    aléatoires. Preprint.

3. de Guzmán, M. Real variable methods in Fourier analysis.
    North-Holland Mathematics Studies 46, Amsterdam, 1981.

4. Jessen, B.; Marcinkiewicz, J.; and Zygmund, A. Note on the
    differentiability of multiple integrals. *Fund. Math.* 25 (1935):
    217-234.

5. Lebesgue, H. Sur l'integration des fonctions discontinues.
    Ann. Sci. Ecole Norm. Sup. 27 (1910):361-450.

6. Marstrand, J. M. A counter-example in the theory of strong
    differentiation. *Bull. London Math. Soc.* 9 (1977):209-211.

7. Melero, B. L. A negative result in differentiation. To be
    published in *Studia Math*.

8. Moriyón, R. On the derivation properties of a class of bases,
    Appendix III. In M. de Guzmán, *Differentiation of Integrals
    in* $\mathbf{R}^n$. Berlin: Springer, 1975.

9. Nagel, A.; Rivière, N. M.; and Wainger, S.   On Hilbert trans-
    forms along curves. *Bull. Amer. Math. Soc.* 80 (1974):106-108.

10. Nagel, A.; Rivière, N. M.; and Wainger, S.   On Hilbert trans-
    forms along curves, II. *Amer. J. Math.* 98 (1976):395-403.

11. Nagel, A.; Rivière, N. M.; and Wainger, S.   A maximal function
    associated to the curve (t, t^2). *Proc. Nat. Acad. Sci.* 73
    (1976):1416-1417.

12. Nikodym, O.   Sur la mesure des ensembles plans dont tous les
    points sont rectilinéairement accessibles. *Fund. Math.* 10
    (1927):116-168.

13. Saks, S.   On the strong derivatives of functions of an
    interval. *Fund. Math.* 25 (1935):235-252.

14. Saks, S.   Théorie de l'integrale. Warszawa, 1933.

15. Stein, E. M., and Wainger, S.   Problems in harmonic analysis
    related to curvature. *Bull. Amer. Math. Soc.* 84 (1978):1239-
    1295.

16. Vitali, G.   Sui gruppi di punti e sulle funzioni di variabile
    reali. *Atti Accad. Sci. Torino* 43 (1908):75-92.

17. Zygmund, A.   A note on the differentiability of multiple
    integrals. *Colloq. Math.* 16 (1967):199-204.

18. Zygmund, A.   On the differentiability of multiple integrals.
    Fund. Math. 23 (1934):143-149.

TANGENTIAL BOUNDARY BEHAVIOR
OF HARMONIC EXTENSIONS OF L^p POTENTIALS

Alexander Nagel , Walter Rudin
University of Wisconsin

Joel H. Shapiro
Michigan State University

1. INTRODUCTION

We describe, mostly without proofs, some results on the boundary
behavior of harmonic functions in classes modeled on the space $\mathfrak{D}$ of
functions harmonic in the open unit disc with finite Dirichlet integral. Detailed proofs will appear elsewhere [6].

For our purposes, $\mathfrak{D}$ is best regarded as the space of Poisson
integrals of functions f square integrable on the unit circle T,
with the additional restriction:

$$\sum_{-\infty}^{\infty} |n| \, |\hat{f}(n)|^2 < \infty, \tag{1}$$

where $\hat{f}(n)$ is the nth Fourier coefficient of f. Such f are somewhat
more regular than "typical" L^2 functions, but still not necessarily
continuous, or even bounded. We study how the additional regularity
of f affects the boundary behavior of its harmonic extension u.

This type of problem was considered by Salem and Zygmund
[7], who showed that the Fourier series of each $f \in L^2(T)$ satisfying (1) converges at each point of T, with the possible exception of

Research partially supported by the National Science
Foundation.

a set of logarithmic capacity zero. This implies that each $u \in \mathcal{D}$
has a radial limit at each point of T, with the possible exception
of such a set. In fact, the same is true for nontangential limits.

Our contribution is to show that if larger classes of
exceptional sets—intermediate between "log-capacity zero" and
"measure zero"—are allowed, then the functions in $\mathcal{D}$ will converge
within regions that meet the unit circle *tangentially*. We describe
the precise relationship between the curvature of the approach
region and the size of the exceptional sets for convergence within
these regions. In particular, we show that each $u \in \mathcal{D}$ has at
almost every $\zeta \in T$ a limit as $z \to \zeta$ through an approach region
making *exponential* contact with T at ζ.

To make these matters precise, consider the following
regions in the open unit disc U. If $c > 0$ and $\gamma \geq 1$, let

$$\mathcal{A}_{\gamma,c}(\varphi) = \left\{ re^{i\theta} \in U : 1 - r > c \left| \sin\left(\frac{\theta - \varphi}{2}\right) \right|^{\gamma} \right\},$$

while if $\gamma > 0$, let

$$\mathcal{E}_{\gamma,c}(\varphi) = \left\{ re^{i\theta} \in U : 1 - r > \exp\left[-c \left| \sin\left(\frac{\theta - \varphi}{2}\right) \right|^{-\gamma} \right] \right\}.$$

Thus $\mathcal{A}_{\gamma,c}(\varphi)$ has *order of contact* γ with T at $e^{i\varphi}$, while $\mathcal{E}_{\gamma,c}(\varphi)$
has *exponential contact*. Note that the regions $\mathcal{A}_{1,c}(\varphi)$ are the
usual nontangential approach regions.

We say a function u defined in U has $\mathcal{A}_{\gamma}$-*limit* L at $e^{i\varphi}$ if
$u(z) \to L$ as $z \to e^{i\varphi}$ within $\mathcal{A}_{\gamma,c}(\varphi_0)$ for every $c > 0$. A similar
definition applies to $\mathcal{E}_{\gamma}$-limits. We can state the classical results
in this language: if $f \in L^2(T)$ and u is the Poisson integral of f,
then u has an $\mathcal{A}_1$-limit at almost every point of T. If, in addition,
f satisfies (1), then u has an $\mathcal{A}_1$-limit at every point of T, with
the possible exception of a set of logarithmic capacity zero. Our
results for the space $\mathcal{D}$ can be stated as follows.

Theorem 1. *Suppose* $u \in \mathfrak{D}$.   *Then*

(a)   *U has an $\mathcal{E}_1$-limit at almost every point of T.*

(b)   *If, in addition, $0 < \beta < 1/2$ and $\gamma = (1 - 2\beta)^{-1}$, then U has an $\mathcal{Q}_\gamma$-limit at every point of T, with the possible exception of a set of $C_{\beta,2}$-capacity zero.*

Here the capacity $C_{\beta,2}$ is the analogue for the unit circle of the corresponding Bessel capacity on R^n (see [5], for example). It coincides with the classical capacity $C_{1-2\beta}$ discussed by Kahane and Salem in [3; Chapter 3, p. 33]. In particular, $C_{1/2,2}$ is the logarithmic capacity, so when $\beta = 1/2$, our result coincides with the previously mentioned one of Salem and Zygmund.

The correct setting for this work is a more general one, motivated by the fact that $\mathfrak{D} = P[K * L^2]$, where P is the Poisson integral for the unit disc,

$$K(\theta) = \left| \sin \frac{\theta}{2} \right|^{-1/2} \sim \sum_{-\infty}^{\infty} (|n| + 1)^{-1/2} e^{in\theta},$$

and $*$ denotes the convolution on T. We prove a generalization of Theorem 1 valid for the classes $P[K * L^p]$ where $1 \leq p < \infty$ and K is a positive, integrable, even function on $[-\pi, \pi]$ that is decreasing on $(0, \pi]$. As a by-product of our work, we answer a question of H. S. Shapiro and A. L. Shields concerning the zeros of holomorphic functions in classes $P[K * L^2]$. These results are stated in detail in the next section.

As the reader has probably guessed, our tangential convergence theorems follow from weak-type estimates on maximal functions associated with our approach regions. For $p > 1$, we use Hansson's strong-type capacity inequality [2] to obtain strong-type maximal estimates, which are in turn crucial to the proof of part (b) of Theorem 1 and its generalizations. We state these results precisely in Secs. 3 and 4. In Sec. 5 we show how the strong-type maximal estimates enter into the proof of Theorem 1, part (b); and in Sec. 6 we discuss Carleson measures for the classes $P[K * L^p]$.

2. TANGENTIAL CONVERGENCE THEOREMS

We prefer to work in the upper half-space R_+^{n+1} of $(n + 1)$-dimensional Euclidean space, instead of the unit disc; so R^n replaces the unit circle as the boundary. We write L^p for $L^p(R^n)$ and denote points of R_+^{n+1} by (x, y) with $x \in R^n$ and $y > 0$. P will denote the Poisson integral for R_+^{n+1}, with $\{P_y : y > 0\}$ the corresponding Poisson kernels on R^n. $*$ will denote convolution on R^n.

Potential Spaces and Dirichlet-Type Spaces

Suppose K is a *kernel* on R^n; that is, K is positive, integrable, radially symmetric, and $K(x)$ decreases as $|x|$ increases. For $1 \leq p < \infty$, let

$$L_K^p = \{K * F : F \in L^p\}$$

denote the space of L^p potentials associated with K, and let

$$h_K^p = P[f] : f \in L_K^p$$

be the corresponding space of harmonic extensions to R_+^{n+1}. Every h_K^p is thus an analogue of the Dirichlet space $\mathfrak{D}$ of Sec. 1. In particular, it follows from Plancherel's theorem that h_K^2 consists of the harmonic extensions of functions $f \in L^2$ satisfying the additional conditions

$$\int_{R^n} |\hat{K}(\lambda)|^{-2} |\hat{f}(\lambda)|^2 d\lambda < \infty \qquad (2)$$

where $\char`^$ denotes the Fourier transform on L^2.

To avoid trivialities, we always assume that $K \notin L^q$, where $p^{-1} + q^{-1} = 1$. Thus L_K^p, and therefore h_K^p, always contains unbounded functions.

Approach Regions

For $(x, y) \in R_+^{n+1}$, let

$$K_y(x) = P[K](x, y) = P_y \star K(x)$$

and set

$$r(y) = r_{K, p}(y) = \|K_y\|_q^{-p/n}, \quad y > 0.$$

For $x_0 \in R^n$ and $\beta > 0$, define

$$\Omega(x_0) = \Omega_{K, \beta}^p(x_0) = \{(x, y) \in R_+^{n+1} : |x - x_0| < \beta r(y)\}.$$

Thus $\Omega(x_0)$ is the region in R_+^{n+1} with spherical cross section of
radius $\beta r(y)$ at height y above R^n. Since $K \notin L^q$, we know that
$r(y) \to 0$ as $y \to 0^+$, so $\Omega(x_0)$ approaches R^n only at the point x_0
(its "vertex"). Moreover, standard estimates of Poisson integrals
show that $y^{-1}r(y) \to 0$ as $y \to 0^+$, so the boundary of $\Omega(x_0)$ actually
approaches R^n *tangentially* at x_0.

We define Ω_K^p *limits* of functions defined on R_+^{n+1} exactly
as in the last section.

Capacity

Following Meyers [5], if $1 < p < \infty$, we define the (K, p)-capacity of
a subset E of R^n as follows. Let $T_{K, p}(E)$ denote those nonnegative
$F \in L^p$ for which $K \star F \geq 1$ everywhere on E. Note that since K and F
are both positive, the convolution $K \star F$ make sense (possibly $+\infty$)
at every point of R^n.

The *capacity* of E is

$$C_{K, p}(E) = \inf\{\|F\|_p^p : F \in T_{K, p}(E)\},$$

where $\|\cdot\|_p$ denotes the L^p norm.

It is easy to check that $C_{K, p}$ is subadditive and monotone
increasing on the subsets of R^n, and that a subset E has capacity
zero if and only if $K \star F \equiv \infty$ on E for some nonnegative F in L^p
[5, Sec. 2].

This last comment shows that the members of the potential space L_K^p are defined and finite at $C_{K,p}$-almost every point of R^n.

Main Results

We can now state our main results on tangential convergence.   In what follows, $1 \le p < \infty$, $p^{-1} + q^{-1} = 1$, $K \notin L^q$, $f \in L_K^p$, and $u = P[f]$.

Theorem 2. *There is a set E of Lebesgue measure zero such that u has Ω_K^p limit $f(x)$ at x for every $x \in R^n \backslash E$.*

Theorem 3. *If, in addition, $p > 1$ and $K = H * G$ where H and G are also kernels, then there is a set E with $C_{H,p}(E) = 0$ such that u has Ω_G^p limit $f(x)$ at x for each $x \in R^n \backslash E$.*

Thus Theorem 3 shows precisely how the degree of tangential convergence influences the size of the exceptional set.

Examples: Bessel Potentials

The most important class of kernels are the Bessel kernels $K = g_\alpha$ defined for $0 < \alpha \le n$ by

$$\hat{g}_\alpha(\lambda) = (1 + |\lambda|^2)^{-\alpha/2}, \quad \lambda \in R^n.$$

The corresponding capacities $B_{\alpha,p}$ make sense for $\alpha p \le n$ and are called *Bessel capacities*. Observe that $g_\alpha * g_\beta = g_{\alpha+\beta}$. Properties of these kernels are worked out in detail in (for example) Meyers [5, Sec. 7, p. 279]. The analogous kernels for the unit circle are the ones considered in Kahane and Salem [3, Chapter 3]:

$$\Phi_\alpha(\theta) = \left| \sin \frac{\theta}{2} \right|^{\alpha-1} \sim \sum (|n| + 1)^{-\alpha} e^{in\theta}$$

for $0 < \alpha < 1$, and

$$\Phi_1(\theta) = -\log \left| \sin \frac{\theta}{2} \right| \sim \sum (|n| + 1)^{-1} e^{in\theta}.$$

As $y \to 0^+$, we have the following asymptotic estimates on the radius $r_{g_\alpha, p}(y) = r_{\alpha, p}(y)$, defined in "Approach Regions" of Sec. 2:

$$r_{\alpha, p}(y) \sim \begin{cases} y^{1-(\alpha p/n)} & \text{if } \alpha p < n \\ (-\log y)^{-1/n(q-1)} & \text{if } \alpha p = n, \, p > 1 \\ (-\log y)^{-1/n} & \text{if } \alpha = n, \, p = 1 \end{cases} \tag{3}$$

where $p^{-1} + q^{-1} = 1$. By analogy with the work of Sec. 1, we define approach regions for $c > 0$: for $\gamma \geq 1$,

$$\mathcal{a}_{\gamma, c} = \{(x, y) \in R_+^{n+1} : y > c|x - x_0|^\gamma\},$$

while if $\gamma > 0$, then

$$\mathcal{E}_{\gamma, c} = \{(x, y) \in R_+^{n+1} \quad y > \exp(-c|x - x_0|^{-\gamma})\}.$$

Then estimate (3) shows that the regions $\Omega_{g_{\alpha, \beta}}^p$ essentially coincide with the classes of regions:

$$\mathcal{a}_{\gamma, c} \text{ if } \alpha p < n, \text{ where } \gamma = n/(n - \alpha p),$$
$$\mathcal{E}_{\gamma, c} \text{ if } \alpha p = n \text{ and } p > 1, \text{ where } \gamma = n(q - 1),$$

and

$$\mathcal{E}_{n, c} \text{ if } \alpha = n \text{ and } p = 1.$$

Thus Theorems 2 and 3 have the following corollaries, which generalize Theorem 1. Here $f \in L_{g_\alpha}^p$ where $\alpha p \leq n$, so $u = P[f] \in h_{g_\alpha}^p$; and "a.e." refers to Lebesgue measure on R^n.

Corollary 1.

   (a)   *If $\alpha p < n$ and $p > 1$, then u has $\mathcal{a}_{n/(n-\alpha p)}$-limit $f(x)$ at a.e. x in R^n.*

   (b)   *If $\alpha p = n$ and $p > 1$, then u has $\mathcal{E}_{n(q-1)}$-limit $f(x)$ at a.e. x in R^n.*

   (c)   *If $\alpha = n$ and $p = 1$, then U has $\mathcal{E}_n$-limit $f(x)$ at a.e. x in R^n.*

Corollary 2.   Suppose in addition to the hypotheses above, that $1 < p < \infty$, $\alpha p \leq n$, and $\alpha = \tau + \kappa$, where τ and κ are positive numbers. Then u has $\alpha_{n/(n-\tau p)}$-limit $f(x)$ at each $x \in R^n$ with the possible exception of a set of $B_{\kappa,p}$ capacity zero.

Remark. If we return to the unit circle and the case $p = 2$, then as noted earlier, the capacity $B_{\alpha,2}$ corresponds to the classical capacity $C_{1-2\alpha}$ of Kahane and Salem [3, Chapter 3] for $0 < \alpha \leq 1/2$, with the understanding that C_0 is logarithmic capacity. Note that the index in Kahane and Salem's capacity refers to the exponent associated with the kernel, while the index α of the Bessel capacity refers to the exponent associated with the Fourier transform of the kernel.

Application to Zeros of h_κ^p Functions

The next result generalizes one proved by H. S. Shapiro and A. L. Shields in the case $n = 1$, $p = 2$ for special kernels and holomorphic functions [8, Theorem 3], and it answers a question posed by them [8, p. 224].

Theorem 4. Suppose $1 \leq p < \infty$ and suppose $(y_j)_1^\infty$ is a sequence of positive numbers with $\sum r_{K,p}^n(y_j) = \infty$. Then there exists a sequence $(x_j)_1^\infty$ in R^n such that no nontrivial function in the class h_K^p vanishes at each point (x_j, y_j) of R_+^{n+1}.

Proof. The hypothesis on (y_j) ensures that we can choose open balls B_j of radius $r_{K,p}(y_j)$ such that each point of R^n lies in infinitely many B_j. Let x_j be the center of B_j and set $z_j = (x_j, y_j) \in R_+^{n+1}$. If $x \in R^n$, then x belongs to some sequence $B_{j_1}, B_{j_2}, \ldots$ of balls; hence $z_{j_1}, z_{j_2}, \ldots$ all belong to $\Omega_{K,1}^p(x)$. So if $u \in h_K^p$ vanishes at each z_j, then it has Ω_K^p limit zero at each x for which it has an Ω_K^p limit. By Theorem 2 this happens for almost every x in R^n. Since u is the Poisson integral of its boundary function, this implies $u \equiv 0$, which completes the proof.

3. TANGENTIAL MAXIMAL FUNCTIONS: WEAK-TYPE INEQUALITIES

If u is a complex valued function defined on R_+^{n+1}, define for $1 \leq p < \infty$ and $\beta > 0$:

$$\mathfrak{M}_{K,p,\beta} u(x_0) = \sup\{|u(x, y)| : (x, y) \in \Omega_{K,\beta}^p(x_0)\}.$$

If $u = P[f]$ we will also write this as $\mathfrak{M}_{K,p,\beta} f(x_0)$. Our fundamental estimate on this tangential maximal function comes from a direct comparison with the following L^p-variant of the Hardy-Littlewood maximal function. For $f \in L^p$ and $x_0 \in R^n$, let

$$M_p F(x_0) = \sup_{r > 0} \left\{ \frac{1}{m(B_r(x_0))} \int_{B_r(x_0)} |F|^p \, dm \right\}^{1/p},$$

where $B_r(x_0)$ is the ball of radius r in R^n, centered at x_0, and m is Lebesgue measure on R^n. Clearly M_p is subadditive, and by the usual Hardy-Littlewood maximal theorem, it is of weak type (p, p). We have:

Lemma. *Suppose* $1 \leq p < \infty$ *and* $\beta > 0$. *Then there exists* $A = A(K, p, \beta) > 0$ *such that if* $f = K * F$ *for* $F \in L^p$, *then*

$$\mathfrak{M}_{K,p,\beta} f(x_0) \leq A M_p F(x_0)$$

for every $x_0 \in R^n$.

This lemma immediately gives the following weak-type estimate, which by standard arguments yields Theorem 2.

Theorem 5. *For each* $1 \leq p < \infty$ *and* $\beta > 0$ *there exists* $A = A(K, p, \beta) < \infty$ *such that for every* $f = K * F$ *with* $F \in L^p$:

$$m\{x \in R^n : \mathfrak{M}_{K,p,\beta} f(x) > \lambda\} \leq \left(\frac{A \|F\|_p}{\lambda} \right)^p$$

for every $\lambda > 0$.

Theorem 5 is "best possible" in the following sense. $\mathfrak{M}_{K,p,\beta}$ is not of weak type (r, r) for any $r < p$. Moreover, if Ω is any region in R_+^{n+1} that approaches R^n only at 0, and $\Omega(x_0) = \Omega + x_0$, then the Ω-maximal function

$$M_\Omega f(x_0) = \sup\{|u(x, y)| : (x, y) \in \Omega(x_0)\}$$

$(u = P[f])$ is of weak type (p, p) on L_K^p only if there is a "slab" in R_+^{n+1}:

$$S = \{(x, y) \in R_+^{n+1} : y < y_0\},$$

and a $\beta > 0$ such that

$$\Omega \cap S \subset \Omega_{K,\beta}^p(0) \cap S.$$

We remark that the lemma is a consequence of the following convolution inequality, whose proof is fairly straightforward.   If $1 \leq p < \infty$, then there exists $A = A(n, p) < \infty$ such that if $F \in L^p$ and K is a nonnegative, radial *decreasing* function on R^n, then

$$|K * F(x)| \leq AM_pF(x_0)\left[|x - x_0|^{n/p}\|K\|_q + \|K\|_1\right]$$

for all x and x_0 in R^n.

The lemma follows from this inequality upon replacing K by K_y and taking the supremum of both sides for $|x - x_0| < r_{K,p}(y)$ and $y > 0$.

4. STRONG-TYPE ESTIMATES

For $1 < p < \infty$ we have the following improvement of Theorem 5.

Theorem 6.   *Suppose $1 < p < \infty$ and $\beta > 0$.   Then there exists $A = A(K, p, \beta) < \infty$ such that if $f = K * F$ for $F \in L^p$, then $\mathfrak{M}_{K,p,\beta}f \in L^p$ and $\|\mathfrak{M}_{K,p,\beta}f\| \leq A\|F\|_p$.*

Since M_p is not of strong type (p, p), this result shows that $\mathfrak{M}_{K,p,\beta}$ is not "equivalent" to M_p if $1 < p < \infty$.   As we will see in Sec. 5, Theorem 6 is the main step in the proof of Theorem 3.

Toward the Proof of Theorem 6

We need the following geometry. For $x_0 \in R^n$, let

$$\Gamma(x_0) = \{(x, y) \in R_+^{n+1} : |x - x_0| < y\}$$

denote the right circular cone in R_+^{n+1} with vertex x_0; and for u continuous on R_+^{n+1}, let

$$Nu(x_0) = \sup\{|u(x, y)| : (x, y) \in \Gamma(x_0)\}$$

be the usual nontangential maximal function. For E an open subset of R^n, let

$$S(E) = R_+^{n+1} \setminus \bigcup_{x \in E} \Gamma(x),$$

the usual "Carleson set" over E, and for F an open subset of R_+^{n+1}, $1 \le p < \infty$, and $\beta > 0$, let

$$J(F) = J_{K, \beta}^p(F) = \{x \in R^n : \Omega_{K, \beta}^p(x) \cap F \ne \emptyset\}.$$

So $J(F)$ is a sort of "shadow" of F on R^n. Then we have the following two results, the first of which is an interesting geometric lower bound for capacity. Assume $1 < p < \infty$ and $\beta > 0$.

Proposition 1. There exists $A = A(K, p, \beta) < \infty$ such that for every open set $E \subset R^n$:

$$m\{J_{K, \beta}^p(S(E))\} \le AC_{K, p}(E).$$

Proposition 2. If u is continuous in R_+^{n+1} and $1 < p < \infty$, then for every $\lambda > 0$:

$$m\{\mathfrak{M}_{K, p, \beta} u > \lambda\} \le AC_{K, p}\{Nu > \lambda\}$$

where $A = A(K, p, \beta) < \infty$.

Proposition 2 follows from Proposition 1 upon observing that if $E = \{Nu > \lambda\}$, then $J_{K, \beta}^p(S(E))$ is an open set containing $\{\mathfrak{M}_{K, p, \beta} u > \lambda\}$. Proposition 1 follows from the weak-type estimate of

Theorem 5 and the fact that if $F \in T_{K,p}(E)$, then (since $K * F \geq 1$ on E) $P[K * F] \geq \lambda_0$ on $S(E)$, where λ_0 depends only on the dimension n and not on E. We describe a similar argument in more detail in the next section.

Finally, we require:

Hansson's Strong-Type Estimate [2, Theorem 2.4]. *If* $1 < p < \infty$, *then there exists* $A = A(K, p) < \infty$ *such that for each* $F \in L^p$:

$$\int_0^\infty C_{K,p}\{|K * F| > \lambda\}d(\lambda^p) \leq A\|F\|_p^p.$$

Special cases of this result have previously been obtained by Adams [1] and Maz'ya [4]. Note that the estimate is a significant improvement over the trivial weak-type estimate

$$C_{K,p}\{|K * F| > \lambda\} \leq \left(\frac{\|F\|_p}{\lambda}\right)^p, \tag{4}$$

which is just a statement of the fact that for every $F \in L^p$ and $\lambda > 0$, the function $|F|/\lambda$ belongs to $T_{K,p}(\{|K * F| > \lambda\})$.

Proof of Theorem 6

Suppose $f = K * F$ for $F \in L^p$, and write $u = P[f]$, $v = P[F]$. Then it is easy to see that $Nu \leq K * Nv$, so

$$\begin{aligned}
\|\mathfrak{M}_{K,p,\beta}u\|_p^p &= \int_0^\infty m\{\mathfrak{M}_{K,p,\beta}u > \lambda\}d(\lambda^p) \\
&\leq A\int_0^\infty C_{K,p}\{Nu > \lambda\}d(\lambda^p) \qquad \text{(Prop. 2)} \\
&\leq A\int_0^\infty C_{K,p}\{K * Nv > \lambda\}d(\lambda^p) \\
&\leq A\|Nv\|_p^p \qquad \text{(Hansson's estimate)} \\
&\leq A\|F\|_p^p,
\end{aligned}$$

where the last inequality follows from the Hardy-Littlewood non-tangential maximal theorem.

5. CAPACITY VS. CURVATURE: PROOF OF THEOREM 3

Recall that Theorem 3 describes how the curvature of our approach regions influences the capacity of the exceptional sets for convergence of h_K^p functions within these regions. Theorem 3 follows by standard arguments from part (b) of the next result, which in turn follows from part (a) and Hansson's estimate.

Theorem 7.   Suppose $1 < p < \infty$, $\beta > 0$, and $K = H \star G$, where H and G are also kernels.

(a)   *Then there exists $A = A(p, \beta, G, H) < \infty$ such that*

$$C_{H,p}\{\mathfrak{M}_{G,p,\beta}u > \lambda\} \leq AC_{K,p}\{Nu > \lambda\}$$

for every u continuous on R_+^{n+1} and every $\lambda > 0$.

(b)   *If, in addition, $u = P[K \star F]$ where $F \in L^p$, then*

$$\int_0^\infty C_{H,p}\{\mathfrak{M}_{G,p,\beta}u > \lambda\}d(\lambda^p) \leq A\|F\|_p^p$$

for A as in part (a).

Note that part (a) above is a capacitary analogue of Proposition 2 of the last section. It follows from a similar analogue of Proposition 1, which we state and prove in detail.

Proposition 3.   Under the hypotheses of Theorem 7 we have for every open subset E of R^n:

$$C_{H,p}\{J_{G,\beta}(S(E))\} \leq AC_{K,p}(E),$$

where $A = A(p, \beta, G, H) < \infty$.

Proof.   Write $M_G = M_{G,\beta,p}$ and $J_G = J_{G,\beta}^p$. Suppose $F \in T_{K,p}(E)$ and write $u = P[K \star F]$ and $v = P[G \star F]$. Since $K \star F \geq 1$ on E, we know that $u \geq \lambda_0$ on $S(E)$, where $\lambda_0 > 0$ is independent of E. Thus

$$J_G(S(E)) \subset \{\mathfrak{M}_G u > \lambda_0\}$$

and so

$$C_{H,p}\{J_G(S(E))\} \leq C_{H,p}\{\mathfrak{M}_G u > \lambda_0\}$$

$$= C_{H,p}\{\mathfrak{M}_G P[H * G * F] > \lambda_0\}$$

$$\leq C_{H,p}\{H * \mathfrak{M}_G P[G * F] > \lambda_0\}$$

$$\leq \left(\frac{\|\mathfrak{M}_G v\|_p}{\lambda_0}\right)^p \qquad (v = P[G * F])$$

$$\leq \left(\frac{A\|F\|_p}{\lambda_0}\right)^p,$$

where the next-to-last inequality follows from the trivial weak-type estimate (4) of Sec. 4, and the last inequality follows from our strong-type estimate Theorem 6. Take the infimum of the right side of the above inequality as F ranges through $T_{K,p}(E)$. The proof is complete.

6. CARLESON MEASURES

We call a finite positive Borel measure μ on R_+^{n+1} a *Carleson measure* for h_K^p if

$$\int |u|^p \, d\mu < \infty$$

for every $u \in h_K^p$. Stengenga [9] has characterized the (K, p) Carleson measures for $1 < p < \infty$ as follows: they are precisely those μ for which there exists $A > 0$ such that

$$\mu(S(E)) \leq AC_{K,p}(E) \qquad (5)$$

for every open subset E of R^n. Stengenga also gives examples that show that if (5) holds merely for open balls E, then μ need not be a Carleson measure.

Using Theorem 6 and standard arguments, we can obtain a more geometric condition on μ that is *sufficient* for it to be a Carleson measure for h_K^p. To state our result, we define for every open subset E of R^n a sort of "(K, p, β)-Carleson region" over E:

$$Q_{K,\beta}^{p}(E) = R_{+}^{n+1} \setminus \bigcup_{x \notin E} \Omega_{K,\beta}^{p}(x).$$

Theorem 8.   Suppose $1 < p < \infty$, $\beta > 0$; and suppose there exists $A < \infty$ such that

$$\mu\{Q_{K,\beta}^{p}(B)\} \leq Am(B)$$

for every open ball B in R_{+}^{n+1}.   Then μ is a Carleson measure for h_{K}^{p}.

However, this condition is *not* necessary.   Using an idea communicated to us by Stengenga, we can show that for each kernel K and $1 < p < \infty$, there exists a Carleson measure μ for h_{K}^{p} that does *not* satisfy the hypothesis of the theorem.

REFERENCES

1.   Adams, D. R.   On the existence of capacitary strong type estimates in R^{n}.   *Ark. Mat.* 14 (1976):125-140.

2.   Hansson, K.   Imbedding theorems of Sobolev type in potential theory.   *Math. Scand.* 45 (1979):77-102.

3.   Kahane, J. P., and Salem, R.   Ensembles parfaits et séries trigonométriques.   *Actualités Sci. Ind.* No. 1301.   Paris: Hermann, 1963.

4.   Maz'ya, V. G.   On capacitary estimetes of strong type for fractional norms.   In Russian.   *Zap. Sem. LOMI Leningrad* 70 (1977): 161-168.

5.   Meyers, N. G.   A theory of capacities for potentials of functions in Lebesgue classes.   *Math. Scand.* 26 (1970):255-292.

6.   Nagel, A.; Rudin, W.; and Shapiro, J. H.   Tangential boundary behavior of functions in Dirichlet-type spaces.   *Annals of Math.* 115 (1982).

7.   Salem, R., and Zygmund, A.   Capacity of sets and Fourier series.   *Trans. Amer. Math. Soc.* 59 (1946):23-41.

8. Shapiro, H. S., and Shields, A. L. On the zeros of functions
    with finite Dirichlet integral, and some related function
    spaces. *Math. Zeit.* 80 (1962):217-229.

9. Stengenga, D. A. Multipliers of the Dirichlet space. *Ill. J.
    Math.* 24 (1980):113-139.

STRONG DIFFERENTIALS IN L^p

G.V. Welland
University of Missouri, St. Louis

In the paper [1], A. Córdoba has shown that a conjecture of
A. Zygmund is true. This conjecture was made in 1935 and roughly
states that for the purpose of differentiation of the integral, the
differentiation basis B_ψ, which consists of rectangles oriented
with sides parallel to a fixed set of coordinate axes and having
side lengths that vary with one of s, t, and $\phi(s, t)$, behaves in
the same way as B_2, the differentiation basis that allows rectangles oriented as in B_ϕ with at most two distinct side lengths. The
function ψ increases monotonically from 0 in both the variables, s
and t. In [3], M. Weiss investigated strong differentiation in L^p.
Among the facts that she established is the following theorem.

*Theorem 1. If $k \geq 1$, $p \geq 1$, and f has a (k, p) differential in E,
it has a $(k, p)''$ differential almost everywhere in E.*

We recommend that the reader see [3 and 4] for many related
results on differentials. The point we will make is that Theorem 1
relies on a lemma that involves the differentiation basis B_2 and,
just as Córdoba has shown in the above case, B_2 can be replaced by
B_ϕ for this case as well. The consequence is that Theorem 1 can
now be stated with $(k, p)''$ differential replaced by $(k, p)\phi$

differential. However, the result of Córdoba is not required and one can follow more established lines of argument.

Definition 1. The function f has a (k, p) differential at x if there is a polynomial $P(t)$ of degree k such that

$$\left(\frac{1}{|Q|} \int_Q |f(x + t) - P(t)|^p \, dt\right)^{1/p} = o(h^k) \text{ as } h \to 0, \tag{1}$$

where Q is a cube in $\mathbf{R}^n$ containing the origin with side length h.

Here and in the following, f is assumed to be in the appropriate Lebesgue class, in this case L^p_{loc}.

Let R be a rectangle containing the origin and have side length s for sides parallel to the axes corresponding to coordinates $x_1, \ldots, x_{j_1}$; side length, t for sides parallel to the axes corresponding to coordinates $x_{j_1+1}, \ldots, x_{j_2}$; and side length $\phi(s, t)$ for sides parallel to the axes corresponding to the coordinates $s_{j_2+1}, \ldots, x_n$. Let ω be the largest of these side lengths.

Definition 2. The function f has a $(k, p)_\phi$ differential at x if there is a polynomial $P(t)$ of degree k such that

$$\left(\frac{1}{|R|} \int_R |f(x + t) - P(t)|^p \, dt\right)^{1/p} = o(w^k) \text{ as } w \to 0, \tag{2}$$

where R is a rectangle described above.

If $\phi(s, t) \equiv s$, then a $(k, p)_\phi$ differential becomes a $(k, p)''$ differential. For what follows, we allow cubes and rectangles like those just described to be translated. The resulting rectangles form the differentiation basis B_ϕ. The lemma from which Theorem 1 can be obtained with $(k, p)_\phi$ replacing $(k, p)''$ is now given. The use of this lemma can be found in [4].

Lemma. There is a positive constant A depending only on the dimension n, j_1, j_2, and $\alpha > 0$ having the following property. Let $g(x) = g(x_1, \ldots, x_n)$ be defined on a cube Q_0, nonnegative and integrable. Let U be the set of points in Q_0 that are contained in a cube Q of side length h such that

$$\int_Q g(\xi)d\xi > h^{n+\alpha}. \tag{3}$$

Let V be the set of points in Q_0 that are contained in a rectangle $R \in B_\phi$ having longest side w such that

$$\int_R g(\xi)d\xi > w^\alpha|R|. \tag{4}$$

Then

$$|V| \leq A|U|. \tag{5}$$

Before we go to the proof, let $r = \phi(s, t)$. A rectangle can have three possible configurations, in the sense that we have that r is between the values of s and t, $r \geq \max(s, t)$ or $r \leq \min(s, t)$. We would like to consider only one situation among these; that is the case $r < t < s$. If we require that $\phi(s, t)$ be continuous and strictly monotone, then the case $t < r < s$ can be described as $\Psi_1(r, s) < r < s$ where Ψ_1 inverts the equation $r = \phi(s, t)$ by solving for t. Similarly, $t < s < r$ can be reduced to $t < \Psi_2(t, r) < r$ where Ψ_2 inverts the equation $r = \phi(s, t)$ by solving for s. In both of these cases, the resulting Ψ_i is also continuous and strictly increasing. Hence, we may assume $\phi(s, t) < t < s$ without loss of generality.

The proof of (5) follows very closely that of [3]. We will be short in describing the proof up to the point where it differs from that of [3].

As in [3], we may assume that there exists a sequence of cubes contained in Q_0 that are nonoverlapping, $\{Q_k\}$, with edge lengths h_1, h_2, ..., such that

$$1 < h_k^{-(n+\alpha)}\int_{Q_k} g(x)dx \leq 2^{n+\alpha}, \quad k = 1, 2, ..., \tag{6}$$

and $g(x) = 0$ almost everywhere on the complement of $\cup Q_k$. Again, as in [3], we have $\cup Q_k \subset U$ and we write $Q_k = Q_k' \times Q_k''$, where Q_k' is the j_1-dimensional cube that has sides parallel to those axes corresponding to the edge lengths s in B_ϕ. Let J_k'' be the cube obtained by expanding Q_k'' about its center by a factor γ, where γ is the smallest odd integer larger than 3 for which

$$(\gamma - 1)^\alpha \geq 2^{n+1+2\alpha} 3^n.$$

Let $\hat{Q}_k$ be Q_k expanded about its center by a factor of 3. Then V_2 is defined to be $V \cap \bigcup_k \hat{Q}_k$ and $V_1 = V - V_2$.   One has

$$|V_2| \leq 3^n |U| \tag{7}$$

and with $V_1^{x''} = \{x' : (x', x'') \in V_1\}$,

$$|V_1| = \int_{Q_0''} |V_1^{x''}| \, dx''. \tag{8}$$

For most of the following, let x'' be fixed.   Then $x' \in V_1^{x''}$ if and only if $(x', x'') \notin \cup \hat{Q}_k$ and there exists an $R = R' \times R''$ with longest edge length s and R' is a j_1-dimensional cube with edge length s such that

$$\int_R g(\xi) \, d\xi > s^{j_1 + \alpha} |R''|, \text{ where } |R''| = t^{j_2} (\phi(s, t))^{n - j_1 - j_2}. \tag{9}$$

By the elementary Vitali covering theorem, there exist a fixed constant $B > 0$ and a finite collection of rectangles R satisfying (9) for which the R''s are disjoint and

$$B \sum |R'| \geq |V_1^{x''}|. \tag{10}$$

Again following [3], we obtain

$$s^{j_1 + \alpha} |R''| \leq 2 \sum_1 \int_{R \cap Q_k} g(\xi) \, d\xi, \tag{11}$$

where $\sum_1$ is the sum over those k for which $x'' \in J_k''$.

Let

$$g_k(\xi'') = \begin{cases} \int_{Q_k'} g(\xi', \xi'') \, d\xi' & \text{if } \xi'' \in Q_k'', \\ 0, & \text{otherwise}, \end{cases}$$

and for a fixed s let

$$g_k^* \ (x'') = \sup\left\{\frac{1}{|R''|}\int_{R''} g_k \ (\xi'')d\xi'' : R'' \text{ with } x'' \ \epsilon \ R''\right\}.$$

For a fixed k there are only 2^{j_1} R's for which $Q_k \cap R \neq \phi$. Let S_k be the set of values of s corresponding to the edge length of R' for which $Q_k \cap R \neq \phi$, where, if an s is repeated it is counted with its multiplicity in S_k. Note that S_k has a cardinality less than or equal to 2^{j_1}. Now we have

$$s^{j_1 + \alpha} \leq 2 \sum_{Q_k \cap R \neq \phi} \frac{1}{|R''|}\int_{R'' \cap Q_k''} g_k \ (\xi'')d\xi''$$

$$= 2 \sum_{Q_k \cap R \neq \phi} \sum_{s \in S_k} \frac{1}{|R''|}\int_{R'' \cap Q_k''} g_k \ (\xi'')d\xi''$$

$$\leq 2 \sum_{Q_k \cap R \neq \phi} \sum_{s \in S_k} g_k^{*(s)}(x'').$$

Hence

$$|R'| = s^{j_1} \leq 2 \sum_{Q_k \cap R \neq \phi} \sum_{s \in S_k} \left(g_k^{*(s)}(x'')\right)^{j_1/j_1 + \alpha}$$

and by (10),

$$|V_1^{x''}| \leq 2B \sum_k \sum_{s \in S_k} \left(g_k^{*(s)}(x'')\right)^{j_1/j_1 + \alpha}. \tag{12}$$

Now for s fixed, the maximal function

$$f^{*(s)}(x, \ y) = \sup_{0 < t}\left\{t^{-j} \ \phi(s, \ t)^{-n + j_1 + j_2}\iint_{|u| < t, \ |v| < \phi(s, t)} |f(x + u, \ y + v)| \, du \, dv\right\} \tag{13}$$

where $u \ \epsilon \ \mathbf{R}^j$ and $v \ \epsilon \ \mathbf{R}^{n - j_1 - j_2}$, satisfies a weak-type inequality, as can be seen from [5, II, pp. 309–310]. The important fact is that the weak-type inequality does not depend on s or ϕ. Then by the results [2], it follows that if Q is a cube of edge length h, in $n - j_1$ dimensions, then

$$\int_Q [f^{*(s)}(x, \ y)]^\delta \, dx \, dy \leq Ch^{(n - j_1)(1 - \delta)/\delta}\int_Q f(x, \ y) \, dx \, dy, \tag{14}$$

where C depends only on δ and $(n - j_1)$.

From (8), (11), (13), and (14), it follows that

$$|V_1| \leq 2B \sum_k \sum_{s \in S_k} \int_{J_k''} \left(g^{*(s)}(x'') \right)^{j_1/j_1 + \alpha} dx''$$

$$\leq CB \sum_k \sum_{s \in S_k} h_k^{(n - j_1)(1 - j_1/j_1 + \alpha)} \left(\int_{Q_k''} g_k(x'') \, dx'' \right)^{j_1/j_1 + \alpha}$$

$$\leq C \sum_k h_k^{(n - j_1)(1 - j_1/j_1 + \alpha)} \, h_k^{(n + \alpha)(j_1/j_1 + \alpha)}$$

$$= C \sum_k h_k^n \leq C|U|.$$

Combining this with (7), (5) follows.

REFERENCES

1. Córdoba, Antonio. Maximal functions, covering lemmas and Fourier multipliers. *Proc. Symp. in Pure Math.* 35, part 1 (1979):29–50.

2. Phillips, K. Maximal theorems of Hardy–Littlewood. *A. M. Monthly*, vol. 76 (June–July 1967).

3. Weiss, M. Strong differentials in L^p. *Studia Math.* 27 (1966): 49–72.

4. Weiss, M. Total and partial differentiation in L^p. *Studia Math.* 25 (1964):103–109.

5. Welland, G. V. Strong differentiation in L^p. *Rev. Acad. Ci. Madrid* 63 (1969):379–390.

6. Zygmund, A. *Trigonometric Series*, vols. I, II. New York: Cambridge Univ. Press, 1959.

PART SEVEN

PARTIAL DIFFERENTIAL EQUATIONS

ON THE LACK OF L∞-ESTIMATES
FOR SOLUTIONS OF ELLIPTIC SYSTEMS
OR EQUATIONS WITH COMPLEX COEFFICIENTS

M. Cwikel
Technion I.I.T., Haifa

E.B. Fabes, C.E. Kenig
University of Minnesota

INTRODUCTION

In this paper we show the lack of uniform L^∞-estimates for solutions to the Dirichlet problem corresponding to strongly coercive second-order elliptic operators defined on a smooth domain $\Omega \subset R^n$. Precisely, our operators are of the form

$$Lu(x) = \operatorname{div}\big(A(x)\nabla u(x)\big),$$

where $A(x) = \big(a_{ij}(x)\big)$ and strong coercivity means there exists $\lambda > 0$ such that for all $\varphi \in C_0^1(\Omega)$,

$$\int \operatorname{Re} \sum_{i,j=1}^{n} a_{ij}(x) D_{x_i}\varphi(x) D_{x_j}\overline{\varphi}(x)\, dx \geq \lambda \int |\nabla\varphi(x)|^2 dx$$

(see [1, p. 143]). In addition, we assume the coefficients to be bounded. We can then speak about the unique solution, u, of the Dirichlet problem

$$Lu = 0, \quad u\big|_{\partial\Omega} = g \tag{1}$$

in the class $H^1(\Omega) = \{u \in L^2(\Omega) : D_{x_i} u \in L^2(\Omega),\ i = 1,\ \ldots,\ n\}$ provided, say, $g \in H^1(\Omega)$. The statement $Lu = 0$ in (1) is in the weak sense; that is, for each $\varphi \in C_0^1(\Omega)$,

$$\sum_{i,j} \int_{\Omega} a_{ij}(x) D_{x_i} u(x) D_{x_j}\overline{\varphi}(x)\, dx = 0,$$

and the condition $u|_{\partial\Omega} = g$ means

$$u - g \in H_0^1(\Omega) = \text{closure of } C_0^1(\Omega) \text{ in } H^1(\Omega).$$

It is well known that when the coefficients of L are real, then the maximum principle holds for solutions of (1); that is, when $g \in H^1(\Omega) \cap C(\bar{\Omega})$,

$$\|u\|_{L^\infty(\Omega)} \leq \underset{\partial\Omega}{\text{Max}} |g|.$$

This fundamental result is crucial in constructing solutions to the Dirichlet problem for arbitrary $g \in C(\partial\Omega)$. By a solution, u, in this case we mean $u \in H_{\text{loc}}^1(\Omega) \cap C(\bar{\Omega})$, $Lu = 0$ in the weak sense, and $u|_{\partial\Omega} = g$ in the classical pointwise sense (see [6]). In fact, an L^∞-estimate similar to that given by the maximum principle is actually necessary if we have unique solvability for arbitrary $g \in C(\partial\Omega)$; for if this is the case, the closed graph theorem implies the existence of a constant, C_L, such that for all $g \in C(\partial\Omega)$ the solution u of $Lu = 0$, $u|_{\partial\Omega} = g$, satisfies

$$\|u\|_{L^\infty(\Omega)} \leq C_L \underset{\partial\Omega}{\text{Max}} |g|. \tag{2}$$

Conversely, the validity of (2) and the existence of solutions to the Dirichlet problem in the class $H^1(\Omega) \cap C(\bar{\Omega})$ when, say, g is smooth implies existence in the class $H_{\text{loc}}^1(\Omega) \cap C(\bar{\Omega})$ for arbitrary continuous boundary data.

When the coefficients of L are continuous in $\bar{\Omega}$, then indeed there exists a (unique) solution of the Dirichlet problem in $H^1(\Omega) \cap C(\bar{\Omega})$ when g is smooth (see [7, Chapter 6]). If, then, in the case of continuous coefficients, one wants to solve the Dirichlet problem for any $g \in C(\partial\Omega)$, we must be able to prove an L^∞-estimate as above. If the coefficients belong to $C^2(\bar{\Omega})$, (2) indeed holds with C_L depending only on the C^2-norm of the coefficients and the coercivity constant λ. However, it is the purpose of this paper to show that an estimate of the type,

$$\|u\|_{L^\infty(\Omega)} \leq C \underset{\partial\Omega}{\text{Max}} |g|, \tag{3}$$

cannot hold uniformly over all operators $L = \text{div}(A\nabla)$ satisfying:

(a) $\displaystyle\sup_{|x-y|<s} |A(x) - A(y)| \leq w(s)$, where $w(s)$ is a given continuous increasing function tending to 0 with s, and

(b) $\text{Re} \int A(\nabla\varphi) \cdot \nabla\bar{\varphi}\, dx \geq \lambda \int |\nabla\varphi|^2 dx$ for all $\varphi \in C_0^\infty(\Omega)$, where λ is a fixed positive number.

The failure of the above L^∞-estimate, (3), will be shown in the case of R^2 and the proof rests upon the following.

Theorem 1. Fix a continuous increasing function $\eta(s)$, $s \geq 0$, *with* $\eta(0) = 0$. *There exists a smooth domain,* Ω, *contained in* $R_+^2 = R \times (0, \infty)$ *and a positive number* ℓ *satisfying*

(i) $\partial\Omega \cap R \supset [-2\ell, 2\ell]$ *and* $\Omega \supset (-2\ell, 2\ell) \times (0, 4\ell)$,

(ii) *if* $\displaystyle\int_0^1 \eta^2(s)\, \frac{ds}{s} < \infty$ *and* $L = \text{div}(A(x, t)\nabla)$ *is a strongly coercive elliptic operator with bounded, uniformly continuous coefficients in* $R_+^2 = R \times (0, \infty)$ *such that*

$$\sup_{\substack{(x,t) \\ 0 \leq t \leq s}} |I - A(x, t)| \leq \eta(s),$$

then the solution, u, *of the Dirichlet problem,* $Lu = 0$ *in* Ω, $u|_{\partial\Omega} = g$, *for* $g \in C_0^\infty(-\ell, \ell)$, *satisfies the estimate*

$$\sup_{0 < t \leq \ell} \int_{-\ell}^{\ell} |u(x, t)|^2 dx \leq c \int |g(x)|^2 dx,$$

where C *is independent of* u *and* g *(see Lemma 2),*

(iii) *if, however,* $\displaystyle\int_0^1 \eta^2(s)\, \frac{ds}{s} = \infty$, *then there exists an elliptic operator,* $L = \text{div}(A(x, t)\nabla)$, *with real coefficients for which*

$$\sup_{\substack{(x,t) \\ 0 \leq t \leq s}} |I - A(x, t)| = \eta(s)$$

and such that the L^p-*estimate*

$$\sup_{0 < t \leq \ell} \int_{-\ell}^{\ell} |u(x, t)|^p dx \leq C_p |g(x)|^p dx \qquad (4)$$

fails for any finite $p \geq 1$ with C_p independent of u and
g (see Lemma 3 and the discussion preceding it).

The negative part of the above result, that is, the lack of an L^p-estimate, is really the consequence of the construction of an elliptic operator L, with real continuous coefficients whose L-harmonic measure (see [2]) is not absolutely continuous with respect to Lebesgue measure on any subinterval of the real line. The relation of this result to the L^∞-estimate, (3), is via interpolation of analytic families of operators. If, in fact, the L^∞-estimate is assumed with C depending only on the coercivity constant λ and the modulus of continuity, $w(s)$, of the coefficients, then the above theorem and an interpolation argument imply the validity of the L^p-estimate, (4), for some finite p and for the operator, L, just mentioned. This, in turn, implies the absolute continuity of the L-harmonic measure with respect to Lebesgue measure on the interval (a, b), a fact that contradicts the known property of L.

Finally, we would like to point out the similarity of our theorem stated above to some of the results of Stein and Zygmund in [9] and of L. Carleson in [4]. In the former work the authors prove the following.

Theorem 2. Let $\eta(s)$ be a function defined in $(0, 1]$ monotonically decreasing to 0 with s and such that

$$\int_0^1 \eta^2(s) \, \frac{ds}{s} < \infty.$$

If for each x belonging to a set $E \subset R$ we have

$$|F(x + s) + F(x - s) - 2F(x)| = 0\{s\eta(s)\} \qquad (s \to 0),$$

not necessarily uniformly in x, then F is differentiable almost everywhere in E.

However, if $\eta(s)$ is a function monotonically decreasing in s satisfying the condition $\dfrac{\eta(2s)}{\eta(s)} \to 1$ for $s \to 0$ and

$$\int_0^1 \eta^2(s) \, \frac{ds}{s} = \infty,$$

then there is a continuous $F(x)$ satisfying for all x the estimate

$$\left| F(x + s) + F(x - s) - 2F(x) \right| \leq s\eta(s)$$

but F is differentiable in a set of measure zero only.

In his paper [4], Carleson proves the following.

Theorem 3. Let $\phi(x)$ be a continuous, stictly increasing function on R such that

$$\frac{\phi(x + s) - \phi(x)}{\phi(x) - \phi(x - s)} = 1 + 0\left(\eta(\sqrt{s})\right) \qquad (s > 0)$$

uniformly on compact sets where

$1°$ $\eta(s)$ is continuous for $s \geq 0$, $\eta(0) = 0$

$2°$ $\eta(s)$ and $\eta(s)s^{-1/2} \to 0$ for $s > 0$.

If

$$\int_0^1 \eta^2(s) \, \frac{ds}{s} < \infty,$$

then ϕ is absolutely continuous and $\phi' \in L^2_{\text{loc}}$.

Moreover, given any $\eta(s)$ satisfying $1°$ and $2°$, if

$$\int_0^1 \eta^2(s) \, \frac{ds}{s} = \infty,$$

then there exists a continuous, strictly increasing $\phi(x)$ for which

$$\frac{\phi(x + s) - \phi(x)}{\phi(x) - \phi(x - s)} = 1 + 0\left(\eta(\sqrt{s})\right),$$

uniformly on compact sets, and ϕ is purely singular.

All the theorems stated above are not merely similar in form. Carleson himself points out the close relation between his result and that of Stein and Zygmund. At the same time, our

construction of an elliptic operator with real continuous coefficients whose "harmonic measure" is singular with respect to Lebesgue
measure is based directly on the results of Carleson (see [2]).

STATEMENT AND PROOF OF MAIN RESULT

Let $A(t, s) = (a_{ij}(t, s))$ denote an $n \times n$ matrix of complex-valued
entries defined on $\overline{R_+^2}$. For such matrix-valued functions, $A(t, s)$,
we set $|A(t, s)| = \max|a_{ij}(t, s)|$ and for each $\delta > 0$ we let

$$w(\delta, A) = \sup|A(t, s) - A(t', s')|$$

$$|(t, s) - (t', s')| < \delta$$

$$(t, s), (t', s') \in \overline{R_+^2}.$$

The main goal is to prove the following.

*Theorem 4.   There exists a smooth $\Omega \subset R_+^2$ so that given $0 < \epsilon < \frac{1}{2}$,
we can find a family of symmetric matrices, $A(t, s; y)$, depending
on the parameter $y \in R$, with the following properties:*

(i) $A(t, s; y)|_{\partial\Omega} = I$, *the identity matrix, and*

$$\sup_{(t,s) \in R_+^2} |A(t, s; y) - I| \leq \epsilon \;\left(\text{in particular,}\right.$$

$$\sup_{(t, s) \in \overline{R_+^2}} |\operatorname{Im} A(t, s; y)| \leq \epsilon, \quad \sup_{(t, s) \in \overline{R_+^2}} |A(t, s; y)| \leq 2,$$

$$\operatorname{Re} \sum_{i, j = 1}^{n} a_{ij}(t, s; y)\xi_i\xi_j \geq \frac{1}{2}|\xi|^2, \text{ and}$$

$$\left|\sum_{i, j = 1}^{2} \int_\Omega a_{ij}(t, s; y)D_iu(t, s)D_j\bar{u}(t, s)\,dt\,ds\right|$$

$$\geq \frac{1}{4}\int_\Omega |\nabla u(t, s)|^2\,dt\,ds$$

for all $u \in C_0^\infty(\Omega)$. As usual, $D_1u = \dfrac{\partial u}{\partial t}$ and $D_2u = \dfrac{\partial u}{\partial s}$),

 (ii)   *there exists an increasing continuous function, $\phi(\delta)$,
        with $\phi(0) = 0$, such that for all y,*

$$w\big(\delta;\ A(\cdot,\ \cdot;\ y)\big)\ \leq\ \phi(\delta),$$

(iii)   the estimate

$$\|u(\cdot,\ \cdot;\ y)\|_{L^\infty(\Omega)}\ \leq\ C\|g\|_{L^\infty(\partial\Omega)}$$

with C independent of y and $g \in C_0^\infty(R^2)$ does not hold for the family of functions, $u(t,\ s;\ y)$, where for each y, $u(t,\ s;\ y)$ is the solution of the Dirichlet problem

$$\mathrm{div}\big(A(t,\ s;\ y)\nabla u(t,\ s;\ y)\big)\ =\ 0\ \text{in}\ \Omega$$

$$u(t,\ s;\ y)\big|_{\partial\Omega}\ =\ g.$$

As mentioned in the Introduction, the proof of the above theorem is accomplished by a contradiction argument.   The existence of a constant, C, making the inequality in (iii) above true for all y, is combined with an L^2-inequality stated in Lemma 2 below to imply, via interpolation, an L^p-estimate, in the form of Lemma 2, with $p < \infty$ for a specific operator, L, with real coefficients.   This L^p-estimate then implies that a Poisson kernel exists for L that at least belongs to $L^1(\partial\Omega)$.   However, for this particular L, it had been previously shown in [2] that such a kernel could not exist. This aspect of our proof is given below in Lemmas 2 and 3 and the discussion in between.

Consider now real-valued functions, $b(t,\ s)$, defined on the square $T = (0,\ \ell) \times (0,\ \ell)$ that are continuous on $\overline{T}$ and that *vanish* on ∂T.   For such functions, b, we define

$$w_*(\delta,\ b)\ =\ \sup_{\substack{s,t \in [0,\ \ell] \\ 0 \leq s \leq \delta}}\ |b(t,\ s)|.$$

Let $p > 1$ and assume

$$\int_0^\ell \frac{W_*(\delta,\ b)^p}{\delta}\ d\delta\ <\ \infty.$$

We consider the strips

$$S = \{z : 0 \leq \mathrm{Re}\ z \leq 1\}\quad \text{and}\quad S_\tau = \{z : -\tau \leq \mathrm{Re}\ z \leq 1 + \tau\}$$

in the complex plane. Pick $q > p$, $0 < \tau < 1$, and for $z \in S_\tau$ set

$$b(t,\ s;\ z)\ =\ |b(t,\ s)|^{p(1-z)+\frac{p}{q}z}(\operatorname{sgn}\ b(t,\ s))\frac{1 + \tilde{\theta}}{1 + z},$$

where $\frac{1}{p} = (1 - \tilde{\theta}) + \frac{\tilde{\theta}}{q}$. The following lemma is easily verified.

Lemma 1. *Let* $b(t,\ s;\ \tilde{\theta})$ *be as above. There exists* $\tau_0 > 0$ *such that*

(i) $b(t,\ s;\ \tilde{\theta})\ =\ b(t,\ s)$,

(ii) $\displaystyle\sup_{z \in S_{\tau_0}} \|b(\cdot,\ \cdot;\ z)\|_{L^\infty(T)}\ \le\ 2\ \max\left\{\|b\|_{L^\infty(T)}^{pq/2},\ \|b\|_{L^\infty(T)}^{p(1+\tau_0)-p\tau_0/q}\right\},$

(iii)   *for each* $z \in S_{\tau_0}$, $b(s,\ t;\ z)$ *is continuous on* $\overline{T}$, *vanishes on* ∂T, *and for all* $z \in S$,

$$w(\delta;\ b(\cdot,\ \cdot;\ z))\ \le\ C\ \max(w(\delta;\ b);\ w(\delta;\ b)^{p/q}),$$

(iv)   *for each* $z = iy$, $-\infty < y < \infty$,

$$w_*(\delta,\ b(\cdot,\ \cdot;\ iy))\ \le\ 2w_*(\delta,\ b)^p,$$

and hence

$$\int_0^\ell w_*(\delta,\ b(\cdot,\ \cdot;\ iy))\frac{d\delta}{\delta}\ \le\ 2\int_0^\ell w_*(\delta,\ b)^p\frac{d\delta}{\delta},$$

(v)   *uniformly for* $(t,\ s) \in \overline{T}$ *we have*

$$\lim_{\zeta \to z} b(t,\ s;\ \zeta)\ =\ b(t,\ s;\ z)$$

for each $z \in S_{\tau_0}$ *and*

$$\lim_{\zeta \to z} \frac{b(t,\ s;\ \zeta) - b(t,\ s;\ z)}{\zeta - z}\ =\ \frac{d}{dz}\ b(t,\ s;\ z)$$

for each z *in the interior of* S_{τ_0}.

Lemma 2. *For any fixed number* ℓ, $0 < \ell < 1$, *let* Ω *denote a smooth bounded domain in* R_+^2 *such that*

$$\partial\Omega \cap R \supset [-2\ell,\ 2\ell] \quad \text{and} \quad \Omega \supset (-2\ell,\ 2\ell) \times (0,\ 4\ell).$$

Given a function $\eta(s)$ *with*

$$\int_0^1 \eta^2(s)\,\frac{ds}{s} < \infty,$$

there exists $\varepsilon_0 > 0$, *depending only on* η, *so that for all matrices* $A(t,\,s)$ *satisfying*

(a) $\omega_*(\delta;\,A - I) \leq \eta(\delta)$,

(b) $|A - I| \leq \varepsilon_0$,

(c) $A = I$ *on* $\overline{R}_+^2 \setminus \left[-\frac{3}{2}\ell,\,\frac{3}{2}\ell\right] \times [0,\,3\ell]$,

we have

$$\left(\int_{-\ell}^{\ell} |u(t,\,s)|^2 dt\right)^{1/2} \leq C\left(\int |g|^2\right)^{1/2} \text{ for all } g \in C_0^\infty(-\ell,\,\ell),$$

where u *solves the Dirichlet problem*

$$\mathrm{div}\bigl(A(t,\,s)\nabla u(t,\,s)\bigr) = 0 \text{ in } \Omega$$

$$u\big|_{\partial\Omega} = g$$

and C *depends only on* η.

Proof.   Set $P_s(t) = \dfrac{1}{\pi}\,\dfrac{s}{t^2 + s^2}$ and $G(t,\,s,\,r) = \log\dfrac{\sqrt{t^2 + (s + r)^2}}{\sqrt{t^2 + (s - r)^2}}$.

Given $g \in C_0^\infty(-\ell,\,\ell)$, extend the solution, u, of the Dirichlet problem described in Theorem 4 to be zero in $\overline{R}_+^2\setminus\overline{\Omega}$. We now can write

$$u(t,\,s) = \int_R P_s(t - r)g(r)\,dr$$

$$+ \int_0^\infty\!\!\int_R \nabla_{y,r}\bigl(G(t - y,\,s,\,r)\bigr)\cdot\bigl(I - A(y,\,r)\bigr)\nabla u(y,\,r)\,dy\,dr.$$

Set $I(g)(t,\,s) = P_s * g(t)$.   From the easy part of the Littlewood-Paley estimates in [8, p. 82],

$$\|I(g)(\cdot,\,s)\|_{L^2(R)} + \|s^{1/2}\nabla I(g)\|_{L^2(R_+^2)} \leq C\|g\|_{L^2(R)}.$$

Now set

$$V(f)(t,\,s) = \int_0^\infty\!\!\int_R DG(t - y,\,s,\,r)f(y,\,r)\,dy\,dr,$$

where D *denotes a partial derivative of* G *with respect to* t *or* r.

First Estimate. Assume $f(y, r) = 0$ for $r \geq 1$. There exists a constant C independent of f and η such that for any $0 < \delta < 1$,

$$\|Vf(\cdot, s)\|_{L^2(R)} \leq C\left[\left(\int_0^\delta \eta^2(s) \frac{ds}{s}\right)^{1/2} \|r^{1/2}\eta(r)^{-1}f(t, r)\|_{L^2(R_+^2)}\right.$$
$$\left. + (\log 1/\delta)^{1/2} \|r^{1/2}f(t, r)\|_{L^2(R_+^2)}\right].$$

Proof. For each s and r,

$$\|DG(\cdot, s, r) \star f(\cdot, r)\|_{L^2(R)} \leq C\|f(\cdot, r)\|_{L^2(R)},$$

where C is an absolute constant independent of s, r, and f. Hence, by Minkowski's inequality,

$$\|Vf(\cdot, s)\|_{L^2(R)} \leq C \int_0^1 \|f(\cdot, r)\|_{L^2(R)} \, dr$$
$$\leq C\left(\int_0^\delta \eta^2(r) \frac{dr}{r}\right)^{1/2} \|r^{1/2}\eta(r)^{-1}f(t, r)\|_{L^2(R_+^2)}$$
$$+ C(\log 1/\delta)^{1/2} \|r^{1/2}f(t, r)\|_{L^2(R_+^2)}.$$

Second Estimate. Again assume $f(y, r) = 0$ for $r \geq 1$. There exists a constant C independent of f and η such that for any $0 < \delta < 1$,

$$\|s^{1/2}DV(f)(t, s)\|_{L^2(R_+^2)} \leq C\left[\left(\int_0^\delta \eta^2(r) \frac{dr}{r}\right)^{1/2} \|r^{1/2}\eta(r)^{-1}f\|_{L^2(R_+^2)}\right.$$
$$\left. + (\log 1/\delta)^{1/2} \|r^{1/2}f\|_{L^2(R_+^2)}\right].$$

Here D denotes the partial derivative with respect either to t or to s.

Proof.

$$s^{1/2}DVf(t, s) = \int_0^\infty \int_R D^2G(t - y, s, r)r^{1/2}f(y, r) \, dy \, dr$$
$$+ \int_0^\infty \int_R D^2G(t - y, s, r)(s^{1/2} - r^{1/2})f(y, r) \, dy \, dr \equiv Af + Bf.$$

(D^2 denotes a general second-order partial.) Since $D^2 \log(t^2 + s^2)$

is a Calderón-Zygmund kernel in R^2, we have

$$\|Af\|_{L^2(R^2_+)} \leq C \|r^{1/2}f\|_{L^2(R^2_+)} .$$

We write

$$Bf(t, s) = \int_0^{s/2} \int_0^\infty D^2G(t - y, s, r)(s^{1/2} - r^{1/2})f(y, r)\,dy\,dr$$

$$+ \int_{s/2}^\infty \int_0^\infty D^2G(t - y, s, r)(s^{1/2} - r^{1/2})f(y, r)\,dy\,dr$$

$$= B_1f + B_2f.$$

Since

$$\int |D^2G(y, s, r)|\,dy \leq \frac{C}{|s - r|},$$

a combination of Minkowski's inequality and Young's theorem for convolution operators gives

$$\|B_2f(\cdot, s)\|_{L^2(R)} \leq C \int_{s/2}^\infty \frac{|s^{1/2} - r^{1/2}|}{|s - r|r^{1/2}} \left(r^{1/2}\|f(\cdot, r)\|_{L^2(R)}\right)dr.$$

An application of Hardy's theorem [5, p. 227] yields

$$\|B_2f\|_{L^2(R^2_+)} \leq C \|r^{1/2}f\|_{L^2(R^2_+)} .$$

Finally,

$$B_1f(t, s) = s^{1/2} \int_0^{s/2} \int_R (D^2G(t - y, s, r) - D^2G(t - y, s, 0))f(y, r)\,dy\,dr$$

$$+ s^{1/2} \int_R D^2G(t - y, s, 0) \left(\int_0^{s/2} f(y, r)\,dr\right)dy$$

$$- \int_0^{s/2} \int_R D^2G(t - y, s, r)r^{1/2}f(y, r)\,dy\,dr$$

$$= B_{11}f + B_{12}f + B_{13}f.$$

Using the mean-value theorem in the r-variable, we have

$$\|B_{11}f(\cdot, s)\|_{L^2(R)} \leq C \frac{1}{s^{3/2}} \int_0^{s/2} \|f(\cdot, r)\|_{L^2(R)} r\,dr.$$

Also

$$\left\| B_{13} f(\cdot, \ s) \right\|_{L^2(R)} \le \frac{c}{s} \int_0^{s/2} \left\| f(\cdot, \ r) \right\|_{L^2(R)} r^{1/2} \, dr.$$

If we once again use Hardy's theorem, we conclude that

$$\left\| B_{11} f \right\|_{L^2(R_+^2)} + \left\| B_{13} f \right\|_{L^2(R_+^2)} \le C \left\| r^{1/2} f \right\|_{L^2(R_+^2)}.$$

In considering $B_{12} f$, observe first that

$$s^{1/2} D^2 G(t, \ s, \ 0) = \frac{\psi(t/s)}{s^{3/2}}, \quad \text{where } \mathfrak{F}(\psi)(\zeta),$$

the Fourier transform of ψ, is bounded by $c|\xi| e^{-c|\xi|}$. Hence

$$\mathfrak{F}(B_{12} f(\cdot, \ s))(\xi) = \frac{\mathfrak{F}(\psi)(\xi s)}{s^{1/2}} \int_0^{s/2} \mathfrak{F}(f(\cdot, \ r))(\xi) \, dr$$

and so

$$\left| \mathfrak{F}(B_{12} f(\cdot, \ s))(\xi) \right| \le \left| \frac{\mathfrak{F}(\psi)(\xi s)}{s^{1/2}} \right| \left[\left(\int_0^\delta \eta^2(r) \, \frac{dr}{r} \right)^{1/2} \right.$$

$$\cdot \left(\int_0^\delta |\mathfrak{F}(\cdot, \ r)(\xi)|^2 \frac{r}{\eta(r)^2} \, dr \right)^{1/2}$$

$$\left. + (\log 1/\delta)^{1/2} \left(\int_\delta^1 |\mathfrak{F}(f(\cdot, \ r))(\xi)|^2 r \, dr \right)^{1/2} \right].$$

Since $\displaystyle\int_0^\infty \left| \frac{\mathfrak{F}(\psi)(\xi s)}{s} \right|^2 \, ds \le C$, independent of ξ, it readily follows that

$$\left\| B_{12} f \right\|_{L^2(R_+^2)} \le C \left[\left(\int_0^\delta \eta^2(r) \, \frac{dr}{r} \right)^{1/2} \left\| r^{1/2} \eta(r)^{-1} f \right\|_{L^2(R_+^2)} \right.$$

$$\left. + (\log 1/\delta)^{1/2} \left\| r^{1/2} f \right\|_{L^2(R_+^2)} \right].$$

Our representation of the solution, u, together with the first and second estimates and the hypotheses concerning the matrix A, yields the inequality

$$\|u(\cdot,\ s)\|_{L^2(R)} + \|r^{1/2}Du\|_{L^2(R_+^2)} \leq C\|g\|_{L^2(R)}$$

$$+ C\left[\left(\int_0^\delta \eta^2(r)\ \frac{dr}{r}\right)^{1/2}\left\|\frac{(I - A)(Du)r^{1/2}}{\eta(r)}\right\|_{L^2(R_+^2)}\right.$$

$$\left. + (\log 1/\delta)^{1/2}\|(I - A)(Du)r^{1/2}\|_{L^2(R_+^2)}\right]$$

$$\leq C\|g\|_{L^2(R)} + C\left[\left(\int_0^\delta \eta^2(r)\ \frac{dr}{r}\right)^{1/2} + (\log 1/\delta)^{1/2}\varepsilon_0\right]\|r^{1/2}Du\|_{L^2(R_+^2)}.$$

It is clear that if we choose first $\delta < 1$ so that

$$C\left(\int_0^\delta \eta^2(r)\ \frac{dr}{r}\right)^{1/2} \leq 1/4$$

and then ε_0 so that

$$C(\log 1/\delta)^{1/2}\varepsilon_0 \leq 1/4,$$

the conclusion of Lemma 2 follows.

Because of the results in [4 and 2], given $\theta(s)$ continuous, increasing with $\theta(0) = 0$, $s^{-1/4}\theta(s)$ decreasing, and

$$\int_0^1 \theta^2(s)\ \frac{ds}{s} = \infty,$$

we can find a real symmetric matrix, $T = I + C$, in the upper half-plane $(t,\ s)$, $s > 0$, such that the entries of C are $O(\theta(s))$ uniformly on compact sets in t, and such that the operator $\tilde{L} = \text{div}(T\nabla)$ has a harmonic measure $w_{\tilde{L},\mathbf{R}_+^2}^P$ strictly singular with respect to Lebesgue measure on $\mathbf{R}$. Here, by the harmonic measure, $w_{\tilde{L},\mathbf{R}_+^2}^P$ we mean the finite Borel measure that gives the representation

$$u(P) = \int_\mathbf{R} g(Q)\,dw_{\tilde{L},\mathbf{R}_+^2}^P(Q)$$

of the solution of the Dirichlet problem for $\tilde{L}$ in $\mathbf{R}_+^2$, with boundary data $g \in C_0(\mathbf{R})$. (The existence of $w_{\tilde{L},\mathbf{R}_+^2}^P$ is guaranteed by the maximum principle, since T is real and symmetric.) Now fix one such θ as above, but with the additional property that

$$\int_0^1 \theta(s)^P\ \frac{ds}{s} < \infty$$

for some $p > 2$, which we also fix from now on.   Set

$$\eta(s) = \theta(s)^p.$$

Given $\varepsilon_1 > 0$, we can find an interval, $[-\ell, \ell]$, such that in the box $[-2\ell, 2\ell] \times [0, 4\ell]$ we have the estimate

$$|I - T(t, s)| \leq \varepsilon_1.$$

Now let $\varphi \in C_0^\infty(R^2)$ satisfy $0 \leq \varphi \leq 1$ and $\varphi \equiv 1$ in $[-\ell, \ell] \times [0, 2\ell]$, $\varphi \equiv 0$ outside $[-3/2\ell, 3/2\ell] \times [0, 3\ell]$.   The matrix

$$A(t, s) = (1 - \varphi(t, s))I + \varphi(t, s)T(t, s)$$

has the properties:

$$|I - A(t, s)| \leq \varepsilon_1, \text{ for all } (t, s) \in \overline{R}_+^2,$$

$A(t, 0) = I$ for all t, $A(t, s) = I$ in $\overline{R}_+^2 \setminus [-3/2\ell, 3/2\ell] \times [0, 3\ell]$, and $A(t, s) = T(t, s)$ on $[-\ell, \ell] \times [0, 2\ell]$.

We choose a smooth domain, $\Omega \subset R_+^2$, such that $\partial\Omega \cap R \supset [-2\ell, 2\ell]$ and $\Omega \supset (-2\ell, 2\ell) \times (0, 4\ell)$ (see Lemma 2).   We set

$$\tilde{L} = \text{div}(T\nabla) \quad \text{and} \quad L = \text{div}(A\nabla).$$

By Theorem 1.4 in [3], $w_{\tilde{L}, \Omega}^p$ is singular with respect to Lebesgue measure on $[-1/2\ell, 1/2\ell]$.   By the same theorem, there exists a closed set $|F| \subset [-1/2\ell, 1/2\ell]$ with $|F| = 0$, but $w_{\tilde{L}, \Omega}^p(F) > 0$. An easy consequence of this fact, and Harnack's principle, is the following.

Lemma 3.   The a priori estimate

$$\left(\int_{-\ell}^{\ell} |u(t, s_0)|^r \, dt\right)^{1/r} \leq C \left(\int |g(t)|^r \, dt\right)^{1/r}$$

cannot hold *for any fixed* $0 < s_0 < 2\ell$, $1 \leq r < \infty$, *with C independent of* $g \in C_0(-\ell, \ell)$ *and* u, *the solution of the Dirichlet problem*

$$Lu = 0 \text{ in } \Omega, \ u|_{\partial\Omega} = g.$$

Set $B(t, s) = I - A(t, s)$ and $b_{ij}(t, s)$ equal to a general entry in the matrix B. As in Lemma 1, define

$$b_{ij}(t, s; z) = |b_{ij}(t, s)|^{p(1-z)+\frac{p}{q}z}(\text{sgn } b(t, s))\frac{1+\tilde{\theta}}{1+z},$$

where $q > p$, $1/p = 1 - \tilde{\theta} + \tilde{\theta}/q$, $z \in S_{\tau_0} = \{z : -\tau_0 \le \text{Re}(z) \le 1 + \tau_0\}$, and the positive number τ_0 has been chosen so that the conclusions of Lemma 1 are valid. Finally, we set

$$B(t, s; z) = (b_{ij}(t, s; z))$$

and

$$A(t, s; z) = I - B(t, s; z).$$

The main properties of the matrix $A(t, s; z)$ are stated in the following lemma, whose proof is immediate.

Lemma 4. _Set_ $Q_\ell = [-\ell, \ell] \times [0, 2\ell]$. _Given_ $\varepsilon_1 > 0$, _there exist_ ℓ _so that_

   (i) $A(t, s; \tilde{\theta}) = A(t, s)$,

   (ii) $\sup\limits_{z \in S_{\tau_0}} |I - A(t, s; z)| \le \varepsilon$,

   (iii)   _for each_ $z \in S_{\tau_0}$, $A(t, s; z)$ _is continuous in_ R_+^2, _equals_ I _on_ R, _and in_ $R_+^2 \backslash Q_{3/2\ell}$ _also_

$$\sup\limits_{z \in S = \{a : 0 \le \text{Re } z \le 1\}} |w(\delta, A(\cdot, \cdot; z))| \le \phi(\delta)$$

   where $\psi(\delta) \to 0$ _as_ $\delta \to 0+$,

   (iv) $\sup\limits_{y \in (-\infty, \infty)} w_*(\delta; I - A(\cdot, \cdot; iy)) \le \eta(\delta)$, _where_ $\eta(s) = \theta(s)^p$

   and $\theta(s)$ _was given following Lemma 2 and satisfies_

$$\int_0^1 \frac{\theta(s)^p}{s} ds < \infty;$$

   in particular, $\int_0^1 \eta(s) \frac{ds}{s} < \infty$,

   (v)   _for each_ (t, s), $A(t, s; z)$ _is analytic in_ S_{τ_0}, _and for_
$$z \in S = \{z : 0 \le \text{Re } z \le 1\},$$

$$\lim_{\zeta \to z} A(t, s; \zeta) = A(t, s; z)$$

and

$$\lim_{\zeta \to z} \frac{A(t, s; \zeta) - A(t, s; z)}{\zeta - z} = \frac{d}{dz} A(t, s; z)$$

with both limits occurring uniformly for $(t, s) \in \overline{R}^2_+$.

The family of matrices mentioned in Theorem 4 is the family $A(t, s; 1 + iy)$. The domain Ω mentioned in the same theorem is the domain Ω described just prior to Lemma 3. The properties of Ω that we need to recall are: $\Omega \subset R^2_+$, $\partial\Omega \cap R \supset [-2\ell, 2\ell]$, and $\Omega \supset (-2\ell, 2\ell) \times (0, 4\ell)$. The number ℓ is chosen so that the conclusions of Lemma 4 hold for a given positive number $\varepsilon_1 < \varepsilon_0$, ε_0 was determined by Lemma 2, and the function $\eta \equiv \theta^p$, and θ was described following Lemma 2.

The family of matrices $A(t, s; 1 + iy)$ satisfies conditions (i) and (ii) of the theorem by construction. By a contradiction argument, we will show that the estimate in part (iii) of the theorem cannot hold in a uniform manner over the family of associated elliptic operators.

Lemma 5. For $z \in S = \{z : 0 \leq Re\ z \leq 1\}$ and $g \in C_0^\infty(-\ell, \ell)$, set $L_z = div(A(t, s, z)\nabla)$ and u_z equal to the solution of the Dirichlet problem

$$L_z u_z = 0 \ in \ \Omega, \ u_z|_{\partial\Omega} = g.$$

There exists a constant C depending only on $\theta(s)$ (see Lemma 4(iv)) such that for all $0 < s < \ell$,

$$\sup_{y \in R} \int_{-\ell}^\ell |u_{iy}(t, s)|^2 dt \leq C \int_{-\ell}^\ell |g(t)|^2 dt.$$

Proof. The proof is an immediate consequence of Lemma 2.

Lemma 6. Let $g \in C_0^\infty(-\ell, \ell)$ and for $z \in S_{\tau_0}$ let u_z denote the solution to the Dirichlet problem

$$L_z u_z \equiv \mathrm{div}\big(A(t,\ s;\ z)\nabla_{t,s} u_z(t,\ s)\big) = 0 \text{ in } \Omega,$$

where $\{A(t,\ s;\ z)\}$ is the family of matrices described in Lemma 4.
If J is an interval $\subset (-\ell,\ \ell)$ and $0 < s_0 < \ell$, then

$$\int_J u_z(t,\ s_0)\,dt$$

is a well-defined bounded analytic function for $z \in \overset{\circ}{S}$, the interior
of S, that is continuous in S.

Proof. We may assume that g is the restriction to $\partial\Omega$ of a function,
which we again denote by g, belonging to $C_0^\infty(R^2)$. Let $v_z \in H_0^1(\Omega)$ be
the solution of the problem

$$L_z v_z = \mathrm{div}\big(A(t,\ s;\ z)\nabla g(t,\ s)\big).$$

Then $u_z(t,\ s) = g(t,\ s) - v_z(t,\ s)$.

Since $v_z \in H_0^1(\Omega)$, the trace theorem [8, pp. 192-193] shows
that

$$\int_J v_z(t,\ s_0)\,dt$$

is well defined. All we have to show is that it is analytic and
bounded in $\overset{\circ}{S}$ and continuous in S. All this follows from the Lax-
Milgram theorem [1].

Proof of Theorem. Consider the family of matrices $\{A(t,\ s,\ z);$
$z \in S_{\tau_0}\}$ described in Lemma 4 and the domain Ω described just prior
to Lemma 3. For each $g \in C_0^\infty(-\ell,\ \ell)$ and $z \in S_{\tau_0}$, let u_z denote the
solution to the Dirichlet problem

$$L_z u_z = \mathrm{div}\big(A(z)\nabla u_z\big) = 0 \text{ in } \Omega$$

$$u_z\big|_{\partial\Omega} = g.$$

Assume

$$\sup_{y \in R}\|u(\cdot,\ \cdot,\ 1 + iy)\|_{L^\infty(\Omega)} \le C\|g\|_{L^\infty(\partial\Omega)} \tag{5}$$

with C independent of g.

Recall that $p > 2$ and $\theta(s)$ have been chosen so that

$$\int_0^1 \frac{\theta^p(s)}{s}\, ds < \infty,$$

while $\int_0^1 \frac{\theta^2(s)}{s}\, ds = \infty$. We also have fixed $q > p$ and have chosen $\tilde{\theta}$ so that

$$\frac{1}{p} = (1 - \tilde{\theta}) + \frac{\tilde{\theta}}{q}.$$

With $\tilde{\theta}$ fixed, choose $r > 2$ satisfying $\frac{1}{r} = \frac{1 - \tilde{\theta}}{2}$. The assumed estimate (5) and the known estimate

$$\sup_y \|u(\cdot,\, s_0;\, iy)\|_{L^2(-\ell,\, \ell)} \leq C\|g\|_{L^2(-\ell,\, \ell)}$$

with C independent of g, given by Lemma 5, imply, via interpolation of analytic families of operators [10], the estimate

$$\|u(\cdot,\, s_0;\, \tilde{\theta})\|_{L^r(-\ell,\, \ell)} \leq C\|g\|_{L^r(-\ell,\, \ell)}$$

with C independent of g. This contradicts the results of Lemma 3 where $L \equiv L_{\tilde{s}}$. This contradiction is due to the assumed validity of the L^{∞}-estimate (5). This concludes the proof of the theorem.

Remark 1. In the above proof of the main theorem, if we replace the assumed estimate

$$\sup_y \|u(\cdot,\, \cdot;\, 1 + iy)\|_{L^{\infty}(\Omega)} \leq C\|g\|_{L^{\infty}(\partial\Omega)} \tag{6}$$

by the weaker inequality

$$\sup_y \|u(\cdot,\, s_0;\, 1 + iy)\|_{BMO[-\ell,\, \ell]} \leq C\|g\|_{L^{\infty}(\partial\Omega)} \tag{7}$$

with C independent of g, we would still arrive at the same contradiction. (BMO$[-\ell,\, \ell]$ = the space of functions of bounded mean oscillation on $[-\ell,\, \ell]$.) Thus the inequality (7) cannot hold with C depending only on a fixed parameter of coercivity and a fixed modulus of continuity of the coefficients.

Remark 2. Interior estimates of the form

$$\|u\|_{L^p(\Omega')} \leq C\|g\|_{L^{\infty}(\partial\Omega)}$$

for some $1 < p < \infty$ and for each $\Omega' \, cc \, \Omega$ also cannot hold with C depending only on Ω', Ω, a fixed parameter of coercivity and a fixed modulus of continuity. This is because for continuous coefficients, solutions of the Dirichlet problem with data g satisfy the estimate

$$\|u\|_{L^\infty(\Omega')} \leq C\|u\|_{L^p(\Omega'')},$$

where $\Omega' \, cc \, \Omega'' \, cc \, \Omega$, and C depends only on Ω', Ω'', Ω, the modulus of continuity of the coefficients and the parameter of coercivity. If we now assume the validity of (8), we would obtain the estimate

$$\|u\|_{L^\infty(\Omega')} \leq C\|g\|_{L^\infty(\partial\Omega)}.$$

In particular for the domain Ω of our theorem, we would have the two estimates

$$\sup_y \|u(\cdot, \, s_0; \, 1 + iy)\|_{L^\infty(-\ell, \, \ell)} \leq C\|g\|_{L^\infty(-\ell, \, \ell)}$$

and

$$\sup_y \|u(\cdot, \, s_0; \, iy)\|_{L^2(-\ell, \, \ell)} \leq C\|g\|_{L^2(-\ell, \, \ell)}$$

$(g \in C_0^\infty(-\ell, \, \ell))$. Interpolation of analytic families of operators would again give

$$\|u(\cdot, \, s_0, \, \tilde{\theta})\|_{L^r(-\ell, \, \ell)} \leq C_{s_0, \, \ell, \, r} \|g\|_{L^r(-\ell, \, \ell)}$$

for some r, $1 < r < \infty$, and $C_{s_0, \ell, r}$ independent of g. This last estimate contradicts Lemma 3.

Remark 3. The L^2 bounds in Lemma 2 also hold for any L^p, $1 < p < \infty$. This can be done by the same technique of proof, using vector-valued L^p-estimates for singular integrals.

REFERENCES

1. Agmon, S. *Lectures on Elliptic Boundary Value Problems.* New York: Van Nostrand, 1965.

2. Caffarelli, L.; Fabes, E.; and Kenig, C. Completely singular
    elliptic-harmonic measures. To appear in the *Ind. Univ.
    Math. J.*, Vol. 30, No. 6 (1981):917-924.

3. Caffarelli, L.; Fabes, E.; Mortola, S.; and Salsa, S. Boundary
    behaviour of nonnegative solutions of elliptic operators in
    divergence form. *Ind. Univ. Math. J.*, Vol. 30, No. 4 (1981):621-639.

4. Carleson, L. On mappings, conformal at the boundary. *J.
    D'Analyse* 19 (1967):1-13.

5. Hardy, G. H.; Littlewood, J. E.; and Polya, G. *Inequalities.*
    2d ed. New York: Cambridge Univ. Press, 1964.

6. Littman, W.; Stampacchia, G.; and Weinberger, H. Regular
    points for elliptic equations with discontinuous coefficients.
    Ann. Sc. Norm. Sup Pisa, Cl. Sci., Ser III, 17 (1963):45-79.

7. Morrey, C. B. *Multiple Integrals in the Calculus of Variations.*
    New York: Springer-Verlag, 1966.

8. Stein, E. *Singular Integrals and Differentiability Properties
    of Functions.* Princeton, N.J.: Princeton Univ. Press, 1970.

9. Stein, E., and Zygmund, A. On the differentiability of func-
    tions. *Studia Math.* 23 (1964):247-283.

10. Zygmund, A. *Trigonometric Series.* 2d ed. New York: Cambridge
    Univ. Press, 1968.

BOUNDARY BEHAVIOR OF
SOLUTIONS TO DEGENERATE ELLIPTIC EQUATIONS

E.B. Fabes, C.E. Kenig
University of Minnesota

D. Jerison
University of Chicago

This is the third and last in a series of articles examining solutions to degenerate elliptic equations in divergence form,

$$Lu(x) = - \sum_{i,j=1}^{n} \frac{\partial}{\partial x_i}(a_{ij}(x)\frac{\partial}{\partial x_j}u(x)) = 0,$$

where the coefficients $a_{ij}(x)$ are real, measurable, symmetric, and satisfy

$$c^{-1}m(x)|\xi|^2 \leq \sum_{i,j=1}^{n} a_{ij}(x)\xi_i\xi_j \leq cm(x)|\xi|^2$$

for suitable $m(x)$. We will be concerned mainly with Fatou-type theorems, but we will also review briefly the aspects of potential theory treated in the previous two articles [6 and 5].

Denote $B(x, r) = \{y \in \mathbf{R}^n : |x - y| < r\}$ and $m(E) = \int_E m(x)\,dx$. The difference between the degenerate elliptic case (where m may vanish or become infinite) and the uniformly elliptic case (where m is constant) is that the function $\theta(x, r) = r^2/m(B(x, r))$ plays the role of r^{2-n}.

The property underlying most of the arguments in this subject is a scale-invariant version of Harnack's inequality. The

All authors were supported in part by the NSF. The second author is an NSF postdoctoral fellow. Univ. of Minnesota Math. Rpt. 81-106.

technique was introduced by L. Carleson [2] (see also J. Moser [11]) and extended by many authors (see [8, 9, 4, 13, 1, and 10]). In the process, there has been a progression to more general contexts. The study of harmonic functions in a Lipschitz domain is reduced by a change of variable to a special case of the study of solutions to a uniformly elliptic equation in a smooth domain. Similarly, the study of harmonic functions in the quasiconformal image of a ball is a special case of the study of solutions to a degenerate elliptic equation in the ball. The present version is more general still. It is only a slight exaggeration to say that it is based on estimates (1) and (2) alone.

We will not need to repeat most of the arguments of preceding work here, since they are the same in our case. Instead we will present an outline and refer the reader to [10] for details. Only one lemma (Lemma 3) requires a different proof.

NOTATIONS

We will assume that the weight $m(x)$ satisfies either Muckenhoupt's condition

$$(A_2) \qquad \sup_B \left(\frac{1}{|B|} \int_B m(x)\,dx \right) \left(\frac{1}{|B|} \int_B m(x)^{-1}\,dx \right) \leq c < \infty,$$

where the supremum is taken over the balls B and $|B| = \int_B dx$, or that $m(x) = |f'(x)|^{1-2/n}$, where f is a global quasiconformal mapping of $\mathbf{R}^n$ and $|f'(x)|$ denotes the absolute value of the Jacobian determinant of f. All functions $m(x)$ of the form $|x|^\alpha$, $\alpha > -n$, satisfy at least one of these conditions. For more details on the nature of the conditions, see [3, 6, and 7]. In particular, it is well known that $m(x)\,dx$ and dx are mutually absolutely continuous and

$$m(B(x, 2r)) \leq cm(B(x, r)) \qquad (doubling\ condition). \qquad (1)$$

Ω will always denote an open, connected, bounded region in $\mathbf{R}^n$. Denote by $H(\Omega)$ [resp. $H_0(\Omega)$] the completion of $\mathbf{C}^\infty(\overline{\Omega})$ [resp. $\mathbf{C}_0^\infty(\Omega)$] in the norm

$$\|\varphi\|_\Omega = \left(\int_\Omega |\varphi|^2 m + \int_\Omega |\nabla\varphi|^2 m\right)^{1/2}.$$

An element u of $H(\Omega)$ need not be a distribution, but

$$\nabla u = \left(\frac{\partial u}{\partial x_1}, \ \ldots, \ \frac{\partial u}{\partial x_n}\right)$$

is still a well-defined n-tuple of elements of $L^2(\Omega, m)$ in an appropriate limiting sense [6, Property 1]. The Dirichlet form $D : H(\Omega) \times H(\Omega) \to \mathbf{R}$ is defined by

$$D(u, v) = \int a_{ij}(x)\frac{\partial u}{\partial x_i}(x)\frac{\partial u}{\partial x_j}(x)\,dx.$$

We will say that $Lu = 0$ in Ω if $u \in H(\Omega')$ and $D(u, v) = 0$ for every $v \in H_0(\Omega')$ and every $\Omega' \subset\subset \Omega$.

ESTIMATES

The first step is the inequality

$$\|\varphi\|_\Omega^2 \leq CD(\varphi, \varphi) \text{ for } \varphi \in \mathbf{C}_0^\infty(\Omega).$$

[See the stronger inequality (2) below.] It follows that $D(u, v)$ is an inner product for the Hilbert space $H_0(\Omega)$ and:

Theorem 1. *There exists a unique, bounded linear mapping $G : H_0(\Omega)^* \to H_0(\Omega)$ such that if $T \in H_0(\Omega)^*$ and $u \in G(T)$, then $D(u, v) = T(v)$ for every $v \in H_0(\Omega)$. [$H_0(\Omega)^*$ denotes the dual space to $H_0(\Omega)$.]*

It is easy to verify that the space $H_0(\Omega)^*$ is identified with the space of distributions $T = f_0 - \mathrm{div}\ \vec{f}$, where $\vec{f} = (f_1, \ldots, f_n)$ and $f_i/m \in L^2(\Omega, m)$, $i = 0, 1, \ldots, n$. The action of T on $H_0(\Omega)$

is given by

$$T(v) = \int_\Omega f_0(x) v(x) \, dx + \sum_{i=1}^{n} \int_\Omega f_i(x) \frac{\partial v}{\partial x_i}(x) \, dx.$$

Some of the main results of [6] are:

Theorem 2. If $Lu = 0$ in Ω, then u has a representative that is Hölder continuous in Ω, that is, there exists $\alpha > 0$ such that for any $\Omega' \subset\subset \Omega$, there exists c such that $|u(x) - u(x')| \le c|x - x'|^\alpha$ for x, x' in Ω'.

We will always take the continuous representative of u.

Harnack's Inequality. If $Lu = 0$ in $B(x, 2r)$ and $u \ge 0$ in $B(x, 2r)$, then $\max\limits_{B(x,r)} u \le c \min\limits_{B(x,r)} u$. ($c$ is independent of x and $r > 0$.)

Strong Maximum Principle. If $Lu = 0$ in Ω and u attains its maximum or minimum in Ω, then u is constant.

The main point of the proof is a kind of fractional integration estimate. There exist $k > 1$, $c < \infty$, such that

$$\left(\frac{1}{|B|} \int_B |\varphi|^{2k} m \right)^{1/2k} \le cr \left(\frac{1}{|B|} \int_B |\nabla\varphi|^2 m \right)^{1/2}, \quad \varphi \in \mathcal{C}_0^\infty(B), \; B = B(x, r). \quad (2)$$

It is crucial that c and k are independent of r and x as well as φ. One can then follow Moser's proof of the theorem of de Georgi and Nash to obtain the scale-invariant version of Harnack's inequality above (with suitable interpretations of max and min) and Hölder continuity.

Let $\varphi \in \mathcal{C}^\infty(\mathbf{R}^n)$ and $T = L\varphi = -\text{div } \vec{f}$, where $\vec{f} = (f_1, \ldots, f_n)$,

$$f_j(x) = \sum_i a_{ij}(x) \frac{\partial \varphi}{\partial x_i}(x).$$

Then T belongs to $H_0(\Omega)^*$ with the action

$$T(v) = \int a_{ij} \frac{\partial \varphi}{\partial x_i} \frac{\partial v}{\partial x_j}.$$

Let $u = \varphi - G(T)$. Then $Lu = 0$ in Ω and u is the (variational) solution to the Dirichlet problem: $Lu = 0$ in Ω, $u = \varphi$ on $\partial\Omega$. The weak maximum principle of Stampacchia [5, 1.18] implies that $\sup_\Omega |u| \leq \max_{\partial\Omega} |\varphi|$. It then follows from the methods of the previous theorems that:

Theorem 3. Suppose that $h \in \mathcal{C}(\partial\Omega)$, $\varphi_j \in \mathcal{C}^\infty(\mathbf{R}^n)$, $\varphi_j \to h$ uniformly on $\partial\Omega$; $Lu_j = 0$ in Ω, $u_j = \varphi_j$ on $\partial\Omega$ in the same given above. Then there exists a unique Hölder continuous function u on Ω such that for any $\Omega' \subset\subset \Omega$, $u_j \to u$ uniformly on Ω'. Moreover, $Lu = 0$ in Ω and $\sup_\Omega |u| \leq \max_{\partial\Omega} |h|$.

We will say that a point $y \in \partial\Omega$ is *regular for Ω* if $\lim_{x \to y} u(x) = h(y)$ for every $h \in \mathcal{C}(\partial\Omega)$.

THE GREEN FUNCTION

A complete description of regular points is based on interior estimates for the Green function. Denote $\Sigma = \{x : |x| < R\}$ for some large value of R. Here are some facts about the Green function proved in [5 and 6].

There is a function $g : \overline{\Sigma} \times \overline{\Sigma} \to [0, \infty]$ such that $g(x, y)$ is Hölder continuous in $\overline{\Sigma} \times \overline{\Sigma} \backslash \Delta$ ($\Delta = \{(y, y) : y \in \overline{\Sigma}\}$); $g(x, y) = g(y, x)$, $g(x, y) = 0$ for $y \in \partial\Sigma$, $g(x, \cdot) \in H(\Sigma \backslash \overline{B}(x, r))$, and $Lg(x, \cdot) = 0$ in $\Sigma \backslash \overline{B}(x, r)$ for any $r > 0$. The class of functions ψm such that $\psi \in L^\infty(\Sigma, dx)$ is dense in $H_0(\Sigma)^*$. Furthermore, $G(\psi m)$ is continuous. Let μ be a finite measure supported in Σ and $u(x) = \int_\Sigma g(x, y) d\mu(y)$. Then

$$\int_\Omega G(\psi m) d\mu = \int_\Omega \psi m u \quad \text{for all } \psi \in L^\infty(\Sigma, dx). \tag{3}$$

Recall the notation $\theta(x, r) = r^2 / m\big(B(x, r)\big)$.

Theorem 4. *Let* $x \in \Sigma$ *and let* $d = \mathrm{dist}(x, \partial\Sigma)$. *If* $|x - y| = r < \frac{1}{2}d$, *then**

$$g(x, y) \simeq \int_r^d \theta(x, s)\frac{ds}{s}.$$

($\simeq$ means that the ratio of the two numbers is bounded above and below by positive constants that depend only on the (A_2) or quasi-conformal dilatation constant of m.)

The proof of Theorem 4 requires the notion of capacity

$$\mathrm{cap}_\Sigma(K) = \inf\{D(\varphi, \varphi) : \varphi \geq 1 \text{ on } K, \varphi \in \mathcal{C}_0^\infty(\Sigma)\}.$$

(For many equivalent definitions, see [5].) Capacity also enters in the Wiener test, of course.

Theorem 5 (Wiener test [5, 5.1]). *Let* $\Omega \subset\subset \Sigma$, $y \in \partial\Omega$ *is regular for* Ω *if and only if either*

(a) $\displaystyle\int_0^R \theta(y, s)\frac{ds}{s} < \infty$ *or*

(b) $\displaystyle\int_0^R \mathrm{cap}_\Sigma(\overline{B}(y, s) \cap \Sigma\backslash\Omega)\theta(y, s)\frac{ds}{s} = \infty.$

Conditions (a) and (b) are mutually exclusive. Condition (a) is equivalent to $\mathrm{cap}_\Sigma(\{y\}) > 0$ and also to $\limsup\limits_{x \to y} g(x, y) < \infty$. In fact, $\lim\limits_{x \to y} g(x, y) = 1/\mathrm{cap}_\Sigma(\{y\})$. Possibility (a) does not arise in the uniformly elliptic case. In keeping with the analogy of $\theta(y, s)$ with S^{2-n}, the "dimension" at points of positive capacity is less than two.

---

*This is very slightly different from the statement in [5], but the proof is the same.

HARMONIC MEASURE

For each $x \in \Omega$, *harmonic measure* (at x) is the measure ω^x on $\partial\Omega$ satisfying $u(x) = \int_{\partial\Omega} h \, d\omega^x$, where u and h are as in Theorem 3.

We call a domain Ω *nontangentially accessible* (abbreviated NTA) provided the following three properties hold.

(i)   There exist constants $A > 1$ and $r_0 > 0$ such that for every r, $0 < r < r_0$ and every $y \in \partial\Omega$, there exists $e \in \Omega$ such that $|y - e| < Ar$ and $B(e, r/A) \subset \Omega$.

(ii) Property (i) for the complement of Ω.

(iii) For every $c > 0$, there exists N such that if $0 < \varepsilon < r_0$, $x, y \in \Omega$, $\text{dist}(x, \partial\Omega) < \varepsilon$, $\text{dist}(y, \partial\Omega) > \varepsilon$, and $|x - y| < c\varepsilon$, then there is a sequence of N points, $x = x_1, x_2, \ldots, x_N = y$, such that $|x_j - x_{j+1}| < \varepsilon/A$ and $B(x_j, 2\varepsilon/A) \subset \Omega$.

From now on Ω will denote an NTA domain.

It is easy to see from their proofs that the properties of the Green function for Σ, including Theorem 4, are valid without change for any NTA domain Ω in place of Σ. A consequence of the Wiener test (which can also be checked more directly) is that all boundary points of an NTA domain are regular. Thus, for every $h \in \mathfrak{C}(\partial\Omega)$,

$$u(x) = \int_{\partial\Omega} h \, d\omega^{x_0} \text{ is continuous in } \overline{\Omega} \text{ and } u|_{\partial\Omega} = h. \qquad (4)$$

Fix $x_0 \in \Omega$ and $\omega = \omega^x$. Harnack's inequality implies that harmonic measure ω^x differs from ω by a bounded factor. Hence, the Lebesgue classes $L^p(\partial\Omega, d\omega^x)$, $1 \leq p \leq \infty$, are equivalent for each value of x. The *kernel function* is defined for almost every (ω) $y \in \partial\Omega$ as the Radon Nikodym derivative $K(x, y) = \frac{d\omega^x}{d\omega}(y)$. A nontangential approach region at $y \in \partial\Omega$ is

$$\Gamma_\alpha(y) = \{x : |x - y| < (1 + \alpha)\text{dist}(x, \partial\Omega)\}.$$

The fact that Ω is an NTA domain implies that $\Gamma_\alpha(y)$ is nontrivial for sufficiently large α. The nontangential maximal function is given by $N(u)(y) = \sup\{|u(x)| : x \in \Gamma_\alpha(y)\}$. The particular value of $\alpha > 0$ turns out not to be important, so for simplicity we will not mention it further.

Here is a Fatou-type theorem.

Theorem 6. Let Ω be an NTA domain. If $f \in L^1(\partial\Omega, d\omega)$ and

$$u(x) = \int_{\partial\Omega} f \, d\omega^x,$$

then $Lu = 0$ in Ω, $\lim\limits_{\substack{x \to y \\ x \in \Gamma_\alpha(y)}} u(x) = f(y)$ for almost every (ω) $y \in \partial\Omega$. Moreover,

$$\omega(\{y \in \partial\Omega : N(u)(y) > \lambda\}) \le c\lambda^{-1} \int_{\partial\Omega} |f| \, d\omega.$$

In addition, if $f \in L^p(\partial\Omega, d\omega)$ for $1 < p \le \infty$, then

$$\|N(u)\|_{L^p(\partial\Omega, d\omega)} \le c_p \|f\|_{L^p(\partial\Omega, d\omega)}.$$

Denote

$$Mf(y) = \sup\left\{\frac{1}{\omega(\Delta)} \int_\Delta |f| \, d\omega : \Delta = B(y, r) \cap \partial\Omega, r > 0\right\}.$$

The proof of Theorem 6 is a well-known argument based on (4) and

$$N(u)(y) \le cMf(y) \text{ for } f \in L^1(\partial\Omega, d\omega), \tag{5}$$

and

$$\omega(\{y \in \partial\Omega : Mf(y) > \lambda\}) \le c\lambda^{-1} \int_{\partial\Omega} |f| \, d\omega. \tag{6}$$

The estimate (6) follows from:

Lemma 1. For any $y \in \partial\Omega$, $r > 0$, $\omega(2\Delta) \le c\omega(\Delta)$, where $\Delta = B(y, r) \cap \partial\Omega$ and $2\Delta = B(y, 2r) \cap \partial\Omega$.

The inequality (5) follows from:

Lemma 2. There exists $\alpha > 0$ such that if $y \in \partial\Omega$, $r > 0$ and $e \in \Omega$ are as in (i) above, then

$$K(e, y') \leq c/\omega(\Delta)(\max(|y'-y|/r)^\alpha, 1),$$

where $\nabla = B(y, r) \cap \partial\Omega$.

Caffarelli et al. [1] observed that the simplest way to prove these two lemmas is by means of a comparison of harmonic measure with the Green function first used by Dahlberg [4] in the context of harmonic functions in Lipschitz domains. The comparison is:

Lemma 3. Let $y \in \partial\Omega$, $e \in \Omega$ be as in (i) above, and let $V = B(y, r) \cap \partial\Omega$. If $x \in \Omega \backslash B(y, 4Ar)$, then

$$g(e, x) \simeq \omega^x(\Delta)\theta(e, r).$$

($\simeq$ means that the ratio of the two sides is bounded above and below by positive constants independent of y, e, and r.)

Lemma 1 follows immediately from Lemma 3 and Harnack's inequality. Lemma 2 depends on Lemma 3, (7), Remark 1 below, and the following consequence of Lemma 3 (proved by Dahlberg [4] for harmonic functions in Lipschitz domains [10, 4.10]).

Boundary Harnack Inequality. There are positive constants c_1 and c_2 such that if $y \in \partial\Omega$, u and v are positive solutions to $Lu = Lv = 0$ in Ω that vanish continuously on $B(y, 2r) \cap \partial\Omega$, and $u(e) = v(e)$ with e as in (i), then $c_1 < u(x)/v(x) < c_2$ for $x \in B(y, r) \cap \Omega$.

Proof of Lemma 3. An easy consequence of Hölder continuity at the boundary ([10, 4.1, 4.2]) is that for $z \in \partial B(e, r/2A)$, $\omega^z(\Delta) > c_0 > 0$. Furthermore, a special case of Theorem 4 is

$$g(e, z) \simeq \theta(e, r) \text{ for } z \in \partial B(e, r/2A). \tag{7}$$

The maximum principle in the domain $\Omega \backslash \bar{B}(e, r/2A)$ implies $g(e, x) \leq c\omega^x(\Delta)\theta(e, r)$ for all $x \in \Omega \backslash B(y, 4Ar)$.

The opposite inequality requires three remarks.   Let $B_1 = B(y, r)$, $B_2 = B(y, 2r)$, $B_3 = B(y, 2Ar)$, $B_4 = B(y, 4Ar)$.

Remark 1. Let u satisfy $Lu = 0$ in $\Omega \cap B_4$. Suppose that u vanishes continuously on $B_4 \cap \partial\Omega$; then $u(x) \leq cu(e)$ for all $x \in B_3 \cap \Omega$.

Remark 1 originates with Carleson [2]. For a proof, see [10, 4.4]. Next we have the well-known inequality [6, 2.4.2].

Remark 2. Let u be as in Remark 1; then

$$\left(\int_{B_2 \cap \Omega} |\nabla u|^2 m \right)^{1/2} \leq c \left(r^{-2} \int_{B_3 \cap \Omega} |u|^2 m \right)^{1/2} .$$

Finally, let $T = -\text{div }\vec{f}$ belong to $H_0(\Omega)^*$. Suppose that the support of T is disjoint from x. Then $G(T)$ is continuous at x. A limiting argument based on (3) with μ equal to the unit mass at x and the fact that $g(x, \cdot) \in H(\Omega \backslash \overline{B}(x, s))$ yields:

Remark 3. $G(T)(x) = T(g(x, \cdot)) = \displaystyle\int_\Omega f_j(z) \frac{\partial}{\partial z_j} g(x, z) dz.$

Let $\varphi \in \mathcal{C}_0^\infty(B_2)$ so that $\varphi \geq 0$, $\varphi = 1$ on B_1, and $|\nabla\varphi| \leq 2r^{-1}$. Let $x \in \Omega \backslash B_4$.   Remark 3 implies

$$\omega^x(\Delta) \leq \int \varphi \, d\omega^x = \varphi(x) - G(L\varphi)(x) = -G(L\varphi)(x)$$

$$= -\int a_{ij}(z) \frac{\partial\varphi}{\partial z_i}(z) \frac{\partial}{\partial z_j} g(x, z) dx.$$

By Schwarz's inequality, Remark 2, Remark 1, and (1),

$$\omega^x(\Delta) \leq c \left(\int_\Omega |\nabla\varphi|^2 m \right)^{1/2} \left(\int_{B_2 \cap \Omega} |\nabla_z g(x, z)|^2 m(z) dz \right)^{1/2}$$

$$\leq c \left(\frac{m(B_2)}{r^2} \right)^{1/2} \left(r^{-2} \int_{B_3 \cap \Omega} |g(x, z)|^2 m(z) dz \right)^{1/2}$$

$$\leq c \left(\frac{m(B_2)}{r^2} \right)^{1/2} \left(\frac{m(B_3)}{r^2} \right)^{1/2} g(x, e) < c\theta(e, r)^{-1} g(x, e).$$

THE KERNEL FUNCTION

A consequence of the results of the previous section is that the kernel function $K(x, y)$ is continuous as a function of $g \in \partial\Omega$ (see [9, 1, and 10]). In addition:

Corollary 1. There is a one-to-one correspondence between positive solutions to $Lu = 0$ in Ω and positive measures on $\partial\Omega$ given by

$$u(x) = \int_{\partial\Omega} K(x, y) \, d\mu(y).$$

Also,

$$\lim_{\substack{x \to y \\ x \in \Gamma_a(y)}} u(x) = h(y) \quad a.e. \ (\omega) \ y \in \partial\Omega,$$

where $\mu = h\omega + \nu$, $h \in L^1(\partial\Omega, d\omega)$, and ν is singular with respect to ω.

Following Jerison and Kenig [10], who treated the uniformly elliptic case, we can show that $K(x, y)$ is actually Hölder continuous as a function of y and deduce Hardy space estimates.

Theorem 7 [10, 7.1]. $g(x, z)/g(x_0, z)$ is Hölder continuous for $s \in \overline{\Omega} \setminus \{x, x_0\}$ and $K(x, y) = \lim_{z \to y} g(x, z)/g(x_0, z)$ for every $y \in \partial\Omega$. Furthermore, there is $\alpha > 0$ and $c < \infty$ such that

$$|K(x, y') - K(x, y)| \leq c \left(\frac{|y - y'|}{|x - y|} \right)^{\alpha} K(x, y)$$

for all $x \in \Omega$, $y, y' \in \partial\Omega$.

The proof is based on the boundary Harnack inequality above. An iteration (analogous to Moser's proof that the Harnack inequality implies Hölder continuity) yields:

Theorem 7'. Under the hypotheses of the statement of the boundary Harnack inequality above, $u(x)/v(x)$ is Hölder continuous in $B(y, r) \cap \overline{\Omega}$.

Theorem 7 is proved in the same way.

A function $a \in L^\infty(\partial\Omega, d\omega)$ is a *p-atom* if $\int a\, d\omega = 0$, a is supported in $\Delta = B(y, r) \cap \partial\Omega$ for some $y \in \partial\Omega$, $r > 0$, and $\sup|a| \leq 1/\omega(\Delta)^{1/p}$. The constant function $a \equiv 1$ is also assumed to be a *p-atom*. Let $p \leq 1$. The atomic Hardy space H_{at}^p is the class of f such that

$$\|f\|_{H_{at}^p} = \inf\{(\Sigma|\lambda_j|^p)^{1/p} : f = \Sigma\lambda_j a_j,\ a_j(x)\ \text{are } p\text{-atoms}\}$$

is finite. A more formal definition requires us to make sense of the sum $\Sigma\lambda_j a_j$ in the dual space to a Hölder class. We will spare the reader the technical details. The estimate in Theorem 7 implies:

Corollary 2 [10, 8.13, 8.14]. *There exists* $p_0 < 1$ *such that if* $p_0 < p \leq 1$ *and* $f \in H_{at}^p$, *then* $u(x) = \int_{\partial\Omega} f\, d\omega^x$ *is a solution to* $Lu = 0$ *in* Ω *and*

$$\|N(u)\|_{L^p(\partial\Omega,\, d\omega)} \simeq \|f\|_{H_{at}^p}.$$

Conversely, if $N(u) \in L^1(\partial\Omega, d\omega)$, *then there is a unique* $f \in H_{at}^1$ *such that* $u(x) = \int_{\partial\Omega} f\, d\omega^x$.

The converse for $p < 1$ is open. To prove it one need only show that $L^1(\partial\Omega, d\omega)$ is dense in the space of boundary values of u such that $N(u) \in L^p(\partial\Omega, d\omega)$.

In closing, we pose the question of whether some of these theorems apply to solutions of nonlinear elliptic equations. In particular, it would be interesting to know if the components of a quasiregular mapping of the ball have such a theory (see [12]).

REFERENCES

1. Caffarelli, L.; Fabes, E.; Mortola, S.; and Salsa, S. Boundary behavior of nonnegative solutions of elliptic operators in divergence form. *Ind. Univ. Math. J.* 30 (1981):621-640.

2. Carleson, L.   On the existence of boundary values for harmonic functions in several variables. *Ark. Mat.* 4 (1962):393-399.

3. Coifman, R., and Fefferman, C.   Weighted norm inequalities for maximal functions and singular integrals. *Studia Math.* 51 (1974):241-250.

4. Dahlberg, B. E. J.   Estimates of harmonic measure. *Arch. for Rat. Mech. and Anal.* 65 (1977):275-288.

5. Fabes, E.; Jerison, D.; and Kenig, C.   The Weiner test for degenerate elliptic equations. To appear in *Ann. Inst. Fourier.*

6. Fabes, E.; Kenig, C.; and Serapioni, R.   The local regularity of solutions of degenerate elliptic equations. *Communications in P.D.E.* 7 (1982):77-116.

7. Gehring, F.   The L^p-integrability of the partial derivatives of a quasiconformal mapping. *Acta Math.* 130 (1973):266-277.

8. Hunt, R., and Wheeden, R.   On the boundary values of harmonic functions. *I.A.M.S.* 132 (1968):307-322.

9. Hunt, R., and Wheeden, R.   Positive harmonic functions on Lipschitz domains. *I.A.M.S.* 147 (1970):507-527.

10. Jerison, D., and Kenig, G.   Boundary behavior of harmonic functions in nontangentially accessible domains. To appear in *Advances in Math.*

11. Moser, J.   On Harnack's theorem for elliptic differential equations. *Comm. Pure Appl. Math.* 14 (1961):577-591.

12. Väisälä, J.   A survey of quasiregular maps in $\mathbf{R}^n$. *Proc. Int. Cong. Math. Helsinki* (1978):685-691.

13. Wu, J. M. G.   Comparisons of kernel functions, boundary Harnack principle and relative Faton theorem on Lipschitz domains. *Ann. Inst. Fourier,* Grenoble, 28.4 (1978):147-167.

SUBELLIPTIC EIGENVALUE PROBLEMS

C. Fefferman
Princeton University
D.H. Phong
Columbia University

This article studies subellipticity and asymptotic eigenvalue distribution of second-order partial differential operators with non-negative principal symbols. We obtain definitive results for self-adjoint operators with real coefficients. Deep theorems on subellipticity and related topics have been obtained by Hörmander [6], Kohn [7], and Rothschild and Stein [10] for sums of squares of vector fields; and by Olenik and Radkevitch [9] for general second-order equations; while subelliptic eigenvalue asymptotics were studied by Menikoff and Sjöstrand [8]. The interest of our work is that it gives sharp necessary and sufficient conditions for operators not assumed to be written as sums of squares. It would be interesting to understand also non-self-adjoint equations and equations with complex coefficients.

To fix the notation, let L be a second-order self-adjoint differential operator on a compact manifold M with smooth measure μ. Assume that in local coordinates

$$L = - \sum_{ij} a^{ij}(x) \frac{\partial^2}{\partial x_i \partial x_j} + \sum_j b_j(x) \frac{\partial}{\partial x_j} + c(x),$$

NSF Grant No. MCS80-03072.
NSF Grant No. MCS78-27119.

where a^{ij}, b_j, c are real and $(a^{ij}(x))$ is positive semidefinite. We want to understand the following:

(A)   When does L satisfy the subelliptic estimates

$$\text{Re}\langle Lu, u \rangle + C\|u\|^2 \geq c\|u\|^2_{(\epsilon)} \quad \text{for } u \in C^\infty(M); \tag{1}$$

$$\|Lu\| + C\|u\| \geq c\|u\|_{(2\epsilon)} \qquad \text{for } u \in C^\infty(M)? \tag{2}$$

(B)   What is the asymptotic behavior as $\lambda \to \infty$ of $N(\lambda, L)$, the number of eigenvalues of L which are $< \lambda$?

These questions are answered in terms of a family of "non-Euclidean balls," which we now define.

A tangent vector $X = \sum_j \gamma_j \dfrac{\partial}{\partial x_j}$ at $x \in M$ is said to be *subunit* for L if $(\gamma_j \gamma_k) \leq (a^{jk}(x))$ as matrices. One easily checks that this notion is independent of the particular coordinate chart. For $x \in M$ and $\rho > 0$, the "ball" $B_L(x, \rho)$ consists of all the points $y \in M$ that can be joined to x by a Lipschitz path $\gamma : [0, \rho] \to M$ for which $\dfrac{d}{dt} \gamma(t)$ is a subunit vector for L at $\gamma(t)$ for almost every t.

By $B_E(x, \rho)$ we denote an ordinary Euclidean ball of radius ρ about x. Note that if $-L$ is the Laplacian for a metric ds^2 on M, then $B_L(x, \rho)$ agrees with the usual ball of radius ρ in ds^2. See also Rothschild and Stein [10], in which suitable non-Euclidean balls play an important role.

Theorem 1.   For L as above, the estimates (1) and (2) are equivalent, and they both hold if and only if

$$B_E(x, \rho) \subseteq B_L(x, C\rho^\epsilon) \text{ for } x \in M, \rho > 0.$$

Theorem 2.   Assume L satisfies (1) and (2). Then for large λ, $N(\lambda, L)$ is bounded above and below by constant multiples of

$$\tilde{N}(\lambda, L) = \int_M \frac{d\mu(x)}{\mu(B_L(x, \lambda^{-1/2}))}.$$

The rest of this paper sketches the proofs of Theorems 1 and 2. We make essential use of results and techniques in [2] and [5], with which we assume the reader is familiar. The new tools needed in our proofs are the following geometric lemmas.

Lemma 1. $B_E(x, \lambda) \subseteq B_L(x, C\lambda^\varepsilon)$ *if and only if*

$$B_E(x, \lambda) \subseteq B_{L - \lambda^{2N}\Delta}(x, C\lambda^\varepsilon),$$

provided $N \geq \varepsilon^{-50}$. *Moreover, these inclusions imply that*

$$\mu(B_L(x, \lambda)) \sim \mu(B_{L - \lambda^{2N}\Delta}(x, \lambda)).$$

Lemma 2. *In* R^{n+1}, *set*

$$L_\lambda = -\left(\frac{\partial}{\partial t}\right)^2 - \sum_{ij} a^{ij}(t, x)\frac{\partial^2}{\partial x_i \partial x_j} - \lambda^{2N}\Delta_x;$$

in R^n, *set*

$$\overline{L}_\lambda = -\sum_{ij} \overline{a^{ij}}(x)\frac{\partial^2}{\partial x_i \partial x_j} - \lambda^{2N}\Delta_x,$$

where

$$\overline{a^{ij}}(x) = \lambda^{-1}\int_{|t|\leq\lambda} a^{ij}(t, x)dt.$$

Then the non-Euclidean balls B_{L_λ}, $B_{\overline{L}_\lambda}$ *about the origin are related by* $B_{L_\lambda}(c\lambda) \subseteq \{|t| \leq \lambda\} \times B_{\overline{L}_\lambda}(\lambda) \subseteq B_{L_\lambda}(C\lambda)$.

Lemma 1 and the terms $\lambda^{2N}\Delta_x$ in Lemma 2 are technicalities; while Lemma 2 contains new geometric information, which is essential both here and in our paper [4].

To prove Lemma 2, we need the following result, which forms an L^∞-analogue of the spectral decomposition theorem in [5].

For

$$L = -\sum_{ij} a^{ij}(x)\frac{\partial^2}{\partial x_i \partial x_j} + \cdots$$

defined on the unit cube Q^0, and for $u \in L^\infty(Q^0)$, define

$$L_K(u) = \inf_{v \in C^1(Q^0)} \left\{ K\|u - v\|^2 + \left\| \sum_{ij} a^{ij}(x) \frac{\partial v}{\partial x_i} \frac{\partial v}{\partial x_j} \right\| + K^{-2N} \|\nabla u\|^2 \right\},$$

where the norms are taken in $L^\infty(Q^0)$.

Lemma 3. $c_N (L + L')_K (u) \le L_K(u) + L'_K(u)$.

We first sketch how Lemmas 1 and 2 yield Theorems 1 and 2
by using [2 and 5]. Next we prove Lemma 1 and reduce Lemma 2 to
Lemma 3. Finally, we sketch the proof of Lemma 3.

Proof of Theorem 1.   It is easy to show that (1) and (2) are equivalent. In fact, (2) asserts that $(L + CI)^2 > \Lambda^{4\varepsilon}$ as self-adjoint
operators on $L^2(M)$. One knows from operator theory that $A^2 \ge B^2$
implies $A \ge B$ for positive operators A, B, so $L + CI \ge \Lambda^{2\varepsilon}$, which
is (1). On the other hand, assume (1), and note that
$\mathrm{Re}\langle Lu, \Lambda^{2\varepsilon} u \rangle \le c_1 \|u\|_{(2\varepsilon)}^2 + 10c_1^{-1} \|Lu\|^2$ for a small c_1 to be picked
later. However, $\mathrm{Re}\langle Lu, \Lambda^{2\varepsilon} u \rangle = \mathrm{Re}\langle L\Lambda^\varepsilon u, \Lambda^\varepsilon u \rangle + \mathrm{Re}\langle \Lambda^\varepsilon [\Lambda^\varepsilon, L]u, u \rangle$.
One computes that $\Lambda^\varepsilon [\Lambda^\varepsilon, L] = T_1 + T_2$ with $T_1 \in S^{2\varepsilon+1}$ skew-adjoint
and $T_2 \in S^{2\varepsilon}$. Therefore,

$$\left| \mathrm{Re}\langle \Lambda^\varepsilon [\Lambda^\varepsilon, L]u, u \rangle \right| = \left| \mathrm{Re}\langle T_2 u, u \rangle \right| \le c_1 \|u\|_{(2\varepsilon)}^2 + 10c_1^{-1} \|u\|^2.$$

Moreover, our assumption (1) implies

$$\mathrm{Re}\langle L\Lambda^\varepsilon u, \Lambda^\varepsilon u \rangle \ge c\|u\|_{(2\varepsilon)}^2 - C\|u\|_{(\varepsilon)}^2 \ge \frac{1}{2} c\|u\|_{(2\varepsilon)}^2 - C'\|u\|^2.$$

Putting these estimates together, we obtain

$$\frac{c}{2}\|u\|_{(2\varepsilon)}^2 - C'\|u\|^2 - c_1 \|u\|_{(2\varepsilon)}^2 - 10c_1^{-1}\|u\|^2 \le c_1\|u\|_{(2\varepsilon)}^2 + 10c_1^{-1}\|Lu\|^2,$$

which implies (2) if we pick $c_1 < \frac{c}{10}$. We are indebted to J. J. Kohn
and E. M. Stein for stimulating conversations on estimate (2).

The hard part of Theorem 1 is to show that (1) is equivalent
to $B_E(x, \lambda) \le B_L(x, C\lambda^\varepsilon)$. To prove this, we formulate a stronger
result, which can be proved by induction on the dimension.

Given a second-order operator L defined on R^n, and large
constants K, S, we say that $L \ge K$ *microlocally in* $|x| < 1$, $|\xi| \sim S$

if $\theta(x, D)^*(L - KI)\theta(x, D) \geq -CI$ for some symbol $\theta \in S^0$ satisfying $\theta(x, \xi) = 1$ for $|x| \leq 1$, $|\xi| \sim S$.

One checks easily that (1) amounts to saying that in local coordinates $L \geq$ (const)K microlocally in $|x| \leq 1$, $|\xi| \sim S$, with $K = S^{2\epsilon}$.

Therefore, Theorem 1 will follow at once from Lemma 1 and the

Main Estimate. *Assume* $K \geq S^\delta$ *and* $N \geq N_{\min}(\delta)$. *Then*

(i) $L \geq$ (const)K *microlocally in* $|x| \leq 1$, $|\xi| \sim S$ *iff*

(ii) $B_E(x, (\text{const})S^{-1}) \subseteq B_{L-K^{-\eta}\Delta}(x, K^{-1/2})$.

So out task is to prove the main estimate. In one variable, the result is trivial; in n dimensions we proceed by induction on n as follows.

We first reduce matters to the case

$$L = -\left(\frac{\partial}{\partial t}\right)^2 + \tilde{L}\left(t, y, \frac{\partial}{\partial y}\right).$$

This is done by localizing. We make a Calderón–Zygmund decomposition of $\{|x| \leq 1\}$ into cubes $\{Q_\nu\}$ of diameters δ_ν, stopping at Q_ν when $\max_{ij} \max_{x \in Q_\nu^*} |a^{ij}(x)| \geq C\delta_\nu^2$. $\left(\text{Recall}\right.$

$$L = -\sum_{ij} a^{ij}(x)\frac{\partial^2}{\partial x_i \partial x_j} + \cdots .\bigg)$$

We may assume all the $\delta_\nu \geq S^{-1}$, since otherwise (i) and (ii) are both false. In each Q_ν, make a change of coordinates from $x \in Q_\nu^*$ to $(t, y) \in$ unit cube, so that L goes over to

$$L_\nu = -\left(\frac{\partial}{\partial t}\right)^2 + \tilde{L}_\nu\left(t, y, \frac{\partial}{\partial y}\right).$$

Now the symbolic calculus of pseudodifferential operators shows that $L \geq$ (const)K microlocally for $|\xi| \sim S$ if and only if each $L_\nu \geq$ (const)K microlocally for $|(\tau, \eta)| \sim S\delta_\nu$. Moreover, under the coordinate change $x \to (t, y)$, $B_E(x, S^{-1})$ goes over to $B_E((t, y), (S\delta_\nu)^{-1})$, while $B_{L + \text{junk}}(x, K^{-1/2})$ goes over to $B_{L + \text{junk}}((t, y), K^{-1/2})$. Thus, the main estimate holds for L if and only if it holds for L_ν. So it is enough to look at

$$L = -\left(\frac{\partial}{\partial t}\right)^2 + L(t, y, \partial y).$$

Also, we may assume $K \leq S^2$, since otherwise (i) and (ii) are again obviously false. Now, however, Lemma 2 shows that in place of $B_{L - K^{-N_\Delta}}((\bar{t}, \bar{y}), K^{-1/2})$, we may take

$$\{|t - \bar{t}| \leq K^{-1/2}\} \times B_{\bar{L}_{\bar{t} - K} - N_\Delta}(\bar{y}, K^{-1/2})$$

in (ii). Here

$$\bar{L}_{\bar{t}}(y, \partial y) = K^{1/2} \int_{|t - \bar{t}| \leq K^{-1/2}} \tilde{L}(t, y, \partial_y) dt.$$

Also, the results in Fefferman and Phong [5] show that (i) is equivalent to saying that $\bar{L}_{\bar{t}} \geq (\text{const})K$ microlocally in $|y| \leq 1$, $|\eta| \sim S$ for each $\bar{t}$. Therefore, our main estimate holds for L, provided it holds for $\bar{L}_{\bar{t}}$ for every $\bar{t}$. Since $\bar{L}_{\bar{t}}$ is an operator in $n - 1$ variables, the induction step is complete. So we know both the main estimate and Theorem 1.

Proof of Theorem 2. Using symbolic calculus and the minimax formula for eigenvalues, one can easily localize the problem by assuming L to be elliptic outside $\{|x| \leq 1\}$. Another application of symbolic calculus localizes the problem still further to the operators L_ν in the proof of Theorem 1. So we may assume

$$L = -\left(\frac{\partial}{\partial t}\right)^2 + \tilde{L}(t, y, \partial y)$$

with L elliptic outside $\{|t|, |y| \leq 1\}$. Since L is subelliptic, one checks that $N(\lambda, L)$ is comparable to $N(\lambda; L - \lambda^{-2N}\Delta)$ for large N depending on the ε in (1). So we may also assume that

$$\tilde{L}(t, y, \partial_y) = -\sum_{ij} a^{ij}(t, y)\frac{\partial^2}{\partial y_j \partial y_k} + \cdots,$$

with $(a^{ij}(t, y)) \geq \lambda^{-2N}(\delta_{ij})$. (This uses Lemma 1.) Now define

$$\bar{L}_{\bar{t}}(y, \partial_y) = \lambda^{+1/2} \int_{|t - \bar{t}| \le \lambda^{+1/2}} \tilde{L}(t, y, \partial_y) dt.$$

Lemma 2 shows that $B_L((t, y), \lambda^{1/2})$ is comparable to $\{|t - \bar{t}| \le \lambda^{-1/2}\} \times B_{\bar{L}_{\bar{t}}}(y, \lambda^{-1/2})$, while the results of Fefferman and Phong [2] show that $N(\lambda, L)$ is comparable to

$$\lambda^{+1/2} \int_{|\bar{t}| \le 1} N(\lambda, \bar{L}_{\bar{t}}) d\bar{t}$$

for large λ. Therefore, the formula

$$N(\lambda, L) \sim \tilde{N}(\lambda, L) = \int_M \frac{d\mu(x)}{\mu(B_L(x, \lambda^{-1/2}))}$$

is easily deduced from the corresponding formula for the $\bar{L}_{\bar{t}}$. Since $\bar{L}_{\bar{t}}$ is an operator in $n - 1$ variables, we can now proceed by induction on the dimension.

The above argument shows that N and $\tilde{N}$ are comparable in the sense that $c\tilde{N}(c\lambda, L) \le N(\lambda, L) \le C\tilde{N}(C\lambda, L)$. A final application of Lemma 2 gives $\tilde{N}(C\lambda, L) \le C'\tilde{N}(\lambda, L)$ and $\tilde{N}(c\lambda, L) \ge c'N(\lambda, L)$, so that $c''\tilde{N}(\lambda, L) \le N(\lambda, L) \le C''\tilde{N}(\lambda, L)$ for large λ.

Proof of Lemma 1. Assuming $B_E(x, \lambda^{N/2}) \subseteq B_{L - \lambda^{2N}\Delta}(x, \lambda)$ for $x \in M$, $\lambda > 0$, we shall prove that

(+) $$B_{L - \lambda^{2N}\Delta}(x, \lambda) \subseteq B_L(x, C\lambda).$$

This easily implies Lemma 1. In fact, using $C\lambda^\varepsilon$ in place of λ and εN in place of N in (+), we obtain $B_{L - \lambda^{2N}\Delta}(x, C\lambda^\varepsilon) \subseteq B_L(x, C'\lambda^\varepsilon)$, so that $B_E(...) \subseteq B_{L - \lambda^{2N}\Delta}(...)$ implies $B_E(...) \subseteq B_L(...)$. The converse implication is trivial, since $B_L(x, C\lambda^\varepsilon) \subseteq B_{L - \lambda^{2N}\Delta}(x, C\lambda^\varepsilon)$. Inclusion (+) also yields the part of Lemma 1 on $\mu(B_L(x, \lambda))$, since $\mu(B_L(x, \lambda)) \le \mu(B_{L - \lambda^{2N}\Delta}(x, \lambda)) \le \mu(B_L(x, C\lambda))$. This implies $\mu(B_{L - c\lambda^{2N}\Delta}(x, c\lambda)) \le \mu(B_L(x, \lambda)) \le \mu(B_{L - \lambda^{2N}\Delta}(x, C\lambda))$.

An application of Lemma 2 shows that the terms on the extreme left and right are comparable, so that indeed (+) implies all of Lemma 1.

We proceed to prove (+).

Given a point $y \in B_{L-\lambda^{2N}\Delta}(x, \lambda)$, we may join x to y by a geodesic $\gamma : [0, \lambda] \rightarrow M$ in the metric

$$ds^2 = \sum_{ij} g_{ij}(x) dx_i dx_j,$$

where

$$(g_{ij})^{-1} = (a^{ij} + \lambda^{2N}\delta_{ij}) > 0, \quad L = -\sum_{ij} a^{ij}(x)\frac{\partial^2}{\partial x_i \partial x_j} + \cdots .$$

Thus $\gamma(0) = x$, $\gamma(\lambda) = y$, and $\frac{d}{dt}\gamma(t)$ is a subunit vector for $L - \lambda^{2N}\Delta$.

We shall perturb $\gamma(t)$ to a broken path $\gamma_\#(t)$ that starts at x and ends very near y, so that $\frac{cd\gamma_\#(t)}{dt}$ is a subunit vector for L.

We construct $\gamma_\#$ using the following elementary remarks.

(a) PERTURBATION OF TANGENT VECTORS

Say that $X = \sum_j \gamma_j \frac{\partial}{\partial x_j}$ is a subunit vector for $L - \lambda^{2N}\Delta$ at x^0, and suppose $|x' - x^0| < c_1\lambda^N$. Then there is a tangent vector $Y = \sum_j \gamma_j' \frac{\partial}{\partial x_j}$ at x' satisfying

(i) $|\gamma_j - \gamma_j'| \leq C\lambda^N$,

(ii) cY is a subunit vector for L at x',

(iii) $\gamma_i' = \sum_j a^{ij}(x')\xi_j'$ with $|\xi'| \leq C\lambda^{-N}$.

For our hypothesis on X is that

$$\left(\sum_j \gamma_1 n_1\right)^2 \leq \sum_{ij}(a^{ij}(x^0) + \lambda^{2N}\delta_{ij})n_i n_j$$

for $\eta \in R^n$. Applying Lemma 4.1 in [5] to the function

$$f(t) = |\eta|^{-2} \sum_{ij}(a^{ij}(x' + t(x^0 - x')) + \lambda^{2N}\delta_{ij})n_i n_j \geq 0$$

yields

$$\sum_{ij} (a^{ij}(x^0) + \lambda^{2N}\delta_{ij})\eta_i\eta_j \leq C\sum_{ij} (a^{ij}(x') + \lambda^{2N}\delta_{ij})\eta_i\eta_j$$
$$+ C|x' - x^0|^2|\eta|^2,$$

so that for $|x^0 - x'| \leq c_1\lambda^N$, we have

$$c\left(\sum_j \gamma_j\eta_j\right)^2 \leq \sum_{ij} (a^{ij}(x') + \lambda^{2N}\delta_{ij})\eta_i\eta_j.$$

Thus $c\widetilde{X}$ is a subunit vector for $L - \lambda^{2N}\Delta$ at x', where $\widetilde{X} = \sum_j \gamma_j\dfrac{\partial}{\partial x_j}$ at x'. Next we rotate the coordinate axes so that $a^{ij}(x')$ is diagonalized: $(a_{ij}(x')) = (\lambda_i\delta_{ij})$, $\lambda_i \geq 0$. Note that conclusions (i), (ii), and (iii) are unaffected by the rotation. Now we know that $c\widetilde{X}$ is a subunit for $L - \lambda^{2N}\Delta$, that is,

$$\sum_i \frac{\gamma_i^2}{\lambda_i + \lambda^{2N}} \leq C.$$

Define $Y = \sum_i \gamma_i'\dfrac{\partial}{\partial x_i}$ at x', where $\gamma_i' = \gamma_i$ if $\lambda_i \geq \lambda^{2N}$, $\gamma_i' = 0$ if $\lambda_i < \lambda^{2N}$. Setting $\xi_i' = \gamma^i/\lambda_i$ for $\lambda_i \geq \lambda^{2N}$, $\xi_i' = 0$ for $\lambda_i < \lambda^{2N}$, we see easily that (i), (ii), and (iii) hold.

(b) ESTIMATES FOR SECOND DERIVATIVES OF HAMILTONIAN PATHS

Say $H = \dfrac{1}{2}\sum_{ij} a^{ij}(x)\xi_i\xi_j$ and initially $|\xi| \leq C\lambda^{-N}$. Hamiltonian's equations

$$\dot{x}^i = \sum_j a^{ij}(x)\xi_j, \qquad \dot{\xi}_1 = -\sum_{ij} \frac{\partial a^{ij}(x)}{\partial x_1}\xi_i\xi_j,$$

$$\ddot{x}^i = \sum_j a^{ij}(x)\dot{\xi}_j + \sum_{j\ell} \frac{\partial a^{ij}(x)}{\partial x_\ell}\dot{x}_\ell\xi_j$$

imply $|\dot{x}| \leq C|\xi|$, $|\dot{\xi}| \leq C|\xi|^2$, $|\ddot{x}| \leq C|\xi|^2$. So if we flow for time $\Delta t \leq c_1\lambda^N$, then $|\dot{\xi}| \leq C'\lambda^{-N}$, $|\ddot{x}| \leq C''\lambda^{-2N}$ along the path.

Now we can construct the broken path $\gamma_\#(t)$ mentioned above. Suppose we have constructed $\gamma_\#(t)$ for $0 \leq t \leq \tau_\kappa$, and that $\gamma_\#(0) = \gamma(0)$, $|\gamma(t) - \gamma_\#(t)| \leq C_+\lambda^N\tau_\kappa$ for $0 \leq t \leq \tau_\kappa$. (The large constant C_+ will be picked later. Note that the assertion is vacuous for $\tau_0 = 0$.) Apply (a) above with $x^0 = \gamma(\tau_\kappa)$, $x' = \gamma_\#(\tau_\kappa)$, and

$X = \frac{d}{dt} \gamma(t) \Big|_{\tau_\kappa}$.   Since λ is small, we have

$$|x^0 - x'| \leq C_+ \lambda^N \tau_\kappa \leq C_+ \lambda^{N+1},$$

so the hypotheses of (a) are satisfied. With Y, ξ' as in (a), we now define $\gamma_\#(t)$ for $\tau_\kappa \leq t \leq \tau_{\kappa+1}$ as the projection onto the x-coordinate of the Hamiltonian curve for

$$H = \frac{1}{2} \sum_{ij} a^{ij}(x) \xi_i \xi_j$$

starting at (x', ξ') for $t = \tau_\kappa$. Since $\frac{d}{dt} \gamma_\# \Big|_{\tau_\kappa} = Y$, and since H is conserved along the path, it follows from (a)(ii) above that $c \frac{d}{dt} \gamma_\#$ is a subunit vector for L throughout $\tau_\kappa \leq t \leq \tau_{\kappa+1}$. The estimates of (b) show that $\left| \frac{d^2}{dt^2} \gamma_\# \right| \leq C\lambda^{-2N}$ for $\tau_\kappa \leq t \leq \tau_{\kappa+1}$, provided $\tau_{\kappa+1} - \tau_\kappa < c_1 \lambda^N$. Similarly, $\left| \frac{d^2}{dt^2} \gamma \right| \leq C\lambda^{-2N}$, while $\left| \frac{d\gamma}{dt}(\tau_\kappa) - \frac{d\gamma_\#}{dt}(\tau_\kappa) \right| = |\overline{X} - \overline{Y}| \leq C\lambda^N$ by (a)(i). So if $\tau_{\kappa+1} - \tau_\kappa = \lambda^{3N}$, then we have for $\tau_\kappa \leq t <\_\tau_{\kappa+1}$ that

$$\left| \frac{d\gamma}{dt} - \frac{d\gamma_\#}{dt} \right| \leq C\lambda^{+N} + C\lambda^{-2N} |t - \tau_\kappa| \leq C'\lambda^N.$$

Our assumption $|\gamma(\tau_\kappa) - \gamma_\#(\tau_\kappa)| \leq C_+ \lambda^N \tau_\kappa$ now implies for $\tau_\kappa \leq t \leq \tau_{\kappa+1}$ that

$$|\gamma(t) - \gamma_\#(t)| \leq C_+ \lambda^N \tau_\kappa + C'\lambda^N (t - \tau_\kappa) \leq C_+ \lambda^N \tau_{\kappa+1},$$

provided we pick $C_+ \geq C'$. Our construction of $\gamma_\#(t)$ is complete, with $\tau_\kappa = \kappa \cdot \lambda^{3N}$.

The proof of (+) is now easy. Given $y \in B_{L-\lambda^{2N}\Delta}(x, \lambda)$ and γ as above, we constructed a broken curve $\gamma_\# : [0, \lambda] \to M$ so that $c \frac{d}{dt} \gamma_\#$ is subunit for L, while $\gamma_\#(0) = x$, $|\gamma_\#(\lambda) - y| \leq C_+ \lambda^{N+1}$. Thus $y_\# = \gamma_\#(\lambda) \in B_L(x, C\lambda)$, while $y \in B_E(y_\#, \lambda^N) \subseteq B_{L-(\lambda^2)^{2N}\Delta}(y^\#, \lambda^2)$. Repeating the process yields a sequence of paths $\gamma_\#^\mu : [t_\mu, t_{\mu+1}] \to M$ with tangent vectors subunit for L, so that $\gamma_\#^0$ joins $x = \gamma_\#^0 (t_0 = 0)$ to $y_\#^1 = \gamma_\#(t_1)$, $\gamma_\#^\mu$ joins $y_\#^\mu$ to $y_\#^{\mu+1}$, $y_\#^\mu \to y$, and $\sum_\mu t_\mu \leq C'\lambda$.

Combining the $\gamma_\#^\mu$ into a single Lipschitz path $\gamma_{\#\#}$, we see that $y \in B_L(x, C'\lambda)$ as required.

Proof of Lemma 2.   Since $(a^{ij}(t, x)) \leq C(\overline{a^{ij}}(x) + \lambda^{2N}\delta_{ij})$ for $|t| \leq \lambda$, we see at once that $B_{L_\lambda}(c\lambda) \subseteq B_{L_\lambda^*}(\lambda) \subseteq \{|t| \leq \lambda\} \times B_{\overline{L}_\lambda}(\lambda)$, where $L_\lambda^* = (\partial/\partial t)^2 + \overline{L}_\lambda(y, \lambda y)$.   So our problem is to show that $\{|t| \leq \lambda\} \times B_{\overline{L}_\lambda}(\lambda) \subseteq B_{L_\lambda}(C\lambda)$.   To see this, let $u(t, y)$ be the distance from (t, y) to the origin induced by L_λ.   Note that $u_\lambda \in \text{Lip}(1)$, since $(a^{ij} + \lambda^{2N}\delta_{ij}) > 0$.   Since $\partial/\partial t$ is a subunit vector for L_λ, we have $t^{-2}|u_\lambda(t, x) - u_\lambda(0, x)|^2 \leq 1$, while also

$$\sum_{ij}(a^{ij}(t, y) + \lambda^{2N}\delta_{ij})\frac{\partial u_\lambda(t, y)}{\partial y_i}\frac{\partial u_\lambda(t, y)}{\partial y_j} \leq 1,$$

where u_λ is differentiable.   By mollifying $u_\lambda(t, \cdot)$ slightly, we obtain $v = v_{\lambda, t} \in C^\infty$, so that

$$t^{-2}\|u_\lambda(0, \cdot) - v\|^2 + \left\|\sum_{ij}a^{ij}(t, \cdot)\frac{\partial v}{\partial y_i}\frac{\partial v}{\partial y_j}\right\| + \lambda^{2N}\|\nabla_y v\|^2 \leq 10,$$

the norms being taken in L^∞.   That is, with

$$L(t) = -\sum_{ij}a^{ij}(t, y)\frac{\partial^2}{\partial y_i \partial y_j},$$

$(L(t))_{\lambda^{-2}}(u_\lambda(0, \cdot)) \leq 10$ for $|t| \leq \lambda$.   Taking $t_\ell = \ell\lambda/N^2$, $\ell = 0, \ldots, N^2$, and applying Lemma 3 repeatedly, we obtain

$$\left(\sum_\ell L(t_\ell)\right)_{\lambda^{-2}}(u(0, \cdot)) \leq C_N,$$

which implies easily $(\overline{L}_\lambda)_{\lambda^{-2}}(u(0, \cdot)) \leq C_N'$.   Thus, for a suitable C^1 function v we have

$$\lambda^{-2}|u_\lambda(0, y) - v(y)|^2 + \sum_{ij}a^{ij}(y)\frac{\partial v(y)}{\partial y_i}\frac{\partial v(y)}{\partial y_j} + \lambda^{2N}|\nabla v(y)|^2 \leq C$$

for all $y \in M$.   In particular, $|v(0)| \leq C\lambda$.   Moreover, let $\gamma : [0, \lambda] \to M$ be a Lipschitz path with $d\gamma/dt$ subunit for $\overline{L}_\lambda$.   Our estimates on v and the definition of subunit vectors together yield

$$|v(\gamma(\lambda)) - v(\gamma(0))| \leq \int_0^\lambda |\langle dv, \dot\gamma(t)\rangle|dt,$$

while

$$\langle dv, \ \dot{\gamma}(t) \rangle^2 \le \sum_{ij} a^{ij}(y) \frac{\partial v}{\partial y_i} \frac{\partial v}{\partial y_j} + \lambda^{2N} |\nabla v(y)|^2 \text{ [with } y = \gamma(t)] \le C.$$

Therefore, $|v(\gamma(\lambda)) - v(\gamma(0))| \le C\lambda$, and it follows that $|v(y)| \le C\lambda$ for $y \in B_{\overline{L}_\lambda}(\lambda)$. The estimate defining v now shows that also $u_\lambda(0, y) \le C'\lambda$ for $y \in B_{\overline{L}}(\lambda)$. Recalling that $|u_\lambda(t, y) - u_\lambda(0, y)| \le |t|$, we obtain $u_\lambda(t, y) \le C''\lambda$ for $(t, y) \in \{|t| \le \lambda\} \times B_{\overline{L}_\lambda}(\lambda)$. By definition of u_λ, this means that $\{|t| \le \lambda\} \times B_{L_\lambda}^-(\lambda) \subseteq B_{L_\lambda}(C\lambda)$, as needed.

Sketch of the Proof of Lemma 3. We use induction on the dimension n. The result being trivial in $n = 0$, we assume Lemma 3 is known in R^{n-1} and deduce successively the following seven lemmas.

Lemma 3.1. *Let* $A = \sum_j a^j(x) \frac{\partial}{\partial x_j}$ *be a first-order operator in* R^{n-1}, *and let* $u, v, w: I \to C^1(R^{n-1})$, *where* I *is an interval of length* $\sim K^{-1/2}$. *Set* $u^0 = \frac{1}{|I|} \int_I u(t) dt$. *Then there exists* $u^+ \in C^1(R^{n-1})$ *for which*

$$K\|u^0 - u^+\|^2 + \|Au^+\|^2 + K^{-2N}\|\nabla_x u^+\|^2 \le C \sup_I \Big\{ K\|u(t) - v(t)\|^2$$

$$+ \left\|\frac{\partial v}{\partial t}\right\|^2 + K\|u(t) - w(t)\|^2$$

$$+ \left\|\left(\frac{\partial}{\partial t} - A\right)w(t)\right\|^2 + K^{-2N}\|\nabla_x w\|^2 \Big\},$$

the norms being taken in L^∞ *(unit cube in* R^{n-1}).

Lemma 3.2. *Set* $I_\ell = \{|t - \ell K^{-1/2}| \le 10K^{-1/2}\}$, $I^+ = $ *union of consecutive* I_ℓ's,

$$L(t) = -\sum_{i,j=1}^{n-1} a^{ij}(t, x) \frac{\partial^2}{\partial x_i \partial x_j},$$

where a^{ij} *is smooth and* $(a^{ij}) \ge K^{-2N}(\delta_{ij})$. *Then given* C^1-*functions* $u(t, x)$ *on* $I^+ \times$ *unit cube,* u_ℓ *and* v_ℓ *on* $I_\ell \times$ *unit cube, we can find* $v \in C^1(I^+ \times $ *cube) so that*

$$K\|u - v\|^2 + \left\|\frac{\partial v}{\partial t}\right\|^2 + \left\|\sum_{ij} a^{ij}(t, x)\frac{\partial v}{\partial x_i}\frac{\partial v}{\partial x_j}\right\|$$

$$\leq C \max_{\ell}\left\{K\|u - v_{\ell}\|^2 + \left\|\frac{\partial v_{\ell}}{\partial t}\right\|^2 + K\|u - u_{\ell}\|^2\right.$$

$$\left. + \left\|\sum_{ij} a^{ij}(t, x)\frac{\partial u_{\ell}}{\partial x_i}\frac{\partial u_{\ell}}{\partial x_j}\right\|\right\}.$$

Here the norms on the left are taken in $L^{\infty}(I^+ \times \text{cube})$, while those on the right are taken in $L^{\infty}(I_{\ell} \times \text{cube})$.

Lemma 3.3. *Set* $A(t) = \sum_{j=1}^{n-1} a^j(t, x)\frac{\partial}{\partial x_j}$ *with* a^j *smooth, and let* I *be an interval of length* $\sim K^{-1/2}$. *Given* C^1-*functions* u, v, w *on* $I \times$ *unit cube, we can find* $u^{\#}$ *so that*

$$K\|u - u^{\#}\|^2 + \left\|\frac{\partial u^{\#}}{\partial t}\right\|^2 + \|A(t)u^{\#}\|^2 + K^{-2N}\|\nabla_x u^{\#}\|^2$$

$$\leq C\left\{K\|u - v\|^2 + \left\|\frac{\partial v}{\partial t}\right\|^2 + K\|u - w\|^2\right.$$

$$\left. + \left\|\left(\frac{\partial}{\partial t} + A(t)\right)w\right\|^2 + K^{-2N}\|\nabla_x w\|^2\right\},$$

the norms being taken in $L^{\infty}(I \times \text{cube})$.

Lemma 3.4. *Let the unit cube in* R^n *be partitioned into subcubes* $\{Q_{\nu}\}$ *of diameters* δ_{ν}, *where* $Q_{\nu}^* \cap Q_{\mu}^* \neq \emptyset$ *implies* $\delta_{\nu} \sim \delta_{\mu}$.

Let

$$L = -\sum_{ij} a^{ij}(x)\frac{\partial^2}{\partial x_i \partial x_j}$$

for smooth functions a^{ij} *satisfying* $K^{-2N}(\delta_{ij}) \leq (a^{ij}(x)) \leq C\delta_{\nu}^2(\delta_{ij})$ *for* $x \in Q_{\nu}$. *Then given* $u \in L^{\infty}$ *and* C^1-*functions* v_{μ} *defined on* Q_{ν}^*, *we have*

$$L_K(u) \leq C \max_{\mu}\left\{K\|u - v_{\mu}\|_{L^{\infty}(Q_{\mu}^*)} + \left\|\sum_{ij} a^{ij}(x)\frac{\partial v_{\mu}}{\partial x_i}\frac{\partial v_{\mu}}{\partial x_j}\right\|_{L^{\infty}(Q_{\mu}^*)}\right\}.$$

Lemma 3.5. *Assume* $L = -\sum_{ij} a^{ij}(x)\frac{\partial^2}{\partial x_i \partial x_j} + \cdots$ *with* $\max_{ij}|a^{ij}| \geq C$, *and set*

$$L' = -\left(\frac{\partial}{\partial x_1}\right)^2 - K^{-2N}\Delta.$$

Then

$$c (L + L')_K (u) \le L_K (u) + L'_K (u).$$

Lemma 3.6. Set

$$L' = -\left(\frac{\partial}{\partial x_1}\right)^2 - K^{-2N}\Delta, \; L = -\sum_{ij} a^{ij} (x) \frac{\partial^2}{\partial x_i \, \partial x_j},$$

where $(a^{ij}(x))$ is smooth and positive semidefinite.   Then

$$c (L + L')_K (u) \le L_K (u) + L'_K (u).$$

Lemma 3.7.   Let

$$L = -\sum_{ij} a^{ij} (x) \frac{\partial^2}{\partial x_i \, \partial x_j}, \; L' = -\sum_{ij} a'^{ij} (x) \frac{\partial^2}{\partial x_i \, \partial x_j},$$

and assume $\max_{ij} |a^{ij}| \ge C$.   Then

$$c (L + L')_K (u) \le L_K (u) + L'_K (u).$$

For the most part, these lemmas are proved in strict analogy with Lemmas 1, 2, 3, 5, 7, 8, and 9 in the proof of the spectral decomposition theorem in [5].   Instead of L^2-norms, one uses L^∞-norms; and for

$$L = -\sum_{ij} a^{ij} (x) \frac{\partial^2}{\partial x_i \, \partial x_j},$$

we work with $\left\| \sum_{ij} a^{ij} (x) \frac{\partial v}{\partial x_i} \frac{\partial v}{\partial x_j} \right\|_{L^\infty}$ in place of $\mathrm{Re}\langle Lv, v \rangle$.

In some ways, the present lemmas are easier than their L^2-analogues in [5]. For instance, where [5] uses the sharp Gårding inequality of [1], we use here merely the trivial estimate

$$\left\| \sum_{ij} b^{ij} (x) \frac{\partial v}{\partial x_i} \frac{\partial v}{\partial x_j} \right\|_{L^\infty} \le \left\| \sum_{ij} a^{ij} (x) \frac{\partial v}{\partial x_i} \frac{\partial v}{\partial x_j} \right\|_{L^\infty}$$

for $(b^{ij}) \le (a^{ij})$. Also, the Fourier integral operators of [5] are now replaced by simple changes of variable $y = \Phi(x)$.

It is now routine for a patient reader to reconstruct the proofs of Lemmas 3.2 through 3.7, given Lemma 3.1.   Moreover,

Lemma 3.7 implies Lemma 3 in R^n, just as Lemma 6.9 implies the spectral decomposition theorem in [5]. Consequently, the proof of Lemma 3 is reduced to Lemma 3.1. Since the L^2-analogue of 3.1 in [5] was proved using the spectral theorem, we have to make a new argument to show (3.1). Since space is limited, we omit all details of Lemmas 3.2-3.7 and Lemma 3.7 → Lemma 3 in R^n. We close our article with the proof of Lemma 3.1. First suppose $A = \partial/\partial x_1$, and let Ω denote the right-hand side of the estimate in Lemma 3.1.

Since $K\|u - v\|^2 + \left\|\dfrac{\partial v}{\partial t}\right\|^2 \leq \Omega$, we have

$$K|u(t + h, x_1, x') - u(t, x_1, x')|^2 \leq C\Omega$$

for $|h| \leq K^{-1/2}$. Also $K\|u - w\|^2 + \left\|\left(\dfrac{\partial}{\partial t} - A\right)w\right\|^2 \leq \Omega$, so that

$$K|u(t + h, x_1 + h, x') - u(t, x_1, x')|^2 \leq C\Omega$$

for $|h| \leq K^{-1/2}$; and $K\|u - w\|^2 + K^{-2N}\|\nabla_x w\|^2 \leq \Omega$ so that

$$K|u(t, x_1, x') - u(t, x_1, y')|^2 \leq C\Omega$$

for $|x' - y'| \leq K^{-N}$. Now take

$$u^+(x_1, x') = \frac{1}{|I|} \int_I \int_{R^{n-1}} \psi(y_1, y')u(t, x_1 - y_1, x' - y')\,dt\,dy_1\,dy',$$

where $\psi(y_1, y')$ is a suitable approximate identity supported in $|y_1| \leq K^{-1/2}$, $|y'| \leq K^{-N}$. One checks easily using the previous estimates that

$$K\|u^0 - u^+\|^2 + \|Au^+\|^2 + K^{-2N}\|\nabla_x u^+\|^2 \leq C\Omega,$$

as needed.

Next we pass to the general case. With $A = \sum\limits_j a^j(x)\dfrac{\partial}{\partial x_j}$, we make a Calderón-Zygmund decomposition of the unit cube in R^{n-1}, stopping at Q_ν with diameter δ_ν if

(i) $\max\limits_{x \in Q_\nu^*} \max\limits_j |a^j(x)| \geq 10\delta_\nu$

or

(ii) $\delta_\nu \leq CK^{-N}$.

In each Q_ν^* we can find a function u_ν^+ so that

$$K|u^0 - u^+|^2 + |Au^+|^2 + K^{-2N}|\nabla_x u_\nu^+|^2 \leq C\Omega \text{ on } Q_\nu^*.$$

For Q_ν arising from (i), this reduces by a change of variable to a slight variant of the case $A = \partial/\partial x_1$, in which the terms $K^{-2N}\|\nabla_x w\|^2$ and $K^{-2N}\|\nabla_x u^+\|^2$ are changed to $K^{-2N}\delta_\nu^{-2}\|\nabla_x w\|^2$ and $K^{-2N}\delta^{-2}\|\nabla_x u^+\|^2$—there is no trouble in adapting our discussion of $A = \partial/\partial x_1$ to this variant. If Q_ν arises from (ii), then we just take $u_\nu^+ = w$.

Now take a partition of unity $1 = \sum_\nu \phi_\nu$ with $\phi_\nu \in C_0^\infty(Q_\nu^*)$, $\|\nabla\phi_\nu\| \leq C\delta_\nu^{-1}$, and set $u^+ = \sum_\nu \phi_\nu u_\nu^+$. Clearly

$$K\|u^0 - u^+\|^2 \leq C\Omega.$$

We also have $Au^+ = \sum_\nu (Au_\nu^+)\phi_\nu + \sum_\nu (u_\nu^+ - u^0)(A\phi_\nu)$, while $|a^j(x)| \leq C\delta_\nu$ and $|\nabla\phi_\nu| \leq C\delta_\nu^{-1}$ in supp $\phi_\nu \subseteq Q_\nu^*$. Therefore,

$$\|Au^+\|^2 \leq C \max\{\|Au^+\|^2 + \|u_\nu^+ - u^0\|^2\} \leq C\Omega.$$

Similarly, $\nabla_x u^+ = \sum_\nu (\nabla_x u_\nu^+)\phi_\nu + \sum_\nu (u_\nu^+ - u^0)(\nabla\phi_\nu)$, while $|\nabla\phi_\nu| \leq \delta_\nu^{-1} \leq K^N$. So

$$K^{-2N}\|\nabla u^+\|^2 \leq C \max\{K^{-2N}\|\nabla_x u_\nu^+\|^2 + \|u_\nu^+ - u^0\|^2\} \leq C\Omega.$$

Thus

$$K\|u^+ - u^0\|^2 + \|Au^+\|^2 + K^{-2N}\|\nabla_x u^+\|^2 \leq C\Omega,$$

proving Lemma 3.1.

REFERENCES

1. Fefferman, C., and Phong, D. H. On positivity of pseudodifferential operators. *Proc. Nat. Acad. Sci.* 75 (1978):4673–4674.

2. Fefferman, C., and Phong, D. H. On the asymptotic eigenvalue distribution of a pseudodifferential operator. *Proc. Nat. Acad. Sci.* 77 (1980):5622–5625.

3. Fefferman, C., and Phong, D. H.   On the lowest eigenvalue of
    a pseudodifferential operator. *Proc. Nat. Acad. Sci.* 76 (1979):
    6055-6056.

4. Fefferman, C., and Phong, D. H.   Symplectic geometry and posi-
    tivity of pseudodifferential operators. *Proc. Nat. Acad. Sci.*
    79 (1982):710-713.

5. Fefferman, C., and Phong, D. H.   The uncertainty principle and
    sharp Gårding inequalities. *Comm. Pure Appl. Math.* 34 (1981):
    285-331.

6. Hörmander, L.   Hypoelliptic second-order differential equa-
    tions. *Acta Math.* 119 (1967):147-171.

7. Kohn, J. J.   Pseudodifferential operators and hypoellipticity.
    Proc. Symp. Pure Math. AMS 23 (1973):61-69.

8. Menikoff, A., and Sjöstrand, J.   Eigenvalues for hypoelliptic
    operators and related methods. *Proc. Int. Cong. Math. Helsinki*
    (1978):797-801.

9. Olenik, O., and Radkevitch, E.   Second-order equations with
    non-negative characteristic form.

10. Rothschild, L. and Stein, E. M.   Hypoelliptic differential
    operators and nilpotent groups. *Acta Math.* 137 (1976):247-320.

SUMMABILITY AND BOUNDS
FOR MULTIPLIERS OF ELLIPTIC OPERATORS ON R^n

David Gurarie
Oregon State University

Mark A. Kon
Boston University

In this paper we consider two related subjects: the properties of kernels of resolvents of elliptic operators on **R**^{n}, and summability theory for these operators. For operators with variable (possibly singular) coefficients, we study summability of eigenfunction expansions using analytic summator functions. This involves consideration of resolvent kernels and yields functional analytic information (for example, closedness, semigroup generation) about the operators themselves. More comprehensive summability results are obtained for constant coefficient operators.

Classical summability theory originated in Fourier analysis [16] and has been studied systematically for Sturm-Liouville expansions [8 and 15] and more recently for eigenfunction expansions of pseudodifferential operators on compact manifolds [13 and 6]. The techniques used for these results cannot be adapted directly to the noncompact situation, or to operators with singular (nonsmooth) coefficients. Here we employ methods that are typical of functional as well as harmonic analysis. We consider the class [*RB*] of integral operators on **R**^{n} whose kernels $K(x, y)$ are bounded by L^1 radial decreasing functions $h(|x - y|)$, and prove that suitable families

Partially supported by NSF Grant No. MCS-800 3407.

of these form approximate identities, in L^p and pointwise. For an elliptic operator A with coefficients in certain L^p classes, the resolvent and analytic multipliers are in [RB], which implies L^p-closedness, essential self-adjointness (if A is symmetric), and other regularity properties for A. In particular, the analytic semigroup $\{e^{-tA^*}\}_{\text{Re } t > 0}$ is continuous (in L_p and on the Lebesgue set) at $t = 0$; this yields convergence at the boundary for solutions of certain Cauchy and Dirichlet problems.

For brevity, most proofs have been sketched or omitted; details of these will appear elsewhere.

1. PRELIMINARIES: CONSTANT COEFFICIENT OPERATORS

Let $S_{1,0}^{m,N}$ denote the classical symbol class of functions $\phi(x, \xi)$ on $\mathbf{R}^n \times \mathbf{R}^n \backslash 0$ with N derivatives in ξ, such that

$$\|\phi\|_{s,\alpha} = \left\| (1 + |\xi|)^{-m+|\alpha|} \partial_\xi^\alpha \phi(x, \xi) \right\|_\infty < \infty$$

for $|\alpha| \le N$, where $\alpha = (\alpha_1, \ldots, \alpha_n)$ is a multiindex, and ∂^α the corresponding partial derivative. Then

$$\phi(x, D)f(x) = \int \phi(x, \xi) e^{-ix \cdot \xi} \hat{f}(\xi) d\xi, \tag{1}$$

where $D^\alpha = i^{-|\alpha|} \partial^\alpha$; henceforth in Fourier integrals $d\xi = (2\pi)^{-n/2} d\lambda$, where λ is Lebesgue measure. For L^p-estimates a more useful form of (1) is

$$\phi(x, D)f(x) = \int K(x, y)f(y) dy,$$

where $K(x, y) = \int \phi(x, \xi) e^{i\xi \cdot (x-y)} d\xi$ is a distribution kernel.

The following (known) radial bounds on $K(x, y)$ are central to our estimates:

If $\phi \in S_{1,0}^{-m,N}$ and $N > n - m$, then

$$|K(x, y)| \le h(|x - y|) = c_{s,t} \|\phi\|_s \begin{cases} |x|^{-s} \ ; & x \le 1 \\ |x|^{-t} \ ; & |x| > 1 \end{cases}, \tag{2}$$

where $n - m < s, t \leq N; s, t > 0;$ and

$$\|\phi\|_s = \sum_{|\alpha| \leq N} \|\phi\|_{s,\alpha}.$$

Consequently, for $1 \leq p \leq \infty$, and $\frac{n}{N} < p < \frac{n}{n-m}$ (if $m \leq n$), and $\frac{n}{N} < p \leq \infty$ (if $m > n$), h is in L^p and $\|h\|_p \leq c_p \|\phi\|_s$.

These bounds follow easily from the fact that symbol class bounds interpolate correctly for fractional derivatives, that is,

$$\left| \Delta_\xi^{s/2} \phi(x, \xi) \right| \leq c_s (1 + |\xi|)^{-m-s} \qquad (0 \leq s \leq N). \tag{3}$$

Similar calculations show, in fact, that for $\phi \in S_{\rho, 0}^{m, N}$ $(0 \leq \rho \leq 1)$, the exponentials s and corresponding L^r classes vary within the limits $\frac{n-m}{\rho} < s \leq N; \frac{n}{N} < r < \frac{n\rho}{n-m}$. An essential part of (2) is the uniformity for $\|\phi\|_s$ bounded. This, together with estimates for symbols singular near 0, will yield general summability for constant coefficient pseudodifferential operators.

Theorem 1. Let $a \in S_1^m$ $(m > 0)$ be a positive, constant coefficient elliptic symbol, and $A = a(D)$ the corresponding operator. Then:

(a) for any $\phi \in S_1^{-m'}$ $(m' > 0)$, $\phi(A)$ is convolution with the kernel $K(x) = (\widehat{\phi \circ a})(-x)$ in $[RB]$.

(b) (summability) if $f \in L^p$, $1 \leq p \leq \infty$, then

$$\phi(\varepsilon A) f(x) \xrightarrow[\varepsilon \to 0]{} \phi(0) f(x)$$

in L^p (for $p < \infty$) and on the Lebesgue set of f.

(c) (resolvent summability) $(\zeta - A)^{-1}$ exists in L^p for $\zeta \notin \mathbf{R}^+$, and has kernel in $[RB]$. We have

$$\zeta(\zeta - A)^{-1} f(x) \xrightarrow[\zeta \to \infty]{} f(x)$$

in L^p $(p < \infty)$ and on the Lebesgue set of f, uniformly in each sector

$$\Omega_\theta = \{ \zeta \in A \,\big|\, |\arg \zeta| \geq \theta \}, \; \theta > 0.$$

Parts (a) and (b) generalize to

$$\phi(\varepsilon A_1, \ldots, \varepsilon A_k) f(x) \xrightarrow[\varepsilon \to 0]{} \phi(0) f(x)$$

for $\phi : \mathbf{R}^k \to \mathbf{C}$ and A_i real elliptic for each i. If $A_j = \frac{1}{i} \frac{\partial}{\partial x_j}$, this is a standard result on summation of the Fourier transform, and (b) easily reduces to this case if A is homogeneous.

In the general case, $\phi \circ a \in S_1^{-mm'}$ by the iterated chain rule, and its symbol class seminorms are estimated by those of ϕ, a, and a^{-1}. Thus, $\phi \circ a(\xi) \in [RB]$. One now considers the "approximation" $a_\varepsilon(\xi) = \varepsilon a(\varepsilon^{-1/m}\xi)$ of $a(\xi)$. In computing $\phi \circ a_\varepsilon$, the standard technique of "cutting off" a neighborhood of 0 takes care of a possible divergence at the origin in the uniform bound on $a_\varepsilon(\xi)$. There are no problems away from the origin, and by the chain and Leibniz rules,

$$D^\alpha [\chi \phi \circ a_\varepsilon] \leq d_\alpha (1 + |\xi|)^{-mm' - |\alpha|},$$

where $\chi(\xi) \in C^\infty$ is 1 for $|\xi| \geq 2$, and 0 for $|\xi| \leq 1$, and d_α depends only on the seminorms of a, a^{-1}, ϕ, and χ. Thus,

$$\phi \circ a_\varepsilon(x) \leq h(|x|) \in L^1.$$

Similarly,

$$\left(\frac{\zeta}{\zeta - a(|\zeta|^{1/m} \xi)} \right)^\wedge (x) \leq g(|x|) \in L^1 \tag{4}$$

with g independent of ζ for $|\zeta| \geq \rho_0 > 0$, $|\arg \zeta| \geq \theta_0 > 0$. The convolution kernels for $\phi(\varepsilon A)$ and $\zeta(\zeta - A)^{-1}$ (in the above region of the ζ-plane) are thus uniformly bounded by L^1-dilations of radial decreasing functions, while their integrals over $\mathbf{R}^n$ converge to 1. The proof is completed by the following, which extends a well-known result (see [12, Chapter 1]).

Proposition 1. If a family $\{K_\varepsilon\} \in [RB]$ is bounded by L^1-dilations of a radially decreasing function $\{\varepsilon^{-n} h(\varepsilon^{-1} |x - y|)\}$ and $\int K_\varepsilon(x, y) dy \to c$ uniformly in x, then for all $f \in L^p$, $K_\varepsilon(f) \to cf$ in L^p $(1 \leq p < \infty)$ and on the Lebesgue set of f $(1 \leq p \leq \infty)$.

The symbol classes of a and ϕ may clearly be relaxed to be $a \in S_1^{m,n+1}$, $\phi \in S_1^{-m',n+1}$.

If $\phi \in S_1^0$, $\phi(\varepsilon A)$ is uniformly bounded in L^p $(1 < p < \infty)$, by Mihlin's theorem [5]. It is not hard to show that $\phi(\varepsilon A)f \xrightarrow[\varepsilon \to 0]{} f$ in L^p if f is in the dense subset $\{f(x) \,|\, \hat{f}(\xi) \in C_K^\infty\}$, so that zero-class summability holds for all $f \in L^p$. However, summability on the Lebesgue set fails in this case: in $\mathbf{R}^1$, let $a(\xi) = \xi$, $\phi(\xi) = |\xi|^i$, and $f(x) = |x|^{-i}/\ell n|x|$. Then $\check{\phi}(x)$ is proportional to the principal value distribution $|x|^{-i-1}$, and $\check{\phi}(x) \star f(x)\Big|_{x=0} = \infty$, even though f is continuous at $x = 0$. Failure on the Lebesgue set does not of course exclude the possibility of convergence a.e.

2. RESOLVENTS AND SUMMABILITY FOR PERTURBED OPERATORS

If $A = A_0 + B$ is a perturbed operator, the resolvent perturbation series

$$(\zeta - A)^{-1} = (\zeta - A_0)^{-1} \sum_{k=0}^{\infty} [B(\zeta - A_0)^{-1}]^k \tag{5}$$

is useful when the terms can be appropriately bounded. Let A_0 be a constant coefficient, positive and homogeneous differential operator of order m, and

$$B = \sum_{|\alpha| < m} b_\alpha(x)D^\alpha.$$

b_α may be defined on quotient spaces $V_\alpha = \mathbf{R}^n/U_\alpha$, with U_α a subspace of $\mathbf{R}^n$. We let $n_\alpha = \dim V_\alpha = n - \dim U_\alpha$, and require $b_\alpha \in L^{r_\alpha} + L^\infty$, for $n_\alpha/r_\alpha + |\alpha| < m$, $|\alpha| < m$. $U_\alpha \neq \{0\}$ is of interest, for example, in Schrödinger operators for Coulomb systems,

$$-\Delta + \sum_{i \neq j} \frac{c_{ij}}{|x_i - x_j|} + \sum_i \frac{c_i}{|x_i|},$$

with $x_i \in \mathbf{R}^3$ $(i = 1, \ldots, N)$, Δ the Laplacian on $\mathbf{R}^{3N}$.

Theorem 2. Let $A = A_0 + B$ be as above, and let

$$d = \max\{n_\alpha/r_\alpha + |\alpha| \,\big|\, |\alpha| < m\} < m.$$

There exists a constant $c > 0$ depending on A such that the resolvent set of A contains $\Omega = \{\zeta = \rho e^{i\theta} \,\big|\, c\rho^{d/m-1} < |\sin \theta/2|^{n+2}\}$ and $(\zeta - A)^{-1} \in [RB]$ for $\zeta \in \Omega$. Moreover, the kernel $\zeta L_\zeta(x, y)$ of $\zeta(\zeta - A)^{-1}$ is bounded as follows:

$$|\zeta L_\zeta(x, y)| \le \frac{c}{|\sin \theta/2|^{n+2}} \left(1 - \frac{c\rho^{d|m-1}}{|\sin(\theta/2)|^{n+2}}\right)^{-1} \times \rho^{n/m} h_{s,t}(\rho^{1/m}|x - y|),$$

where

$$h_{s,t}(x) = \begin{cases} |x|^{-s}; & |x| \le 1 \\ |x|^{-t}; & |x| > 1 \end{cases}.$$

The proof requires the notion of (p, q)-mixed convolution with respect to a decomposition

$$\mathbf{R}^n = U \oplus V = \{(x', x'') \,\big|\, x' \in U, \ x'' \in V\},$$

with $\dim V = n'$, $\dim U = n''$. Define

$$f *_{(p,q)} g = \left(\int_U \left(\int_V |f(y' + y'')g(x' + x'' - y' - y'')|^p \, dy'\right)^{q/p} dy''\right)^{1/q}.$$

After a rather lengthy calculation, one obtains:

Proposition 2. Let $\mathbf{R}^n = U_i \oplus V_i$ ($1 \le i \le k$) be a sequence of partitions; let $\dim V_i = n_i$, $\dim U_i = n_i'$. If sequences of reals $\{s_i\}_1^k$, $\{t_i\}_1^k$, $\{p_i\}_1^k$, $\{q_i\}_1^k$ satisfy $t_i > n$;

$$\max\{s_i, \min(s_1, \ldots, s_{i-1})\} \le \frac{n_i}{p_i} + \frac{n_i'}{p_i'},$$

then the estimate

$$(h_{s_0,t_0} * h_{s_1,t_1}) * \cdots * h_{s_k,t_k} \le c_1^k h_{s',t'}$$

holds, where $s' = \min\{s_i\}$, $t' = \min\{t_i\}$ and c_1 is a constant. Above, the ith $*$ denotes a (p_i, q_i) mixed convolution.

The theorem is proved by bounding each term of (5) in [RB]. Note that

$$B(\zeta - A_0)^{-1} = \sum_{|a| < m} b_a(x) D^a (\zeta - A_0)^{-1}$$

contains terms $\phi a(D)$, whose symbol $\phi_a(\xi) = \xi^a (\zeta - a(\xi))^{-1}$, where $a(\xi)$ is the symbol of A_0. Let $K_a(x)$ be the corresponding convolution kernels. Then

$$L_k = (\zeta - A_0)^{-1} [B(\zeta - A_0)]_k = \sum_{\alpha_1, \ldots, \alpha_k} L_{\alpha_1, \ldots, \alpha_k}.$$

Each term in the last sum has kernel

$$L_{\alpha_1, \ldots, \alpha_k}(x, y) = K_0 * b_{\alpha_1} K_{\alpha_1} * \cdots * b_{\alpha_k} K_{\alpha_k}$$

obtained from multiplying out the left-hand side. Above, operations go from right to left and K_k has argument $x - y$. We can assume that $b_a \in L^{r_a}$, as the argument is the same for $r_a \le r \le \infty$.

Application of a multiple Hölder inequality, two basic interpolation inequalities for multiplication and convolution of functions in L^r classes, and appropriate radial bounds yield (we let $b_i = b_{\alpha_i}$, and so on)

$$|L_{\alpha_1, \ldots, \alpha_k}(x, y)| \le \prod_i \| b_i \|_{r_{\alpha_i}} \frac{c_2}{|\sin \theta/2|^{n+2}} |\zeta|^{n/m-1} \left(\frac{c_2}{|\sin \theta/2|^{n+2}} \zeta^{d/m-1} \right)^k$$

$$\times \left(h_{s_0, t} *_{(p_1, 1)} h_{s_1, t} *_{(p_2, 1)} \cdots *_{(p_k, 1)} h_{s_k, t} \right) (|\zeta|^{1/m} |x - y|)$$

$$\equiv \prod_i \| b_i \|_{r_{\alpha_i}} H_{\zeta, k} (|x - y|),$$

where $d = \max\left(\frac{n_a}{r_a} + |a|\right)$, $s_j > n - m - |a|$, and $t > n$. According to Proposition 2, the multiple convolution is bounded by $c_1^k h_{s, t} (|\zeta|^{1/m} |x - y|)$, where $s = \min s_j \le s_0$; hence

$$|L_k(x, y)| \le \left(\sum_\alpha \| b_\alpha \|_{r_\alpha} \right)^k H_{\zeta, k}.$$

The geometric series $\sum L_k(x, y)$ converges (and hence is in [RB]) if

$$\frac{c_2}{|\sin \theta/2|^{n+2}}|\zeta|^{d/m-1} < 1;$$

the kernel is then estimated by summing the series.

Note that if A_0 is not positive and instead has range in a sector $D = \{\zeta \,|\, \theta_1 \leq \arg \zeta \leq \theta_2\}$, the theorem still holds with the replacement of $\sin \theta/2$ by a slightly more complicated function; the same will hold for later statements.

3. APPLICATIONS OF THEOREM 2

The radial bounds of Theorem 2 provide information on the smoothing properties of $(\zeta - A)^{-1}$, and additional global bounds on the kernel at infinity, giving uniform estimates of L^p, Sobolev, and Hölder bounds. These estimates extend to analytic functions $\phi(A)$ via Cauchy integration, yielding summability results and other interesting information about A and $\phi(A)$.

a. The Operator A

It follows immediately from the theorem that the L^p spectrum $(1 \leq p \leq \infty)$ of A is contained in

$$\Omega' = \left\{ \zeta = \rho e^{i\theta} \,\middle|\, \rho^{d/m-1} \geq \frac{|\sin \theta/2|^{n+2}}{c} \right\}$$

which is a "U-shaped" domain about $\mathbf{R}^+$. Consequently, A is closeable in all L^p. Furthermore, the norm of the resolvent has maximum decrease in all directions $\theta \neq 0$. If A_0 has range in D as above, Ω' encloses D in a similar way.

If A is formally self-adjoint, then A admits a self-adjoint extension in L^2; indeed A is clearly bounded from below. Such results have previously been obtained when $A = -\Delta + V(x)$ in the theory of Schrödinger operators. An argument similar to Theorem 2 yields a priori estimates for the pair (B, A_0) in certain L^p spaces.

Precisely, for $1 \leq p \leq \min\{r_\alpha\}$, $\|Bf\|_p \leq \epsilon \|A_0 f\|_p + c_\epsilon \|f\|_p$. If, in particular, $\min\limits_\alpha\{r_\alpha\} \geq 2$ and A is formally self-adjoint, the Kato-Rellich theorem [7] applies to show that A is essentially self-adjoint on $\mathscr{L}_m^2$ or any essential domain of A_0 (c.f. [11, Chapter 10]).

b. Resolvent Summability

This has been studied in the context of regular ([14]) and singular [3 and 10] Sturm-Liouville expansions, with

$$A = \frac{-d^2}{dx^2} + q(x).$$

For eigenfunction expansions on $\mathbf{R}^n$, we have

Theorem 3. *If $A = A_0 + B$ is as in Theorem 2 and $f \in L^p$, then $\zeta(\zeta - A)^{-1} f(x) \xrightarrow[\zeta \to \infty]{} f(x)$ uniformly for $|\arg \zeta| \geq \theta_0 > 0$, in L^p-norm $(1 \leq p < \infty)$ and on the Lebesgue set of f $(1 \leq p \leq \infty)$.*

By Theorem 2 and Proposition 1, what needs to be proved is that $\int \zeta L_\zeta(x, y)\,dy \xrightarrow[\zeta \to \infty]{} 1$ uniformly for $|\arg \zeta| \geq \theta_0$; this can be done using decompositions like those in Theorem 2.

c. Summability for Analytic Multipliers

If $\phi_\epsilon(\zeta)$ is analytic in a neighborhood of Ω', $\phi_\epsilon(A)$ can be defined by

$$\frac{1}{2\pi i} \int_\Gamma \phi_\epsilon(\zeta)(\zeta - A)^{-1} d\zeta,$$

where Γ is a contour containing Ω'. Such calculi for unbounded operators are treated in [4, Chapters 5 and 15]; one such calculus involves multipliers

$$\phi_\epsilon(\zeta) = \int_0^\infty e^{-\zeta x} da(x),$$

where $a(x)$ is a function of strong bounded variation on $\mathbf{R}^+$. In order that $\phi_\varepsilon(A)$ be in $[RB]$, it suffices that

$$\Gamma \subset \{\zeta \mid |\arg \zeta| \geq \theta_0 > 0\},$$

and

$$\int_\Gamma \left| \frac{\phi_\varepsilon(\zeta)}{\zeta} \right| |d\zeta| < \infty. \tag{6}$$

Writing down the obvious L^1 radial bound by integrating the resolvent bounds shows summability in L^p and pointwise for analytic dilations $\phi_\varepsilon(A) = \phi(\varepsilon A)$, via techniques of Theorem 3.

Let A be as before and $\phi_t(A) = e^{-tA^s}$ for $s \in \mathbf{R}$. A^s will then be well defined if Γ does not enclose the origin (for example, when A has spectrum in the right half-plane). Cauchy integration of $e^{-t\zeta^s}$ gives

Theorem 4. *A^s generates an analytic semigroup $\{e^{-tA^s}\}_{\mathrm{Re}\, t > 0}$ in all L^p spaces for which A^s is defined as above. Furthermore, we have L^p ($1 \leq p < \infty$) and Lebesgue convergence $e^{-tA^s}f(x) \xrightarrow[t \to 0]{} f(x)$ uniformly in each sector $\{t \mid |\arg t| \leq \theta_0 < \pi/2\}$.*

For $s = 1$, this is a statement of Abel summability and is also equivalent to convergence at the boundary $u(x, t) \xrightarrow[t \to 0]{} f(x)$ in the generalized heat equation $u_t(x, t) = Au(x, t)$, with initial value $u(x, 0) = f(x)$.

Remark. The radial bounds in Theorem 2 can be sharpened to yield optimal (in the sense of being the best generic) radial bounds for $\zeta L_\zeta(x, y)$. These in turn integrate to give bounds on the kernel of $\phi(A)$, and via Hölder's inequality provide information about the (p, q) type of $\phi(A)$.

d.   Dirichlet Problems on the Half-Space

Let $A = A_0 + \sum_{|\alpha| < 2m} b_\alpha(x) D^\alpha$ be an elliptic operator as before, this time with sufficiently smooth $b_\alpha \in \mathcal{L}_s^{r_\alpha}$ ($s \geq m$), $\frac{n_\alpha}{r_\alpha} + |\alpha| < 2m$

$(n_\alpha = \dim V_\alpha)$, where $\mathscr{L}_s^{r_\alpha} = (1 - \Delta)^{-s/2} L^{r_\alpha}$ is an L^{r_α}-Sobolev space. Consider the generalized Dirichlet problem for the elliptic operator $\dfrac{\partial^{2m}}{\partial t^{2m}} + (-1)^m A$

$$\left(\frac{\partial^{2m}}{\partial t^{2m}} - (-1)^m A\right) u(x, t) = 0; \quad \left(\frac{\partial}{\partial t}\right)^j u\bigg|_{t=0} = f_j(x); \qquad (7)$$

$$(j = 0, \ldots, m - 1),$$

with $f_j \in \mathscr{L}_{m-1-j}^p$. Let $\omega = e^{2\pi i/2m}$ or $e^{\pi i/2m}$, depending on whether m is odd or even. By considering the general decaying solution

$$u = \sum_{\text{Re } \omega^j > 0} e^{-t\omega^j A^{1/2m}} g_j,$$

we can formally solve for g_j at $t = 0$ and obtain

$$u(x, t) = \sum_{\text{Re } \omega^j > 0} e^{-t\omega^j A^{1/2m}} \sum_{0 \le k \le m-1} \omega_{jk} A^{-k/2m} f_k, \qquad (8)$$

where $\{\omega_{jk}\}$ is the inverse of the Wandermond matrix

$$W(\{\omega^j \,|\, \text{Re } \omega^j > 0\}).$$

To make (8) rigorous we need well-defined fractional power semigroups e^{-tA^s} ($s \in \mathbf{R}$, $t \ge 0$). Theorem 4 supplies them, provided A has spectrum in the right half-plane. Henceforth, we assume that A is formally self-adjoint and positive, or constant coefficient and positive. In the first case, the L^2 and $\mathscr{L}_s^2$ ($s \ge 0$) spectrum of A is in $\mathbf{R}^+$, and in the second the same is true in $\mathscr{L}_s^p$ ($1 \le p < \infty$).

Theorem 5. In either of these cases, (7) has a unique solution (8) that is regular $(u(x, t) \in \mathscr{L}_s^p(\mathbf{R}_+^{n+1})$ *for all* $s)$, *and approaches its boundary values, that is,* $\partial_t^k u(x, t) \xrightarrow[t \to 0]{} f_k(x)$ $(0 \le k \le 2\ell)$ *in* $\mathscr{L}_{2\ell-k}^p$ *and on the Lebesgue set of* f_k.

Above, $p = 2$ in the first case, and $1 < p < \infty$ in the second.

Nontangential convergence can be proved using standard Hardy-Littlewood maximal function techniques.

REFERENCES

1. Calderón, A. P., and Zygmund, A.   On singular integrals.
    Acta Math. 88 (1952):289-309.

2. Carmona, R.   Regularity properties of Schrödinger and Dirichlet
    semigroups.   *J. Func. Anal.* 35 (1980):215-229.

3. Diamond, H.; Kon, M. A.; and Raphael, L.   Stable summation
    methods for a class of singular Sturm-Liouville expansions.
    Proc. AMS 81 (1981):279-286.

4. Hille, E., and Phillips, R. S.   *Functional Analysis and Semi-
    groups.*   AMS Colloq. Publ. Vol. 31, 1957.

5. Hörmander, L.   Estimates for translation invariant operators
    in L^p-spaces.   *Acta Math.* 104 (1960):93-139.

6. Hörmander, L.   On the Riesz means of spectral functions and
    eigenfunction expansions for elliptic differential operators.
    Recent Advances in the Basic Sciences, Yeshiva Univ. Conf.,
    Nov. 1966, pp. 155-202.

7. Kato, T.   *Perturbation Theory for Linear Operators.*   New York:
    Springer-Verlag, 1976.

8. Levitan, B. M., and Sargsjan, I. S.   *Introduction to Spectral
    Theory.*   Providence, R.I.: A.M.S., 1975.

9. Nagel, A., and Stein, E. M.   *Lectures on Pseudo-Differential
    Operators.*   Princeton, N.J.: Princeton Univ. Press, 1979.

10. Raphael, L. A.   A generalized Stieltjes transform summability
    method and summing Sturm-Liouville expansions.   Preprint.

11. Reed, M., and Simon, B.   *Methods of Modern Mathematical
    Physics*, vol. 2.   New York: Academic Press, 1975.

12. Stein, E., and Weiss, G.   *Introduction to Fourier Analysis on
    Euclidean Spaces.*   Princeton, N.J.: Princeton Univ. Press, 1971.

13. Taylor, M.   Fourier integral operators and harmonic analysis
    on compact manifolds.   *AMS Proc. of Symposia in Pure Math.* 32
    (1979):115-136.

14. Tikhonov, A. N.   Stable methods for the summation of Fourier
    series.   *Soviet Math. Dokl.* 5 (1964):641-644.

15. Titschmarsh, E. C. *Eigenfunction Expansions*. 2d ed. New
     York: Cambridge Univ. Press, 1959.

16. Zygmund, A. *Trigonometric Series*. 2d ed. New York:
     Cambridge Univ. Press, 1959.

ON THE DIRICHLET PROBLEM
FOR HIGHER-ORDER EQUATIONS

Lars Inge Hedberg
University of Stockholm

The Dirichlet problem for higher-order elliptic partial differential equations is usually treated only for regions with smooth
boundaries. The purpose of this expository paper is to discuss the
problem for more general regions, and to present some new results,
mostly without proofs. For simplicity, we treat the equation
$\Delta^m u = 0$, $m \geq 2$, although the results carry over to much more general equations. Here Δ^m means the iterated Laplacian; $\Delta^2 u = \Delta(\Delta u)$,
and so on.

     The classical Dirichlet problem can be formulated in the
following way.

Problem 1. Let $G \subset \mathbf{R}^d$ be a bounded region, and suppose that ∂G has
a tangent hyperplane at each of its points. Let g_0, g_1, ..., g_{m-1}
be given continuous functions on ∂G. Find a function
$u \in C^{m-1}(\bar{G}) \cap C^{2m}(G)$ such that $\Delta^m u = 0$ in G, and $\partial^k u/\partial n^k = g_k$ on
∂G for $0 \leq k \leq m - 1$. (Here $\partial^k/\partial n^k$ denotes the derivative in the
exterior normal direction.)

     To discuss the problem for more general regions, it is
convenient to reformulate it.

Supported by the Swedish Natural Science Research Council.

Problem 2. Let $G \subset \mathbf{R}^d$ be an arbitrary bounded region. Let
$g \in C^{m-1}(\overline{G})$ be given. Find $u \in C^{m-1}(\overline{G}) \cap C^{2m}(G)$ such that $\Delta^m u = 0$
in G, and $D^\alpha(u - g)\big|_{\partial G} = 0$ for all multiindices α, $0 \le |\alpha| \le m - 1$.

In this generality the problem is far from being solved.
The highly developed theory of the Laplace equation breaks down com-
pletely because of the failure of the maximum principle. The fol-
lowing theorem is typical for what is known. See, for example,
S. Agmon [3 and 2, Sec. 9], L. Nirenberg [20], and C. Miranda [19,
Sec. 52].

Theorem 1. *Let $G \subset \mathbf{R}^d$ be a bounded region with C^∞ boundary, and
let $g \in C^\infty(\overline{G})$. Then Problem 1 (and 2) has a unique solution u, and
$u \in C^\infty(\overline{G})$.*

General criteria for the solvability of Problem 2 are
unknown. To the author's knowledge, the only result in this direction is a sufficient condition of Wiener-type valid for $m = 2$ and
$d \le 7$ dimensions, due to V. G. Maz'ja [16 and 17].

Because of the difficulty of the classical Dirichlet problem,
it is natural that various generalized formulations were considered
long before Theorem 1 was proved.

Let $G \subset \mathbf{R}^d$ be an arbitrary open set, let $H^m(G)$ denote the
closure of $C^m(G)$ in the norm $\|\cdot\|_m$ defined by

$$\|f\|_m^2 = \sum_{0 \le |\alpha| \le m} \int_G |D^\alpha f|^2 \, dx,$$

and denote by $H_0^m(G)$ the closure in $H^m(G)$ of $C_0^\infty(G)$, the test functions with support in G.

Equivalently, $H^m(G)$ consists of those functions in $L^2(G)$
whose distribution derivatives $D^\alpha f$ belong to $L^2(G)$ for all multi-
indices α with $|\alpha| \le m$. That a function f belongs to $H_0^m(G)$ is
interpreted as meaning that f and $D^\alpha f$, $|\alpha| \le m - 1$, have boundary

values zero, in a generalized sense. Thus, the following "generalized Dirichlet problem" is natural.

Problem 3. Let $G \subset \mathbf{R}^d$ be open and bounded, and let $g \in H^m(G)$ be
given. Find $u \in H^m(G)$ such that $\Delta^m u = 0$ in G, and $u - g \in H_0^m(G)$.

Here $\Delta^m u = 0$ is interpreted in the sense of distributions,
but by the well-known lemma of Weyl and Schwartz, this implies
that $u \in C^\infty(G)$, and thus $\Delta^m u = 0$ in the ordinary sense.

Using projections in Hilbert space, one easily proves the
following theorem.

Theorem 2. *Problem 3 always has a unique solution.*

This result is, of course, very satisfactory through its
simplicity and generality, but on the other hand, it is difficult
to understand in which way the boundary values are really taken.
What does it mean that $u - g \in H_0^m(G)$ if, for example, G is a d-
dimensional ball with a $(d - k)$-dimensional manifold removed from
its interior?

Generalizing earlier results of K. Friedrichs [9], S. L.
Sobolev studied this problem in the case when G is bounded by a
finite union of manifolds of arbitrary codimension. His fundamental paper [23] appeared in 1937, but it was translated into
English only in 1963. Sobolev's results also appear in his mono-
graph [22] from 1950, which was also translated in 1963.

Let $f \in H^m(G)$, and let M be a compact piece of a $(d - k)-$
dimensional smooth manifold in G. Here $1 \leq k \leq d$. Let f_h, $h > 0$,
be the averaged function, obtained by taking the convolution with
a smooth approximate identity with support in a ball with radius h.
Then Sobolev proved an imbedding theorem of which the following is
a special case.

Theorem 3. *Suppose* $k < 2m$. *Then* $\lim\limits_{h \to 0} f_h(x) = \tilde{f}(x)$ *exists a.e. on* M *in the* $(d - k)$*-dimensional sense, and* $\lim\limits_{h \to 0} \|f_h - \tilde{f}\|_{L^2(M)} = 0$. *In particular,* $f_h(x)$ *converges pointwise, and* $\tilde{f}(x)$ *is continuous if* $d < 2m$. *The condition* $k < 2m$ *is sharp.*

We shall denote the restriction of $\tilde{f}$ to M by $f|_M$. It is called the trace of f on M.

Now suppose that ∂G consists of a finite union of smooth manifolds, not forming cusps. We write

$$\partial G = \bigcup_{k=1}^{d} M_k,$$

where each M_k is a finite union of $(d - k)$-manifolds. Then Sobolev proved, by letting the above M approach the boundary, that $f|_{M_k}$ is also well defined provided $k < 2m$.

Since $D^\alpha f \in H^{m - |\alpha|}(G)$ if $f \in H^m(G)$, it follows that $D^\alpha f|_{M_k}$ is well defined, provided $k < 2(m - |\alpha|)$, that is, $|\alpha| < m - \dfrac{k}{2}$. Thus, the Dirichlet problem can be formulated in the following way.

Problem 4. Let $G \subset \mathbf{R}^d$ be open and bounded, and let

$$\partial G = \bigcup_{k=1}^{d} M_k$$

satisfy the above assumptions. Let $g \in H^m(G)$. Find $u \in H^m(G)$ such that $\Delta^m u = 0$ in G and $D^\alpha(u - g)|_{M_k} = 0$ for $1 \leq k < 2(m - |\alpha|)$, and $0 \leq |\alpha| \leq m - 1$.

It is clear from Theorem 2 that solutions always exist. In fact, it is easily seen that the above traces vanish if $u - g \in H_0^m(G)$. What is not evident is that this solution is the only one, but this was proved by Sobolev.

Theorem 4. *Problem 4 always has a unique solution. In fact,* $f \in H_0^m(G)$ *if and only if* $f \in H^m(G)$ *and* $D^\alpha f|_{M_k} = 0$ *for* $1 \leq k < 2(m - |\alpha|)$, *and* $0 \leq |\alpha| \leq m - 1$.

Example. Consider the biharmonic equation $\Delta^2 u = 0$ in $G \subset \mathbf{R}^3$. Then $G = M_1 \cup M_2 \cup M_3$, where M_1 is the two-dimensional part of the boundary. Let, for example, G be a ball with a one-dimensional interval (M_2) and a finite number of points (M_3) removed from its interior. Let $u \in H^2(G)$ be a solution. By Theorem 3, and the remarks following it, $u|_{M_k}$ is defined for $k = 1, 2, 3$, and $\nabla u|_{M_1}$ is defined. Theorem 4 implies that $u = 0$ in G if $u|_{M_k} = 0$, $k = 1, 2, 3$, and $\nabla u|_{M_1} = 0$.

One difficulty in extending Sobolev's theorem to more general regions lies in the definition of the trace. This problem can be overcome through the idea of corrected, or precisely defined, functions, which appeared, at least for H^1, in the 1950s in the work of Deny [6, Chapter 4] and Deny and Lions [8]. From now on, we shall consider the more general Sobolev spaces $H^{m,p}(G)$, $1 < p < \infty$, $(H^{m,2} = H^m)$ defined as the closure of $C^m(G)$ in the norm $\|\cdot\|_{m,p}$ defined by

$$\|f\|_{m,p}^p = \sum_{0 \le |\alpha| \le m} \int_G |D^\alpha f|^p \, dx.$$

The extension of the theory of corrected functions to this context is due to V. G. Maz'ja and V. P. Havin [18].

Let $f \in H^{m,p}(\mathbf{R}^d)$. By the well-known imbedding theorem of Sobolev, f is continuous for $mp > d$ (after correction on a set of measure zero), but f can in general not be so corrected if $mp \le d$. The natural way of measuring the deviation from continuity is by means of an (m, p)-capacity, which we shall now define.

If K is compact, we set

$$C_{m,p}(K) = \inf\{\|\varphi\|_{m,p}^p; \ \varphi \in C_0^\infty, \ \varphi \ge 1 \text{ on } K\}.$$

$C_{m,p}$ is then extended to arbitrary sets as an outer capacity in the usual way; that is, for open G,

$$C_{m,p}(G) = \sup\{C_{m,p}(K); \ K \text{ compact}, \ K \subset G\}$$

and for arbitrary E, $C_{m,p}(E) = \inf\{C_{m,p}(G); \ G \text{ open}, \ G \supset E\}$.

There is an equivalent approach that is sometimes more
convenient. By a theorem of A. P. Calderón [5], a function f
belongs to $H^{m,\,p}(\mathbf{R}^d)$ for $1 < p < \infty$ if and only if $f = G_m * g$ for
some $g \in L^p$, where G_m is the so-called Bessel kernel, defined as
the inverse Fourier transform of $\hat{G}_m(\xi) = (1 + |\xi|^2)^{-m/2}$. There is
also equivalence of norms, that is, $A^{-1}\|f\|_{m,p} \leq \|g\|_{0,p} \leq A\|f\|_{m,p}$.
One can now define $(m,\,p)$-capacity at once for arbitrary sets by
setting

$$C_{m,\,p}(E) = \inf\{\|g\|_{0,\,p}^p ;\ g \in L^p,\ g \geq 0,\ G_m * g \geq 1 \text{ on } E\}.$$

The definition makes sense, since $G_m * g(x)$ is defined
($\leq +\infty$) at all points, and one can show that this definition is
equivalent to the previous one.

There are precise comparison theorems between $(m,\,p)$-capa-
cities and Hausdorff measure (see [18]). Here we only quote the
following result.

Theorem 5. *Let M_k be a $(d - k)$-dimensional smooth manifold.* *Then*
$C_{m,\,p}(M_k) = 0$ *if and only if $k \geq mp$.*

Remark. It follows from Theorems 3 and 5 that if $f \in H^m(\mathbf{R}^d)$ and
M_k is a $(d - k)$-manifold, then $D^\alpha f|_{M_k}$ is defined if $C_{m-|\alpha|,\,2}(M_k) > 0$.

Let

$$f(x) = G_m * g(x) = \int_{\mathbf{R}^d} G_m(x - y)g(y)\,dy,\ g \in L^p,$$

so that $f(x)$ is well defined wherever $G_m * |g|(x) < \infty$. Let E_λ
denote the set where either $f(x)$ is undefined or $|f(x)| > \lambda$. Since
$|f(x)| \leq G_m * |g|(x)$, it follows from the second definition of
capacity that

$$C_{m,\,p}(E_\lambda) \leq \frac{1}{\lambda^p} \int_{\mathbf{R}^d} |g|^p\,dx.$$

In particular, $f(x)$ is well defined outside a set of zero $(m,\,p)$-
capacity, that is, $(m,\,p)$-quasieverywhere $[(m,\,p) - \text{q.e.}]$. Let Mf
denote the Hardy-Littlewood maximal function of f, that is,

$$Mf(x) = \sup_{B \ni x} \frac{1}{|B|} \int_B |f(y)| \, dy,$$

where the supremum is taken over all balls B containing x. It is not hard to prove that Mf satisfies a similar inequality (D. R. Adams [1]), that is, there is a constant A such that

$$C_{m,p}(\{x; \, Mf(x) > \lambda\}) \le \frac{A}{\lambda^p} \|f\|_{m,p}^p.$$

Let $\{\chi_h\}_{h>0}$ be a C^∞ approximate identity; that is, $\chi_h = h^{-d} \chi_1\left(\frac{x}{h}\right)$, $\chi_1 \ge 0$, $\int \chi_1 dx = 1$, and χ_1 has support in the unit ball. It follows in a well-known way from the above inequality that

$$\lim_{h \to 0} f * \chi_h (x) = \tilde{f}(x)$$

exists (m, p)-q.e. Moreover, $\tilde{f}(x) = f(x)$ a.e., and for any $\varepsilon > 0$, there is an open set G such that $C_{m,p}(G) < \varepsilon$, and $\tilde{f}\big|_{G^c}$ is continuous on G^c (the complement of G).

Such a function $\tilde{f}$ is called an (m, p)-quasicontinuous representative of f. By an extension of a theorem of Wallin [25] (see also Deny [7]), such representatives are unique up to sets of zero (m, p)-capacity. In particular, the potential $\int G_m(x - y) g(y) dy$ is such a representative.

If a quasicontinuous representative of f is chosen, f is said to be corrected, or precisely defined.

We now define the trace $f\big|_E$ of an element f in $H^{m,p}(\mathbf{R}^d)$ simply as the restriction to E of any (m, p)-quasicontinuous representative of f. Similarly, $D^\alpha f\big|_E$ is defined by taking the restriction to E of an $(m - |\alpha|, p)$-quasicontinuous representative of $D^\alpha f$. Here $D^\alpha f$ is the derivative in the sense of distributions.

In a manuscript from 1968, which appeared in print only in 1977 (see B. W. Schulze and G. Wildenhain [21, IX.5.1]), B. Fuglede proposed the following formulation of the Dirichlet problem.

Problem 5. Let $G \subset \mathbf{R}^d$ be open and bounded. Let $g \in H^m(\mathbf{R}^d)$. Find $u \in H^m(\mathbf{R}^d)$ such that $\Delta^m u = 0$ in G and $D^\alpha(u - g)\big|_{\partial G} = 0$, $0 \le |\alpha| \le m - 1$.

It is again easy to see that a solution exists. In fact,
by Theorem 2 there is a solution u such that $u - g \in H_0^m(G)$, and it
is easy to prove that this implies that $D^\alpha(u - g)\big|_{\partial G} = 0$,
$0 \leq |\alpha| \leq m - 1$. The problem is again to prove that this solution
is the only one, that is, that pathologic null solutions do not
arise.

A bounded open G has this uniqueness property if and only
if ∂G has a property that by analogy with harmonic analysis Fuglede
called the $2m$-spectral synthesis property. More generally, we shall
say that a closed set F admits (m, p)-spectral synthesis if every
f in $H^{m, p}(\mathbf{R}^d)$ that satisfies $D^\alpha f\big|_F = 0$ for $0 \leq |\alpha| \leq m - 1$ belongs
to $H_0^{m, p}(F^c)$. Thus, the Dirichlet problem in G has a unique solu-
tion if and only if ∂G admits $(m, 2)$-spectral synthesis.

The following theorem was proved in [13] after a partial
result had been proved in [14].

*Theorem 6. Problem 5 always has a unique solution. In fact, every
closed set $F \subset \mathbf{R}^d$ admits (m, p)-spectral synthesis, provided only
$p > 2 - \frac{1}{d}$.*

Remark. Whether the result is true for all $p > 1$ remains an open
problem. (Added in proof: The result is true. See the remark following Theorem 7.)

For the proof, more references, and other applications, we
refer to [13 and 14]. See also the survey [12]. Here we only give
a couple of remarks concerning the proof.

The proof depends on the idea of a thin set. A set $E \subset \mathbf{R}^d$
is called (k, p)-thin at x if $kp \leq d$, and

$$\sum_{n=1}^{\infty} \left[2^{n(d-kp)} C_{k, p}(E \cap B_n(x)) \right]^{\frac{1}{p-1}} < \infty.$$

$[B_n(x)$ is the ball $\{y; \ |y - x| \leq 2^{-n}\}.]$ For $k = 1$, $p = 2$, this
agrees with the classical definition.

The importance of this definition is the following generalized Kellogg property.

Theorem 7. *If E' is the subset of E where E is (k, p)-thin, then $C_{k,p}(E') = 0$, provided $p > 2 - \dfrac{k}{d}$.*

Remark. If the restriction on p could be removed, the extension of Theorem 6 to $p > 1$ would follow. (Added in proof: T. Wolff has done this. See [26].)

Let $f \in H^{m,p}(\mathbf{R}^d)$, and suppose that $D^\alpha f|_F = 0$ for all α, $0 \leq |\alpha| \leq m - 1$. Assume that $p > 2 - \dfrac{1}{d}$, so that the Kellogg property holds for $C_{k,p}$, $k = 1, \ldots, m$. Set

$$\left\{ \frac{1}{|B_n(x)|} \int_{B_n(x)} |f|^p \, dy \right\}^{1/p} = [f]_n(x),$$

and

$$\sum_{|\alpha|=m} |D^\alpha f| = |\nabla^m f|.$$

If $x \in F$, one can estimate $[f]_n(x)$ in terms of $C_{k,p}(F \cap B_n(x))$, $1 \leq k \leq m$, and using the Kellogg property, one can prove the following lemma.

Lemma 1. *Under the above assumptions, there are disjoint sets $E_0, E_1, \ldots, E_m$ such that*

$$F = \bigcup_{k=0}^{m} E_k,$$

and

(a) $\displaystyle \sum_{n=1}^{\infty} \left\{ \frac{2^{-nm}[\nabla^m f]_n(x)}{[f]_n(x)} \right\}^{p'} = \infty$ *for all* $x \in E_0$, $\left(p' = \dfrac{p}{p-1} \right)$;

(b) $\displaystyle \liminf_{n \to \infty} \frac{2^{-n(m-k)}[G_k * |\nabla^m f|]_n(x)}{[f]_n(x)} > 0$ *for all* $x \in E_k$,

   and $C_{k,p}(E_k) = 0$, *for* $k = 1, 2, \ldots, m - 1$;

(c) $C_{m,p}(E_m) = 0$.

Suppose for simplicity that the sets E_k are compact, so that one can treat them separately. Using the divergent series in (a), one can construct a C^∞-function χ_0 so that $0 \le \chi_0 \le 1$, $\chi_0 = 0$ on a neighborhood of E_0, and $\|f - f\chi_0\|_{m,q} < \frac{\varepsilon}{2}$, for some arbitrarily given $\varepsilon > 0$.

Then one can use the fact that $C_{1,p}(E_1) = 0$ to construct a suitable function χ_1 so that $0 \le \chi_1 \le 1$, $\chi_1 = 0$ on a neighborhood of E_1, $\int |\nabla \chi_1|^p dx$ is small, and use (b) to show that

$$\|f\chi_0 - f\chi_0\chi_1\|_{m,p} < \frac{\varepsilon}{2^2}.$$

One proceeds step by step, until one has found $\chi_0, \ldots, \chi_m$ so that $\chi = \chi_0\chi_1\cdots\chi_m = 0$ on a neighborhood of F, and $\|f - f\chi\|_{m,p} < \varepsilon$. It follows that $f \in H_0^{m,p}(F^c)$.

We shall illustrate the result by showing that it contains the uniqueness theorem of Sobolev (Theorem 4).

Let G be bounded as before by a finite union of manifolds not forming cusps,

$$\partial G = \bigcup_{k=1}^{d} M_k, \text{ dim } M_k = d - k.$$

Let $f \in H^m(G)$.

In Theorem 6, f is required to belong to $H^m(\mathbf{R}^d)$, but this difference is only apparent. In fact, by a well-known extension theorem (see Stein [24]), a function f in $H^m(G)$ can be extended across M_1 if it is assumed that M_1 satisfies a local Lipschitz condition.

That f can be extended to $H^m(\mathbf{R}^d)$ follows from the following lemma, which is probably well known.

Lemma 2.   Let $f \in H^{m,p}(G\backslash F)$, where F is a closed set with zero $(d - 1)$-dimensional Hausdorff measure.   Then $f \in H^{m,p}(G)$.

Proof.   We can assume without loss of generality that $G = \mathbf{R}^d$. By assumption, almost every line parallel to the x_i-axis is disjoint from F. In fact, the perpendicular projection of F on the hyperplane $x_i = 0$ has $(d - 1)$-dimensional measure zero.

Let $f \in H^{m, p}(\mathbf{R}^d \setminus F)$. Then f and $D^\alpha f$, $|\alpha| \leq m$, are defined almost everywhere in $\mathbf{R}^d$, so $D^\alpha f \in L^p(\mathbf{R}^d)$, $0 \leq |\alpha| \leq m$. It is enough to prove that

$$\int_{\mathbf{R}^d} \varphi D^\alpha f \, dx = (-1)^{|\alpha|} \int_{\mathbf{R}^d} f D^\alpha \varphi \, dx \text{ for all } \varphi \in C_0^\infty.$$

First let $|\alpha| = 1$. By Fubini's theorem,

$$\int_{\mathbf{R}^d} \varphi \frac{\partial f}{\partial x_i} \, dx = \int_{\mathbf{R}^{d-1}} dx' \int_{-\infty}^{\infty} \varphi \frac{\partial f}{\partial x_i} \, dx_i$$

$(dx' = dx_1 \ldots dx_{i-1} dx_{i+1} \ldots dx_d)$. For almost every $a' \in \mathbf{R}^{d-1}$, the line $x' = a'$ lies in $\mathbf{R}^d \setminus F$. Thus, by a well-known property of Sobolev spaces, f is absolutely continuous on almost all such lines. By integration by parts,

$$\int_{\mathbf{R}^{d-1}} dx' \int_{-\infty}^{\infty} \varphi \frac{\partial f}{\partial x_i} \, dx_i = -\int_{\mathbf{R}^{d-1}} dx' \int_{-\infty}^{\infty} f \frac{\partial \varphi}{\partial x_i} \, dx_i = -\int_{\mathbf{R}^d} f \frac{\partial \varphi}{\partial x_i} \, dx.$$

The lemma follows by induction.

Now let $f \in H^m(G)$, where

$$\partial G = \bigcup_{k=1}^{d} M_k,$$

and suppose that $D^\alpha f|_{M_k} = 0$, in the sense of Sobolev, for $1 \leq k < 2(m - |\alpha|)$, $0 \leq |\alpha| \leq m - 1$. It is easily seen that $D^\alpha f|_{M_k} = 0$ in the sense of corrected functions. But by Theorem 5, $C_{m-|\alpha|, 2}(M_k) = 0$ for $k \geq 2(m - |\alpha|)$. It follows that $D^\alpha f|_{\partial G} = 0$, $0 \leq |\alpha| \leq m - 1$, in the sense of corrected functions, and thus Theorem 6 applies.

The formulation in Problem 5 of the Dirichlet problem is less general than the generalized Dirichlet problem (Problem 3) in the sense that the boundary values are supposed to be given through a function g that is defined in all of $\mathbf{R}^d$, or at least across ∂G. One can remove this restriction by formulating a "fine Dirichlet problem."

Let $f \in H^{m, p}(\mathbf{R}^d)$, and suppose that f is corrected. Then one can prove (Fuglede [11]) that for (m, p)-quasievery point x, the set

$\{y; |f(y) - f(x)| \geq \varepsilon\}$ is (m, p)-thin at x for all $\varepsilon > 0$. One says that (m, p)-fine $\lim\limits_{y \to x} f(y) = f(x)$, or one introduces a fine topology and says that f is (m, p)-finely continuous at x.

The fine Dirichlet problem can be formulated as follows.

Problem 6. Let $G \subset \mathbf{R}^d$ be open and bounded, and let $g \in H^m(G)$. Find $u \in H^m(G)$ so that $\Delta^m u = 0$ in G, and

$$(m - |\alpha|, 2)\text{-fine} \lim_{G \ni x \to x_0} D^\alpha(u(x) - g(x)) = 0$$

for

$$(m - |\alpha|, 2)\text{-q.e. } x_0 \in \partial G, \ 0 \leq |\alpha| \leq m - 1.$$

It is clear from the above that a solution exists. Using Theorem 6, T. Kolsrud [15] has proved the following theorem.

Theorem 8. *Problem 6 always has a unique solution. In fact, a function f in $H^{m, p}(G)$ belongs to $H_0^{m, p}(G)$ if and only if*

$$(m - |\alpha|, p)\text{-fine} \lim_{G \ni x \to x_0} D^\alpha f(x) = 0$$

for $(m - |\alpha|, 2)$-q.e. $x_0 \in \partial G, \ 0 \leq |\alpha| \leq m - 1$, provided $p > 2 - \dfrac{1}{d}$. (Added in proof: The theorem is true for $p > 1$. See [26].)

This characterization of $H_0^1(G)$ is due to Deny and Lions [8, Theorem 5.1], who deduced it from a uniqueness theorem for harmonic functions of Brelot [4]. See also Fuglede [10, Theorem 9.1]. Here the procedure is the opposite one; the uniqueness theorem is proved using a characterization of $H_0^m(G)$.

REFERENCES

1. Adams, D. R. Maximal operators and capacity. *Proc. Amer. Math. Soc.* 34 (1972):152-156.

2. Agmon, S. *Lectures on Elliptic Boundary Value Problems*. New York: Van Nostrand, 1965.

3. Agmon, S. Maximum theorems for solutions of higher order elliptic equations. *Bull. Amer. Math. Soc.* 66 (1960):77-80.

4. Brelot, M. Sur l'allure des fonctions harmoniques et sous-
    harmoniques à la frontière. *Math. Nachr.* 4 (1950-51):298-307.

5. Calderón, A. P. Lebesgue spaces of differentiable functions
    and distributions. *Proc. Symp. Pure Math.* 4 (1961):33-49.

6. Deny, J. Les potentiels d'énergie finie. *Acta Math.* 82 (1950):
    107-183.

7. Deny, J. Théorie de la capacité dans les espaces fonction-
    nels. *Séminaire de Théorie du Potentiel* 9 (1964-65), No. 1.

8. Deny, J., and Lions, J.-L. Les espaces du type de Beppo Levi.
    Ann. Inst. Fourier (Grenoble) 5 (1953-54):305-370.

9. Friedrichs, K. Die randwert- und eigenwertprobleme aus der
    theorie der elastischen platten. *Math. Ann.* 98 (1928):205-247.

10. Fuglede, B. Finely harmonic functions. *Lecture Notes in
    Mathematics* 289. New York: Springer-Verlag, 1972.

11. Fuglede, B. Quasi topology and fine topology. *Séminaire de
    Théorie du Potentiel* 10 (1965-66), No. 12.

12. Hedberg, L. I. Spectral synthesis and stability in Sobolev
    spaces, Euclidean harmonic analysis. Proc. Univ. of Maryland,
    1979. *Lecture Notes in Mathematics* 779. New York: Springer-
    Verlag, 1980, pp. 73-103.

13. Hedberg, L. I. Spectral synthesis in Sobolev spaces, and
    uniqueness of solutions of the Dirichlet problem. *Acta Math.*
    147 (1981):237-264.

14. Hedberg, L. I. Two approximation problems in function spaces.
    Ark. Mat. 16 (1978):51-81.

15. Kolsrud, T. A uniqueness theorem for higher order elliptic
    partial differential equations. To appear in *Math.
    Scand.*

16. Maz'ja, V. G. Behavior of solutions to the Dirichlet problem
    for the biharmonic operator at the boundary point. Technische
    Hochschule Karl-Marx-Stadt, Sektion Mathematik, *Wissenschaft-
    liche Informationen* 10 (1979):1-16.

17. Maz'ja, V. G. On the behavior near the boundary of solutions
    of the Dirichlet problem for the biharmonic operator. *Dokl.
    Akad. Nauk SSSR* 235:6 (1977):1263-1266.

18. Maz'ja, V. G., and Havin, V. P.   Non-linear potential theory.
     Uspehi Mat. Nauk 27:6 (1972):67-138.

19. Miranda, C.   Partial differential equations of elliptic type.
     2d ed.   *Ergebnisse der Math.*, vol. 2.   New York: Springer-
     Verlag, 1970.

20. Nirenberg, L.   Remarks on strongly elliptic partial differential
     equations.   *Comm. Pure Appl. Math.* 8 (1955):648-674.

21. Schulze, B.-W., and Wildenhain, G.   *Methoden der Potential-
     theorie für Elliptische Differentialgleichungen beliebiger
     Ordnung.*   Berlin: Akademie-Verlag, 1977.

22. Sobolev, S. L.   Applications of functional analysis in mathe-
     matical physics.   *Izd. LGU*, Leningrad, 1950.   (English trans-
     lation, *Amer. Math. Soc.*, Providence, R.I., 1963).

23. Sobolev, S. L.   On a boundary value problem for polyharmonic
     equations.   *Mat. Sb.* (N.S.) 2 (44) (1937):467-499.   [English
     translation, *Amer. Math. Soc. Translations* (2) 33 (1963):1-40.]

24. Stein, E. M.   *Singular Integrals and Differentiability Proper-
     ties of Functions.*   Princeton, N.J.: Princeton Univ. Press,
     1970.

25. Wallin, H.   Continuous functions and potential theory.   *Ark.
     Mat.* 5 (1963):55-84.

26. Hedberg, L. I., and Wolff, T. H.   Thin sets in nonlinear potential
     theory.   To appear.

L^p-ESTIMATES FOR A SINGULAR HYPERBOLIC EQUATION

Jeff E. Lewis
University of Illinois at Chicago

Cesare Parenti
University of Bologna

Let A be a self-adjoint operator on a Hilbert space H with domain $D(A)$ and compact resolvent. We study the singular hyperbolic equation

$$tu'(t) - (iA + bI)u(t) = f(t) \text{ on } R^+, \tag{1}$$

with b real.

As a simple model for (1), consider the scalar equation

$$tu'(t) - (ic + b)u(t) = f(t) \in C_0^\infty(R^+). \tag{2}$$

Equation (2) has two solutions given by

$$F_{b+ic}f(t) = \int_0^t (t/s)^{b+ic}f(s)\,ds/s,$$

$$B_{b+ic}f(t) = -\int_t^\infty (t/s)^{b+ic}f(s)\,ds/s. \tag{3}$$

If supp $f \subset (T_1, T_2)$, then $F_{b+ic}f(t) = 0$ on $(0, T_1)$ and $B_{b+ic}f(t) = 0$ on (T_2, ∞); moreover,

$$F_{b+ic}f(t) - B_{b+ic}f(t) = t^{b+ic}\tilde{f}(-(b+ic)),$$

where $\tilde{f}(z) = \int_0^\infty t^{z-1}f(t)\,dt$ is the Mellin transform of f.

This work was completed while the second author was visiting the University of Illinois at Chicago. Both authors have been partially supported by the Italian C.N.R., gruppo G.N.A.F.A.

By Hardy's inequality,

if $1/p + a + b < 0$, then

$$\|t^a F_{b+ic} f; \ L^p(0, T)\| \le C \|t^a f; \ L^p(0, T)\|,$$ (5)

if $1/p + a + b > 0$, then

$$\|t^a B_{b+ic} f; \ L^p(T, \infty)\| \le C \|t^a f; \ L^p(T, \infty)\|.$$ (6)

The constant C in (5) or (6) may be chosen as $C = |1/p + a + b|^{-1}$.

For $I = R^+$, $(0, T)$, or (T, ∞), and $t^a f(t) \in L^p(I; H)$, a distribution $u \in \mathscr{D}'(I; H)$ is an $L^{p, a}$ solution of

$$Pu = tu' - (iA + bI)u = f \text{ on } I$$ (7)

if

$$t^a u \in L^p(I; H),$$

    (8)

$$\int_0^\infty (u, P^*v) dt = \int_0^\infty (f, v) dt, \text{ for all } v \in C_0^\infty(I; D(A)).$$

The formal adjoint of P is

$$P^*v = -\{tv' - (iA - (1 + b)I)v\}.$$ (9)

We now give our main result.

Theorem. Let $t^a f \in L^p(R^+; H)$. Then:

(a) If $1/p + a + b < 0$, then for any T, $0 < T \le \infty$, there is a unique $L^{p, a}$ solution of $Pu = f$ on $(0, T)$; the solution $u = Ff$ satisfies

$$\|t^a u; \ L^p(0, T; H)\| \le C \|t^a f; \ L^p(0, T; H)\|.$$ (10)

(b) If $1/p + a + b > 0$, then for any T, $0 \le T < \infty$, there is a unique $L^{p, a}$ solution of $Pu = f$ on (T, ∞); the solution $u = Bf$ satisfies

$$\|t^a u; \ L^p(T, \infty; H)\| \le C \|t^a f; \ L^p(T, \infty; H)\|.$$ (11)

(c) The constant C in (10) or (11) may be taken as

$$C = |1/p + a + b|^{-1}.$$ (12)

Proof. Since A has compact resolvent, denote the eigenvalues of A as $\{c_k\}$ with a corresponding complete orthonormal basis of eigenfunctions $\{y_k\}$. If

$$f(t) = \sum_{\text{finite}} f_k(t)y_k, \quad f_k \in C_0^\infty(R^+),$$

define

$$Ff(t) = \Sigma F_{b+ic_k} f_k y_k, \tag{13}$$

$$Bf(t) = \Sigma B_{b+ic_k} f_k y_k. \tag{14}$$

Obviously, $PFf = f$ and $PBf = f$ on R^+. Since

$$t^{a-ic} F_{b+ic} f(t) = \int_0^t (t/s)^{a+b}(s^{a-ic} f(s))ds/s,$$

we have that

$$|t^a Ff(t)|_H^2 = \Sigma|t^a F_{b+ic_k} f_k|^2 = \Sigma|F_{a+b}(s^{a-ic} f_k)|^2.$$

If $1/p + a + b < 0$, using the bound $C = |1/p + a + b|^{-1}$ for F_{a+b} on $L^p(0, T)$ and the Littlewood-Paley lemma of Zygmund [2, vol. 2, Lemma 2.10], we obtain that $\|t^a Ff; L^p(0, T; H)\|$ is dominated by C times

$$\|(\Sigma|t^{a-ic_k} f_k|^2)^{1/2}; L^p(0, T)\| = \|t^a f; L^p(0, T; H)\|. \tag{15}$$

A similar argument yields (11) for Bf.

To prove the uniqueness, suppose that $1/p + a + b < 0$ and that u is an $L^{p,a}$ solution of $Pu = 0$ on $(0, T)$. Fix $c = c_k$ and for any $g \in C_0^\infty(0, T)$ such that

$$\tilde{g}(-(ic - (1 + b))) = 0, \tag{16}$$

define $v_k = (B_{ic-(1+b)} g(t))y_k$. By (16) and (4), $v_k \in C_0^\infty(0, T)$ and $P^* v_k = -gy_k$. Since $Pu = 0$ on $(0, T)$,

$$\int_0^T (u, y_k)g(t)dt = 0. \tag{17}$$

Since $1/p + a + b < 0$, the linear functional

$$g \to \tilde{g}(1 + b - ic)$$

is not continuous on $L^{q',-a}(0, T)$, $1/p + 1/q = 1$; hence functions

$g(t)$ satisfying (16) are dense in $L^{q,-a}(0, T)$ and (17) holds for all $g \in C_0^\infty(0, T)$. Hence $(u, y_k) = 0$ for every k.

Uniqueness in the case $1/p + a + b > 0$ is handled in a similar manner.

Remarks.

1. When $p = 2$ the equation (1) on $(0, T)$ and many generaliza-
    tions were studied by Alinhac and Baouendi [1] by the
    method of energy inequalities.

2. Under the change of variables $s = \ln(1/t)$, equation (1)
    becomes

    $$-(d/ds)u - (iA + bI)u = f, \tag{18}$$

    and inequality (10) becomes for $1/p + a + b < 0$

    $$\|e^{-as}u; \, L^p(S, \infty; H)\| \le C\|e^{-as}f; \, L^p(S, \infty; H)\|$$

    for solutions of (18).

REFERENCES

1. Alinhac, S., and Baouendi, M. S. Uniqueness for the charac-
    teristic Cauchy problem and strong unique continuation for
    higher order partial differential inequalities. *Amer. J. of
    Math.* 102 (1980):179-217.

2. Zygmund, A. *Trigonometric Series.* New York: Cambridge Univ.
    Press, 1959.

MIXED NORM ESTIMATES
FOR THE KLEIN-GORDON EQUATION

Bernard Marshall
McGill University

Let $u(x, t)$ be the solution of the Klein-Gordon equation

$$\begin{cases} u_{tt} - \Delta_x u + u = 0 \\ u(x, 0) = 0 \qquad u_t(x, 0) = f(x) \end{cases}$$

where $x \in \mathbf{R}^n$, $t > 0$. For what values of q and r does there exist α such that

$$\| (1 + t)^{-\alpha} u \|_{q, r} \equiv \left(\int_0^\infty ((1 + t)^{-\alpha} \| u(\cdot, t) \|_q)^r \, dt \right)^{1/r} \le C \| f \|_2, \quad (1)$$

and what is the smallest possible value of α? Thus $\alpha - \dfrac{1}{r}$ is a measure of the decay of the solution u.

Theorem 1.

(i) *There can exist an estimate of the form (1) only if* $\left(\dfrac{1}{q}, \dfrac{1}{r} \right)$ *is in the region R determined by*

$$0 \le \frac{1}{r} \le 1, \quad \frac{1}{2} - \frac{3}{2n} \le \frac{1}{q} \le \frac{1}{2}, \quad \frac{n}{q} + \frac{1}{r} \ge \frac{n - 2}{2}.$$

(ii) *For* $\left(\dfrac{1}{q}, \dfrac{1}{r} \right)$ *in R, if (1) holds, then* $\alpha \ge \alpha_0 \left(\dfrac{1}{q}, \dfrac{1}{r} \right)$, *where* α_0 *is defined by*

$$\alpha_0 \left(\frac{1}{q}, \frac{1}{r} \right) = \max \left\{ 0, \frac{n}{2q} - \frac{n}{4} + \frac{1}{r}, \frac{1}{r} - \frac{1}{2} \right\}.$$

This research was supported by NSERC Grant u0074.

To describe the corresponding positive result, let R_0 be the region in the $\left(\frac{1}{q}, \frac{1}{r}\right)$ plane defined by

$$0 \le \frac{1}{r} \le 1, \quad \frac{1}{2} - \frac{3}{2n} < \frac{1}{q} \le \frac{1}{2}, \quad \frac{n}{q} + \frac{1}{r} > \frac{n-2}{2},$$

together with the points satisfying

$$\frac{n}{q} + \frac{1}{r} = \frac{n-2}{2}, \quad \frac{1}{r} \le \frac{1}{2} - \frac{1}{n+1}. \tag{2}$$

Note that the closure of R_0 is R. These regions in Figure 1 have vertices P_0, P_1, P_4, P_7, P_5.

The coordinates of the points in Figure 1 are as follows:

$$P_0 = \left(\frac{1}{2}, 0\right) \qquad\qquad P_1 = \left(\frac{1}{2} - \frac{1}{n}, 0\right) \qquad\qquad P_2 = \left(\frac{1}{2} - \frac{1}{n}, \frac{1}{2}\right)$$

$$P_3 = \left(\frac{1}{2} - \frac{1}{n-1}, \frac{1}{2}\right) \qquad P_4 = \left(\frac{1}{2} - \frac{3}{2n}, \frac{1}{2}\right) \qquad P_5 = \left(\frac{1}{2}, 1\right)$$

$$P_6 = \left(\frac{1}{2} - \frac{1}{n}, 1\right) \qquad\qquad P_7 = \left(\frac{1}{2} - \frac{3}{2n}, 1\right)$$

Theorem 2. For $\left(\frac{1}{q}, \frac{1}{r}\right)$ in R_0, if $\alpha > \alpha_0\left(\frac{1}{q}, \frac{1}{r}\right)$, then

$$\| (1 + t)^{-\alpha} u \|_{q, r} \le C \| f \|_2.$$

For $\alpha = \alpha_0$, the operator $Tf = u$ may or may not be bounded. This can be seen quickly in the case

$$\| t^{-1/2} u \|_{2, 2} = \int_{\mathbf{R}^n} \int_0^\infty |\hat{f}(\xi)|^2 (|\xi|^2 + 1)^{-1} \int_0^\infty \sin^2 (t\sqrt{1 + |\xi|^2}) t^{-1} dt \, d\xi = \infty,$$

where we used the Plancherel theorem and the fact that

$$\widehat{T_t f}(\xi) = (1 + |\xi|^2)^{-1/2} \sin(t\sqrt{1 + |\xi|^2}) \hat{f}(\xi).$$

Of particular importance in applications is the quadrilateral in R where $\alpha_0 = 0$. Define

$$Q = \left\{ \left(\frac{1}{q}, \frac{1}{r}\right) \in R_0 : 0 \le \frac{1}{r} \le \frac{1}{2}, \frac{1}{2} - \frac{1}{n} - \frac{1}{nr} \le \frac{1}{q} < \frac{1}{2} - \frac{1}{nr} \right\}.$$

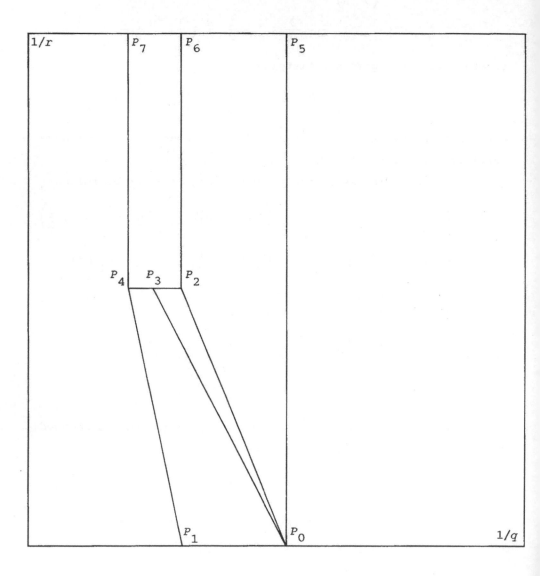

Figure 1

The vertices of Q are P_0, P_1, P_2, P_4. The closure of Q is the region where $\alpha_0 = 0$.

Theorem 3. If $\left(\dfrac{1}{q}, \dfrac{1}{r}\right) \in Q$, then $\|u\|_{q,r} \leq C\|f\|_2$. If $\left(\dfrac{1}{q}, \dfrac{1}{r}\right) \notin \bar{Q}$, then $Tf = u$ is not bounded from L^2 to $L^r(L^q)$.

These estimates are a consequence of estimates involving derivatives of u. Let $\mathscr{I}$ be a triangle

$$\mathscr{I} = \left\{\left(\frac{1}{q}, \frac{1}{r}\right) : 0 \leq r \leq \frac{1}{2}, \ \frac{1}{2} - \frac{2}{(n-1)r} < \frac{1}{q} < \frac{1}{2} - \frac{2}{nr}\right\}.$$

The vertices of $\mathscr{I}$ are P_0, P_2, P_3.

Theorem 4. If $\left(\dfrac{1}{q}, \dfrac{1}{r}\right) \in \mathscr{I}$ and $\gamma < \dfrac{1}{2} + \dfrac{1}{q} - \dfrac{1}{r}$, then

$$\| (I - \Delta)^{\gamma/2} u\|_{q,r} \leq C\|f\|_2. \tag{3}$$

When $q = \infty$, the L^∞-norm is replaced by the BMO norm. Also, by Sobolev's theorem, if $\left(\dfrac{1}{q}, \dfrac{1}{r}\right) \in Q - \mathscr{I}$ and $0 \leq \gamma < 1 - \dfrac{n}{2} + \dfrac{n}{q} + \dfrac{1}{r}$, then (3) holds.

In the case $r = \infty$, the estimate (1) takes the form

$$\|u(\cdot, t)\|_q \leq Ct^\alpha \|f\|_p.$$

The estimates of this type have been determined completely in [3].

In the case of $q = r$, $\alpha = 0$, Strichartz [8] has shown that $(I - \Delta)^{1/4}T$ is bounded from $L^2(\mathbf{R}^n)$ to $L^q(\mathbf{R}^{n+1})$ if

$$2 + \frac{4}{n} \leq q \leq 2 + \frac{4}{n-1}.$$

This result is stronger than Theorem 4, which states that $(I - \Delta)^\sigma T$ is bounded if $\sigma < \dfrac{1}{4}$ and $2 + \dfrac{4}{n} < q < 2 + \dfrac{4}{n-1}$. Thus, in Theorem 4, for $q \leq r$ we can have $\mathscr{I}$ closed and $\gamma = \dfrac{1}{2} + \dfrac{1}{q} - \dfrac{1}{r}$. In [8], Strichartz used duality to take advantage of a device of Tomas [9] for handling the restrictions of Fourier transforms. We follow the same approach for Theorem 4 but use a different analytic family of operators, $(I - \Delta)^z T$, to control the

different behavior in the x- and t-variables. The estimates
needed, then, are those in [3].

The stimulus for studying these estimates came from the
scattering theory of various nonlinear variants of the Klein-Gordon
equation. In particular, the estimates of Theorem 3 have been
used in Strauss [7] in the study of nonlinearities of the form
$(V * u^2)u$.

1. NEGATIVE RESULTS (THEOREM 1)

Proof of Theorem 1. *Part (i)a.* There are three estimates to be
proved:

$$\frac{1}{q} \leq \frac{1}{2}; \quad \frac{1}{2} - \frac{3}{2n} \leq \frac{1}{q}; \quad \frac{n}{q} + \frac{1}{r} \geq \frac{n-2}{2}.$$

Fix $\ell > 1$ and let $k = (\ell^2 - 1)^{1/2}$. Let $f \in L^2$ be such that
$f = 1$ in $|x| < 1$ and let $g(x) = \cos(kx_1)$ and $f_s(x) = s^{n/2} f(sx)$ for
s small. If $u(x, t)$ denotes the solution with $u(x, 0) = 0$,
$u_t(x, 0) = g(x)f_s(x)$, then $u_t(x, 0) = \cos(kx_1)$ for $|x| < s^{-1}$ and
so in the domain of dependence,

$$T_t(g \cdot f_s)(x) = \ell^{-1} \sin \ell t \cos kx_1 \text{ for } A = \{x : |x| < s^{-1} - t\}.$$

Consider $\frac{\pi}{2} \leq \ell t \leq \frac{3\pi}{4}$ and $s^{-1} > 2t$. Then

$$\|t^{-\alpha} u\|_{q, r}^r \geq s^{nr/2} \int_{\pi/4\ell}^{3\pi/4\ell} \left| \frac{\sin \ell t}{\ell t^\alpha} \right|^r \left(\int_A |\cos kx_1|^q \, dx \right)^{r/q} dx$$

$$\geq Cs^{nr/2} \left(\int_{|x| < (2s)^{-1}} |\cos kx_1|^q \, dx \right)^{r/q}$$

$$\geq Cs^{nr\left(\frac{1}{2} - \frac{1}{q}\right)} \to \infty \text{ as } s \to 0$$

whenever $q < 2$. Since $\|gf_s\|_2 \leq \|f_s\|_2 = \|f\|_2$, this shows that
$t^{-\alpha}T$ is unbounded for $\frac{1}{q} > \frac{1}{2}$.

Part (i)b.   Let $f(x) = \sin \frac{1}{\sigma}(|x| - 1)$ for $1 \le |x| \le M$ and $f(x) = 0$ otherwise.   Then $\|f\|_2 \le CM^{n/2}$.   We calculate $T_t f = K_t * f$ by using the explicit formulas for K_t in [1, pp. 693-694].   Suppose that $1 < t < 1 + s$ and $|x| \le C\sigma$.   In odd dimensions, the main term of $K_t * f(x)$ is

$$\left| t^{-\frac{n-1}{2}} \left(\frac{\partial}{\partial t} \right)^{\frac{n-1}{2}} X_t * f(x) \right| \ge C t^{\frac{n-1}{2}} \sigma^{-\frac{n-3}{2}},$$

where X_t is the characteristic function of $\{|x| \le t\}$.   Similarly, in all dimensions

$$|T_t f(x)| \ge C t^{\frac{n-1}{2}} \sigma^{-\frac{n-3}{2}} \qquad |x| \le C\sigma$$

for $1 < t < 1 + \sigma$ and σ sufficiently small.   Therefore,

$$\|t^{-\alpha} Tf\|_{q,r} \ge \left(\int_1^M \left(t^{-\alpha} C t^{\frac{n-1}{2}} \sigma^{\frac{n}{q}-\frac{n-3}{2}} \right)^r dt \right)^{1/r}$$
$$\ge C\sigma^{\frac{n}{q}-\frac{n-3}{2}} \left(1 + M^{-\alpha+\frac{n-1}{2}+\frac{1}{r}} \right).$$

By letting $\sigma \to 0$ or $M \to \infty$, this shows that $t^{-\alpha} T$ is unbounded if either

$$-\alpha + \frac{n-1}{2} + \frac{1}{r} - \frac{n}{2} > 0 \quad \text{or} \quad \frac{n}{q} - \frac{n-3}{2} < 0,$$

that is, $\alpha < \frac{1}{r} - \frac{1}{2}$ or $\frac{1}{q} < \frac{n-3}{2n}$.

Part (i)c.   From Lemma 10, p. 434 of [3], there exist f, a, b, c_t such that

$$|T_t f_s(x)| \ge c_t s^{-1/2} \text{ for } t - b/s < |x| < t - a/s, \ s \ge 1,$$

where $f_s(x) = s^{n/2} f(sx)$.   If we consider only $\frac{1}{2} \le t \le 1$, then c_t can be replaced by an absolute constant.   Consider s fixed for the moment and let g be the characteristic function of the region $A = \{x : 1 - b_1/s \le |x| \le 1 - a_1/s\}$ and let $v(x, t) = f_s(x)$ if $t \in A_2 = \{t : 1 - b_2/s < t < 1 - a_2/s\}$ and $v(x, t) = 0$ otherwise. Choose the constants a_1, a_2, b_1, b_2 so that if $x \in A$ and $t \in A_2$, then $t - b/s < |x| < t - a/s$ and so we can use the estimate from [3]. $\langle\langle\,,\,\rangle\rangle$ will denote the inner product over $\mathbf{R}^{n+1}$ and $\langle\,,\,\rangle$ will be the inner product over $\mathbf{R}^n$. Since $T_t g = u(\cdot,\, t)$ is self-adjoint,

$$\langle\langle t^{-\alpha} Tg,\, v\rangle\rangle = \int_{A_2} t^{-\alpha}\langle T_t g,\, f_s\rangle\, dt$$

$$= \int_{A_2}\int_A t^{-\alpha}(T_t f_s)\, dx\, dt \geq Cs^{-2}s^{-1/2}.$$

Note above that if f is chosen to be real-valued, then $T_t f_s$ is always positive or always negative and so $T_t f_s$ could be replaced by $|T_t f_s|$. But because

$$\|v\|_{q',\,r'} = Cs^{\frac{n}{2}-\frac{n}{q'}-\frac{1}{r'}}\ \text{and}\ \|g\|_2 = Cs^{-1/2},$$

we have

$$\|t^{-\alpha}T\| \geq Cs^{-\frac{n}{q}-\frac{1}{r}+\frac{n-2}{2}}.$$

Letting $s \to \infty$ shows that $t^{-\alpha}T$ is not bounded for $\frac{n}{q} + \frac{1}{r} < \frac{n-2}{2}$.

Part (ii)a. $(\alpha < 0)$. From [3, p. 434] there exist $f,\ a,\ b,\ c$ such that the measure of the set of x that satisfy

$$|T_t f(x)| \geq ct^{-n/2} \quad\text{and}\quad at < |x| < bt$$

is at least ct^n for $t \geq 1$.

Suppose g is the characteristic function of $A = \{x : s + as < |x| < s + bs\}$ and $v(x,\, t) = f(x)$ for $t \epsilon A_2 = \{t : s + d < t < s + e\}$ and $v(x,\, t) = 0$ otherwise. Then

$$|\langle\langle t^{-\alpha} Tg,\, v\rangle\rangle| \geq \int_{A_2}\int_A |T_t f| t^{-\alpha} dt \geq Cs^{-\alpha+\frac{n}{2}}$$

and $\|v\|_{q',\,r'} = C,\ \|g\|_2 = Cs^{n/2}$. Therefore,

$$\|t^{-\alpha} Tg\| \geq Cs^{-\alpha}\|g\|_2 \text{ for all } s \geq 1.$$

Hence $t^{-\alpha}T$ is not bounded if $\alpha < 0$.

Part (ii)b. $\left(\alpha < \frac{n}{2q} - \frac{n}{4} + \frac{1}{r}\right)$. This estimate follows an argument of A. W. Knapp in [8]. Let Q be the rectangle centered at

$(0, 1) \in \mathbf{R}^n \times \mathbf{R}$ whose sides are parallel to the axes. The first n sides have length $\sqrt{\varepsilon}$ and the length parallel to the t-axis is ε. Let v be the inverse Fourier transform of the characteristic function of Q. Let f be the function on $\mathbf{R}^n$ whose Fourier transform is the characteristic function of the cube centered at the origin with sides of length $\sqrt{\varepsilon}$ parallel to the axes.

Then by the Plancherel theorem on $\mathbf{R}^{n+1}$,

$$\langle\langle Tf, v \rangle\rangle = \langle\langle \widehat{Tf}, \hat{v} \rangle\rangle = \int \chi_Q d\mu \geq C\varepsilon^{n/2},$$

where $d\mu$ is the measure on $S = \{(\xi, \tau) : |\xi|^2 + 1 = \tau^2\}$ given by $d\mu = (1 + |\xi|^2)^{-1} d\xi_1 \ldots d\xi_n$. But f and v are essentially equal to

$$\prod_{j=1}^{n} \frac{\sin(\sqrt{\varepsilon}\, x_j)}{x_j} \quad \text{and} \quad \frac{\sin \varepsilon t}{t} \prod_{j=1}^{n} \frac{\sin(\sqrt{\varepsilon}\, x_j)}{x_j}$$

and so $\|f\|_2 \leq C\varepsilon^{n/4}$ and $\|t^\alpha v\|_{q', r'} \leq C\varepsilon^{\frac{n}{2} - \frac{1}{q} + \frac{1}{r} - \alpha}$. Therefore,

$$\|t^{-\alpha} t\| \geq \langle\langle Tf, v \rangle\rangle / \|f\|_2 \|t^\alpha v\|_{q', r'} \geq C\varepsilon^{\frac{n}{4} - \frac{n}{2} + \frac{1}{q} - \frac{1}{r} + \alpha}.$$

By letting $\varepsilon \to 0$, we see that $t^{-\alpha} T$ is unbounded for

$$\frac{n}{4} - \frac{n}{2q} - \frac{1}{r} + \alpha < 0.$$

This completes the proof of Theorem 1.

2. THE POSITIVE RESULTS

The proofs of Theorems 2, 3, and 4 will be given together. If estimates can be obtained at the vertices P_0 to P_7, then the rest of the results follow by interpolation. Since the multipliers for T_t and $\sqrt{I - \Delta}\, T_t$, namely, $(1 + |\xi|^2)^{-1/2} \sin(t\sqrt{1 + |\xi|^2})$ and $\sin(t\sqrt{1 + |\xi|^2})$, are bounded, then T and $\sqrt{I - \Delta}\, T$ map L^2 to $L^\infty(L^2)$. Also, by Sobolev's theorem, T maps L^2 to $L^\infty(L^q)$, where $\frac{1}{q} = \frac{1}{2} - \frac{1}{n}$. This takes care of P_0 and P_1.

The estimates for P_5, P_6, P_7 follow from those of P_0, P_2, P_4, respectively, by Hölder's inequality, since

$$\| (1 + t)^{-\alpha - a} u \|_{q, 1} \leq C \| (1 + t)^{-\alpha} u \|_{q, r_1}$$

so long as $(1 + t)^{-a} \in L^{r_1'} (dt)$.

Observe now that Theorem 4 implies Theorem 3 by Sobolev's theorem. The result of Theorem 3 is stronger than that of Theorem 2 in Q because the former claims boundedness of $\alpha = \alpha_0 = 0$, whereas the latter is for $\alpha > \alpha_0$. Theorem 3 gives the estimates near P_2 and P_4, which complete the proof of Theorem 2. The estimates (2) on the boundary of Q follow by Sobolev's theorem from Strichartz' results [8].

Therefore, it remains only to prove Theorem 4 for

$$\left\{ 0 < \frac{1}{r} \leq \frac{1}{2}, \; \frac{1}{2} - \frac{2}{(n - 1)r} < \frac{1}{q} < \frac{1}{2} - \frac{2}{nr} \right\}.$$

If $T_t f = u(\cdot, t) = K_t * f$ and $D_* = (I - \Delta)^{1/2}$, then it was shown in [3, Lemma 1, p. 421] that for any $0 \leq a \leq \frac{1}{2}$, if $\nu = a + (n - 1)/2$, then

$$\| D_*^{-\nu} K_t \|_{\text{BMO}} \leq C t^{-\nu} \quad \text{for } t \geq 1.$$

For small t,

$$\| D_*^{-\nu} K_t \|_{\text{BMO}} \leq C t^{-\frac{n-1}{2}} \quad \text{for } t \leq 1.$$

Since $D_*^{-\nu} T_t$ is a bounded operator from H^1 to L^∞ with norm $\leq C t^{-\frac{n-1}{2}} (1 + |t|)^{-a}$, and $D_* T_t$ is bounded from L^2 to L^2 with norm ≤ 1, then by interpolation, $D_*^{-b} T_t$ is bounded from $L^{q'}$ to L^q with norm $\leq C t^{\varepsilon - 1} (1 + |t|)^{-\pi}$. The parameters are given by

$$\frac{1}{q'} = \frac{1}{2} + \frac{1 - \varepsilon}{n - 1}, \; b = \left(\frac{n + 1 + 2a}{n - 1} \right)(1 - \varepsilon) - 1, \; \pi = \frac{1 - \varepsilon}{n - 1} 2a. \quad (4)$$

Let $k(t)$ be the norm of $D_*^{-b} T_t$ as an operator from $L^{q'}$ to L^q. Since $|k(t)| \leq C t^{\varepsilon - 1} (1 + |t|)^{-\pi}$, then $k(t) \in L^R (dt)$ if

$$(1 - \varepsilon)R < 1 \quad \text{and} \quad R(1 + \pi - \varepsilon) > 1. \quad (5)$$

Let dv be the measure on $|\xi|^2 + 1 = \tau^2$ determined by

$$dv = (1 + |\xi|^2)^{-(b+1)/2}\, d\xi_1 \cdots d\xi_n.$$

If $\widehat{dv}$ denotes the Fourier transform of dv in both space and time variables, then $\widehat{dv}(x,\ t) = D_*^{-b}K_t(x)$. The next restriction theorem follows an argument of P. Tomas [9].

Lemma. *Under the conditions* (4), (5), $0 \le a \le 1/2$, *and* $R = r/2$, *for all* $F \in L^{r'}(L^{q'})$ *we have*

$$\int |\hat{F}|^2 dv \le C\|F\|_{q',\,r'}^2.$$

Proof. To avoid confusion, $\wedge$ and $*$ will denote the Fourier transform and convolution in the n space variables while $\sim$ and $\#$ will denote the same operators in $(n+1)$ variables.

First we show that the hypotheses imply that $\|\widetilde{dv}\ \#\ F\|_{q,\,r} \le C\|F\|_{q',\,r'}$ where the convolution is in $(n+1)$ variables. Let $F_s(x) = F(x,\ s)$. Then

$$\|\widetilde{dv}\ \#\ F\|_{q,\,r} = \left\| \int D_*^{-b} K_{t-s}\ *\ F_s\, ds \right\|_{q,\,r}$$

$$\le \left(\int \left(\int k(t-s)\|F_s\|_{q'}\, ds \right)^r dt \right)^{1/r}.$$

If $h(s) = \|F_s\|_{q'}$ then the last term is $\|k * h\|_r$, which is dominated by $\|k\|_R \|h\|_{r'}$ since

$$\frac{1}{r} = \frac{1}{R} + \frac{1}{r'} - 1 \quad \text{or} \quad R = \frac{r}{2}.$$

This shows that

$$\|\widetilde{dv}\ \#\ F\|_{q,\,r} \le C\|h\|_{r'} = C\|F\|_{q',\,r'}.$$

The proof of the lemma is completed with an application of the Plancherel theorem,

$$\int |\tilde{F}|^2 dv = \int \bar{\tilde{F}}\ \widetilde{F\ \#\ dv} \le C\|F\|_{q',\,r'}\|F\ \#\ \widetilde{dv}\|_{q,\,r} \le C\|F\|_{q',\,r'}^2.$$

By duality, the lemma is equivalent to

$$\|\widetilde{Fdv}\|_{q,\,r} \le C\left(\int |F|^2 dv \right)^{1/2}. \tag{6}$$

If $u(x, t)$ is the solution of the Klein-Gordon equation with initial velocity $f(x)$ and $\xi_* = (1 + |\xi|^2)^{1/2}$, then $\hat{u}(\xi, t) = \xi_*^{-1} \sin(t\xi_*)\hat{f}(\xi)$. Define $F(\xi, \tau) = \hat{f}(\xi)$. Since the Fourier transform with respect to t of $\xi_*^{-1-b} \sin(t\xi_*)\hat{f}(\xi)$ is $\widehat{f dv} = Fdv$, then $D_*^{-b}u = \widetilde{Fdv}$. By the Plancherel theorem,

$$\int |F|^2 dv = \int |\hat{f}(\xi)|^2 (1 + |\xi|^2)^{-(b+1)/2} d\xi = \|D_*^{-(b+1)/2}f\|_2^2.$$

Therefore, (6) becomes $\|D_*^{-b}u\|_{q,r} \leq C\|D_*^{-(b+1)/2}f\|_2$. Equivalently,

$$\|D_*^{\gamma}u\|_{q,r} \leq C\|f\|_2,$$

where $\gamma = (1 - b)/2$.

Now to complete the proof of Theorem 4, we sort out the values of the various parameters. Use (4) to write b, ε, π in terms of n, q, a. Then

$$\gamma = \frac{1}{2}(1 - b) = 1 - \frac{1}{2}(n + 1 + 2a)\left(\frac{1}{2} - \frac{1}{q}\right). \tag{7}$$

The inequalities in (5) become

$$\frac{n-1}{2}\left(\frac{1}{2} - \frac{1}{q}\right) < \frac{1}{r} < \frac{1}{2}(n - 1 + 2a)\left(\frac{1}{2} - \frac{1}{q}\right)$$

or

$$\frac{1}{2} - \frac{2}{(n-1)r} < \frac{1}{q} < \frac{1}{2} - \frac{2}{r(n - 1 + 2a)}. \tag{8}$$

If $\frac{1}{q}$ equals the right-hand endpoint of (8), then $\gamma = \frac{1}{2} + \frac{1}{q} - \frac{1}{r}$. Thus (7) and (8) are equivalent to

$$\gamma < \frac{1}{2} + \frac{1}{q} - \frac{1}{r}, \quad \frac{1}{2} - \frac{2}{(n-1)r} < \frac{1}{q} < \frac{1}{2} - \frac{2}{nr}.$$

Also $R = r/2 \geq 1$ implies $\frac{1}{r} < \frac{1}{2}$.

In the case $q = \infty$, we have shown that

$$\|D_*^{\gamma}u\|_{BMO,r} \leq C\|f\|_2$$

for $\gamma < 1 - \frac{1}{2}(n + 1 + 2a)\left(\frac{1}{2} - \frac{1}{q}\right)$. But because $\|D_*^{\gamma}u\|_{\infty,r} \leq C\|D_*^{\gamma+\varepsilon}u\|_{BMO,r}$ for any $\varepsilon > 0$, we can replace BMO by L^{∞} in all estimates above. This completes the proof of Theorem 2.

REFERENCES

1. Courant, R., and Hilbert, D. *Methods of Mathematical Physics*, vol. 2. New York: Interscience, 1962.

2. Littman, W. The wave operator and L^p norms. *J. Math. Mech.* 12 (1963):55–68.

3. Marshall, B.; Strauss, W.; and Wainger, S. $L^p - L^q$ estimates for the Klein-Gordon equation. *J. Math. Pures et Appl.* 59 (1980):417–440.

4. Peral, J. C. L^p estimates for the wave equation. *J. Func. Anal.* 36 (1980):114–145.

5. Segal, I. E. Space time decay for solutions of wave equations. *Adv. in Math.* 22 (1976):305–311.

6. Strauss, W. A. Everywhere defined wave operators. In *Nonlinear Evolution Equations*, M. G. Crandall, ed. New York: Academic Press, 1978, pp. 85–102.

7. Strauss, W. A. Nonlinear scattering theory at low energy, I. *J. Func. Anal.* 41 (1981):110–133. II. 43 (1981):281–293.

8. Strichartz, R. S. Restrictions of Fourier transforms to quadratic surfaces . . . . *Duke Math. J.* 44 (1977):705–714.

9. Tomas, P. A restriction theorem for the Fourier transform. *Bull. Amer. Math. Soc.* 81 (1975):477–478.

PART EIGHT

OTHER TOPICS RELATED TO HARMONIC ANALYSIS

SCATTERING AND INVERSE
SCATTERING RELATED TO EVOLUTION EQUATIONS

R. Beals
R.R. Coifman
Yale University

It is now well known that many important nonlinear evolution equations are closely related to spectral problems for ordinary differential equations [1, 7, 8]. In principle, the former problems can be linearized and solved by solving associated scattering and inverse scattering problems for the ordinary differential equations. These latter problems are well understood for the second-order (Sturm-Liouville) case [3] and for certain first-order systems when the Gelfand-Levitan-Marcenko theory is applicable [1]. We summarize here a different approach to scattering and inverse scattering that applies to general first-order systems and contributes to the study of higher-order problems. We also discuss the question of the exact scope of the inverse scattering method for evolution equations and obtain a complete solution to a model case.

1. FIRST-ORDER SYSTEMS

Consider a system of the form

$$Du - Qu = f, \qquad D = d/dx, \tag{1}$$

where $Q = (q_{ij}(x))$, $u = (u_1, \ldots, u_n)^t$, $f = (f_1, \ldots, f_n)^t$. We assume that the matrix-valued function Q satisfies

$$q_{ij} \in L^1(\mathbf{R}), \qquad q_{ii} \equiv 0, \tag{2}$$

and denote the set of such Q again by $L^1(\mathbf{R})$. Consider an associated spectral problem

$$Du - Qu = zJu, \qquad J = \operatorname{diag}(\lambda_1, \ldots, \lambda_n), \tag{3}$$

where the λ_j are distinct complex constants. We seek a fundamental matrix solution $\Psi(x, z)$,

$$D\Psi = Q\Psi + zJ\Psi, \qquad \det \Psi \neq 0. \tag{4}$$

If $Q \equiv 0$ then (4) has the standard solution $\exp xzJ$. If Q has compact support then (4) has a unique solution Ψ such that

$$\Psi(x, z) = e^{xzJ} \quad \text{if } x \ll 0. \tag{5}$$

Then for each $z \in C$ there is a matrix $S(z)$ such that

$$\Psi(x, z) = e^{xzJ} S(z) \quad \text{if } x \gg 0. \tag{6}$$

Associated evolution equations may be described as follows. Let μ be a constant diagonal matrix and assume Q is a (matrix-valued) test function. Then there is an asymptotic expansion

$$\Psi(x, z)\mu\Psi(x, z)^{-1} \sim \sum_{k=0}^{\infty} \mu_{k-1}(x) z^{-k}, \quad z \to \infty, \tag{7}$$

where the coefficients depend on Q, $\mu_j = \mu_j(Q)$. *Formally*, if Q depends on a parameter and evolves according to

$$\frac{\partial Q}{\partial t} = [J, \mu_j(Q)], \tag{8}$$

then the matrix S of (6) evolves according to

$$\frac{\partial}{\partial t} S(z, t) = z^j[\mu, S(z, t)]. \tag{9}$$

Thus the initial value problem for the (generally) nonlinear problem (8) may be solved exactly if we can recover Q from the asymptotic data S. This formal procedure as described so far is well known, but there are several obstructions. First, S as defined will make sense in general only for compactly supported Q, and compact support is generally incompatible with (8). If $J + J^* = 0$, then $S(\xi)$ can be defined for all *real* ξ and all $Q \in L^1(\mathbf{R})$, but the map $Q \mapsto S$ is not injective, while for general J the map $Q \mapsto S(z)$ may not extend to $L^1(\mathbf{R})$ for any z.

A well-known transformation of the problem seeks Ψ of the form

$$\Psi(x, z) = \mathcal{M}(x, z)e^{xzJ}, \qquad \mathcal{M}(\cdot, z) \text{ bounded.} \tag{10}$$

Thus

$$D\mathcal{M} = Q\mathcal{M} + z[J, \mathcal{M}], \qquad \mathcal{M}(\cdot, z) \text{ bounded,} \tag{11}$$

and we normalize by requiring

$$\mathcal{M}(x, z) \to I \text{ as } x \to -\infty. \tag{12}$$

Let Σ be the following union of lines in $\mathbf{C}$:

$$\Sigma = \{z \in \mathbf{C} : \operatorname{Re}(z\lambda_j - z\lambda_k) = 0, \text{ some } j \neq k\}. \tag{13}$$

Theorem 1. *Given* $Q \in L^1(\mathbf{R})$, *the problem* (12), (13) *has a unique solution for every* $z \in \mathbf{C} \backslash \Sigma$ *except for a countable set. For* $x \in \mathbf{R}$, *the function* $\mathcal{M}(x, \cdot)$ *is meromorphic on* $\mathbf{C} \backslash \Sigma$ *with poles at* $\{z_i\}$. *Moreover,*

$$\lim_{z \to \infty} \mathcal{M}(x, z) = I. \tag{14}$$

Definition. $Q \in L^1(\mathbf{R})$ *is* generic *is the associated* $\mathcal{M}$ has only simple poles, different columns have different poles, and for each sector Ω of $\mathbf{C} \backslash \Sigma$, $\mathcal{M}$ has a continuous extension to $\overline{\Omega} \backslash \{z_1, \ldots, z_n\}$.

Proposition. *The set of generic* Q *is a dense open set in* $L^1(\mathbf{R})$.

If Q is generic, $\mathcal{M}(x, \cdot)$ may be considered as a matrix-valued distribution on $\mathbf{C}$. The distribution derivative $\partial\mathcal{M}/\partial\bar{z}$, which has support in $\Sigma \cup \{z_1, \ldots, z_N\}$, also satisfies (11). Formally, this implies that

$$D\left(\mathcal{M}^{-1}\frac{\partial\mathcal{M}}{\partial\bar{z}}\right) = \left[J, \mathcal{M}^{-1}\frac{\partial\mathcal{M}}{\partial\bar{z}}\right], \tag{15}$$

so that $\mathcal{M}^{-1}\partial\mathcal{M}/\partial\bar{z}$ must have the form

$$\mathcal{M}^{-1}\frac{\partial\mathcal{M}}{\partial\bar{z}} = W_x(z) \equiv e^{xzJ}W(z)e^{-xzJ}. \tag{16}$$

We interpret W as a density on Σ, and an isolated set of residues $W(z_j)$, and call it the *spectral transform* of Q.

Theorem 2. *The map $Q \to W$ is injective.*

In fact, if Q_1 and Q_2 yield W, then one may justify the following formal calculation and apply Liouville's theorem:

$$\frac{\partial}{\partial\bar{z}}(\mathcal{M}_1\mathcal{M}_2^{-1}) = \frac{\partial\mathcal{M}_1}{\partial\bar{z}}\mathcal{M}_2^{-1} - \mathcal{M}_1\mathcal{M}_2^{-1}\frac{\partial\mathcal{M}_2}{\partial\bar{z}}\mathcal{M}_2^{-1} \tag{17}$$

$$= \mathcal{M}_1\{W_x - W_x\}\mathcal{M}_2^{-1} \equiv 0.$$

To determine Q from $\mathcal{M}$, we use (14) and (16) to see that

$$\mathcal{M}(x, z) = I + \frac{1}{2\pi i}\int_C (\zeta - z)^{-1}\frac{\partial\mathcal{M}}{\partial\bar{z}}(x, \zeta)d\zeta \wedge d\bar{\zeta} \tag{18}$$

so

$$\mathcal{M}(x, z) = I + \frac{1}{2\pi i}\int_\Sigma (\zeta - z)^{-1}\mathcal{M}(x, \zeta)W_x(\zeta)d\zeta \tag{19}$$

$$+ \Sigma\frac{1}{z_j - z}\mathcal{M}(x, \zeta_j)W_x(\zeta_j).$$

Then

$$Q(x) = \left[J, \frac{1}{2\pi i}\int_\Sigma \mathcal{M}(x, \zeta)W_x(\zeta)d\zeta + \Sigma\mathcal{M}(x, \zeta_j)W_x(\zeta_j)\right]. \tag{20}$$

Not all matrix-valued functions W on Σ can arise as spectral transforms, even those which are test functions with small norm: there are necessary and sufficient algebraic conditions. These conditions involve the shape of the matrix W on the various rays of Σ (relative to an appropriate ordering of the basis of eigenvectors of J) as well as the possibility of solving a certain finite linear system of equations. Here again there exists a class

S of generic scattering data for which the inverse problem is always solvable and which is dense in the class of all scattering transforms (see [2]). When $W \equiv 0$ on Σ, one sees from (19) and (20) that as functions of x, $\mathcal{M}$ and Q are rational functions of exponentials. Special cases give the familiar, soliton, multi-soliton, and breather potentials [6].

In the special case $n = 2$ and $J + J^* = 0$, the restriction of the spectral transform W to $\Sigma = R$ is essentially the "reduced scattering data" of [1]. In all cases it plays the same role in evolution as the ill-defined function S. Thus the evolution (8) corresponds to

$$\frac{\partial W}{\partial t} = z^j [\mu, W], \qquad (21)$$

which has the solution

$$W(z, t) = e^{tz^j \mu} W(z, 0) e^{-tz^j \mu}. \qquad (22)$$

It follows that a necessary condition for stability of (8) in any reasonable Banach space is the boundedness of the exponentials occurring in (22), for $z \in \Sigma$, $z \to \infty$, when W satisfies the algebraic constraints mentioned above. [These constraints are preserved by (22).]

2. HIGHER-ORDER EQUATIONS

Consider an eigenvalue problem

$$\left(D^n + \sum_{j=0}^{n-2} q_j(x) D^j \right) u = z^n u. \qquad (23)$$

This can at least formally be reduced in various ways to a first-order system of the type above, with $J = \text{diag}(\alpha, \alpha^2, \ldots, \alpha^{n-1}, 1)$, $\alpha = \exp \frac{1}{n} 2\pi i$.

Through such a reduction or by a more direct argument, one is led to construct eigenfunctions

$$\Psi_j (x, z) = \mathcal{M}_j(x, z) \exp(\alpha^j xz) \tag{24}$$

via an integral equation for $\mathcal{M} = (\mathcal{M}_1, \mathcal{M}_2, \ldots, \mathcal{M}_n)$:

$$\mathcal{M}(x, z) = (1, 1, \ldots, 1) + \frac{1}{2\pi i} \int_\Sigma (\zeta - z)^{-1} \mathcal{M}(x, \zeta) W_x(\zeta) d\zeta \tag{25}$$

$$+ \text{ (sum over singularities)}$$

where $W_x(z)$ has the form (16) and W is subject to the symmetry requirement

$$W_{j+1, k+1}(z) = W_{j, k}(\alpha z) \tag{26}$$

where we take $W_{n+1, k} = W_{1, k}$, $W_{j, n+1} = W_{j, 1}$. For W a test function of small norm and, for simplicity, no singularities, the solution of (25) is unique and gives eigenfunctions for (23), where, of course, the potentials q_j depend on W. The direct problem in this case is more delicate than in the case of a first-order system because of the difficulties in constructing normalized eigenfunctions as the eigenvalue parameter tends to zero. It appears, however, that the picture is essentially the same as in the case of a first-order system.

3. THE SCOPE OF THE INVERSE SCATTERING METHOD

The results described here are joint work with B. Dahlberg. A considerable amount is known about inverse scattering in more than one dimension, but there is as yet no systematic way to associate tractable evolution equations. The possibility of doing so in one dimension may be due to the particularly simple way in which eigenfunctions may be generated. For example, for any small test function w, the solutions of the integral equations

$$\Psi(x, z) = e^{ixz}\left\{1 + \int_R \frac{e^{ix\zeta} \Psi(x, \zeta)}{z + \zeta} w(\zeta) d\zeta\right\}$$

are eigenfunctions of

$$(D^2 + z^2)\Psi = q\Psi, \qquad D = 1/i \ d/dx, \tag{28}$$

where q depends on w. This is a version of (25) when $n = 2$, and the connection between evolutions of w and evolutions of q can be established directly from (27).

As a general model of this in the scalar case we can ask for a function or measure Φ and a function Q of two variables such that for all small test functions w, the solutions of

$$\Psi(x, z) = e^{ixz}\left\{1 + \int \Phi(x, z, \zeta)\Psi(x, \zeta)w(\zeta)d\zeta\right\} \tag{29}$$

are the solutions of the pseudodifferential equation

$$Q(D, z)\Psi(x, \zeta) = q(x)\Psi(x, \zeta) \tag{30}$$

with q depending on w. Under some regularity assumptions, the problem reduces to the case when $\Phi(x, z, \cdot)$ is a density of the form $e^{ix\zeta}p(z, \zeta)$ and the pair Q, p satisfy the functional equation

$$[Q(\xi + \zeta + z, z) - Q(\xi, \zeta)]p(z, \zeta) \equiv r(\xi, \zeta) \tag{31}$$

for some function r.

The functional equation (31) has three types of solution, which may be normalized as follows. The first is a trivial family in which $Q(\xi, z)$ is a function of $\xi - z$ and one may, therefore, take $p \equiv 1$. This leads, for example, to the Cole-Hopf transformation and has Burger's equation as an associated evolution. The second is the Sturm-Liouville case

$$Q(\xi, z) = \xi^2 - z^2, \ p(z, \zeta) = (z + \zeta)^{-1}, \tag{32}$$

which has KdV as an associated evolution.

Finally, there is a remarkable one-parameter family of solutions giving pseudodifferential operators $Q_\delta(D, z)$, $\delta \neq 0$, where

$$Q_\delta(\xi, z) = \xi - z \ \coth 2\delta z + z \ \text{csch} \ 2\delta z \cdot e^{-2\delta\xi}. \tag{33}$$

An associated evolution is

$$\frac{\partial u}{\partial t} + \delta^{-1} \frac{\partial u}{\partial x} + 2u \frac{\partial u}{\partial x} = 2(\cosh \ \delta D)u.$$

Despite the artificial appearance of this last family of pseudo-differential equations, it occurs in the physics literature in connection with waves in a finite depth fluid [4], and the associated eigenvalue problem (30), (33) has been treated in [5].

REFERENCES

1. Ablowitz, M. J.; Kaup, D. J.: Newell, A. C.; and Segur, H. The inverse scattering transform—Fourier analysis for nonlinear problems. *Studies Appl. Math.* 53 (1974):249-315.

2. Beals, R., and Coifman, R. R. Scattering and inverse scattering for first-order systems. To appear.

3. Deift, P., and Trubowitz, E. Inverse scattering on the line. *Comm. Pure Appl. Math.* 32 (1979):121-253.

4. Joseph, R. I. Solitary waves in a finite depth fluid. *J. Physics* A 10' (1977):L225-L227.

5. Kodama, Y.; Satsuma, J.; and Ablowitz, M. J. Nonlinear intermediate long-wave equation: Analysis and method of solution. *Phys. Rev. Letters* 46 (1981):687-690.

6. Lamb, G. L., Jr. *Elements of Soliton Theory*. New York: Wiley, 1980.

7. Newell, A. C. The inverse scattering transform. In Bullough and Caudrey, eds., *Solitons*. New York: Springer-Verlag, 1980, pp. 177-242.

8. Zakharov, V. E. The inverse scattering method. In Bullough and Caudrey, eds., *Solitons*. New York: Springer-Verlag, 1980, pp. 243-286.

NORM INEQUALITIES FOR HOLOMORPHIC FUNCTIONS OF SEVERAL COMPLEX VARIABLES

Jacob Burbea
University of Pittsburgh

1. INTRODUCTION

We consider the n-dimensional complex space $\mathbf{C}^n$ with the usual inner product

$$z \cdot \overline{\zeta} = z_1 \overline{\zeta}_1 + \cdots + z_n \overline{\zeta}_n; \qquad |z|^2 = z \cdot \overline{z}.$$

The unit ball Δ_n of $\mathbf{C}^n$ is then the set of all $z \in \mathbf{C}^n$ with $|z| < 1$. The class of all holomorphic functions with domain Δ_n shall be denoted by $H(\Delta_n)$. It is a linear space over $\mathbf{C}$. The following standard notation is employed: if $\alpha = (\alpha_1, \ldots, \alpha_n) \in \mathbf{Z}_+^n$ (that is, an n-tuple of nonnegative integers), then $|\alpha| = \alpha_1 + \cdots + \alpha_n$ and $\alpha! = \alpha_1! \ldots \alpha_n!$. The notation $a_\alpha z^\alpha$ is an abbreviation for the monomial

$$a_{\alpha_1 \ldots \alpha_n} z_1^{\alpha_1} \ldots z_n^{\alpha_n}; \qquad a_{\alpha_1 \ldots \alpha_n} \in \mathbf{C}, \ z \in \mathbf{C}^n.$$

Any $f \in H(\Delta_n)$ can be represented as

$$f(z) = \sum_{\alpha \geq 0} a_\alpha z^\alpha, \ z \in \Delta_n,$$

where the power series converges locally uniformly in Δ_n. We may therefore regard $H(\Delta_n)$ as the span of all z^α, $\alpha \geq 0$, in the uniform topology of Δ_n.

We shall introduce a "derivation-operator" $\partial_n : H(\Delta_n) \to H(\Delta_n)$ as follows: we require ∂_n to be linear on $H(\Delta_n)$, and we specify its action on z^α, $\alpha \geq 0$. If $\alpha = 0$, we let

$$\partial_n(z^0) = \partial_n(1) = 0. \tag{1}$$

On the other hand, if $\alpha \geq 1$, we let α_{k_1}, $\ldots$, α_{k_m}, $1 \leq k_1 \leq \cdots$
$\leq k_m \leq n$ $(1 \leq m \leq n)$ be the nonzero (≥ 1) components of α. Thus
$z^\alpha = z_{k_1}^{\alpha_{k_1}} \ldots z_{k_m}^{\alpha_{k_m}}$. We then define

$$\partial_n(z^\alpha) = \sqrt{\alpha_{k_1} \ldots \alpha_{k_m} (|\alpha|)_{n-m+1}}\ z_{k_1}^{\alpha_{k_1}^{-1}} \ldots z_{k_m}^{\alpha_{k_m}^{-1}}, \tag{2}$$

where, for a real $q > 0$,

$$(q)_0 = 1; \quad (q)_m = q(q+1) \ldots (q+m-1), \ m \geq 1.$$

This derivation-operator ∂_n is the n-dimensional generalization of
the usual derivative $\partial \equiv \partial_1$ on $H(\Delta_1)$. In fact, for $f \in H(\Delta_1)$,
$\partial f = f'$.

By $d\sigma_n(z)$ we denote the Lebesgue measure of $\mathbf{C}^n \equiv \mathbf{R}^{2n}$ and by
$ds_n(z)$ we denote the Euclidean measure on the boundary $\partial\Delta_n$ of Δ_n.
For a real $q > 0$, we introduce the "(n, q)-Dirichlet space"

$$D_{n,q} = \left\{ f \in H(\Delta_n) : f(0) = 0, \ \|f\|_{D_{n,q}} < \infty \right\},$$

where

$$\|f\|_{D_{n,q}}^2 = \frac{1}{\pi^n}\ \frac{2}{(n+1)q} \int_{\Delta_n} |\partial_n f(z)|^2 d\sigma_n(z). \tag{3}$$

It will be verified later that $D_{n,q}$ is a functional Hilbert space
of holomorphic functions in Δ_n with the reproducing kernel

$$k_{n,q}(z, \bar{\zeta}) = -\frac{(n+1)q}{2} \log(1 - z \cdot \bar{\zeta}); \qquad z, \zeta \in \Delta_n. \tag{4}$$

Another related functional Hilbert space is the so-called
(generalized) "(n, q)-Hardy space" $H_{n,q}$. This is the space of all
holomorphic functions g in Δ_n normed by $\|g\|_{n,q}$. The norm $\|g\|_{n,q}$
for arbitrary $q > 0$ is represented by means of a suitable weighted
ℓ_2-norm that will be described later (see also Burbea [1] for the
case $n = 1$). When, however, $q > \dfrac{2n}{n+1}$, this norm can also be
realized as the following weighted L_2-norm:

$$\|g\|_{n,\,q}^2 = \frac{1}{\pi^n}\left(\frac{n+1}{2}\,q - n\right)_n \int_{\Delta_n} |g(z)|^2\,(1 - |z|^2)^{\frac{n+1}{2}(q-2)}\,d\sigma_n(z). \tag{5}$$

On the other hand, the limiting case of $q = \dfrac{2n}{n+1}$ is given by

$$\|g\|_{n,\,\frac{2n}{n+1}}^2 = \frac{(n-1)!}{2\pi n}\int_{\partial\Delta_n} |g(z)|^2\,ds_n(z), \tag{6}$$

where in the last integral, g stands for the nontangential boundary values of the holomorphic function $g(z)$ in Δ_n. Thus, $H_{n,\,\frac{2n}{n+1}}$ is the usual "Hardy-Szegö space," $H_{n,\,2}$ is the customary "Bergman space," and $H_{n,\,q}$, $q > \dfrac{2n}{n+1}$, is the "Bergman-Selberg space" (see [3]). The reproducing kernel of $H_{n,\,q}$, $q > 0$, is, as we also shall see later,

$$K_{n,\,q}(z,\,\bar{\zeta}) = (1 - z\cdot\bar{\zeta})^{-\frac{(n+1)q}{2}}\,; \qquad z,\,\zeta \in \Delta_n,$$

and we note the relationship $K_{n,\,q}(z,\,\bar{\zeta}) = \exp[k_{n,\,q}(z,\,\bar{\zeta})]$, $z,\,\zeta \in \Delta_n$.

The main purpose of this paper is in the establishment of the following theorem.

Theorem. *Let $q > 0$ and let $f \in D_{n,\,q}$. Then*

$$\|\exp f\|_{n,\,q}^2 \le \exp\|f\|_{D_{n,\,q}}^2.$$

Equality holds if and only if f is of the form $f = k_{n,\,q}(\cdot,\,\bar{\zeta})$ for some $\zeta \in \Delta_n$.

The inequality and the sufficient condition in the equality statement of this theorem constitute a simple consequence of a more abstract theorem in the general theory of reproducing kernels (see Burbea [1] for details). The essence, therefore, of this theorem is in the necessity of the equality statement. In the special case that $n = 1$ and $q \ge 1$, a proof of the theorem was given by Saitoh [2]. His proof, however, is difficult and rather involved. In [1], Burbea has extended the result of Saitoh [2] to cover the case of $0 < q < 1$ with a proof that is substantially easier and shorter than that of [2]. The method of proof in Burbea [1] conceals in it the main ingredients for establishing the present theorem. To do so, however, it is found necessary to derive certain multinomial and combinatorial identities that are of some interest in their own right. Apart from that, the proof given in this paper is tailored after that of Burbea [1] for the case $n = 1$.

We should also remark that it is possible to extend the theorem to cover the case where Δ_n is replaced by one of the classical Cartan domains once some minor modifications in the present proof are made. However, we shall not pursue this point here.

2. WEIGHTED l_2-SPACES

We first note the following two elementary identities: For $q > 0$ and α, $\beta \in \mathbf{Z}_+^n$, we have

$$\frac{1}{\pi^n} \int_{\Delta_n} z^\alpha \bar{z}^\beta \ (1 - |z|^2)^{q-1} d\sigma_n(z) \ = \ \frac{\alpha!}{(q)_{|\alpha|+n}} \delta_{\alpha\beta} \tag{8}$$

and

$$\frac{1}{2\pi^n} \int_{\partial\Delta_n} z^\alpha \bar{z}^\beta ds_n(z) \ = \ \frac{\alpha!}{(1)_{|\alpha|+n}} \delta_{\alpha\beta}, \tag{9}$$

where $\delta_{\alpha\beta}$ is the Kronecker delta. With the aid of (8) and the derivation operator of (1) and (2), the space $D_{n,\,q}$ can be realized as the weighted ℓ_2-space described below. Let $f \in D_{n,\,q}$ so that

$$f(z) = \sum_{\alpha \geq 1} a_\alpha z^\alpha, \ z \in \Delta_n.$$

By (1)-(3) and (8),

$$\|f\|_{D_{n,\,q}}^2 \ = \ \frac{1}{\pi^n} \ \frac{2}{(n+1)q} \sum_{\alpha \geq 1} |a_\alpha|^2 \alpha_{k_1} \ldots \alpha_{k_m} |\alpha|_{n-m+1} \int_{\Delta_n} z^{\alpha-1} \bar{z}^{\alpha-1} d\sigma_n(z)$$

$$= \ \frac{2}{(n+1)q} \sum_{\alpha \geq 1} |a_\alpha|^2 \alpha_{k_1} \ldots \alpha_{k_m} |\alpha|_{n-m+1} \frac{(\alpha-1)!}{(1)_{n+|\alpha|-m}}$$

$$= \ \frac{2}{(n+1)q} \sum_{\alpha \geq 1} |a_\alpha|^2 \alpha! \ \frac{(|\alpha|)_{n-m+1}}{(|\alpha|+n-m)!},$$

where here $\alpha_{k_1}\ldots\alpha_{k_m}$, $1 \leq k_1 \leq \cdots \leq k_m \leq n$ $(1 \leq m \leq n)$ are the nonzero components of $\alpha \geq 1$. It follows that

$$\|f\|^2_{D_{n,q}} = \frac{2}{(n+1)q} \sum_{\alpha \geq 1} |a_\alpha|^2 \frac{\alpha!}{(|\alpha|-1)!}. \tag{10}$$

From (10) we deduce that $D_{n,q}$ is a functional Hilbert space of holomorphic functions f in Δ_n with $f(0) = 0$. The reproducing kernel of this space is

$$k_{n,q}(z,\bar\zeta) = \frac{(n+1)q}{2} \sum_{\alpha \geq 1} \frac{(|\alpha|-1)!}{\alpha!} z^\alpha \bar\zeta^\alpha$$

$$= \frac{(n+1)q}{2} \sum_{\alpha \geq 1} \frac{1}{\alpha!}\left(\sum_{m=|\alpha|} \frac{1}{m} m!\, z^\alpha \bar\zeta^\alpha\right)$$

$$= \frac{(n+1)q}{2} \sum_{m=1}^\infty \frac{1}{m}\left(\sum_{|\alpha|=m} \frac{m!}{\alpha!} z^\alpha \bar\zeta^\alpha\right)$$

$$= \frac{(n+1)q}{2} \sum_{m=1}^\infty \frac{1}{m}(z\cdot\bar\zeta)^m = -\frac{(n+1)q}{2}\log(1-z\cdot\bar\zeta).$$

This is in agreement with (4). In a similar fashion, the space $H_{n,q}$ can be realized as follows. Let $g \in H(\Delta_n)$ with

$$g(z) = \sum_{\alpha \geq 0} b_\alpha z^\alpha, \quad z \in \Delta_n.$$

For $q > 0$, we write

$$Q_n \equiv \frac{(n+1)q}{2} \tag{11}$$

and

$$\|g\|^2_{n,q} \equiv \sum_{\alpha > 0} |b_\alpha|^2 \frac{\alpha!}{(Q_n)_{|\alpha|}}. \tag{12}$$

The space $H_{n,q}$ can then be defined as

$$H_{n,q} \equiv \left\{g \in H(\Delta_n) : \|g\|_{n,q} < \infty\right\},$$

and therefore $H_{n,q}$ is a functional Hilbert space of holomorphic functions in Δ_n whose reproducing kernel is

$$k_{n,q}(z,\bar\zeta) = \sum_{\alpha \geq 0} \frac{(Q_n)_{|\alpha|}}{\alpha!} z^\alpha \bar\zeta^\alpha = \sum_{m=0}^\infty \frac{(Q_n)m}{m!}\left(\sum_{|\alpha|=m} \frac{m!}{\alpha!} z^\alpha \bar\zeta^\alpha\right)$$

$$= \sum_{m=0}^{\infty} \frac{(Q_n)_m}{m!} (z \cdot \bar{\zeta})^m = (1 - z \cdot \bar{\zeta})^{-Q_n},$$

which is in agreement with (7). By virtue of identities (8) and (9), it is easily verified that for $Q_n \geq n$, (11) and (12) reduce to (5) and (6).

3. SOME COMBINATORIAL IDENTITIES

For a real $q > 0$ and $m \in \mathbf{Z}_+$, we use the notation

$$\frac{(q)_m}{m!} = \frac{\Gamma(q+m)}{\Gamma(m+1)\Gamma(q)} = \binom{q+m-1}{m} = (-1)^m \binom{-q}{m}, \qquad (13)$$

and thus, for $m = 0$,

$$\binom{q-1}{0} = \binom{-q}{0} = 1.$$

Formula (13) may, of course, be extended to include the case where q is any real number. A basic combinatorial identity is the "Vandermonde convolution formula"

$$\binom{q}{m} = \sum_{k=0}^{m} \binom{q-p}{m-k}\binom{p}{k}; \quad q, p \in \mathbf{R}, \ m \in \mathbf{Z}_+. \qquad (14)$$

This formula is easily derived from the identity $(1 + x)^q = (1 + x)^{q-p}(1 + x)^p$. From (13) and (14) we deduce that

$$\binom{q}{m} = \binom{q-m+1+m-1}{m} = (-1)^m \binom{-q+m-1}{m}$$

$$= (-1)^m \sum_{k=0}^{m} \binom{-q+m-1+p}{m-k}\binom{-p}{k}$$

$$= (-1)^m \sum_{k=0}^{m} (-1)^{m-k} \binom{q-m+1-p+m-k-1}{m-k}(-1)^k\binom{p+k-1}{k},$$

and therefore,

$$\binom{q}{m} = \sum_{k=0}^{m} \binom{q-p-k}{m-k}\binom{p+k-1}{k}; \qquad q, p \in \mathbf{R}, \ m \in \mathbf{Z}_+. \qquad (15)$$

Another consequence of (14) is the identity

$$\binom{q + m}{m} = \sum_{k=0}^{m} \binom{q + k - 1}{k}; \quad q \in \mathbf{R}, \ m \in \mathbf{Z}_+. \tag{16}$$

This is obtained as follows. By (13) and (14), with $p = -q$,

$$\binom{q + m}{m} = (-1)^m \binom{-q - 1}{m} = (-1)^m \sum_{k=0}^{m} \binom{-q - 1 + q}{m - k}\binom{-q}{k}$$

$$= (-1)^m \sum_{k=0}^{m} \binom{-1}{m - k}\binom{-q}{k}$$

$$= (-1)^m \sum_{k=0}^{m} (-1)^{m-k} \binom{1 + m - k - 1}{m - k}(-1)^k \binom{q + k - 1}{k}$$

and (16) follows.

Identity (16) admits a multinomial extension described below. For $\alpha, \beta \in \mathbf{Z}_+^n$ with $\beta \leq \alpha$, we define

$$\binom{\alpha}{\beta} \equiv \binom{\alpha_1}{\beta_1}\binom{\alpha_2}{\beta_2} \cdots \binom{\alpha_n}{\beta_n} = \frac{\alpha!}{\beta!\,(\alpha - \beta)!} \tag{17}$$

Then

$$\binom{q + |\alpha|}{|\alpha|} = \sum_{0 \leq \beta \leq \alpha} \binom{\alpha}{\beta}\binom{|\alpha|}{|\beta|}^{-1}\binom{q + |\beta| - 1}{|\beta|}; \quad q \in \mathbf{R}, \ \alpha \in \mathbf{Z}_+^n. \tag{18}$$

For $n = 1$, (18) clearly reduces to (16), while for $n > 1$, (18) may be established as follows. From (13) and (17), we have

$$\sum_{0 \leq \beta \leq \alpha} \binom{\alpha}{\beta}\binom{|\alpha|}{|\beta|}^{-1}\binom{q + |\beta| - 1}{|\beta|} = \frac{\alpha!}{|\alpha|!} \sum_{0 \leq \beta \leq \alpha} \frac{|\beta|!\,(|\alpha| - |\beta|)!}{\beta!\,(\alpha - \beta)!}\binom{q + |\beta| - 1}{|\beta|}$$

$$= \frac{\alpha!}{|\alpha|!} \sum_{\beta_1 = 0}^{\alpha_1} \binom{q + \beta_1 - 1}{\beta_1} \sum_{\beta_2 = 0}^{\alpha_2} \binom{\alpha_1 - \beta_1 + \alpha_2 - \beta_2}{\alpha_2 - \beta_2}\binom{q + \beta_1 + \beta_2 - 1}{\beta_2} \cdots$$

$$\sum_{\beta_n = 0}^{\alpha_n} \binom{\alpha_1 - \beta_1 + \cdots + \alpha_n - \beta_n}{\alpha_n - \beta_n}\binom{q + \beta_1 + \cdots + \beta_n - 1}{\beta_n},$$

which by a successive application of (15) becomes

$$= \frac{\alpha!}{|\alpha|!} \sum_{\beta_1 = 0}^{\alpha_1} \binom{q + \beta_1 - 1}{\beta_1}\binom{q + \alpha_1 + \alpha_2}{\alpha_2} \cdots \binom{q + \alpha_1 + \alpha_2 + \cdots + \alpha_n}{\alpha_n}$$

$$= \frac{\alpha_1!}{|\alpha|!} \frac{\Gamma(q + \alpha_1 + \cdots + \alpha_n + 1)}{\Gamma(q + \alpha_1 + 1)} \sum_{\beta_1 = 0}^{\alpha_1} \binom{q + \beta_1 - 1}{\beta_1}.$$

Finally, using identities (13) and (16), we see that (18) follows.

4. AN INEQUALITY

Let $\alpha \in \mathbf{Z}_+^n$, $\alpha \neq 0$, and let α_k, $1 \le k \le n$, be a nonzero component of α. We then write

$$\alpha - 1_k = (\alpha_1, \ldots, \alpha_k - 1, \ldots, \alpha_n); \ \alpha_k \ge 1, \ 1 \le k \le n.$$

Let

$$f(z) = \sum_{\alpha \ge 1} a_\alpha z^\alpha, \ z \in \Delta_n, \tag{19}$$

and let

$$g(z) = \exp[f(z)] = \sum_{\alpha \ge 0} b_\alpha z^\alpha \tag{20}$$

be its exponential transform. For an arbitrary k, $1 \le k \le n$, (20) shows that

$$\frac{\partial}{\partial z_k} g(z) = g(z) \frac{\partial}{\partial z_k} f(z)$$

and therefore, by (19) and (20),

$$\sum_{\alpha \ge 0} \alpha_k b_\alpha z^{\alpha - 1k} = \left(\sum_{\alpha \ge 0} b_\alpha z^\alpha\right)\left(\sum_{\alpha \ge 0} \alpha_k a_\alpha z^{\alpha - 1}k\right)$$

$$= \sum_{\substack{\alpha \ge 0 \\ \alpha_k \ge 1}} \left(\sum_{\beta \le \alpha} \beta_k a_\beta b_{\alpha - \beta}\right) z^{\alpha - 1}k.$$

This shows that

$$b_\alpha = \frac{1}{\alpha_k} \sum_{\beta \le \alpha} \beta_k a_\beta b_{\alpha - \beta}; \ \alpha_k \ge 1 \ (b_0 = 1). \tag{21}$$

In view of (10)-(12) and (19) and (20), the theorem is equivalent to the following sharp inequality.

Lemma. Let the notation of (19) and (20) apply and let $q > 0$. Define

$$Q_n \equiv \frac{(n + 1)q}{2}.$$

Then

$$\sum_{\alpha \geq 0} \frac{\alpha!}{(Q_n)_{|\alpha|}} |b_\alpha|^2 \leq \exp\left\{ \frac{1}{Q_n} \sum_{\alpha \geq 1} \frac{\alpha!}{(|\alpha| - 1)!} |a_\alpha|^2 \right\} \tag{22}$$

if the right-hand side is finite. Equality holds if and only if

$$a_\alpha = Q_n \frac{(|\alpha| - 1)!}{\alpha!} \zeta^\alpha, \ \alpha \geq 1; \ b_\alpha = \frac{(Q_n)_{|\alpha|}}{\alpha!} \zeta^\alpha, \quad \alpha \geq 0,$$

for some $\zeta \epsilon \Delta_n$.

Proof. For $r = (r_1, \ldots, r_n) \epsilon [0, 1)^n$, define

$$A(r) = \frac{1}{Q_n} \sum_{\alpha \geq 1} \frac{\alpha!}{(|\alpha| - 1)!} |a_\alpha|^2 r^\alpha, \ B(r) = \sum_{\alpha \geq 0} \frac{\alpha!}{(Q_n)_{|\alpha|}} |b_\alpha|^2 r^\alpha.$$

Let k, $1 \leq k \leq n$, be fixed. By (21) and the Cauchy-Schwarz inequality,

$$|b_\alpha|^2 = \frac{1}{\alpha_k^2} \left| \sum_{\substack{\beta \leq \alpha \\ \beta_k \geq 1}} \beta_k \left(\frac{\beta! (\alpha - \beta)!}{(|\beta| - 1)! \beta_k (Q_n)_{|\alpha - \beta|}} \right)^{1/2} \right.$$

$$\left. \times a_\beta b_{\alpha - \beta} \left(\frac{(|\beta| - 1)! \beta_k (Q_n)_{|\alpha - \beta|}}{\beta! (\alpha - \beta)!} \right)^{1/2} \right|$$

$$\leq \frac{1}{\alpha_k^2} \left(\sum_{\substack{\beta \leq \alpha \\ \beta_k \geq 1}} \frac{\beta_k \beta!}{(|\beta| - 1)!} |a_\beta|^2 \frac{(\alpha - \beta)!}{(Q_n)_{|\alpha - \beta|}} |b_{\alpha - \beta}|^2 \right)$$

$$\times \left(\sum_{\substack{\beta \leq \alpha \\ \beta_k \geq 1}} \frac{\beta_k (|\beta| - 1)!}{\beta!} \frac{(Q_n)_{|\alpha - \beta|}}{(\alpha - \beta)!} \right).$$

On the other hand, by changing to variables

$$\gamma = \alpha - 1_k \geq 0, \qquad \delta = \gamma - \beta,$$

we obtain

$$\sum_{\substack{\beta \leq \alpha \\ \beta_k \geq 1}} \frac{\beta_k (|\beta| - 1)!}{\beta!} \frac{(Q_n)_{|\alpha - \beta|}}{(\alpha - \beta)!} = \sum_{0 \leq \delta \leq \gamma} \frac{(|\gamma| - |\delta|)!}{\delta! (\gamma - \delta)!} (Q_n)_{|\delta|},$$

which, by virtue of (13), (17), and (18), is equal to

$$\frac{|\gamma|!}{\gamma!} \sum_{0 \leq \delta \leq \gamma} \binom{\gamma}{\delta} \binom{|\gamma|}{|\delta|}^{-1} \binom{Q_n + |\delta| - 1}{|\delta|}$$

$$= \frac{|\gamma|!}{\gamma!} \binom{Q_n + |\gamma|}{|\gamma|} = \frac{1}{\gamma!} (Q_n + 1)_{|\gamma|}.$$

Therefore,

$$\sum_{\beta \le \alpha} \frac{\beta_k (|\beta| - 1)!}{\beta!} \frac{(Q_n)_{|\alpha - \beta|}}{(\alpha - \beta)!} = \frac{\alpha_k}{\alpha!} \frac{(Q_n)_{|\alpha|}}{Q_n} \tag{23}$$

and

$$\frac{\alpha_k \alpha!}{(Q_n)_{|\alpha|}} |b_\alpha|^2 \le \frac{1}{Q_n} \sum_{\substack{\beta \le \alpha \\ \beta_k \ge 1}} \frac{\beta_k \beta!}{(|\beta| - 1)!} |a_\beta|^2 \frac{(\alpha - \beta)!}{(Q_n)_{|\alpha - \beta|}} |b_{\alpha - \beta}|^2. \tag{24}$$

The last inequality shows that

$$\frac{\partial}{\partial r_k} B(r) \le B(r) \frac{\partial}{\partial r_k} A(r)$$

or that

$$\frac{\partial}{\partial r_k} \log B(r) \le \frac{\partial}{\partial r_k} A(r); \qquad 1 \le k \le n, \tag{25}$$

as formal power series. Note that if the right-hand side of (22) is finite, then $A(r)$ converges for $r \in [0, 1)^n$, and therefore, by (25), $B(r)$ also converges for $r \in [0, 1)^n$. Let $\rho \in [0, 1)$ and choose a path $r_k = r_k(t)$, where $r(t)$ is increasing in $t \in [0, 1]$ and $r_k(0) = 0$, $r_k(1) = \rho$ for $1 \le k \le n$. Since $A(0, \ldots, 0) = 0$ and $B(0, \ldots, 0) = 1$, it follows from (25) that

$$\log B(\rho, \ldots, \rho) = \int_0^1 \frac{d}{dt} \log B[r(t)] dt$$

$$\le \int_0^1 \frac{d}{dt} A[r(t)] dt = A(\rho, \ldots, \rho)$$

and (22) follows by letting $\rho \to 1 - 0$. In view of (25), equality in (22) holds if and only if for any $r \in [0, 1)^n$ and any k, $1 \le k \le n$,

$$\frac{\partial}{\partial r_k} B(r) = B(r) \frac{\partial}{\partial r_k} A(r),$$

which is equivalent to having equality in (24) for every component $\alpha_k \ge 1$. This is obviously equivalent to an existence of $\lambda_\alpha \in \mathbf{C}$ so that

$$\beta_k a_\beta b_{\alpha - \beta} \left(\frac{\beta! (\alpha - \beta)!}{(|\beta| - 1)! \beta_k (Q_n)_{|\alpha - \beta|}} \right)^{1/2} = \lambda_\alpha \left(\frac{(|\beta| - 1)! \beta_k (Q_n)_{|\alpha - \beta|}}{\beta! (\alpha - \beta)!} \right)^{1/2}$$

for all $\beta \le \alpha$ and $\beta_k \ge 1$. Therefore,

$$a_\beta b_{\alpha - \beta} = \lambda_\alpha \frac{(|\beta| - 1)! (Q_n)_{|\alpha - \beta|}}{\beta! (\alpha - \beta)!}, \quad 1 \le \beta \le \alpha. \tag{26}$$

Putting $\beta = \alpha$ in (26) results in

$$\lambda_\alpha = \frac{\alpha!}{(|\alpha| - 1)!} \, a_\alpha, \quad \alpha \geq 1. \tag{27}$$

Also, by (21), (24), and (26),

$$b_\alpha = \frac{\lambda_\alpha}{\alpha_k} \sum_{\substack{\beta \leq \alpha \\ \beta_k \geq 1}} \frac{\beta_k(|\beta| - 1)!}{\beta!} \frac{(Q_n)_{|\alpha - \beta|}}{(\alpha - \beta)!} = \lambda_\alpha \frac{(Q_n)_{|\alpha|}}{\alpha! Q_n},$$

and therefore,

$$\lambda_\alpha = \frac{\alpha! Q_n}{(Q_n)_{|\alpha|}} \, b_\alpha, \quad \alpha \geq 1. \tag{28}$$

Relations (27) and (28) give

$$\frac{1}{(|\alpha| - 1)!} \, a_\alpha = \frac{Q_n}{(Q_n)_{|\alpha|}} \, b_\alpha, \quad \alpha \geq 1. \tag{29}$$

For $1 \leq k \leq n$, we denote by e_k the kth unit vector $(0, \ldots, 1_k, \ldots, 0) \in \mathbf{Z}_+^n$, and we define $\zeta = (\zeta_1, \ldots, \zeta_n) \in \mathbf{C}^n$ by

$$\zeta_k = Q_n^{-1}\bar{a}_{e_k}, \quad 1 \leq k \leq n. \tag{30}$$

We shall show that the only solution of (29) subject to (21) is

$$a_\alpha = Q_n \frac{(|\alpha| - 1)!}{\alpha!} \, \bar{\zeta}^\alpha, \quad \alpha \geq 1. \tag{31}$$

and therefore, $b_\alpha = [(Q_n)_{|\alpha|}/\alpha!]\bar{\zeta}^\alpha$, $\alpha \geq 0$. Indeed, for $\alpha \geq 1$ with a component $\alpha_k \geq 1$, $1 \leq k \leq n$, we have from (21) and (29),

$$\frac{1}{(|\alpha| - 1)!} \, a_\alpha = \frac{Q_n}{(Q_n)_{|\alpha|}} \frac{1}{\alpha_k}\left[\alpha_k a_\alpha + \sum_{\beta < \alpha} \beta_k a_\beta b_{\alpha - \beta}\right],$$

and therefore,

$$\left[\frac{1}{(|\alpha| - 1)!} - \frac{Q_n}{(Q_n)_{|\alpha|}}\right]a_\alpha = \frac{1}{(Q_n)_{|\alpha|}} \frac{1}{\alpha_k}\sum_{\beta < \alpha} \beta_k \frac{(Q_n)_{|\alpha - \beta|}}{(|\alpha - \beta| - 1)!} \, a_\beta a_{\alpha - \beta}. \tag{32}$$

To prove (31), we use induction on the weight $|\alpha| \geq 1$. For $|\alpha| = 1$, (31) holds true by virtue of the definition (30). We assume that (31) is true for α, $1 \leq |\alpha| < m$. When $|\alpha| = m$, there is a component $\alpha_k \geq 1$, $1 \leq k \leq n$, of α so that (23) and (32) apply. By the induction assumption, (23), and (32),

$$\left[\frac{1}{(|\alpha|-1)!} - \frac{Q_n}{(Q_n)_{|\alpha|}}\right]a_\alpha = \frac{1}{(Q_n)_{|\alpha|}} \frac{1}{\alpha_k} Q_n^2 \bar{\zeta}^\alpha \sum_{\beta<\alpha} \frac{\beta_k(|\beta|-1)!}{\beta!} \frac{(Q_n)_{|\alpha-\beta|}}{(\alpha-\beta)!}$$

$$= \frac{1}{(Q_n)_{|\alpha|}} \frac{1}{\alpha_k} Q_n^2 \bar{\zeta}^\alpha \left[\frac{\alpha_k}{\alpha!} \frac{(Q_n)_{|\alpha|}}{Q_n} - \frac{\alpha_k}{\alpha!}(|\alpha|-1)!\right]$$

$$= \left[\frac{1}{(|\alpha|-1)!} - \frac{Q_n}{(Q_n)_{|\alpha|}}\right]Q_n \frac{(|\alpha|-1)!}{\alpha!} \bar{\zeta}^\alpha.$$

Since the common factor in the square bracket is positive, (31)
follows at once. Also, since the right-hand side of (22) is finite,
$\zeta = (\zeta_1, \ldots, \zeta_n)$ of (30) must belong to Δ_n. This concludes the
proof of the lemma.

The proof of the theorem is now complete.

REFERENCES

1. Burbea, J. A Dirichlet norm inequality and some inequalities
    for reproducing kernel spaces. *Proc. Amer. Math. Soc.* 83 (1981):
    279-285.

2. Saitoh, S. Some inequalities for analytic functions with finite
    Dirichlet integral on the unit disc. *Math. Ann.* 246 (1979):69-
    77.

3. Selberg, A. *Automorphic Functions and Integral Operators.*
    Seminars on Analytic Functions, Institute for Advanced Study,
    Princeton, N.J., 1957, vol. 2, pp. 152-161.

ON THE RADON TRANSFORM
AND SOME OF ITS GENERALIZATIONS

A.P. Calderón
University of Chicago

1. INTRODUCTION

Let $f(x)$ be a function on the n-dimensional Euclidean space $\mathbf{R}^n$ and let N denote a k-dimensional affine subspace of $\mathbf{R}^n$, $0 < k < n$. The k-plane transform of f is the function of N defined by

$$\mathscr{R}_k(f)(N) = \int_N f \, d\sigma,$$

where $d\sigma$ denotes the ordinary measure on N. For $k = n - 1$ this is the well-known Radon transform. Our goal is to obtain existence, integrability, regularity, and support properties of such transforms. Some of the results presented here have been obtained previously or independently by D. M. Oberlin and E. M. Stein [8], S. W. Drury [4 and 5], K. T. Smith and D. C. Solmon [9 and 10], and R. Strichartz [11].

2. STATEMENT OF RESULTS

Before we proceed, we shall describe the k-plane transforms in terms of the group $\mathcal{O}_n$ of proper rotations of $\mathbf{R}^n$ around the origin. Let us

The author was partially supported by a grant from the National Science Foundation.

fix a subspace M of $\mathbf{R}^n$ through the origin and of dimension $n - k$, and for $y \in M$, let N_y be the orthogonal complement of M through y. Thus, if $\omega \in \mathcal{O}_n$ and $N = \omega(N_0 + y) = \Phi(\omega, y)$,

$$\mathcal{R}_k(f)(N) = \int_{\omega N_y} f(x)\, d\sigma = \int_{N_y} f(\omega^{-1}x)\, d\sigma = F_k(\omega, y)$$

is a function on the Cartesian product $\mathcal{O}_n \times M$, which we assume to be equipped with the product of the measure on M and the normalized Haar measure $d\omega$ on $\mathcal{O}_n$. The set of subspaces N carries a measure dN that is invariant under the action of rigid motions of $\mathbf{R}^n$ and is defined by

$$\int F(N)\, dN = \int_{\mathcal{O}_n} d\omega \int_M F[\omega(N_0 + y)]\, dy.$$

This measure is clearly invariant under rotations. That it is also invariant under translations ω follows readily by observing that if P denotes the orthogonal projection on M, then

$$\int F(N + w)\, dN = \int_{\mathcal{O}_n} d\omega \int_M F[\omega(N_0 + y + P\omega^{-1}w)\, dy$$

$$= \int_{\mathcal{O}_n} d\omega \int_M F[\omega(N_0 + y)]\, dy = \int F(N)\, dN.$$

The measure dN is uniquely determined by its invariance under rigid motions and the fact that the measure of the set of all N intersecting a ball of measure 1 is 1. Such an invariant measure we call normalized.

We shall also use the natural topology of the set of subspaces N, which is the strongest that makes the map $(\omega, y) \to N = \Phi(\omega, y)$ continuous.

Theorem 1 (D. C. Solomon [10]). *Let $f(x) \geq 0$ and $\alpha < n - k$. Then $F_k(\omega, y)$ is well defined almost everywhere and $F_k(\omega, y)(1 + |y|)^{-\alpha}$ is integrable over $\mathcal{O}_n \times M$ if and only if $f(x)(1 + |x|)^{-\alpha}$ is integrable over $\mathbf{R}^n$ and the ratio of the corresponding integrals is between two positive constants.*

Suppose now that $F_k(\omega, y)$ is well defined almost everywhere for $|y| \le \varepsilon$, $\varepsilon > 0$, and integrable there; then $f(x)(1 + |x|)^{-n+k}$ is integrable. Conversely, if $f(x)(1 + |x|)^{-n+k}$ is integrable, then $F_k(\omega, y)$ is well defined almost everywhere and if $\varphi(t)$ is positive and decreasing in $(0, \infty)$,

$$\int_{\mathcal{O}_n} d\omega \int F_k(\omega, y)(1 + |y|)^{-n+k} \varphi(|y|) \, dy$$

$$\le c \left[\int_{\mathbf{R}^n} f(x)(1 + |x|)^{-n+k} \, dx \right] \left[\int_0^\infty \varphi(t) \frac{dt}{t} \right].$$

Remark. The result implies in particular that $F_k(\omega, y)$ is well defined almost everywhere if $f \in L^p$, $p < n/k$. However, this is false if $p = n/k$, as the example

$$f(x) = (1 + |x|)^{-k} [\log(1 + |x|)]^{-1}$$

shows.

Theorem 2. Let $f(x) \in L^{p_1} \cap L^{p_2}$, $1 < p_1 \le p_2 \le 2$, $p_1 < n/k$. Then $F_k(\omega, y)$ is a well-defined function of y for almost all ω and its Fourier transform with respect to y is a locally integrable function $\hat{F}_k(\omega, z)$ for almost all ω. Thus, the operator L

$$[(L \, F_k)(\omega, y)]^{\wedge} = m(x)\hat{F}_k(\omega, z),$$

where $m(z)$ is a positively homogeneous function of complex degree $s + it$, $s \ge 0$, is well defined for almost all ω. Now let $p_1 \le p \le p_2$, $p > 1$, and let $m(z)$ be infinitely differentiable in $|z| > 0$ and its degree of homogeneity be such that

$$k - \frac{n}{p} < s < k - \frac{k}{p}.$$

Then

$$\int_{\mathcal{O}_n} \| (L \, F_k)(\omega, y) \|_r^{p'} \, d\omega \le c \|f\|_p^{p'}$$

where the norm in the integral is the norm of $L^r(M)$, $p' = p/(p-1)$ and

$$\frac{1}{r} = \frac{1}{p} \frac{n}{n-k} - \frac{k-s}{n-k}.$$

In the extreme case $p = 1$ and $s = 0$, we have the following substitute result. If f belongs to the Hardy space $H^1(\mathbf{R}^n)$, then $F_k(\omega, y)$ and $(L\,F_k)(\omega, y)$ are well defined for every ω and

$$\left\| (L\,F_k)(\omega, y) \right\|_{H^1(M)} \leq c\|f\|_{H^1(\mathbf{R}^n)}.$$

Now if $s < k - n/p_2$ [which requires that $p_2 > n/(k - s)$], then $(L\,F_k)(\omega, y)$,is bounded for almost all ω and

$$\left\| L\,F_k \right\|_\infty^u d\omega \leq c\left(\|f\|_{p_1} + \|f\|_{p_2} \right)^u, \quad u = n/(n - k + s).$$

Finally, if $s < k - n/p_2 < s + 1$, then $(L\,F_k)(\omega, y)$ satisfies a uniform Hölder condition of order $k - n/p_2 - s$ in y for almost all ω.

Remark. If $k - n/p_2 > 0$ and we specialize $m(x)$ to be monomial, we find that F_k has bounded derivatives of all orders h, $0 \leq h < k - n/p_2$, in y for almost all ω and derivatives of order h, $h < k - n/p_2 < h + 1$, satisfying a Hölder condition of order $k - n/p_2 - h$ in y. In general, F_k has derivatives with respect to y of orders less than $k - k/p_2$ in the corresponding classes described in the theorem. What happens for $p > 2$ remains an open problem. However, if $f(x)$ is the characteristic function of the unit ball, then

$$F_k(\omega, y) = c(1 - |y|^2)^{k/2}, \quad |y| \leq 1,$$

and if k is even, this function does not have functions as derivatives of order higher than $k/2$.

Theorem 3. Let C be a closed subset of $\mathbf{R}^n$ with the following property: for every x not in C, there is a subspace L_x through x of dimension $k + 1$, $(k < n)$, such that no $(k + 1)$-dimensional subspace L in a neighborhood of L_x intersects C. Let $f(x)(1 + |x|)^{-n+k}$ be integrable and suppose that

$$\mathscr{R}_k(f)(N) = 0$$

for almost all N not intersecting C. Then $f(x)$ vanishes almost everywhere outside C.

Remark. If C is convex and bounded, the theorem asserts that if $k < n - 1$ and $\mathscr{R}_k(fg)\,(N) = 0$ for almost every N not intersecting C, then f vanishes almost everywhere outside C. This is no longer true if $k = n - 1$. In fact, if $k = n - 1$, then for every α, $\alpha > n$, there exists a continuous function $f(x)$, $|f(x)| \leq (1 + x)^{-\alpha}$, such that $\mathscr{R}_k(f)\,(N) = 0$ for every N not intersecting a ball and the support of f is unbounded. For positive results in the case $k = n - 1$, see [6 and 7].

3. PROOF OF THEOREM 1

Let M and N_y be as in the beginning of Sec. 2, and let P denote the orthogonal projection on M. Then we have

$$\int_M F_k(\omega, y)(1 + |y|)^{-\alpha}\,dy = \int_M (1 + |y|)^{-\alpha} \int_{N_y} f(\omega^{-1}x)\,d\sigma\,dy$$

$$= \int_{\mathbf{R}^n} (1 + |Px|)^{-\alpha} f(\omega^{-1}x)\,dx$$

$$= \int_{\mathbf{R}^n} (1 + |P\omega x|)^{-\alpha}\,d\omega\,dx.$$

Integrating with respect to ω and interchanging the order of integration, we obtain

$$\int_{M \times \mathcal{O}_n} F_k(\omega, y)(1 + |y|)^{-\alpha}dy\,d\omega = \int_{\mathbf{R}^n} f(x) \int_{\mathcal{O}_n} (1 + |P\omega x|)^{-\alpha}\,d\omega\,dx.$$

But the inner integral in the last expression is nothing but the mean value of $(1 + |Px|)^{\alpha}$ over the sphere of radius $|x|$, which an elementary calculation shows to be between two positive constants times $(1 + |x|)^{-\alpha}$. To prove the second part of the theorem, we argue as above, replacing the function $(1 + |y|)^{-\alpha}$ by $\chi(|y|/t)$, where χ is the characteristic function of the interval $[0, 1]$ and t is a positive real number, and obtain

$$\int_{\mathcal{O}_n} d\omega \int_{|y| \leq t} F_k(\omega, y)\,dy = \int_{\mathbf{R}^n} f(x) \int_{\mathcal{O}_n} \chi\left[\frac{|P\omega x|}{t}\right]d\omega\,dx.$$

Again, the inner integral on the right is the mean value of the function $\chi(|Px|)$ over the sphere of radius $|x|/t$, which is readily seen to be of the order of

$$\left(1 + \frac{|x|}{t}\right)^{-n+k}.$$

Thus, the integrability of

$$f(x)\left(1 + \frac{|x|}{t}\right)^{-n+k}$$

is equivalent to that of $F_k(\omega, y)$ in $|y| \le t$. But the integrability of

$$f(x)\left(1 + \frac{|x|}{t_1}\right)^{-n+k}$$

is equivalent to that of

$$f(x)\left(1 + \frac{|x|}{t_2}\right)^{-n+k}, \quad t_1 > 0, \ t_2 > 0,$$

whence setting $t_1 = \varepsilon$, $t_2 = 1$, the first assertion follows. Finally, suppose that $f(x)(1 + |x|)^{-n+k}$ is integrable. Then if $\varphi(t)$ is positive and decreasing, we have

$$\int_{O_n} d\omega \int_M F_k(\omega, y)(1 + |y|)^{-n+k} \varphi(|y|) dy$$

$$= -\int_{O_n} d\omega \int_0^\infty \left[\int_{|y| \le t} F_k(\omega, y) dy\right] d(1 + t)^{-n+k}\varphi(t)$$

$$\le -c \int_0^\infty \left[\int_{R^n} f(x)\left(1 + \frac{|x|}{t}\right)^{-n+k} dx\right] d(1 + t)^{-n+k}\varphi(t)$$

$$= c(n - k)\int_{R^n} f(x)\int_0^\infty (1 + t)^{-n+k}\left(1 + \frac{|x|}{t}\right)^{-n+k-1}\frac{|x|}{t^2}\varphi(t) dt.$$

But, as is readily verified,

$$(1 + t)^{-n+k}\left(1 + \frac{|x|}{t}\right)^{-n+k-1}\frac{|x|}{t} \le c(1 + |x|)^{-n+k},$$

whence, substituting in the preceding integral, the desired result follows.

4. PROOF OF THEOREM 2

First let us consider the case when f is a function in $H^1(\mathbf{R}^n)$.
An atom (see [1]) is a bounded function with support in a ball and
integral equal to 0. The norm of the atom is the least upper bound
of $|f|$ times the measure of the ball. Every sum of atoms with
finite sum of norms is a function in H^1 with norm less than or equal
to a constant (depending on the choice of the norm for H^1) times
the sum of the norms of the atoms. Conversely, every function in
H^1 is a sum of atoms with sum of norms less than or equal to a
constant times the norm of the function. Now, if f is an atom,
so is $F_k(\omega, y)$ as a function of y for every ω, and the norm of F
as an atom is less than or equal to a constant times the norm of f
as an atom. This is readily seen and is left to the reader to
verify. From this there follows by addition that if f is in $H^1(\mathbf{R}^n)$,
then $F_k(\omega, y)$ is in $H^1(M)$ for every ω and

$$\|F_k(\omega, y)\|_{H^1} \leq c\|f\|_{H^1}.$$

Assume now that the Fourier transform of f is infinitely differentiable, has compact support, and vanishes near the origin. For z a
point of M, we have

$$\hat{F}_k(\omega, z) = \int_M F_k(\omega, y)\exp 2\pi i\,(y \cdot z)\,dy$$

$$= \int_M \exp 2\pi i\,(y \cdot z)\int_{N_y} f(\omega^{-1}x)\,d\sigma$$

$$= \int_{\mathbf{R}^n} f(\omega^{-1}x)\exp 2\pi i\,(x \cdot z)\,dx$$

$$= \int_{\mathbf{R}^n} f(x)\exp 2\pi i\,(x \cdot \omega z)\,dx = \hat{f}(\omega z).$$

Thus, $\hat{F}_k(\omega, z) = \hat{f}(\omega z)$. For a complex number ξ, $\xi = s + it$, let
D^ξ be the operator defined on functions F in M with Fourier transform vanishing near the origin by

$$(D^\xi F)^\wedge = |z|^\xi \hat{F}(z).$$

Then for $s = k/2$, we have

$$\|D^\xi F_k\|_2^2 = \| |z|^{k/2} \hat{f}(\omega z)\|_2^2,$$

and integrating with respect to ω, we obtain

$$\int_{\mathcal{O}_n} \|D^\xi F_k\|_2^2 d\omega = \int_{\mathcal{O}_n} d\omega \int_M |z|^k |\hat{f}(\omega z)|^2 dz = \int_{\mathcal{O}_n} d\omega \int_{\omega^{-1}M} |x|^k |\hat{f}(x)|^2 d\sigma.$$

The last expression is the integral of $|\hat{f}|$ with respect to a certain measure μ in $\mathbf{R}^n$ that clearly is rotation invariant and homogeneous of degree n in the sense that

$$t^n \int g(tx) d\mu = \int g(x) d\mu, \quad t > 0.$$

Now, every measure with these properties is a multiple of the ordinary measure dx. Consequently,

$$\int_{\mathcal{O}_n} \|D^\xi F_k\|_2^2 d\omega = c \int |\hat{f}|^2 dx = c\|f\|_2^2.$$

Now let $s = 0$. Then, since $|z|^{it}$ is a multiplier for $H(M)$ with norm less than or equal to $c(1 + |t|)^\alpha$ for some constants c and α (see [3, Theorem 4.6]), we find that

$$\|D^\xi F_k\|_1 \le c\|D^\xi F_k\|_{H^1(M)} \le c(1 + |t|)^\alpha \|F_k\|_{H^1(M)}$$

$$\le c(1 + |t|)^\alpha \|f\|_{H^1(\mathbf{R}^n)}.$$

Thus the operator taking f to the function $(1 + \xi)^{-\alpha} D^\xi F_k$ in $M \times \mathcal{O}_n$ has the property of being continuous with respect to the norms of $H^1(\mathbf{R}^n)$ and $\sup_\omega \|F(\omega, y)\|_1$ for $s = 0$, and the norms of $L^2(M \times \mathcal{O}_n)$ for $s = k/2$, uniformly in t. Thus, by interpolation (see, for example, [3, Theorem 3.4, and 2, Theorem 13.6]), there follows that

$$\int_{\mathcal{O}_n} \|D^{k/p'} F_k\|_p^{p'} d\omega \le c\|f\|_p^{p'}, \quad 1 < p < 2.$$

Now let p, r, and s be as in the first part of the theorem. Then

$$k/p' - s > 0 \quad \text{and} \quad 1/p > (k/p' - s)/(n - k),$$

and since $D^{-(k/p'-s)}$ is fractional integration of order $k/p' - s$, by Sobolev's theorem we have

$$\left\| D^s F_k \right\|_r = \left\| D^{-(k/p'-s)} D^{k/p'} F_k \right\|_r < c \left\| D^{k/p'} F_k \right\|_p ,$$

and substituting above, we obtain

$$\int_{\mathcal{O}_n} \left\| D^s F_k \right\|_r^{p'} d\omega \leq c \| f \|_p^{p'} .$$

Now, since $m(z) |z|^{-s}$ is a multiplier for $L^r(M)$ (we have $r > p > 1$),

$$\left\| L \, F_k \right\|_r \leq c \left\| D^s F_k \right\|_r .$$

Combining this with the preceding inequality, we obtain the first inequality of the theorem for functions f whose Fourier transforms are infinitely differentiable, have compact support, and vanish near the origin. Such functions are dense in the intersection of any two L^p spaces, $p > 1$. To see this, we observe first that bounded functions with compact support have this property, and if $(\phi_\varepsilon)^\wedge = \hat{\phi}(\varepsilon z)$ where $\hat{\phi}$ is infinitely differentiable, has compact support, and equals 1 near the origin, and if g is bounded and has compact support, then, by the maximal theorem, the convolution $g * \phi_\varepsilon$ will be majorized by a function belonging to every L^p, $p > 1$, and will converge to g as $\varepsilon \to 0$ and to 0 as $\varepsilon \to \infty$ in the norm of every L^p, $p > 1$. Thus, $g * \phi_\varepsilon - g * \phi_\delta$ will converge to g in every such norm as $\varepsilon \to 0$ and $\delta \to \infty$, and its Fourier transform will be infinitely differentiable, have compact support, and will vanish near the origin. Thus, by passage to the limit, we obtain the first inequality of the theorem in the general case.

To prove our assertion in the case $s < k - n/p_2$, we need the following result, which we state without proof.

Lemma. Let F be a function on M belonging to $L^{q_1} \cap L^{q_2}$, $1 < q_1 < q_2 \leq \infty$, and let

$$\frac{s}{n-k} = \frac{v}{q_1} + \frac{1-v}{q_2}, \quad 0 < v < 1.$$

Then $D^{-s}F$ is bounded and

$$\|D^{-s}F\|_\infty \le c\|F\|_{q_1}^v \|F\|_{q_2}^{1-v}.$$

If, in addition,

$$\frac{1}{q_2} < \frac{s}{n-k} < \frac{1}{q_2} + \frac{1}{n-k}$$

then $D^{-s}F$ satisfies a Hölder condition of order $s - (n-k)/q_2$.

Returning to our proof, since $1 < p_1 \le p_2 \le 2$, $p_1 < n/k$, and $s \ge 0$, we must have $p_2 > p_1$, for if $p_2 = p_1$, then $p_2 < n/k$ and $k - n/p_2 < 0$. Now choose q_1, q_2, $p_1 \le q_1 < q_2 \le p_2$, and σ so that

$$k - \frac{k}{q_1} > \sigma, \quad 1 + k - \frac{n}{q_1} > \sigma > k - \frac{n}{q_2} > s > k - \frac{n}{q_1}.$$

This is clearly possible. Then from the last inequalities above, we obtain

$$\frac{1}{r_2} = \frac{1}{q_2}\frac{n}{n-k} - \frac{k-\sigma}{n-k} < \frac{\sigma-s}{n-k} < \frac{1}{q_1}\frac{n}{n-k} - \frac{k-\sigma}{n-k} = \frac{1}{r_1} \tag{1}$$

from the third $0 < r_2 < \infty$, and from the second $r_1 < 1$. Now since

$$k - \frac{n}{q_i} < \sigma < k - \frac{k}{q_i}, \quad i = 1, 2,$$

and $f \in L^{p_1} \cap L^{p_2} \subset L^{q_1} \cap L^{q_2}$; from the first part of the theorem it follows that

$$\int \|D^{\sigma-s}L\, F_k(\omega, y)\|_{r_i}^{q_i'} \le c\|f\|_{q_i}^{q_i'}, \tag{2}$$

and from the lemma and (1),

$$\|L\, F_k(\omega, y)\|_\infty \le c\|D^{\sigma-s}L\, F_k(\omega, y)\|_{r_1}^v \|D^{\sigma-s}L\, F_k(\omega, y)\|_{r_2}^{1-v}, \tag{3}$$

where

$$\frac{\sigma-s}{n-k} = \frac{v}{r_1} + \frac{1-v}{r_2}, \quad 0 < v < 1,$$

which, as follows from (1), is equivalent to

$$\frac{n}{n-k+s}\frac{1}{q_1'}v + \frac{n}{n-k+s}\frac{1}{q_2'}(1-v) = 1.$$

Thus, setting

$$u = \frac{n}{n - k + s}, \ w = \frac{n}{n - k + s} \frac{1}{q_1'} v, \ 1 - w = \frac{n}{n - k + s} \frac{1}{q_2'} (1 - v),$$

applying Hölder's inequality to (3), we obtain

$$\int_{O_n} \|LF_k (\omega, \ y)\|_\infty^\mu d\omega \le c \left[\int_{O_n} \|D^{\sigma - s} LF_k (\omega, \ y)\|_{r_1}^{\frac{uv}{w}} d\omega \right]^w$$

$$\times \left[\int \|D^{\sigma - s} LF_k (\omega, \ y)\|_{r_2}^{\frac{u(1 - v)}{1 - w}} d\omega \right]^{1 - w};$$

and from (2), since $\frac{uv}{w} = q_2'$, $\frac{u(1 - v)}{1 - w} = q_2'$, and $\|f\|_{q_i} \le \|f\|_{p_1} +$
$\|f\|_{p_2}$, $i = 1, 2$, there follows that

$$\int_{O_n} \|LF_k (\omega, \ y)\|_\infty^\mu d\omega \le c (\|f\|_{p_1} + \|f\|_{p_2})^\mu .$$

To prove the last assertion of the theorem we argue as above, now
setting $q_2 = p_2$. If r_2 is as in (1), there will follow that

$$\frac{1}{r_2} + \frac{1}{n - k} > \frac{\sigma - s}{n - k},$$

and applying the lemma, one concludes that $LF_k (y, \ \omega)$, as a function
of y, satisfies a Hölder condition of order

$$\sigma - s - \frac{n - k}{r_2} = k - \frac{n}{p_2} - s$$

whenever $\|D^{\sigma - s} LF_k (\omega, \ y)\|_{r_1}$ and $\|D^{\sigma - s} LF_k (\omega, \ y)\|_{r_2}$ are finite, that is,
for almost all ω. This concludes the proof of Theorem 2.

5. PROOF OF THEOREM 3

First we shall prove a special case of our theorem, namely, when
$k = n - 1$. In this case C must be empty and the theorem asserts
that if $f(x) (1 + |x|)^{-1}$ is integrable and f has vanishing integral
on almost all N, then $f = 0$ almost everywhere. If f were integrable,
then, as was indicated at the beginning of the proof of Theorem 2,
it would follow that $\hat{f} = 0$, which would imply the desired conclusion. In the general case we must proceed otherwise.

Assume first that the Fourier transform $\hat{f}$ of f coincides with an infinitely differntiable function in $|x| \neq 0$. Let ω, M, N_y, N_0 be as in Sec. 2, and $\varphi(y)$, $\psi(z)$, respectively, be functions in the Schwartz classes of M and its orthogonal complement through the origin N_0, with $\psi(0) = 1$. Set

$$\eta_\varepsilon(x) = \varphi(y)\psi(\varepsilon z), \quad x = y + z.$$

Then we have

$$\int_{\mathbf{R}^n} f(\omega^{-1}x)\eta_\varepsilon(x)\,dx = \int_{M \times N_0} f[\omega^{-1}(y + z)]\varphi(y)\psi(\varepsilon z)\,dy\,dz.$$

Now, according to Theorem 1, $f[\omega^{-1}(y + z)]\varphi(y)$ is integrable in $M \times N_0 = \mathbf{R}^n$ for almost all ω. For such an ω, we have

$$\lim_{\varepsilon \to 0} \int_{M \times N_0} f[\omega^{-1}(y + z)]\varphi(y)\psi(\varepsilon z)\,dy\,dz$$

$$= \int_{M \times N_0} f[\omega^{-1}(y + z)]\varphi(y)\,dy\,dz = \int_M \varphi(y)\int_{N_y} f(\omega^{-1}x)\,d\sigma\,dy,$$

and according to our hypotheses, for almost all ω the inner integral in the last expression vanishes for almost all y. Thus

$$\lim_{\varepsilon \to 0} \int_{\mathbf{R}^n} f(\omega^{-1}x)\eta_\varepsilon(x)\,dx = 0$$

for almost all ω. Suppose now that the Fourier transform $\hat{\varphi}(y)$ of $\varphi(y)$ vanishes near the origin. Then $\hat{\eta}_\varepsilon(x) = \hat{\varphi}(y)\varepsilon^{-n}\hat{\psi}(\varepsilon^{-1}z)$, $x = y + z$, also vanishes near the origin, and since $\hat{f}(x)$ was assumed to coincide with an infinitely differentiable function in $|x| \neq 0$, if $\hat{\varphi}(y)$ and $\hat{\psi}(z)$ have compact support, we have

$$\lim_{\varepsilon \to 0} \int f(\omega^{-1}x)\eta_\varepsilon(x)\,dx = \lim_{\varepsilon \to 0} \int \hat{f}(\omega^{-1}x)\hat{\eta}_\varepsilon(-x)\,dx$$

$$= \lim_{\varepsilon \to 0} \int_{M \times N_0} \hat{f}[\omega^{-1}(y + z)]\hat{\varphi}(-y)\varepsilon^{-n}\hat{\psi}(-\varepsilon^{-1}z)\,dy\,dz$$

$$= \int_M \hat{f}(\omega^{-1}y)\hat{\varphi}(-y)\,dy.$$

Thus, for almost every ω, the last integral vanishes whenever $\hat{\varphi}(y)$ has compact support and vanishes near the origin. Since $\hat{f}$

coincides with a continuous function in $|x| \neq 0$, this clearly
implies that $\hat{f}$ is a distribution supported at the origin and f is a
polynomial. But $f(x)(1 + |x|)^{-1}$ is integrable and this is possible
only if $f = 0$.

Now let us pass to the case when $f(x)(1 + |x|)^{-1}$ is merely
assumed to be integrable. Let A denote a linear transformation of
$\mathbf{R}^n$ and I the identity transformation. Let $\eta(A)$ be an infinitely
differentiable function of A with compact support, that is, a C_0^{∞}-
function of the entries a_{ij} of the matrix describing A in terms of
a coordinate system $(x_1, \ldots, x_n)$ in $\mathbf{R}^n$, such that

$$\int \eta(A)\, dA = 1$$

where $dA = \prod (da_{ij})$. Let

$$f_{\varepsilon}(x) = \int f[(I + A)x]\varepsilon^{-n^2}\eta(\varepsilon^{-1}A)\, dA, \quad \varepsilon > 0.$$

If A is sufficiently small, that is, the a_{ij} are sufficiently small,
then $I + A$ is invertible, $f[(I + A)x]$ is well defined,
$f[(I + A)x](1 + |x|)^{-1}$ is integrable, and the integral of
$f[(I + A)x]$ on subspaces N of dimension $n - 1$ vanishes for almost
all N (we leave the proof of this to the reader). Thus, if ε is
sufficiently small, the same is true for $f_{\varepsilon}(x)$. Now, as we shall
see below, for sufficiently small ε, $\hat{f}_{\varepsilon}$ coincides with an infinitely
differentiable function in $|x| \neq 0$, and this, as we have shown,
implies that $f_{\varepsilon} = 0$. Now, from the definition of f_{ε}, it follows
that for every $\varphi(x)$ in C_0^{∞},

$$\int f(x)\varphi(x)\, dx = \lim_{\varepsilon \to 0} \int f_{\varepsilon}(x)\varphi(x)\, dx = 0.$$

Thus, we conclude that $f = 0$ almost everywhere.

To prove that $\hat{f}_{\varepsilon}$ coincides with an infinitely differentiable
function in an open set $\mathcal{O}$ will suffice to show that for every compact set C contained in $\mathcal{O}$ and every monomial differential operator
$(\partial/\partial x)^{\alpha}$, there is a constant c such that if φ is a testing function
with support in C, then

$$\left| \left\langle \hat{f}_\epsilon, \left(\frac{\partial}{\partial x}\right)^\alpha \varphi \right\rangle \right| \leq c \|\varphi\|_\infty .$$

Let us take a coordinate system $(x_1, \ldots, x_n)$ in $\mathbf{R}^n$ and let the linear transformation $\bar{x} = Ax$ be given by $\bar{x}_i = \Sigma a_{ij}x_j$. If B is a linear transformation and ψ a function on $\mathbf{R}^n$, let us write

$$\psi(Bx) = (\psi \circ B)(x) .$$

Then if B_t denotes the transpose of B,

$$\hat{\psi} \circ B^{-1} = (\det B)(\psi \circ B_t)\hat{} .$$

Thus, denoting $\epsilon^{-n^2}\eta(\epsilon^{-1}A)$ simply by $\eta_\epsilon(A)$, we have

$$\langle \hat{f}_\epsilon, \varphi \rangle = \int f[(I + A)x]\hat{\varphi}(x)\eta_\epsilon(A)\,dA\ dx$$

$$= \int f(x)\hat{\varphi}[(I + A)^{-1}x]\det(I + A)^{-1}\eta_\epsilon(A)\,dA\ dx$$

$$= \int \langle \hat{f}, \varphi \circ (I + A_t) \rangle \eta_\epsilon(A)\,dA$$

$$= \langle \hat{f}, \int [\varphi \circ (I + A_t)]\eta_\epsilon(A)\,dA \rangle$$

and more generally,

$$\left\langle \hat{f}_\epsilon, \left(\frac{\partial}{\partial x}\right)^\alpha \varphi \right\rangle = \left\langle \hat{f}, \int \left[\left(\left(\frac{\partial}{\partial x}\right)^\alpha \varphi\right) \circ (I + A_t)\right]\eta_\epsilon(A)\,dA \right\rangle .$$

Now, if φ has support in a compact set C, then the preceding integral is a function ψ of x with support in a compact set C_ϵ and

$$|\langle \hat{f}, \psi \rangle| \leq c \sup_{|\beta| \leq M} \left\| \left(\frac{\partial}{\partial x}\right)^\beta \psi \right\|_\infty$$

for some M.

Assume now that C is contained in $|x_1| > \delta > 0$. Then differentiating under the integral sign in

$$\left(\frac{\partial}{\partial x}\right)^\beta \psi = \left(\frac{\partial}{\partial x}\right)^\beta \int \left(\frac{\partial}{\partial x}\right)^\alpha \varphi \circ (I + A_t)(x)\eta_\epsilon(A)\,dA ,$$

using the identity

$$\left(\frac{\partial}{\partial x_j} \xi\right)[(I + A_t)x] = \frac{1}{x_1}\frac{\partial}{\partial a_{1j}} \xi[(I + A_t)x] ,$$

and integrating by parts with respect to a_{1j}, we find that

$$\left(\frac{\partial}{\partial x}\right)^\beta \psi = \int x_1^{-|\beta|} \varphi[(I + A_t)x] (D\eta_\varepsilon)(A) dA,$$

where D is a differential operator with polynomial coefficients
acting on the function η_ε of the variables a_{ij}. Thus, since for
ε sufficiently small and A in the support of $\eta_\varepsilon(A)$, the support
of $\varphi[(I + A_t)x]$ is also contained in $|x_1| > \delta > 0$, we find that
for small ε,

$$\left\|\left(\frac{\partial}{\partial x}\right)^\beta \psi\right\|_\infty \le c\|\varphi\|_\infty,$$

where c depends on β, δ, and η_ε, but not on φ. Thus, for small ε,

$$\left|\left\langle \hat{f}_\varepsilon, \left(\frac{\partial}{\partial x}\right)^\alpha \varphi \right\rangle\right| = |\langle \hat{f}, \psi \rangle| \le \sup_{|\beta| \le M} \left\|\left(\frac{\partial}{\partial x}\right)^\beta \psi\right\|_\infty \le c\|\varphi\|_\infty$$

and we conclude that $\hat{f}_\varepsilon$ coincides with an infinitely differentiable
function in $|x_1| > \delta > 0$. Since the same is true in $|x_j| > \delta > 0$,
the validity of our theorem in the special case is established.

Now let us pass to the general case. Since $f(x)(1 + |x|)^{-n+k}$
is integrable, the function $f(x)(1 + |x|)^{-1}$ is integrable on almost
every subspace L of dimension $k + 1$. Let $x \notin C$, L_x a subspace of
dimension $k + 1$ through x and not intersecting C, U a neighborhood
of L_x such that $L \in U$ implies that L does not intersect C. Let dL
denote the normalized invariant measure on the set of $(k + 1)$-
dimensional subspaces of $\mathbf{R}^n$, dN the normalized invariant measure on
the set of k-dimensional subspaces of $\mathbf{R}^n$, and for each L, $d_L N$ the
normalized invariant measure on the set of k-dimensional subspaces
of L. Let $F(N)$ be a continuous function with compact support on
the set of k-dimensional subspaces of $\mathbf{R}^n$ and

$$G(L) = \int_{\{N|N \subset L\}} F(N) d_L N.$$

Then $G(L)$ is continuous and

$$\int G(L) dL = \int dL \int_{\{N|N \subset L\}} F(N) d_L N = \int F(N) dN.$$

To see this, one merely has to observe that the iterated integral
above gives a linear functional of F that is invariant under the

action of rigid motions of $\mathbf{R}^n$ and therefore must coincide with a multiple of

$$\int F(N)\, dN.$$

That is actually coincides with this integral follows by setting

$$F(N) = \mathscr{R}_k(g)(N) = \int_N g\, d\sigma,$$

where g is now a continuous function in $\mathbf{R}^n$ with compact support and nonvanishing integral, in which case, as is readily verified,

$$\int F(N)\, dN = \int_{\mathbf{R}^n} g(x)\, dx$$

$$G(L) = \int_{\{N\,|\,N\,\cap\,L\}} F(N)\, d_L N = \int_L g\, d\sigma$$

$$\int G(L)\, dL = \int_{\mathbf{R}^n} g(x)\, dx.$$

Now, by passage to the limit through monotone sequences of functions, we find that if $F(N)$ is measurable, nonnegative, then

$$G(L) = \int_{\{N\,|\,N\,\cap\,L\}} F(N)\, d_L N$$

is well defined for almost all L and

$$\int G(L)\, dL = \int F(N)\, dN.$$

Now let $V = \{N\,|\,N \subset L,\ L \in U\}$ and $\chi_v(N)$ the characteristic function of this set. Let

$$F(N) = \chi_v(N)\left|\int_N f\, d\sigma\right| = \chi_v(N)\,|\,\mathscr{R}_k(f)(N)\,|.$$

Since if $N \in V$, then N does not intersect C, our assumption on $\mathscr{R}_k(f)(N)$ implies that $F(N)$ vanishes almost everywhere, and, consequently, so does

$$G(L) = \int_{\{N\,|\,N\,\subset\,L\}} F(N)\, d_L N.$$

Thus, for almost every L in U, $f(x)(1 + |x|)^{-1}$ is integrable and the integral of f vanishes on almost every subspace of L of codimension 1. But, as we have shown, this implies that f vanishes almost everywhere on L, and from this follows that f vanishes almost everywhere in a neighborhood of x.   Thus our assertion is established.

REFERENCES

1. Calderón, A. P.   An atomic decomposition of distributions in parabolic H^p spaces. *Adv. in Math.* 25 (1977):216-225.

2. Calderón, A. P.   Intermediate spaces and interpolation, the complex method. *Studia Math.* 24 (1964):113-190.

3. Calderón, A. P., and Torchinsky, A.   Parabolic maximal functions associated with a distribution, II. *Adv. in Math.* 214 (1977):101-171.

4. Drury, S. W. L^p estimates for the X-ray transform.   To appear.

5. Drury, S. W.   Some remarks on k-plane transforms.   Preprint.

6. Helgason, S.   The Radon transform on Euclidean spaces, compact two-point homogeneous spaces and Grassmann manifolds. *Acta Math.* 113 (1965):153-180.

7. Ludwig, D.   The Radon transform on Euclidean spaces. *Comm. Pure Appl. Math.* 23 (1966):49-81.

8. Oberlin, D. M., and Stein, E. M.   Mapping properties of the Radon transform.   Preprint.

9. Smith, K. T., and Solmon, D. C.   Lower dimensional integrability of L^2 functions. *J. Math. Anal. and Appl.* 51 (1975):539-549.

10. Solmon, D. C.   A note on k-plane integral transforms. *J. Math. Anal. and Appl.* 71 (1979):351-358.

11. Strichartz, R. L^p estimates for Radon transforms in Euclidean and non-Euclidean spaces.   Preprint.

AN ADDITION THEOREM FOR HEISENBERG HARMONICS

Charles F. Dunkl
University of Virginia

1. INTRODUCTION

The subelliptic Laplacian on the Heisenberg group has been studied by G. Folland, P. Greiner, and E. Stein [3, 6, 7]. Greiner [6] proposed the study of harmonic polynomials as a way of attacking the Dirichlet problem for the ball. This led him to a class of generalized ultraspherical polynomials and a situation analogous to ordinary spherical harmonics in $\mathbf{R}^N$. In this paper, after a brief description of the general situation (Greiner's paper discusses only H_1), an addition theorem is obtained that explicitly shows the effect of translation on harmonic polynomials and also leads to the harmonic polynomial expansion of an arbitrary translate of the fundamental solution. It is shown that this series converges (in an appropriate sense) in the largest ball avoiding the singularity. The disc polynomials (some spherical functions of the unitary group) are important in the analysis.

The Heisenberg group H_N is the space $\mathbf{C}^N \times \mathbf{R}$ furnished with the group operation $(z, t) \cdot (w, s) := (z + w, s + t + 2 \operatorname{Im}\langle z, w \rangle)$, where $N = 1, 2, \ldots$, and

$$\langle z, w \rangle := \sum_{j=1}^{N} z_j \overline{w}_j .$$

Two important functions on H_N are the norm $\|(z,\ t)\| := (t^2 + |z|^4)^{1/4}$
and $S(z,\ t) := t + i|z|^2$.

Two groups of automorphisms of H_N are the unitary group
$U(N)$ (on $\mathbf{C}^N$) acting by $U(z,\ t) := (Uz,\ t)$ and the dilation group
($\{r > 0\}$) acting by $r : (z,\ t) \mapsto (rz,\ r^2 t)$.

If p is a polynomial function on H_N (in the variables z_j,
$\bar{z}_j$, t, $1 \leq j \leq N$), say p is homogeneous of degree n if
$p(rz,\ r^2 t) = r^n p(z,\ t)$, $r > 0$.

The left-invariant tangent fields are spanned by

$$Z_j := \frac{\partial}{\partial z_j} + i\bar{z}_j \frac{\partial}{\partial t}, \ \bar{Z}_j := \frac{\partial}{\partial \bar{z}_j} - iz_j \frac{\partial}{\partial t}, \ 1 \leq j \leq N, \ T := \frac{\partial}{\partial t},$$

and the subelliptic Laplacian is

$$L := -\frac{1}{2} \sum_{j=1}^{N} (Z_j \bar{Z}_j + \bar{Z}_j Z_j)$$

(see Folland and Stein [3]). We will be concerned with the operator
$L_\gamma := L + i\gamma T$, for $\gamma \ \epsilon \ \mathbf{C}$, which arises for certain values of γ when
applying $\square_b$ to forms [3]. Say a function on H_N (or some open sub-
set) is L_γ-harmonic if it is twice differentiable and annihilated
by L_γ. Note that L_γ applied to a homogeneous polynomial decreases
the degree by 2. Also L_γ commutes with the action of $U(N)$. Greiner
found the L_γ-harmonic homogeneous polynomials for $N = 1$; here we
will extend his ideas to $N \geq 2$ [the main difference is due to $U(N)$
being non-Abelian for $N \geq 2$].

When the space of polynomials on $\mathbf{C}^N$ is split as a $U(N)$-
module, the following irreducible components arise.

Definition 1. For k, $\ell \geq 0$ let $V_{k\ell} := \left\{ p(z) : p \text{ is a homogeneous} \right.$
polynomial on $\mathbf{C}^N$ of bidegree $(k,\ \ell)$, and

$$\sum_{j=1}^{N} \frac{\partial^2}{\partial z_j\, \partial \bar{z}_j}\, p = 0 \left. \right\},$$

[bidegree $(k,\ \ell)$ means $p(cz) = c^k \bar{c}^\ell p(z)$ for $c \ \epsilon \ \mathbf{C}$].

Definition 2. The generalized ultraspherical polynomial of degree n and index (α, β) (for $\alpha, \beta \in \mathbf{C}$) is

$$C_n^{(\alpha, \beta)}(\zeta) := \sum_{j=0}^{n} \frac{(\alpha)_j \, (\beta)_{n-j}}{j! \, (n-j)!} \, \bar{\zeta}^j \, \zeta^{n-j} \quad (\zeta \in \mathbf{C}),$$

where the shifted factorial $(x)_m$ is defined by

$$(x)_0 = 1, \quad (x)_{m+1} = (x)_m \, (x + m), m = 0, 1, 2, \ldots .$$

We are using the notation of Gasper [4], who found an orthogonality relation. However, this relation uses all values of ζ on $|\zeta| = 1$, whereas in our context Im $\zeta \geq 0$ always. Also the (finite) linear span of $\{C_n^{(\alpha, \beta)} : n \geq 0\}$ is not an algebra when $\alpha \neq \beta$, so this is not a family of polynomials in the usual sense.

Proposition 1. Let $p \in V_{k\ell}$; then $L_\gamma(p(z) C_n^{(\alpha + \ell, \, \beta + k)}(t + i|z|^2)) = 0$, where $\alpha := (N - \gamma)/2$, $\beta := (N + \gamma)/2$; and every L_γ-harmonic polynomial is a linear combination of such terms (all values of k, ℓ, n).

This expansion for a left translate of the basic L_γ-harmonic function $C_n^{(\alpha, \beta)}(t + i|z|^2)$ is exactly the addition theorem. Specifically, we want the expansion of $C_n^{(\alpha, \beta)}(S((-w, -s) \cdot (z, t)))$ [note that $S((-w, -s) \cdot (z, t)) = t - s + i(|w|^2 + |z|^2 - 2\langle z, w \rangle)$]. To expand functions in $\langle z, w \rangle$ (for fixed w) in terms of $\{V_{k\ell}\}$, the spherical functions for $U(N)/U(N-1)$, namely, the disc polynomials, are needed (see Sec. 4 for more details).

Definition 3. For k, $\ell \geq 0$, $\lambda > -1$, $\xi \in \mathbf{C}$,

$$R_{k\ell}^{(\lambda)}(\xi) := \frac{(\lambda + 1)_{k+\ell}}{(\lambda + 1)_k \, (\lambda + 1)_\ell} \xi^k \bar{\xi}^\ell \, _2F_1(-k, \, -\ell; \, -\lambda - k - \ell; \, 1/\xi\bar{\xi}).$$

Also, for z, $w \in \mathbf{C}^N$, let

$$F_{k\ell}(z, w) := (|z| \, |w|)^{k+\ell} R_{k\ell}^{(N-2)}(\langle z, w \rangle / (|z| \, |w|)).$$

For fixed $w \neq 0$, the polynomial $z \mapsto F_{k\ell}(z, w)$ is in $V_{k\ell}$. Also, $|F_{k\ell}(z, w)| \leq (|z||w|)^{k+\ell}$ with equality at $z = w$ [see (14) and (16)].

Theorem 1 (Addition formula). For $\alpha + \beta = N$,

$$
C_n^{(\alpha, \beta)}(S((-w, -s) \cdot (z, t))) = \sum_{k+\ell \leq n} (2|w||z|)^{k+\ell} R_{k\ell}^{(N-2)}(\langle z, w \rangle / |z||w|)
$$

$$
\cdot \frac{i^{\ell-k}(\alpha)_\ell (\beta)_k (N - 1)_k (N - 1)_\ell}{k!\ell!(N - 1)_{k+\ell}}
$$

$$
\cdot \sum_{m=0}^{n-k-\ell} \frac{(N + k + \ell)_{n-k-\ell}}{(N + k + \ell)_m (N + k + \ell)_{n-k-\ell-m}}
$$

$$
\cdot C_m^{(\alpha+\ell, \beta+k)}(t + i|z|^2) C_{n-k-\ell-m}^{(\alpha+\ell, \beta+k)}(-s + i|w|^2).
$$

Folland and Stein [3] showed that a certain multiple of the function $\psi_\gamma(z, t) := (-t + i|z|^2)^{-\alpha}(-t - i|z|^2)^{-\beta}$ is the fundamental solution for L_γ (provided $\pm\gamma$ avoids the values N, $N + 2$, $N + 4$, ..., and $\gamma = \beta - \alpha$, $N = \alpha + \beta$). Thus $L_\gamma\psi = 0$ except at $(0, 0) \in H_N$.

Remark on Branch Points. Whenever a power function with a branch point is used, we will choose a branch cut to lie in some half-plane (Im ζ > or < 0) as appropriate, and the determination that is positive for all large positive arguments.

Theorem 2. *The L_γ-harmonic polynomial expansion of a translate of ψ_γ is*

$$
\psi_\gamma((-w, -s) \cdot (z, t)) = \sum_{k, \ell \geq 0} (2|z||w|)^{k+\ell} R_{k\ell}^{(N-2)}(\langle z, w \rangle / |z||w|)
$$

$$
\cdot \frac{i^{\ell-k}(\alpha)_\ell (\beta)_k (N - 1)_k (N - 1)_\ell}{k!\ell!(N - 1)_{k+\ell}}
$$

$$
\cdot \sum_{m=0}^{\infty} \frac{m!}{(N + k + \ell)_m} C_m^{(\alpha+\ell, \beta+k)}(t + i|z|^2) C_m^{(\alpha+\ell, \beta+k)}(s + i|w|^2)
$$

$$
\cdot (s - i|w|^2)^{-\beta-k-m}(s + i|w|^2)^{-\alpha-\ell-m}.
$$

For each k, ℓ the m-sum converges absolutely for $\|(z, t)\| < \|(w, s)\|$, and the (k, ℓ)-sum converges in the same region in the Abel or Cesaro $(C, 2N - 1)$ methods applied to the partial sums $\sum_{k+\ell \leq n}$ (see Theorem 5).

2. THE L_γ-HARMONIC POLYNOMIALS AND THE ADDITION THEOREM

Since the space of L_γ-harmonic polynomials is a $U(N)$-module, it can be decomposed into copies of $V_{k\ell}$, with k, $\ell \geq 0$, so that every L_γ-harmonic is a sum of terms like $p(z)g_1(z, t)$ where $p \in V_{k\ell}$ and g_1 is invariant under $U(N)$. Clearly $g_1(z, t) = g(|z|^2, t)$ for some function g.

Lemma 1. $L_\gamma = -\Delta - |z|^2 \left(\dfrac{\partial}{\partial t}\right)^2 + i\left(\gamma + \displaystyle\sum_{j=1}^{N}\left(z_j\dfrac{\partial}{\partial z_j} - \bar{z}_j\dfrac{\partial}{\partial \bar{z}_j}\right)\right)\dfrac{\partial}{\partial t}$, *where*

$$\Delta : = \sum_{j=1}^{N} \frac{\partial^2}{\partial z_j \partial \bar{z}_j}.$$

Further, if $p \in V_{k\ell}$ *and* g *is twice differentiable, then*

$$L_\gamma\left(p(z)g(|z|^2, t)\right) = p(z)\left(-s\left(\frac{\partial^2}{\partial s^2} + \frac{\partial^2}{\partial t^2}\right) - (k + \ell + N)\frac{\partial}{\partial s}\right.$$
$$\left. + i(k - \ell + \gamma)\frac{\partial}{\partial t}\right)g(s, t),$$

where $s = |z|^2$.

Proof. The calculation uses the product formula for Δ, the fact that $\Delta p = 0$, Euler's formula

$$\sum_{j=1}^{N} z_j \frac{\partial}{\partial z_j} p(z) = kp(z),$$

and the similar $\bar{z}$-result.

To find polynomial solutions for g, we use Greiner's trick, and let $g(|z|^2, t) = g_2(t + i|z|^2)$.

Lemma 2. Let $p \in V_{k\ell}$; then $L_\gamma\left(p(z)g_2(t + i|z|^2)\right) = 0$ if and only if g_2 satisfies

$$\left((\zeta - \bar{\zeta})\frac{\partial^2}{\partial\zeta\partial\bar{\zeta}} - (\alpha + \ell)\frac{\partial}{\partial\zeta} + (\beta + k)\frac{\partial}{\partial\bar{\zeta}}\right)g_2(\zeta) = 0$$

$(\alpha + \beta = N, \; \beta - \alpha = \gamma)$.   *The polynomial solutions for g_2 are arbitrary linear combinations of $C_n^{(\alpha + \ell, \; \beta + k)}(\zeta)$, $n = 0, 1, 2, \ldots$.*

Proof.   The change of variable $\zeta = t + i|z|^2$ leads to this equation.   Polynomial solutions can be split by degree of homogeneity, and the substitution

$$g_2(\zeta) = \sum_{j=0}^{n} a_j \bar{\zeta}^j \zeta^{n-j}$$

leads to the two-term recurrence

$$(j + 1)(\beta + k + n - j - 1)a_{j+1} - (n - j)(\alpha + \ell + j)a_j = 0,$$

whose unique solution is

$$a_j = c \; \frac{(\alpha + \ell)_j (\beta + k)_{n-j}}{j!(n - j)!},$$

some constant c.

Another solution for g_2 is given by $(c - \bar{\zeta})^{-\alpha - \ell}(c - \zeta)^{-\beta - k}$, some constant c.   To obtain the addition formula, we first deduce the general form of the expansion and then find the coefficients by a Lie technique.   Note that $C_n^{(\alpha, \; \beta)}(S((-w, -s) \cdot (z, t)))$ is homogeneous of degree $2n$, in the sense that the replacement of (z, w, t, s) by (rz, rw, r^2t, r^2s) multiplies the value by r^{2n} $(r > 0)$.   Further, if this function is to be expanded as a function of z, for fixed w, with respect to $U(N)$, then the terms must be invariant under the subgroup fixing w, hence functions of $\langle z, w \rangle$, $|z|$, $|w|$ only (namely, the $F_{k\ell}$ functions; see Definition 3).   Since L_γ is left-invariant, each term in the $U(N)$-expansion must be L_γ-harmonic (in z), so the multipliers of $F_{k\ell}(z, w)$ must be multiples of $C_m^{(\alpha + \ell, \; \beta + k)}(t + i|z|^2)$ $(0 \leq m \leq n - k - \ell)$.

On the other hand, for fixed (z, t), the left side of Theorem 1 is a left translate of the function :

$$(s, w) \mapsto C_n^{(\alpha, \beta)}\left(S\left((s, \cdot w)^{-1}\right)\right) = C_n^{(\alpha, \beta)}(-s + i|w|^2)$$

$$= (-1)^n C_n^{(\beta, \alpha)}(s + i|w|^2),$$

which (by interchanging α and β, or γ and $-\gamma$) can similarly be expanded in a series of

$$F_{\ell k}(w, z)C_m^{(\beta + k, \alpha + \ell)}(s + i|w|^2) = F_{k\ell}(z, w)(-1)^m C_m^{(\alpha + \ell, \beta + k)}(-s + i|w|^2).$$

Thus

$$C_n^{(\alpha, \beta)}\left(S\left((-w, -s) \cdot (z, t)\right)\right) = \sum_{k + \ell + m \leq n} A_{k\ell m}^n F_{k\ell}(z, w)$$

$$\cdot C_m^{(\alpha + \ell, \beta + k)}(t + i|z|^2)C_{n - k - \ell - m}^{(\alpha + \ell, \beta + k)}(-s + i|w|^2)$$

(1)

with the constants $A_{k\ell m}^n$ to be determined.

To do this, we will use two differential operators that commute with left translation by a fixed $(w, s) \in H_N$. Accordingly, when such an operator is applied to the right-hand side of identity (1), the same sum will be obtained regardless of whether the operator acts on the (z, t) or the (w, s) variables. Equating the two results will give recurrence relations for $A_{k\ell m}^n$.

The operators are T, defined by

$$Tf(z, t) := \frac{\partial}{\partial \theta} f\left((0, \theta) \cdot (z, t)\right)\Big|_{\theta = 0} \quad \text{(that is, } \partial/\partial t\text{)},$$

and R, defined by

$$Rf(z, t) := \frac{\partial}{\partial \theta} f\left((\theta w, 0) \cdot (z, t)\right)\Big|_{\theta = 0}$$

[for f differentiable, $(z, t) \in H_N$, θ real].

Equating the two results of applying T to (1), we obtain

$$\sum_{k + \ell + m \leq n} A_{k\ell m}^n F_{k\ell}(z, w)(N + n - m - 1)C_m^{(\alpha + \ell, \beta + k)}(t + i|z|^2)$$

$$\cdot C_{n - k - \ell - m - 1}^{(\alpha + \ell, \beta + k)}(-s + i|w|^2) = \sum_{k + \ell + m \leq n} A_{k\ell m}^n F_{k\ell}(z, w)$$

$$\cdot (N + k + \ell + m - 1)C_{m - 1}^{(\alpha + \ell, \beta + k)}(t + i|z|^2)C_{n - k - \ell - m}^{(\alpha + \ell, \beta + k)}(-s + i|w|^2)$$

[using $\partial/\partial t = \partial/\partial \zeta + \partial/\partial \overline{\zeta}$, when $\zeta = t + i|z|^2$ and the identity (9)]. This leads to the recurrence

$$A^n_{k\ell, m+1} = \frac{(N + n - m - 1)}{(N + k + \ell + m)} A^n_{k\ell m}$$

with solution

$$A^n_{k\ell m} = A^n_{k\ell 0} \frac{(N + k + \ell)_{n-k-\ell}}{(N + k + \ell)_m (N + k + \ell)_{n-k-\ell-m}} \tag{2}$$

(note $A^n_{k\ell, n-k-\ell} = A^n_{k\ell 0}$).

We use R to get relations among the $A^n_{k\ell 0}$ coefficients.   We already know $A^n_{000} = 1$ [set $(w, s) = (0, 0)$].   Since

$$(-w, -s) \cdot ((\theta w, 0) \cdot (z, t)) = ((\theta - 1)w, -s) \cdot (z, t),$$

we apply $\partial/\partial\theta$ to the right-hand side of (1) with $(-w, -s)$ replaced by $((\theta - 1)w, -s)$ and obtain

$$\sum_{k+\ell+m\leq n} A^n_{k\ell m} C^{(\alpha+\ell, \beta+k)}_m (t + i|z|^2) F_{k\ell}(z, w)(-(k + \ell) + R)$$

$$C^{(\alpha+\ell, \beta+k)}_{n-k-\ell-m}(-s + i|w|^2)$$

(setting $\theta = 0$).   It will suffice to use the $m = n - k - \ell$ terms. On the other hand, we have:

Lemma 3. $RF_{k\ell}(z, w)C^{(\alpha+\ell, \beta+k)}_m(t + i|z|^2)$

$$= \Big(2i(\beta + k)(N + k - 1)F_{k+1, \ell} C^{(\alpha+\ell, \beta+k+1)}_{m-1} \tag{3}$$

$$- 2i(\alpha + \ell)(N + \ell - 1)F_{k, \ell+1} C^{(\alpha+\ell+1, \beta+k)}_{m-1}$$

$$+ (N + k + \ell + m - 1)|w|^2 (k F_{k-1, \ell} C^{(\alpha+\ell, \beta+k-1)}_m$$

$$+ \ell F_{k, \ell-1} C^{(\alpha+\ell-1, \beta+k)}_m)\Big)/(N + k + \ell - 1),$$

with arguments (z, w) and $S(z, t)$ for the F and C functions, respectively.

Proof.   We use differential, recurrence, and contiguity relations for F- and C-functions.   First we note that

$$RF(z) = \sum_{j=1}^N \left(w_j \frac{\partial}{\partial z_j} + \bar{w}_j \frac{\partial}{\partial \bar{z}_j}\right)F(z),$$

so

$$RF_{k\ell}(z, w) = k|w|^2 F_{k-1, \ell}(z, w) + \ell|w|^2 F_{k, \ell-1}(z, w),$$

where we use $R\langle z, w\rangle = R\langle w, z\rangle = |w|^2$, $R(|z|^2) = 2\operatorname{Re}\langle z, w\rangle$, and a direct computation with the definition of $F_{k\ell}$ (see Definition 3).

Let $\zeta := t + i|z|^2$; then

$$Rg(\zeta) = 2i\left(\langle z, w\rangle\frac{\partial}{\partial\zeta} - \langle w, z\rangle\frac{\partial}{\partial\overline{\zeta}}\right)g(\zeta).$$

Using these relations, the recurrence for $\langle z, w\rangle F_{k\ell}$ (18) and $\frac{\partial}{\partial\zeta} C_m^{(\alpha, \beta)}(\zeta) = \beta C_{m-1}^{(\alpha, \beta+1)}(\zeta)$ (8), we obtain the term $F_{k+1, \ell}$ with coefficient as in (3) and the term $F_{k, \ell-1}$ with coefficient

$$\ell|w|^2(C_m^{(\alpha+\ell, \beta+k)}(\zeta) + (2i|z|^2(\beta + k)/(N + k + \ell - 1))C_{m-1}^{(\alpha+\ell, \beta+k+1)}(\zeta)),$$

which equals

$$\ell|w|^2((N + k + \ell + m - 1)/(N + k + \ell - 1))C_m^{(\alpha+\ell-1, \beta+k)}(\zeta)$$

by (10). A similar computation applies to $\langle w, z\rangle\frac{\partial}{\partial\overline{\zeta}} C_m^{(\alpha, \beta)}(\zeta)$, $F_{k, \ell+1}$, and $F_{k-1, \ell}$.

Theorem 3. *The coefficient in* (1),

$$A_{k\ell m}^n = \frac{2^{k+\ell}i^{\ell-k}(\alpha)_\ell(\beta)_k(N-1)_k(N-1)_\ell(N+k+\ell)_{n-k-\ell}}{(N-1)_{k+\ell} k! \ell! (N+k+\ell)_m(N+k+\ell)_{n-k-\ell-m}},$$

$$(k + \ell + m \leq n).$$

Proof. We equate the coefficients of $F_{k\ell}(z, w)C_{n-k-\ell}^{(\alpha+\ell, \beta+k)}(t + i|z|^2)$ in the two expressions, which result from applying R to (1). This gives the recurrence

$$\frac{2i\big((\beta + k - 1)(N + k - 2)A_{k-1,\ell, n-k-\ell+1} - (\alpha + \ell - 1)(N + \ell - 2)A_{k, \ell-1, n-k-\ell+1}\big)}{(N + k + \ell - 2)}$$

$$= -(k + \ell)A_{k, \ell, n-k-\ell}$$

[left side from (z, t), right from (w, s)]. We make the substitution (arrived at by judicious guessing)

$$A_{k,\,\ell,\,n-k-\ell} = \frac{2^{k+\ell}\bar{z}^{\ell-k}(\alpha)_\ell\,(\beta)_k\,(N-1)_k\,(N-1)_\ell}{(N-1)_{k+\ell}\,(k+\ell)!}C_{k\ell}$$

(for $k+\ell \leq n$) and obtain the equation $C_{k-1,\,\ell}+C_{k,\,\ell-1}=C_{k\ell}$.
Since $A_{00n}=A_{000}=C_{00}=1$, the solution is obviously
$C_{k\ell}=\binom{k+\ell}{\ell}$. This, together with (2), proves the theorem.

3. THE FUNDAMENTAL SOLUTION OF L_γ

We will use a trick that exploits the addition formula to derive an
expansion for an arbitrary translate of the L_γ-harmonic function
$\psi_\gamma(z,\,t) = (-\bar{S}(z,\,t))^{-\alpha}(-S(z,\,t))^{-\beta}$.

Lemma 4. For $m = 0,\,1,\,2,\,\ldots,\,r$ real,

$$(r-\bar{\zeta})^{-\alpha-m}(r-\zeta)^{-\beta-m}C_m^{(\alpha,\,\beta)}(r-\bar{\zeta})r^{m+N}m!$$

$$= \sum_{n=0}^{\infty}r^{-n}(N+n)_m C_n^{(\alpha,\,\beta)}(\zeta),$$

(4)

*with absolute convergence in $|z| < |r|$, and for fixed r the func-
tion is L_γ-harmonic ($\zeta = t + i|z|^2$).*

Proof. We apply T^m to both sides of the generating function

$$(r-\bar{\zeta})^{-\alpha}(r-\zeta)^{-\beta}r^N = \sum_{n=0}^{\infty}r^{-n}C_n^{(\alpha,\,\beta)}(\zeta)$$

(in $|\zeta| < |r|$, r real), where $T = \partial/\partial\zeta + \partial/\partial\bar{\zeta}$. We have

$$TC_n^{(\alpha,\,\beta)}(\zeta) = (N+n-1)C_{n-1}^{(\alpha,\,\beta)}(\zeta),$$

and inductively,

$$TC_m^{(\alpha,\,\beta)}(r-\bar{\zeta})(r-\bar{\zeta})^{-\beta-m}(r-\zeta)^{-\alpha-m} = (r-\bar{\zeta})^{-\alpha-m-1}(r-\zeta)^{-\beta-m-1}$$

$$\cdot\,\{((r-\bar{\zeta})(\beta+m)+(r-\zeta)(\alpha+m))C_m^{(\alpha,\,\beta)}(r-\bar{\zeta})$$

$$-(r-\zeta)(r-\bar{\zeta})(N+m-1)C_{m-1}^{(\alpha,\,\beta)}(r-\bar{\zeta})\}$$

$$= (m+1)(r-\bar{\zeta})^{-\alpha-m-1}(r-\zeta)^{-\beta-m-1}C_{m+1}^{(\alpha,\,\beta)}(r-\bar{\zeta})$$

by (12).

If ζ is replaced by $r - \zeta_1$ in Lemma 4, then absolute convergence holds in $|\zeta_1 + r| < |r|$, which is possible only if $\operatorname{sgn} r = -\operatorname{sgn}(\operatorname{Re} \zeta_1) \neq 0$.

The argument will use three regions, all subsets of $H_N \times H_N$,

$$\Omega := \{((z, t), (w, s)) : \|(z, t)\| < \|(w, s)\|\}$$

$$\Omega' := \{((z, t), (w, s)) : \|(z, t)\|^2 + 2|z||w| < \|(w, s)\|^2\}$$

$$\Omega'' := \{((z, t), (w, s)) : \|(z, t)\|^2 + 2|z||w| < s/2\}$$

in decreasing order. We will use an auxiliary variable in Ω'' and extend the result to Ω' and Ω, in turn, by analytic continuation.

Let

$$B_{k\ell} := \frac{2^{k+\ell} i^{\ell-k} (\alpha)_\ell (\beta)_k (N-1)_k (N-1)_\ell}{k! \ell! (N-1)_{k+\ell}}.$$

Lemma 5. *Suppose $((z, t), (w, s)) \in \Omega''$ and $\alpha > 0$, $\beta > 0$ (that is, $-N < \gamma < N$). Then there exists r such that*

$$r - |r - s + i|w|^2| > s/2 \quad and \quad |S((w, r - s) \cdot (z, t))| < r;$$

and

$$\psi_\gamma((-w, -s) \cdot (z, t)) = \sum_{k, \ell \geq 0} F_{k\ell}(z, w) B_{k\ell}$$

$$\cdot \sum_{m=0}^{\infty} \frac{1}{(N+k+\ell)_m} C_m^{(\alpha+\ell,\ \beta+k)}(t + i|z|^2) r^{-(k+\ell+m+N)} \tag{5}$$

$$\cdot \sum_{j=0}^{\infty} r^{-j} (N+k+\ell+j)_m C_j^{(\alpha+\ell,\ \beta+k)}(r - s + i|w|^2),$$

with absolute convergence.

Proof. Fix a point in Ω''. Since $s > 0$, $\lim_{r \to \infty} (r - |r - s + i|w|^2|) = s$ ($r > 0$), there exists an r with $r - |r - s + i|w|^2| > s/2$, and so $|S((-w, r-s) \cdot (z, t))| \leq |r - s + i|w|^2| + \|(z, t)\|^2 + 2|z||w| < r$ (by definition of Ω''). Using (4) with $m = 0$, we have

$$\psi_\gamma((-w, -s) \cdot (z, t)) = (r - \bar{S}((-w, r - s) \cdot (z, t)))^{-\alpha}$$

$$(r - S((-w, r - s) \cdot (z, t)))^{-\beta}$$

$$= \sum_{n=0}^{\infty} r^{-n-N} C_n^{(\alpha,\ \beta)}(S((-w, r - s) \cdot (z, t)))$$

(by the addition theorem)

$$= \sum_{n=0}^{\infty} r^{-n-N} \sum_{k+\ell+m \atop n} A_{k\ell m}^{n} F_{k\ell}\,(z,\;w)C_{m}^{(\alpha+\ell,\;\beta+k)}(t+i|z|^2)$$

$$\cdot\; C_{n-k-\ell-m}^{(\alpha+\ell,\;\beta+k)}(r-s+i|w|^2)$$

$$= \sum_{k,\;\ell \geq 0} F_{k\ell}\,(z,\;w)\sum_{m=0}^{\infty} C_{m}^{(\alpha+\ell,\;\beta+k)}(t+i|z|^2)r^{-(k+\ell+m+N)}$$

$$\cdot\; \sum_{j=0}^{\infty} r^{-j} A_{k\ell m}^{j+k+\ell+m} C_{j}^{(\alpha+\ell,\;\beta+k)}(r-s+i|w|^2),$$

which is the sum (5). The interchange of summation needs to be justified. The sum is dominated by

$$\sum_{k,\;\ell,\;m,\;j} (|z||w|)^{k+\ell}|B_{k\ell}|\frac{(N+k+\ell)_{m+j}}{m!j!}r^{-(k+\ell+m+j+N)}$$

$$\|(z,\;t)\|^{2m}|r-s+i|w|^2|^j$$

[see (13)]. The sums over j, then m, are negative binomial series, giving the sum

$$\sum_{k,\;\ell} (|z||w|)^{k+\ell}|B_{k\ell}|\,(r-|r-s+i|w|^2|-\|(z,\;t)\|^2)^{-N-k-\ell}$$

(the last term is positive by construction). Finally,

$$|B_{k\ell}| \leq K(k+\ell)^M 2^{k+\ell},$$

where K and M are constants, so the $(k,\;\ell)$ series converges absolutely.

Lemma 6. $\psi_\gamma((-w,\;-s)\cdot(z,\;t)) = \sum_{k,\;\ell \geq 0} F_{k\ell}\,(z,\;w)B_{k\ell}\sum_{m=0}^{\infty}\frac{m!}{(N+k+\ell)_m}$

$$\cdot\; C_{m}^{(\alpha+\ell,\;\beta+k)}(t+i|z|^2)C_{m}^{(\alpha+\ell,\;\beta+k)}(s+i|w|^2) \tag{6}$$

$$\cdot\; (s+i|w|^2)^{-\alpha-\ell-m}(s-i|w|^2)^{-\beta-k-m},$$

with absolute convergence in Ω'.

Proof. With the variables as in Lemma 5, sum (6) follows immediately from (5), because the sum over j is done by Lemma 4 with

$\zeta = r - s + i|w|^2$ (appropriate α, β). We claim that the right-hand side of (6) converges uniformly and absolutely on compact subsets of Ω' and γ in any compact subset of **C**. Indeed, let M be an integer so that $|\alpha| + |\beta| \leq N + M$. The m-sum in (6) is dominated term-wise by the series

$$\sum_{m=0}^{\infty} \frac{((|\alpha| + |\beta| + k + \ell)_m)^2}{(N + k + \ell)_m m!} \| (z, t) \|^{2m} \| (w, s) \|^{-2(m+k+\ell)} W,$$

which in turn is dominated by the series

$$W\| (w, s) \|^{-2(k+\ell)} {}_2F_1 (N + M + k + \ell, N + M + k + \ell; N + k + \ell; x)$$

$$= W\| (w, s) \|^{-2(k+\ell)} (1 - x)^{-N-2M-k-\ell} {}_2F_1 (-M, -M; N + k + \ell; x)$$

(by Euler's transformation), where

$$W = | (s + i|w|^2)^{-\alpha} (s - i|w|^2)^{-\beta} |$$

and

$$x = (\| (z, t) \| / \| (w, s) \|)^2.$$

The last series terminates and has positive terms only, so

$${}_2F_1 (-M, -M; N + k + \ell; x) < {}_2F_1 (-M, -M; N + k + \ell; 1)$$

$$= \frac{(N + M + k + \ell)_M}{(N + k + \ell)_M},$$

which is bounded in $k + \ell$. Since $\log|B_{k\ell}| = 0((k + \ell)\log 2)$ [see (5)], the (k, ℓ)-sum converges absolutely if

$$2|z||w| < \| (s, w) \|^2 - \| (t, z) \|^2.$$

Consider both sides of (6) as functions of the variables s, t, $|z|$, $|w|$, ξ, $\bar{\xi}$ (where $\xi := \langle z, w \rangle / |z||w|$), γ. Any compact subset of $\{t^2 + |z|^4 < s^2 + |w|^4, |\xi| \leq 1, \gamma \in \mathbf{C}, s, t, |z|, |w|$ real$\}$ can be fattened slightly to allow complex values and still preserve the uniform and absolute convergence in the sum as above. Thus, by analytic continuation, the identity (6) holds throughout Ω', $\gamma \in \mathbf{C}$.

Theorem 4. *The expansion (6) holds throughout* Ω *in the disc poly-*
nomial sense; in particular, for fixed $|z|$, t, $|w|$, s,

$$\psi_\gamma((-s, -w) \cdot (t, z)) = \lim_{r \to 1^-} \sum_{k, \ell \geq 0} r^{k+\ell} F_{k\ell}(z, w) B_{k\ell} G_{k\ell}(t + i|z|^2, s + i|w|^2),$$

where $G_{k\ell}$ *denotes the value of the m-sum in (6). The Cesaro*
(C, 2N - 1) sums in the $(k + \ell)$*-index also converge.*

Proof. Let $\Psi(\xi; |z|, t, |w|, s) := (s - t + i|z|^2 + i|w|^2 - 2i|z||w|\bar{\xi})^{-\alpha}(s - t - i|z|^2 - i|w|^2 + 2i|z||w|\xi)^{-\beta} = \psi_\gamma((-w, -s) \cdot (z, t))$ (where $\xi = \langle z, w \rangle / |z||w|$). Then Ψ is a smooth
function in ξ provided $(|z|, t) \neq (|w|, s)$, and as such, it has a
disc polynomial expansion [see (19) and (20)]

$$\sum_{k, \ell \geq 0} h_{k\ell}^{(N-2)} \hat{\Psi}_{k\ell}(|z|, t, |w|, s) R_{k\ell}^{(N-2)}(\xi),$$

where

$$\hat{\Psi}_{k\ell} = \frac{(N-1)}{\pi} \int_D \Psi(\xi; |z|, t, |w|, s) \cdot \overline{R_{k\ell}(\xi)} (1 - |\xi|^2)^{N-2} \, dm(\xi).$$

In Lemma 6 we showed that

$$\hat{\Psi}_{k\ell} = (2|z||w|)^{k+\ell} i^{\ell-k} \frac{(\alpha)_\ell (\beta)_k}{(N)_{k+\ell}} G_{k\ell}$$

holds in Ω'. But $\hat{\Psi}_{k\ell}$ and $G_{k\ell}$ are analytic in the variables $|z|$, t,
$|w|$, s in the domain Ω (by an argument similar to that of Lemma 6)
and agree in Ω', so by analytic continuation agree in Ω. The
convergence follows from Theorem 5.

Greiner studied the L_γ-harmonics primarily because of their
hoped-for applicability to the Dirichlet problem (which is solvable
with $\gamma = 0$, for the ball $\|(z, t)\| \leq 1$, Gaveau [5]). He pointed out
for a particular example $[(w, s) = (i, 0), N = 1]$ that the expansion
(6) arranged by degree of homogeneity of $\|(z, t)\|$ (first sum over
$k + \ell + 2m = n$ then n) converges only in the region Ω' [an argument
based on power series in $\|(z, t)\|$]. However, the fact that the m-
series, for each k, ℓ, converges absolutely in Ω still supports the

conjecture that the functions $\{C_n^{(\alpha,\ \beta)}(\zeta) : n = 0,\ 1,\ 2,\ \ldots\}$ are a complete set on $|\zeta| = 1$, Im $\zeta \geq 0$, for most values of α and β.

4. SOME RESULTS ON $C_n\alpha\beta$ AND THE DISC POLYNOMIALS

The generating function for the $C_n^{(\alpha,\ \beta)}$-functions is

$$(1 - r\bar{\zeta})^{-\alpha} (1 - r\zeta)^{-\beta} = \sum_{n=0}^{\infty} r^n C_n^{(\alpha,\ \beta)}(\zeta),$$

for $|r\zeta| < 1$. These hold (all can be directly verified from Definition 2):

$$\left((\zeta - \bar{\zeta})\frac{\partial^2}{\partial\zeta\partial\bar{\zeta}} - \alpha\frac{\partial}{\partial\zeta} + \beta\frac{\partial}{\partial\bar{\zeta}}\right)C_n^{(\alpha,\ \beta)}(\zeta) = 0; \tag{7}$$

(i) $\quad \dfrac{\partial}{\partial\zeta} C_n^{(\alpha,\ \beta)}(\zeta) = \beta C_{n-1}^{(\alpha,\ \beta+1)}(\zeta),$

(ii) $\quad \dfrac{\partial}{\partial\bar{\zeta}} C_n^{(\alpha,\ \beta)}(\zeta) = \alpha C_{n-1}^{(\alpha+1,\ \beta)}(\zeta);$
$$\tag{8}$$

$$\left(\frac{\partial}{\partial\zeta} + \frac{\partial}{\partial\bar{\zeta}}\right)C_n^{(\alpha,\ \beta)}(\zeta) = (\alpha + \beta + n - 1)C_{n-1}^{(\alpha,\ \beta)}(\zeta); \tag{9}$$

$$C_n^{(\alpha,\ \beta+1)}(\zeta) - (\zeta - \bar{\zeta})\frac{\alpha}{\alpha + \beta} C_{n-1}^{(\alpha+1,\ \beta+1)}(\zeta) = \frac{\alpha + \beta + n}{\alpha + \beta} C_n^{(\alpha,\ \beta)}(\zeta); \tag{10}$$

$$C_n^{(\alpha+1,\ \beta)}(\zeta) + (\zeta - \bar{\zeta})\frac{\beta}{\alpha + \beta} C_{n-1}^{(\alpha+1,\ \beta+1)}(\zeta) = \frac{\alpha + \beta + n}{\alpha + \beta} C_n^{(\alpha,\ \beta)}(\zeta); \tag{11}$$

$$(\zeta(\beta + n) + \bar{\zeta}(\alpha + n))C_n^{(\alpha,\ \beta)}(\zeta) - (\alpha + \beta + n - 1)\zeta\bar{\zeta}C_{n-1}^{(\alpha,\ \beta)}(\zeta)$$
$$= (n + 1)C_{n+1}^{(\alpha,\ \beta)}(\zeta); \tag{12}$$

(i) $\quad C_n^{(\alpha,\ \beta)}(1) = \dfrac{(\alpha + \beta)_n}{n!},$

(ii) $\quad \left|C_n^{(\alpha,\ \beta)}(\zeta)\right| \leq \dfrac{(|\alpha| + |\beta|)_n}{n!}|\zeta|^n.$
$$\tag{13}$$

The disc polynomials $R_{k\ell}^{\lambda}$ (see Definition 3) satisfy (for $\lambda > -1$):

$$R_{k\ell}^{(\lambda)}(1) = 1; \tag{14}$$

$$\bar{R}_{k\ell}^{(\lambda)}(\xi) = R_{\ell k}^{(\lambda)}(\xi); \tag{15}$$

$$\left| R_{k\ell}^{(\lambda)}(\xi) \right| \leq 1, \text{ for } |\xi| \leq 1 \text{ and } \lambda \geq 0; \tag{16}$$

$$R_{k\ell}(\xi) = \frac{\ell! \xi^{k-\ell}}{(\lambda + 1)_\ell} P_\ell^{(\lambda, k-\ell)}(2|\xi|^2 - 1) \tag{17}$$

[the Jacobi polynomial, for $k \geq \ell$; use (15) for $\ell \geq k$];

$$\xi R_{k\ell}^{(\lambda)}(\xi) = \left((\lambda + k + 1) R_{k+1, \ell}^{(\lambda)}(\xi) + \ell R_{k, \ell-1}^{(\lambda)}(\xi) \right) / (\lambda + k + \ell + 1) \tag{18}$$

and a similar relation for $\bar{\xi} R_{k\ell}^{(\lambda)}(\xi)$ [use (15)], (Boyd [1]). They form an orthogonal basis for $L^2(D, \mu_\lambda)$, where $D := \{\xi \in \mathbf{C}: |\xi| \leq 1\}$ and $d\mu_\lambda(\xi) := ((\lambda + 1)/\pi)(1 - |\xi|^2)^\lambda dm(\xi)$ (dm is Lebesgue measure in $\mathbf{R}^2$); μ_λ is a probability measure. The orthogonality relation is:

$$\int_D R_{k\ell}^{(\lambda)} \bar{R}_{mn}^{(\lambda)} d\mu_\lambda = \delta_{km} \delta_{\ell n} / h_{k\ell}^{(\lambda)}$$

$$\text{where } h_{k\ell}^{(\lambda)} := \frac{k + \ell + \lambda + 1}{\lambda + 1} \frac{(\lambda + 1)_k (\lambda + 1)_\ell}{k! \ell!}.$$

For $\lambda = N - 2$, $N = 2, 3, \ldots$, they are spherical functions for $U(N)/U(N - 1)$ (Ikeda [8], see also Koornwinder [10]) and $h_{k\ell}^{(N-2)} = \dim V_{k\ell}$ (see Definition 1). Folland [2] used them to expand the Poisson-Szegö kernel for the ball in $\mathbf{C}^N$.

Each $f \in L^1(D, \mu_\lambda)$ has a formal series

$$\sum_{k, \ell \geq 0} h_{k\ell}^{(\lambda)} \hat{f}_{k\ell} R_{k\ell}^{(\lambda)}, \text{ where } \hat{f}_{k\ell} = \int_D f \bar{R}_{k\ell}^{(\lambda)} d\mu_\lambda. \tag{20}$$

Two summation processes are suggested by the interpretation of the unit sphere in $\mathbf{C}^N$ as the unit sphere in $\mathbf{R}^{2N}$ and the fact that $\sum_{k+\ell=n} V_{k\ell}$ is the space of spherical harmonics of degree n on $\mathbf{R}^{2N}$. The following is proved by using the corresponding result for ultraspherical series of index $\lambda + 1$ and the usual convolution methods; the Cesaro summability was established by Kogbetliantz [9, p. 169].

Theorem 5. For $f \in C(D)$, define the functions

$$P_r f(\xi) : = \sum_{n=0}^{\infty} r^n \sum_{k+\ell=n} h_{k\ell}^{(\lambda)} \, \hat{f}_{k\ell} \, R_{k\ell}^{(\lambda)} \, (\xi), \; 0 \le r < 1,$$

$$S_m^{(\delta)} \, f(\xi) : = \sum_{n=0}^{m} \frac{(-m)_n}{(-\delta - m)_n} \sum_{k+\ell=n} h_{k\ell}^{(\lambda)} \, \hat{f}_{k\ell} \, R_{k\ell}^{(\lambda)} \, (\xi),$$

$m = 1, 2, \ldots, \delta \ge 2\lambda + 3$. *Then $P_r f \to f$ and $S_m^{(\delta)} f \to f$ uniformly on D as $r \to 1-$ and $m \to \infty$, respectively, and if $f \ge 0$, then so are $P_r f$ and $S_m^{(\delta)} f$.*

REFERENCES

1. Boyd, J. Orthogonal polynomials on the disc. M.A. Thesis, Univ. of Virvinia, 1972.

2. Folland, G. B. Spherical harmonic expansion of the Poisson-Szegö kernel for the ball. *Proc. Amer. Math. Soc.* 47 (1975): 401-408.

3. Folland, G. B., and Stein, E. M. Estimates for the $\overline{\partial}_b$ complex and analysis on the Heisenberg group. *Comm. Pure and Appl. Math.* 27 (1974):429-522.

4. Gasper, G. Orthogonality of certain functions with respect to complex valued weights. To appear.

5. Gaveau, B. Principe de moindre action, propagation de chaleur et estimées sous elliptiques sur certains groupes nilpotents. *Acta Math.* 139 (1977):95-153.

6. Greiner, P. C. Spherical harmonics on the Heisenberg group. *Canad. Math. Bull.* 23 (1980):383-396.

7. Greiner, P. C., and Stein, E. M. *Estimates for the $\overline{\partial}$-Neumann Problem. Mathematical Notes.* Princeton, N.J.: Princeton Univ. Press, 1977.

8. Ikeda, M. On spherical functions for the unitary group, I, II, III. *Mem. Fac. Engrg. Hiroshima Univ.* 3 (1967):17-75.

9.   Kogbetliantz, E.   Recherches sur la sommabilité de séries
     ultrasphériques par la methode des moyennes arithmétiques.
     J.de Math . (9)3 (1924):107-187.

10. Koornwinder, T.   Two-variable analogues of the classical
     orthogonal polynomials.   In *Theory and Application of Special
     Functions*, R, Askey, Ed.   New York: Academic Press, 1975,
     pp. 435-495.

SUR LES DIFFEOMORPHISMES DU
CERCLE DE NOMBRE DE ROTATION DE TYPE CONSTANT

M.R. Herman

Centre de Mathématiques de l'Ecole Polytechnique, Palaiseau

1. INTRODUCTION

Pour presque tout nombre α $\mathbf{T}^1$ - (**Q/Z**) (presque tout pour la
mesure de Lebesgue $m = d\theta$), si f est un difféomorphisme du cercle
$\mathbf{T}^1 = \mathbf{R/Z}$ de classe $C^{2+\varepsilon}$, $\varepsilon > 0$, si $\rho(f) = \alpha$, et si f est $C^{2+\varepsilon}$-
proche de la translation $R_\alpha : \mathbf{T}^1 \to \mathbf{T}^1$, $R_\alpha(x) = x + \alpha$, alors f est
C^1-conjugué à R_α. Le difféomorphisme f laisse invariant une mesure
de densité de classe C^0, strictement positive (voir [4, Appendice]).

De plus, si f est de classe C^3, il n'y a pas de restriction de proximité (voir [4, IX]).

J. Hawkins et K. Schmidt [3] ont montré que, si α n'est pas
un nombre de type constant [i.e., $\underset{p/q}{\mathrm{Inf}}(q^2|\alpha - (p/q)|) = 0$, où
$p/q \in \mathbf{Q}$, $(p, q) = 1$] et si α est un nombre irrationnel, alors il
existe un difféomorphisme f de classe C^2 vérifiant $\rho(f) = \alpha$; f est
aussi C^2-proche de R_α qu'on veut; et de plus f ne laisse pas de
mesure σ-finie invariante absolument par rapport à la mesure de
Lebesgue.

Y. Katznelson [7] a montré que, pour tout module de continu-
ité w différent de Lipschitz (Lip$_1$) et tout nombre $\alpha \in \mathbf{T}^1$ - (**Q/Z**)

(vérifiant l'hypothèse suivante: si $\alpha = [a_0, a_1, \ldots]$, où les a_i sont les quotients partiels du développement en fraction continue de α, alors il existe une infinité de $a = a$, avec $a_i > 15$), il existe un difféomorphisme f de classe C^1 du cercle, $Df > 0$, $Df \in \text{Lip}(w)$, f est C^{1+w}-proche de R_α, $\rho(f) = R_\alpha$, et telle que f ne laisse pas invariant de mesure σ-finie absolument continue par rapport à la mesure de Lebesgue. (Pour le théorème de Denjoy en classe C^{1+w}, voir [4, X]).

Nous proposons de démontrer que si α est un nombre de type constant, alors il existe $\varepsilon > 0$, tel que si f est un difféomorphisme de classe C^2 du cercle vérifiant $\|f - R_\alpha\|_{C^2} < \varepsilon$ et $\rho(f) = \alpha$, alors f laisse invariant une mesure strictement positive de densité L^1 équivalente à la mesure de Lebesgue.

Ceci est équivalent à ce que f soit conjugué à R_α par un homéomorphisme du cercle h qui soit un homéomorphisme absolument continu: si $f = h^{-1} \circ R_\alpha \circ h$, alors

$$h(x) = \int_0^x d\mu(t),$$

où μ est l'unique mesure de probabilité de $\mathbf{T}^1$ invariante par f [si $\rho(f) = a \notin \mathbf{Q}/\mathbf{Z}$, μ est unique et sans atome, et on a $h_\star\mu = m$].

La mesure μ est donc la dérivée au sens des distributions de la fonction à variation bornée h.

On peut "raisonnablement" conjecturer qu'en général pour la catégorie de Baire un difféomorphisme de classe C^2 de nombre de rotation α de type constant n'est pas C^1-conjugué à la rotation R_α (cf. [4, XIII 4.6]).

Ce travail repose de façon essentielle sur le lemme d'Yves Meyer [8], et je désire lui exprimer ma très grande reconnaissance d'avoir démontré ce lemme. Je tiens aussi à le remercier de m'avoir signalé le résultat de Peter Jones [6].

2. RAPPEL SUR LES ESPACES DE SOBOLEV

2.1

Soient r un entier, $r \geq 1$, et $p > 1$ $(p \in \mathbf{R})$. On considère
l'espace de Sobolev $W^{r, p} = \{\varphi \in C^{r-1}(\mathbf{T}^1) \mid D\varphi \in L^p\}$, où $D\varphi$ est la
dérivée de φ au sens des distributions et $L^p = L^p(\mathbf{T}^1, d\theta, \mathbf{R})$, et
$d\theta$ est la mesure de Lebesgue (ou de Haar), de $\mathbf{T}^1 = \mathbf{R}/\mathbf{Z}$. On note
aussi la mesure de Lebesgue par m $(m = d\theta)$.

$$\text{Si } \varphi \in L^p, \text{ on note } \|\varphi\|_{L^p} = \left(\int_0^1 |\varphi(\theta)|^p d\theta\right)^{1/p}.$$

2.2

Par Zygmund [11, VI 3.8, p. 242], si $\varphi \in W^{1,p}$,

(a) φ est absolument continue (φ est donc à variation bornée
sur $\mathbf{T}^1$),

(b)   La série de Fourier de

$$\varphi(x) = \sum_{k \in \mathbf{Z}} a_k e^{2\pi i k x}$$

vérifie $\sum |a_k| < +\infty$ (i.e., $\varphi \in$ classe A),

(c) $\varphi \in C^{1/p'}(\mathbf{T}^1)$ avec $(1/p) + (1/p') = 1$, en effet, on a

$$|\varphi(x + h) - \varphi(x)| \leq \int_x^{x+h} |D\varphi(t)| dt \leq \left(\int_0^1 |D\varphi(t)|^p dt\right)^{1/p} h^{1/p'}$$

et donc

$$|\varphi|_{1/p'} = \sup_{x \neq y} \frac{|\varphi(x) - \varphi(y)|}{|x - y|^{1/p'}} \leq \|D\varphi\|_{L^p}.$$

2.3

Par [10, V], on a (i) $\Leftrightarrow$ (ii):

(i) $\varphi \in L^p$ vérifie $\|\varphi(x + t) - \varphi(x)\|_{L^p} \leq t$, $t > 0$,

(ii) $\varphi \in W^{1/p}$ et $\|D\varphi\|_{L^p} \leq 1$.

2.4

On fixe $q > 1$. Soit

$$K_q = \left\{ \varphi \in L^q \mid \int_0^1 \varphi(x)\,dx = 1, \quad \|D\varphi\|_{L^q} \leq 2, \quad \varphi \geq 0 \right\}$$

($\varphi \geq 0$ veut dire Lebesgue presque partout). Pour la topologie faible induite par la dualité L^q, $L^{q'}\left(\frac{1}{q} + \frac{1}{q'} = 1\right)$, K_q est faiblement compact convexe et faiblement métrisable. Noter que

$$\varphi \geq 0 \Leftrightarrow \forall\, \eta \in L^{q'}, \ \eta \geq 0, \ \text{on a} \ \int_0^1 \varphi\eta \, d\theta \geq 0.$$

2.5

A chaque $\varphi \in K_q$ on associe

$$h(x) = \int_0^x \varphi(t)\,dt.$$

L'application continue h de $\mathbf{R}$ dans $\mathbf{R}$ induit une application continue du cercle $\mathbf{T}^1$ dans lui-même, monotone non décroissante, homotope à l'identité et vérifiant $h(0) = 0$. En effet, on a $h(x + 1) = h(x) + 1$ puisque

$$\int_0^1 \varphi(t)\,dt = 1.$$

2.6

Lemme 1. Soit $(\varphi_i)_{i \in \mathbf{N}} \subset K_q$ *une suite convergeant faiblement vers* φ. *Soit* (h_i) *la suite d'applications continues associées à la suite* (φ_i). *Alors, si* $i \to +\infty$, $h_i \to h$ *pour la topologie de la convergence uniforme dans* $C^0(\mathbf{T}^1, \mathbf{T}^1)$.

Démonstration. Puisque la suite (φ_i) est faiblement convergente si $i \to +\infty$, $h_i \to h$ pour la topologie de la convergence simple, i.e.,

$$\forall\, x \in \mathbf{R}, \quad \int_0^x \varphi_i(t)\,dt \to \int_0^x \varphi(t)\,dt.$$

Par 2.2(c), la suite $(h_i)_i$ est équicontinue et la résultat suit.

2.7

Lemme 2. Soient $(\varphi_i)_i \subset K_q$ *une suite faiblement convergente vers* φ *et* $(\psi_i)_i \subset L^\infty$ *une suite convergente dans* L^∞ *vers* ψ *pour la topologie de la norme* L^∞. *La suite* $(\psi_i \cdot \varphi_i) \in L^q$ *converge faiblement dans* L^q *vers* $\varphi\psi$.

Démonstration. Soit $\eta \in L^{q'}$. On doit montrer que, si $i \to +\infty$,

$$\int_0^1 (\psi_i \varphi_i \eta)\,(x)\,dx \to \int_0^1 (\psi\varphi\eta)\,(x)\,dx.$$

On a

$$\left| \int_0^1 \psi_i \varphi_i \eta \, dx - \int_0^1 \psi \varphi_i \eta \, dx \right| \le \|\psi - \psi_i\|_{L^\infty} \|\varphi_i \eta\|_{L^1}$$

$$\le \|\psi - \psi_i\|_{L^\infty} \|\varphi_i\|_{L^q} \|\eta\|_{L^{q'}} \le \|\psi - \psi_i\|_{L^\infty} 2\|\eta\|_{L^{q'}},$$

et donc si $i \to +\infty$, $\|\psi - \psi_i\|_{L^\infty} \|\varphi_i \eta\|_{L^1} \to 0$.

Puisque $\psi\eta \in L^{q'}$ est une fonction fixée, et comme $\varphi_i \to \varphi$ faiblement dans L^q on a, si $i \to +\infty$,

$$\int_0^1 \psi\varphi_i \eta \, dx \to \int_0^1 \psi\varphi\eta \, dx. \quad \blacksquare$$

2.8

Soit la rotation $R_\alpha : \mathbf{T}^1 \to \mathbf{T}^1$, $R_\alpha(\theta) = \theta + \alpha$. L'application $\varphi \in L^q \to \varphi \circ R_\alpha \in L^q$ est une *isométrie* de L^q *faiblement continue* pour la topologie faible induite par la dualité entre L^q et $L^{q'}$.

3. RAPPELS SUR BMO

3.1

On désigne par BMO ou aussi BMO($\mathbf{T}^1$) l'espace des fonctions $\varphi \in L^1(\mathbf{T}^1)$ tel que

$$\sup_J \frac{1}{|J|} \int_J |\varphi(x) - \varphi_J|\,dx = \|\varphi\|_{\text{BMO}} < +\infty,$$

où J est un intervalle de $\mathbf{T}^1$, $|J|$ sa longueur et

$$\varphi_J = \frac{1}{|J|} \int_J \varphi(x)\,dx.$$

Pour la norme $\|\varphi\|_* = \|\varphi\|_{L^1} + \|\varphi\|_{BMO}$, BMO est un espace de Banach. Comme $\|C\|_{BMO} = 0$, si C est une constante, $\|\ \|_{BMO}$ est seulement une seminorme.

Si $\varphi \in$ BMO vérifie $\int_0^1 \varphi(t)\,dt = 0$, on a $\|\varphi\|_{L^1} \leq \|\varphi\|_{BMO}$.

3.2

Le théorème suivant est de John et Nirenberg [5].

Théorème 1. Il existe des constantes $b > 0$, $B > 0$, telles que, si $\varphi \in$ BMO,

$$\kappa = \|\varphi\|_{BMO}$$

verifie $b' \leq \kappa^{-1}b$, alors $e^{b'|\varphi - \varphi_I|} \in L^1$ (où $I = [0, 1]$) et on a

$$\int_0^1 e^{b'|\varphi - \varphi_I|}dx \leq 1 + \frac{Bb'\kappa}{b - \kappa b'}.$$

3.3

Il suit que si $\varphi_I = 0$, $\kappa^{-1}b > b'$, on a

$$\int_0^1 e^{b'\varphi}\,dx \leq 1 + \frac{Bb'\kappa}{b - \kappa b'},$$

$$\int_0^1 e^{-b'\varphi}\,dx \leq 1 + \frac{Bb'\kappa}{b - \kappa b'}.$$

Il en résulte que si $q \geq 1$ est fixé, alors, pour tout $\varepsilon > 0$, il existe $\eta(q) > 0$ tel que, si φ vérifie $\|\varphi\|_{BMO} \leq \eta(q)$, $\int_0^1 \varphi(x)\,dx = 0$, on ait

$$\int_0^1 e^{q\varphi}\,dx \leq 1 + \varepsilon, \quad \int_0^1 e^{-q\varphi}\,dx \leq 1 + \varepsilon.$$

4. UNE LEMME REMARQUABLE DE REGULARITE (D'YVES MEYER)

4.1

Soit α un nombre de type constant, il existe $\gamma > 0$ tel que pour tout $p/q \in \mathbf{Q}$, $(p,q) = 1$, on ait $|\alpha = (p/q)| \geq \gamma q^{-2}$. La constante de Markoff de α est $\underset{p/q}{\mathrm{Inf}}(q^2|\alpha - (p/q)|) = \gamma$. Par le principe de Dirichlet, on a $\gamma < 1$.

Le lemme suivant est dû à Yves Meyer [8].

4.2

Lemme 3. Soit $p > 1$, alors il existe une constante C_p telle que, si α est un nombre de type constant, γ étant la constante de Markoff de α, pour tout $\varphi \in W^{1,p}$ vérifiant $\int_0^1 \varphi(x)\,dx = 0$, il existe un unique $\eta \in \mathrm{BMO},^*$ $\int_0^1 \eta(x)\,dx = 0$, tel que l'on ait

$$\eta - \eta \circ R_\alpha = \varphi,$$

$$\|\eta\|_{\mathrm{BMO}} \leq C_p \gamma^{-1} \|D\varphi\|_{L^p}.$$

Remarques.

(1) La constante C_p ne dépend pas de α ni de γ, mais seulement de p.

(2) Si $p \geq 2$, il est évident que $\eta \in L^2$; ce qui est remarquable, c'est qu'on a beaucoup mieux.

Le lemme d'Yves Meyer résulte de [8] et du lemme suivant:

Lemme 4. Soit α un nombre de type constant, de constante de Markoff γ, et $p > 1$. Il existe une constante C_p^1 dépendant seulement de p, telle que, si $N \geq 1$, on ait

---

*On peut aussi montrer que η appartient à la classe $\exp(\lambda x^2)$: il existe $\lambda > 0$ tel que l'on ait $\int \exp(\lambda|\eta|^2)\,dx < +\infty$. Je remercie Yves Meyer de m'avoir aussi communiqué ce résultat.

$$\left(\sum_{N \leq |n| \leq 2N} \frac{1}{|n(1 - e^{2\pi i n\alpha})|^p} \right)^{1/p} \leq C_p^1 \gamma^{-1}.$$

Démonstration. Par la même démonstration [4, IX 6.7.1], on a

$$\sum_{0 < |n| \leq N} \frac{1}{|1 - e^{2\pi i n\alpha}|^p} \leq C_p'' \gamma^{-p} N^p,$$

ou C_p'' est une constante ne dépendant que de $p > 1$ (cette inégalité a été remarquée semble-t-il pour la première par Rüssmann [9] dans le cas $p = 2$). D'où:

$$\sum_{N \leq |n| \leq 2N} \frac{1}{|n(1 - e^{2\pi i n\alpha})|^p} \leq \frac{1}{N^p} \sum_{N \leq |n| \leq 2N} \frac{1}{|1 - e^{2\pi i n\alpha}|^p}$$

$$\leq \frac{1}{N^p} \sum_{0 < |n| \leq 2N} \frac{1}{|1 - e^{2\pi i n\alpha}|^p} \leq C_p'' \gamma^{-p} \frac{(2N)^p}{N^p} = 2^p C_p'' \gamma^{-p}.$$

5. DIFFEOMORPHISMES DE CLASSE SOBOLEV

5.1

Soient r un entier $r \geq 1$ et $D^r(\mathbf{T}^1) = \{ f \in \text{Diff}^r + (\mathbf{R}) \mid f - \text{Id} \in C^r(\mathbf{T}^1) \}$. $D^r(\mathbf{T}^1)$ est le groupe revêtement universel du groupe des difféomorphismes de classe C^r du cercle préservant l'orientation. Pour la C^r-topologie, c'est un groupe topologique.

5.2

Soient $r \geq 2$ et $p \geq 1$, on définit

$$D^{r, p}(\mathbf{T}^1) = \{ f \in D^{r-1}(\mathbf{T}^1) \mid f - \text{Id} \in W^{r, p} \}.$$

Comme $D^{r, p}(\mathbf{T}^1) \subset D^{r-1}(\mathbf{T}^1) \subset D^1(\mathbf{T}^1)$ et en utilisant la formule de Leibnitz et de Faa-di Bruno, on vérifie que $D^{r, p}(\mathbf{T}^1)$ est un groupe (c'est même un groupe topologique si on munit $W^{r, p}$ de sa topologie naturelle d'espace de Banach). On a $D^r(\mathbf{T}^1) \subset D^{r, p}(\mathbf{T}^1)$.

5.3

A chaque $f \in D^{r, p}(\mathbf{T}^1)$ est associé son nombre de rotation $\alpha = \rho(f) \in \mathbf{R}$ et l'application $f \to \rho(f)$ est continue (voir [4, II]).

5.4

Si $f \in D^{2, p}(\mathbf{T}^1)$ vérifie $\rho(f) = \alpha \notin \mathbf{Q}$, alors par le théorème de Denjoy, qui s'applique puisque $f \in D^1(\mathbf{T}^1)$ et que la fonction Df est à variation bornée (et même absolument continue par 2.2), alors f est C^0-conjugué par $h \in D^0(\mathbf{T}^1) = \{f \in \text{Homéo}_+(\mathbf{R}), \ f - \text{Id} \in C^0(\mathbf{T}^1)\}$ à la translation $R_\alpha : x \to x + \alpha$. Soit $f = h \circ R_\alpha \circ h^{-1}$. De plus, h est unique si on impose $h(0) = 0$ (si on a $h_1 \circ R_\alpha \circ h_1^{-1} = h \circ R_\alpha \circ h^{-1}$, alors $h_1^{-1} = R_\lambda \circ h^{-1}$, $\lambda \in \mathbf{R}$).

5.5

Si on n'impose pas de restriction sur $\alpha \notin \mathbf{Q}/\mathbf{Z}$, en général (pour la catégorie de Baire), même si $f \in D^\infty(\mathbf{T}^1)$, h est un homéomorphisme singulier pour la mesure de Lebesgue m: il existe $B \subset \mathbf{R}$ un borélien $m(B) = 0$ tel que $m(\mathbf{R} - h(B)) = 0$ (voir [4, XII]).

5.6

Soit $h \in D^0(\mathbf{T}^1)$ un homéomorphisme.

Définition. *L'homeomorphisme h est appelé homéomorphisme absolument continu si les fonctions $h : [0, 1] \to \mathbf{R}$ et $h^{-1} : [0, 1] \to \mathbf{R}$ sont des fonctions absolument continues.*

Puisque h et h^{-1} sont à variations bornées sur $[0, 1]$, il suffit de voir que si B est un borélien de mesure de Lebesgue nulle [i.e., $m(B) = 0$], alors $m(h(B)) = m(h^{-1}(B)) = 0$.

En utilisant ceci, on vérifie que les homéomorphismes absolument continus forment un groupe.

On a d'autre part que, si h est un homéomorphisme tel que $h|_{[0, 1]}$ soit absolument continue, en général h^{-1} n'est pas absolument continue.

Exemple. Soient $K \subset [0, 1]$ un ensemble de Cantor de mesure de Lebesgue $m(K) > 0$ et φ une fonction de classe C^∞ sur $[0, 1]$ vérifiant:

$$\begin{cases} \varphi(x) > 0 & \text{si } x \notin K \\ \varphi(x) = 0 & \text{si } x \in K \end{cases}$$

$$\int_0^1 \varphi(x)\, dx = 1.$$

On pose

$$h(x) = \int_0^x \varphi(t)\, dt.$$

L'homéomorphisme $h : [0, 1] \to [0, 1]$ est une application de classe C^∞ mais h^{-1} n'est pas absolument continue.

5.7

Remarques.

(1) Soit $h \in D^0(\mathbf{T}^1)$ un homéomorphisme tel que l'application $h : [0, 1] \to \mathbf{R}$ soit absolument continue. Si $Dh(t) > 0$ pour presque tout $t \in [0, 1]$, alors $h^{-1} : [0, 1] \to \mathbf{R}$ est une fonction absolument continue. Il suit que h est un homéomorphisme absolument continu. L'hypothèse est vérifiée, par exemple, si $\log Df \in L^1$. (En effet, si h^{-1} n'était pas absolument continue, alors il existerait un ensemble de Cantor $K \subset [0, 1]$, $m(K) > 0$ tel que l'on ait $m(h(K)) = 0$ et donc $Dh(x) = 0$ pour presque tout $x \in K$.)

(2) Si $h \in D^0(\mathbf{T}^1)$ est un homéomorphisme, tel que l'application $h : [0, 1] \to \mathbf{R}$ soit singulière par rapport à m, il en est de même pour h^{-1} (une application à variation bornée est dite singulière si sa dérivée au sens de distributions

est une mesure singulière par rapport à la mesure de
Lebesgue).

(3) Si f est un difféomorphisme de classe C^1 vérifiant
$\rho(f) = \alpha \notin \mathbf{Q}/\mathbf{Z}$, alors f laisse invariant une unique mesure
de probabilité μ. On décompose la mesure $\mu = \mu_{ab} + \mu_{sing}$
(la décomposition étant unique). Puisque f est de classe
C^1, il suit que les mesures (≥ 0) μ_{ab} et μ_{sing} sont invari-
antes par f. De l'unicité de la mesure μ on a $\mu_{sing} = 0$ ou
bien $\mu_{ab} = 0$.

Il résulte de $h \circ f = R_\alpha \circ h$, où $h(x) = \int_0^x d\mu(t)$, que si h
n'est pas absolument continue alors h est singulière pour
la mesure de Lebesgue.

Si h est absolument continue, en général, la mesure μ n'est
pas équivalente à la mesure de Lebesgue (par exemple, un
contre-exemple de Denjoy, de classe C^1, construit en [4, X-3]
laissant invariant un ensemble de Cantor de mesure de Lebesgue
strictement positive obtenu en perturbant une rotation R_α).
Si f est un difféomorphisme de classe C^2, $\rho(f) = \alpha \notin \mathbf{Q}/\mathbf{Z}$,
alors par le théorème de Denjoy, h est un homéomorphisme
(voir [4, VI]). Par la m-ergodicité de f, si la fonction
$h|_{[0,1]}$ est absolument continue, alors la mesure μ est équivalente à la mesure de Lebesgue (cf. [4, VII.1, XII.1]). Il en
résulte que h est un homéomorphisme absolument continu.

6. THEOREME PRINCIPAL

On se propose de démontrer le théorème suivant:

6.1

Théorème 2.   Soient $p > 1$ fixé, et α un nombre de type constant de Markoff γ.   Il existe $\varepsilon_0 > 0$ (ne dépendant que de p et pas de α), tel que si $f \in D^{2, \, p}(\mathbf{T}^1)$ vérifie

  (a) $\left\| D^2 (f - R_\alpha) \right\|_{L^p} \leq \varepsilon_0 \gamma,$

  (b) $\rho(f) = \alpha,$

alors f est conjugué par un homéomorphisme $h \in D^0(\mathbf{T}^1)$ à R_α avec $h - \mathrm{Id} \in W^{1, \, q}$ et $h^{-1} - \mathrm{Id} \in W^{1, \, q}$, $q > 1$.

6.2

Remarques.

  (1)   Il suit que h est un homéomorphisme absolument continu.

  (2)   En fait, on a mieux; h appartient au groupe définit par Peter Jones [6]: le groupe des homéomorphismes f (absolument continus) de $\mathbf{T}^1$ qui préservent BMO (i.e., si $\varphi \in$ BMO, alors $\varphi \circ \mathbf{f} \in$ BMO).   Ceci équivaut à $Df \in \bigcup_{p > 1} A_p$.

  (3)   En utilisant 4.2 et [4] (en particulier le théorème fondamental de [4]), on peut montrer que si $f \in D^{r, \, p}(\mathbf{T}^1)$ ($r \in \mathbf{N}$, $r \geq 3$, et $p > 1$) vérifie $\rho(f) = \alpha$ est un nombre de type constant, alors $f = h \circ R_\alpha \circ h^{-1}$ où $h \in D^{r-1, \, p}(\mathbf{T}^1)$.

  (4)   Par une démonstration analogue à celle de 6.1 en utilisant 4.2 (et si $p = 2$, la remarque 2 de 4 suffit), on peut démontrer très simplement une version locale de la remarque précédente.

6.3

Démonstration.

① Soit $1 < p_1 < p$. On pose $p/p_1 = r > 1$, $r' = r/(r - 1)$, et
$q = \left(p - (1/r)\right)r'$. On a $p > r$ et comme $q = (p-1)/(r-1)$,
$q > 1$. Soit K_q l'ensmeble défini en 2.4.

② Par 2.5, à $\varphi \in K_q$ on associe $h(x) = \int_0^x \varphi(t)\,dt$. Nous allons
construire une application $\Phi : K_q \to K_q$.

③ On se fixe f vérifiant $\|D^2f\|_{L^p} < \varepsilon\gamma$, $\varepsilon > 0$ petit. Il en
résulte que $\|Df - 1\|_{C^0} < 1/2$, si $\varepsilon < 1/2$ est assez petit
et comme $\gamma < 1$ ceci ne dépend pas de γ. On a donc $Df > 0$.
Soit
$$\lambda(f, h) = -\int_0^1 \log Df \circ h(t)\,dt; \ \lambda(f, h) \in \mathbf{R}.$$

L'application $(h, f) \to \lambda(f, h)$ est continue pour la C^0-topologie.
La fonction $\log Df \circ h + \lambda(f, h)$ est d'intégrale nulle et
$\log Df \circ h + \lambda(f, h) \in W^{1, p_1}$, pour cela on utilise les lemmes
suivants:

Lemme 5. *Si $f \in D^{2, p}(\mathbf{T}^1)$, alors $\log Df \in W^{1, p}$.*

Démonstration. Par [1], on a
$$\log Df(x) - \log Df(0) = \int_0^x \frac{D^2f}{Df}(t)\,dt.$$

(En effet, $\log Df$ est à variation bornée puisque la fonction log
est Lipschitzienne sur [min Df, max Df] et de plus, si A est un
ensemble borélien de mesure de Lebesgue nulle, alors l'ensemble
$\log Df(A)$ est mesure de Lebesgue nulle. Il suit que $\log Df$ est
absolument continue.) Comme $1/Df \in L^\infty$ et que $D^2f \in L^p$, on a
$D \log Df \in W^{1, p}$.

Lemme 6. Soit $h(x) = \int_0^x \varphi(t)\,dt$, avec $\varphi \in L^1$, $\varphi \geq 0$, $\int_0^1 \varphi(t)\,dt = 1$.
Soit $\eta \in L^1(\mathbf{T}^1)$, alors $Dh\eta \circ h \in L^1$, et on a

$$\int_0^1 \eta(t)\,dt = \int_0^1 \eta(h(t))Dh(t)\,dt.$$

Démonstration. [1]. ∎

④ On a bien $\log Df \circ h + \lambda(f, h) \in W^{1, p_1}$ puisque

$$\int_0^1 |(D \log Df) \circ hDh|^{p_1}\,dx = \int |((D \log Df) \circ h)^{p_1} Dh^{\frac{1}{r}}|Dh^{p_1 - \frac{1}{r}}\,dx$$

$$\underset{\text{inégalité de Hölder}}{\leq} \left(\int |((D \log Df) \circ h)^{p_1 r} Dh|\,dx\right)^{\frac{1}{r}} \left(\int Dh^{(p_1 - \frac{1}{r})r'}\,dx\right)^{\frac{1}{r'}}.$$

Or,

$$\int |(D \log Df) \circ h|^{p_1 r} Dh\,dx = \int |D \log Df|^{p_1 r}\,dx = \int |D \log Df|^p\,dx.$$

On obtient finalement

$$\|(D \log Df) \circ hDh\|_{L^{p_1}} \leq \|D \log Df\|_{L^p} \left(\int Dh^{(p_1 - \frac{1}{r})r'}\,dx\right)^{\frac{1}{r'p_1}},$$

$$\boxed{\|(D \log Df \circ h)Dh\|_{L^{p_1}} \leq \|D \log Df\|_{L^p} \|Dh\|_{L^q}^{q/(r'p_1)}.}$$

⑤ Comme $\|Dh\|_{L^q} \leq 2$ et qu'on suppose p et p_1 fixés, par 4.2 il existe $\varphi_1 \in$ BMO vérifiant $\int_0^1 \varphi_1(x)\,dx = 0$ et

$$\log Df \circ h + \lambda(f, h) = \varphi_1 \circ R_\alpha - \varphi_1 \qquad (+)$$

et on a

$$\boxed{\|\varphi_1\|_{\text{BMO}} \leq C_1\gamma^{-1}\|D \log Df\|_{L^p}}$$

où C_1 est une constante ne dépendant que de p et p_1. Si on remplace φ_1 par $\varphi_1 + c$ ($c \in \mathbf{R}$), alors $\varphi_1 + c$ vérifie encore (+).

⑥ Soit $0 < \eta < 1/2$, par 3.3 et ⑤, il existe $\varepsilon > 0$ tel que si $\|D \log Df\|_{L^p} \leq \varepsilon\gamma$, alors $e^{\varphi_1} \in L^q$, $e^{-\varphi_1} \in L^q$,

$$\int_0^1 e^{q\varphi_1}\,dx \leq 1 + \eta,$$

$$\int_0^1 e^{-\varphi_1}\,dx \leq 1 + \eta.$$

Par l'inégalité de Hölder, il en résulte

$$\int_0^1 e^{\varphi_1}\, dx \geq \left(\int_0^1 e^{-\varphi_1}\, dx\right)^{-1} \geq (1 + \eta)^{-1};$$

soit

$$\left(\int_0^1 e^{\varphi_1}\, dx\right)^{-1} \leq 1 + \eta.$$

(7) On pose $\eta_0 = \mathrm{Inf}\left(\dfrac{1}{2},\ \eta_1\right)$ où $\eta_1 > 0$ vérifie $(1 + \eta_1)^{\frac{q+1}{q}} = 2$. On détermine $0 < \varepsilon_0 < \dfrac{1}{2}$ comme en (6) et associée à η_0. La constante ε_0 ne dépend que de p, p_1 et du théorème de John et Nirenberg.

(8) Si f vérifie $\|D \log Df\|_{L^p} \leq \varepsilon_0 \gamma$ et si $\varphi \in K_q$, on pose

$$\Phi(\varphi) = \left(\int_0^1 e^{\varphi_1}\, dx\right)^{-1} e^{\varphi_1}.$$

On a $\Phi(\varphi) \in K_q$ où encore $\Phi : K_q \to K_q$ puisque $\Phi(\varphi) \geq 0$,

$$\int_0^1 \Phi(\varphi)\, dx = 1$$

et

$$\|\Phi(\varphi)\|_{L^q} \leq (1 + \eta)^{\frac{q+1}{q}} \leq 2.$$

(9) Nous avons ainsi défini une application $\Phi : K_q \to K_q$ telle que, $\Phi(\varphi) = \psi$ vérifie:

$$\boxed{\psi \circ R_\alpha = e^{\lambda(f,\ h)} Df \circ h\psi}.$$

(*)

(10) Nous allons maintenant montrer que *l'application Φ est continue pour la topologie faible sur K_q*.

En effet, si la suite (φ_i) tend faiblement vers $\varphi \in K_q$, alors il faut montrer que la suite $(\Phi(\varphi_i))_i$ tend faiblement vers $\Phi(\varphi)$. Puisque $\Phi(\varphi_i) \in K_q$ et que K_q est faiblement compact métrisable, il suffit démontrer que toute valeur d'adhérence $\tilde{\psi}$ de la suite $(\Phi(\varphi_i))_i$ est égale à $\Phi(\varphi)$, ou ce qui revient au même si $\varphi_i \to \varphi$ et $\psi_i = \Phi(\varphi_i) \to \tilde{\psi}$, alors on a $\Phi(\varphi) = \tilde{\psi}$. Or par (*), on a

$$\psi_i \circ R_\alpha = e^{\lambda(f,\ h_i)} Df \circ h_i \psi_i.$$

Par 2.6-2.8, si $i \to +\infty$, on a

$$\tilde{\psi} \circ R_\alpha = e^{\lambda(f, h)} Df \circ h\tilde{\psi}.$$

Or $\tilde{\psi} \geq 0$, $\int_0^1 \tilde{\psi}\, dx = 1$, et donc $\tilde{\psi} \neq 0$. Par l'ergodicité de la rotation R_α on obtient $\tilde{\psi} = \Phi(\varphi)$.

(11)   Par le théorème de Schauder-Tychonoff [2, p. 456] l'application Φ a un point fixe φ. En fait, par la démonstration de (10), φ est > 0 et de plus φ est unique. Il en résulte: $\varphi = Dh$ avec

$$h(x) = \int_0^x Dh(t)\, dt;$$

h est un homéomorphisme puisque $Dh > 0$ (presque partout).

On a la relation

$$Dh \circ R_\alpha = e^{\lambda(f, h)} Df \circ hDh.$$

(12)   Puisque $\int_0^1 Dh \circ R_\alpha\, dx = 1$ et que

$$\int_0^1 e^{\lambda(f, h)} Df \circ hDh\, dx = e^{\lambda(f, h)},$$

on a $e^{\lambda(f, h)} = 1$, et donc

$$Dh \circ R_\alpha = D(f \circ h)$$

d'où $h \circ R_\alpha = f \circ h + a$, avec $a \in \mathbf{R}$.

Puisque $\rho(f) = \alpha$ est irrationnel par [4, III.4], on a $a = 0$. On a finalement:

$$f = h \circ R_\alpha \circ h^{-1}.$$

(13)   Puisque $Dh > 0$ (presque partout), h est un homéomorphisme absolument continu (voir 5.7). On a de plus

$$Dh \in L^q \quad \text{et} \quad Dh^{-1} \in L^q,$$

en effet

$$\int_0^1 (Dh^{-1})^q\, dx = \int_0^1 \left(\frac{1}{Dh} \circ h^{-1}\right)^q dx = \int_0^1 \left(\frac{1}{Dh}\right)^{q-1} dx$$

et il suffit d'appliquer (6) pour conclure.

Le théorème est ainsi démontré. ∎

7. UNE CONSEQUENCE

Soient $\alpha \in \mathbf{R}$, et $p > 1$, on note

$$F_\alpha^{2,p} = \{f \in D^{2,p}(\mathbf{T}^1) \mid \rho(f) = \alpha\}$$

et

$$F_\alpha^\infty = \{f \in D^\infty(\mathbf{T}^1) \mid \rho(f) = \alpha\}.$$

Par [4, III], $F_\alpha^{2,p}$ est un sous-ensemble fermé de $D^{2,p}(\mathbf{T}^1)$ pour la C^0-topologie. Par le même raisonnement qu'en [4, III.4], si $\alpha \notin \mathbf{Q}$, F_α^∞ est dense dans $F_\alpha^{2,p}$ pour la $W^{2,p}$-topologie.

Proposition. Si α est un nombre de type constant, alors l'ensemble $V = \{f \in F_\alpha^{2,p} \mid f$ est conjugué par un homéomorphisme $h \in D^0(\mathbf{T}^1)$ qui est un homéomorphisme absolument continu$\}$, contient un ouvert dense dans $F_\alpha^{2,p}$.

Démonstration. Puisque α est de type constant, par le théorème fondamental de [4], tout $f \in F_\alpha^\infty$ est C^∞-conjugué à R_α et la proposition suit du théorème 6 par conjugaison C^∞ et de la densité de F_α^∞ dans $F_\alpha^{2,p}$.

REFERENCES

1. Bourbaki, N. Integration. Chap. 5, Sec. 6, No. 5.

2. Dunford, N., and Schwartz, J. T. *Linear Operators*, part I. New York: John Wiley & Sons, 1957.

3. Hawkins, J., and Schmidt, K. On C^2-diffeomorphisms of the circle which are of type III$_1$. *Invent. Math.* 66 (1982):511-518.

4. Herman, M. Sur la conjugaison différentiable des difféomorphismes du cercle à des rotations. *Pub. I.H.E.S.* 49 (1979):5-233.

5. John, F., and Nirenberg, L. On functions of bounded mean oscillation. *Comm. Pure Appl. Math.* 14 (1961):415-426.

6. Jones, P. W. Homeomorphisms of the line which preserve BMO.
    A paraître. *Comm. Math. Helv.*

7. Katznelson, Y. Smooth mappings of the circle without sigma-
    finite invariant measures. *Symposia Math.* 22:363-369.

8. Meyer, Y. Sur un problème de Michael Herman. Preprint, Centre
    de Mathématiques de l'Ecole Polytechnique, 1980.

9. Russmann, H. On optimal estimates for the solutions of linear
    differential equations of the circle. *Celestial Mechanics* 14
    (1976):33-37.

10. Stein, E. M. *Singular Integrals and Differentiability Proper-
    ties of Functions.* Princeton, N.J.: Princeton Univ. Press,
    1970.

11. Zygmund, A. *Trigonometric Series.* 2d ed. New York: Cambridge
    Univ. Press, 1968.

SUR UN PROBLÈME DE MICHAEL HERMAN

Yves Meyer

Ecole Polytechnique, Plateau de Palaiseau

Soient $T = R/Z$ et $\alpha \notin Q/Z$ un nombre irrationnel.

Nous nous proposons d'étudier la régularité des solutions $f : T \rightarrow C$ de l'équation

$$f(x + \alpha) - f(x) = g(x) \tag{1}$$

en fonction de celle du second membre $g(x)$.

Désignons, pour $p \in (1, +\infty)$, par $L_1^p = L_1^p(T)$ l'espace de Banach des fonctions g appartenant à $L^p(T)$ et dont la dérivée g' (prise au sens des distributions) appartient aussi à L^p.

Théorème 1. Les quatre propriétés suivantes sont équivalentes:

    (i)   *il existe $\delta > 0$ tel que pour tout $p \in Z$ et tout $q \geq 1$ on ait*

$$|\alpha - p/q| \geq \delta/q^2 ; \tag{2}$$

    (ii)   *il existe $C > 0$ tel que, pour toute fonction $g \in L_1^2$ d'intégrale nulle, (1) ait une solution $f \in L^2$ telle que $\|f\|_2 \leq C \|g'\|_2$;*

Laboratoire associé au C. N. R. S. n° 169

(iii) *il existe $C > 0$ tel que, pour toute fonction $g \in L_1^2$*

d'intégrale nulle, toute solution f de (1) vérifie

$$\| f \|_{BMO} \leq C \| g' \|_2 ;$$

(iv)   *pour tout $p > 1$, il existe une constante C_p telle que*
   $\| f \|_{BMO} \leq C_p \| g' \|_p$ *pour toute solution f de (1) telle*
   que $g \in L_1^p$.

Naturellement, (iv) $\Rightarrow$ (iii) $\Rightarrow$ (ii). Vérifions que (ii) $\Rightarrow$ (i)
ce qui est facile puis démontrons (i) $\Rightarrow$ (iv). Pour vérifier que
(ii) $\Rightarrow$ (i), on teste (ii) sur $f(x) = e^{2\pi i q x}$ et l'on obtient
$1 \leq C 2\pi q |e^{2\pi i q \alpha} - 1|$ ce qui est équivalent à (i). Montrons que
(i) $\Rightarrow$ (iv). Nous allons nous passer du théorème de Fefferman-Stein
sur la dualité entre H^1 et BMO et de la caractérisation due à E.
Stein de H^1 par la fonction de Littlewood-Paley. La preuve qui
suit est plus "élémentaire." Elle repose sur le Theoreme 2 suivant
(d'intérêt indépendant) dont voici les notations. Pour $f \in L^2(\mathbf{T})$
et tout $k \geq 0$, on pose $t_k(x) = \sum_{2^k \leq |j| < 2 \cdot 2^k} c_j e_j(x)$ où $c_j = \hat{f}(j)$
et $e_j(x) = e^{2\pi i j x}$. On a donc $f(x) = a + \sum_{k \geq 0} f_k(x)$.

Théorème 2.   [2]   *Avec les notations précédentes, il existe une*
constante $C > 0$ telle que, pour toute série de Fourier, on ait

$$\| f \|_{BMO} \leq C \| (\sum_{k \geq 0} |f_k(x)|^2)^{1/2} \|_\infty . \tag{3}$$

Naturellement, on peut alors se demander si la norme L^∞ du second
membre de (3) peut être remplacée par la norme BMO. Il n'en est
rien à cause du lemme suivant.

Lemme 1.   *Pour toute fonction continue $\varphi : \mathbf{T} \to \mathbf{C}$, on a*

$$\| \varphi \|_\infty = \lim_{N \to +\infty} \| e_N \varphi \|_{BMO} \quad (\text{où } e_N(x) = e^{2\pi i N x}).$$

C'est immédiat et laissé au lecteur.

Si φ est une somme trigonométrique finie, on teste (3) sur
$f(x) = e_N(x) \varphi(x)$, $N = 2^k + k$. Alors, avec les notations du

Théorème 2, $f = f_k$ et il en résulte que dans le second membre de (3), la norme L^∞ ne peut être remplacée par la norme BMO.

La preuve du Théorème 2 découle presque immédiatement du lemme suivant.

Lemme 2. Il existe un constante $C > 0$ telle que pour tout entier $m \geq 0$ et tout intervalle $I \subset \mathbf{T}$ de longueur $\geq 2^{-m}$, on ait, pour toute série (2),

$$\left(\frac{1}{|I|} \int_I \Big| \sum_{k \geq m} f_k \Big|^2 \, dx \right)^{1/2} \leq C \, \Big\| \big(\sum_{k \geq m} |f_k(x)|^2 \big)^{1/2} \Big\|_\infty . \qquad (4)$$

Pour démontrer le Lemme 2, on observe que si (4) est satisfaite pour un couple (m, I) et une constante C, il en est, a fortiori, de même si m est remplacé par $m_1 \geq m$. Cette remarque permet de supposer $2^{-m} \leq |I| < 2.2^{-m}$. De plus on peut évidemment se ramener au cas où

$$I = [-2^{-m}, \, 2^{-m}] .$$

On définit le noyau de Fejer $K_N(x) = \frac{1}{N} \big| 1 + \cdots + e_{N-1}(x) \big|^2 = \frac{1}{N} \frac{\sin^2 \pi N x}{\sin^2 \pi x}$. On a, si $N = 2^{m-1}$ et $x \in I$, $K_N(x) \geq \frac{4}{\pi^2} N$. En d'autres termes, en appelant Π_I la fonction indicatrice de I, on a $\frac{1}{|I|} \Pi_I \leq C K_N$ lorsque $N = 2^{m-1}$ et $2^{-m} \leq |I| < 2.2^{-m}$, I étant centré en 0. On pose alors $A_N(x) = N^{-1/2}(1 + \cdots + e_{N-1}(x))$ et l'on a

$$\frac{1}{|I|} \int_I \Big| \sum_{k \geq m} f_k \Big|^2 \, dx \leq C^2 \Big\| \sum_{k \geq m} A_N f_k \Big\|_2^2 .$$

Or les fréquences du produit $A_N f_k$ appartiennent à $E_k = \{ 2^k - 2^{m-1} \leq |j| < 2.2^k + 2^{m-1} \}$ et ces ensembles E_k sont deux à deux disjoints lorsque $k \in 3\mathbf{N}$ (ou $k \in 3\mathbf{N} + 1$ ou $k \in 3\mathbf{N} + 2$). Il en résulte que

$$\left\| \sum_{3\ell \geq m} A_N f_{3\ell} \right\|_2^2 = \sum_{3\ell \geq m} \int_{\mathbb{T}} K_N(x) \left| f_{3\ell}(x) \right|^2 dx$$

$$\leq \sup_{x \in \mathbb{T}} \sum_{3\ell \geq m} \left| f_{3\ell}(x) \right|^2 \int_{\mathbb{T}} K_N(x) \, dx$$

$$= \left\| \sum_{3\ell \geq m} \left| f_{3\ell}(x) \right|^2 \right\|_\infty .$$

Puisque les inégalités analogues sont vraies si $k = 3\ell + 1$ (ou $k = 3\ell + 2$), le Lemme 2 est démontré. La preuve du Théorème 2 est alors calquée sur la démonstration directe du fait que $\sum_0^\infty a_n \cos 2^n x \in \text{BMO}$ lorsque $a_n \in \ell^2$. On suppose que $2^{-m} < |I| <$ 2.2^{-m} et l'on écrit

$$f(x) = \sum_{k \geq 0} f_k(x) = S_m(x) + R_m(x) \quad \text{où } R_m(x) = \sum_{k \geq m+1} f_k(x) .$$

Soit x_0 le centre de I. Montrons que

$$\frac{1}{|I|} \int_I |f(x) - S_m(x_0)|^2 \, dx \leq C \left\| \left(\sum_0^\infty |f_k(x)|^2 \right)^{1/2} \right\|_\infty^2 , \tag{5}$$

où C est une constante absolue. On a, d'une part, $|S_m(x) - S_m(x_0)|$ $\leq |x - x_0| \|S_m'\|_\infty$. Or $\|S_m'\|_\infty \leq \|f_0'\|_\infty + \cdots + \|f_m'\|_\infty \leq 2\|f_m\|_\infty + \cdots +$ $2^{m+1} \|f_m\|_\infty$ (grâce à la célèbre inégalité de S. Bernstein) et cette somme ne dépasse pas

$$2^{m+2} \left\| \left(\sum_0^\infty |f_k(x)|^2 \right)^{1/2} \right\|_\infty .$$

Finalement, $|S_m(x) - S_m(x_0)| \leq 4 \left\| \left(\sum_0^\infty |f_k(x)|^2 \right)^{1/2} \right\|_\infty$. En ce qui concerne $\frac{1}{|I|} \int_I |R_m(x)|^2 \, dx$, le Lemme 2 donne l'estimation cherchée ce qui achève de démontrer le Théorème 2.

Revenons maintenant au Théorème 1. On utilise le lemme suivant.

Lemme 3.   Soit $(\mu_j)_{j \in \mathbf{Z}}$ une suite bornée de nombres complexes. Supposons qu'il existe une constante $C > 0$ et $p \in (1, 2)$ tels que, pour tout $k \geq 0$, on ait $\sum_{2^k \leq |j| < 2.2^k} |\mu_j|^p \leq c^p$.

Alors l'opérateur M défini par $M(e_j) = \mu_j e_j$ se prolonge en un opérateur linéaire continu de $L^p(\mathbb{T})$ dans BMO($\mathbb{T}$).

En effet, si $f \in L^p(\mathbb{T})$, on a $\sum_{k \geq 0} \|f_k\|_p^2 \leq c_p^2 \|f\|_p^2$ ([1]). Ecrivons $f_k(x) = \sum_{2^k \leq |j| < 2.2^k} c_j e_j(x)$. Il vient, en appelant q l'exposant conjugué de p,

$$\left(\sum_{2^k \leq |j| < 2.2^k} |c_j|^q\right)^{1/q} \leq \|f_k\|_p \quad \text{et donc}$$

$$\sum_{2^k \leq |j| < 2.2^k} |c_j \| \mu_j| \leq c \|f_k\|_p .$$

On pose $g_k(x) = \sum_{2^k \leq |j| < 2.2^k} c_j \mu_j e_j(x)$ et l'on a

$$\|g_k\|_\infty \leq \sum_{2^k \leq |j| < 2.2^k} |c_j \| \mu_j| \leq c \|f_k\|_p$$

qui entraîne

$$\sum_0^\infty \|g_k\|_\infty^2 \leq c_p'^2 \|f\|_p^2$$

et, a fortiori, $\quad \left\|\left(\sum_0^\infty |g_k(x)|^2\right)^{1/2}\right\|_\infty \leq c_p' \|f\|_p .$

Dans le cas qui nous intéresse $\mu_j = \dfrac{1}{j(e^{2\pi i j \alpha} - 1)}$. Lorsque $2^k \leq j < 2.2^k$ les distances mutuelles entre les points $e^{2\pi i j \alpha}$ dépassent $\delta 2^{-k} (\delta > 0)$ et il en résulte de façon très simple que $\sum_{2^k \leq j < 2.2^k} |\mu_j|^p \leq c_p$ pour tout $p > 1$. Il en est de même pour $\mu_{-j} = -\bar\mu_j$.

RÉFÉRENCES

1. Zygmund, A. *Trigonometric Series*. 2d ed. New York: Cambridge Univ. Press, 1968.

2. Remarquons qu'avec les notations du Théorème 2, la condition

$(*)$ $\frac{1}{|I|} \int_I (|f_k(x)|^2 + |f_{k+1}(x)|^2 + \cdots) dx \leq C$, satisfaite

pour tous les intervalles I de longueur 2^{-k}, est suffisante pour que $f \in$ BMO. Cette condition est nécessaire si les f_k sont définies de façon "raisonnable"; par exemple si

$\hat{f}_k(j) = \hat{f}(j) \Psi\left(\dfrac{j}{2^k}\right)$ où Ψ est continûment dérivable et est nulle

hors de [1/3, 3]. La condition $(*)$ est la version dyadique de la caractérisation de BMO par les mesures de Carleson.

MINIMAL AND MAXIMAL METHODS
OF INTERPOLATION OF BANACH SPACES

Svante Janson
Uppsala University

Most theorems on interpolation of linear operators can be stated in the following form, where X_0, Y_0, X_1, Y_1, X, Y are some normed (or quasinormed) linear spaces such that $X \subset X_0 + X_1$ and $Y \subset Y_0 + Y_1$. (All inclusions in this paper are assumed to be continuous.)

If T is a linear operator that maps X_0 into Y_0 and X_1 into Y_1 boundedly, then T maps X into Y boundedly.

For simplicity, we will only consider Banach spaces in this paper. $X = Y$ means only that the spaces are isomorphic, that is, the norms are equivalent. Also, we will only consider linear operators, although some results, for example, the Marcinkiewicz interpolation theorem [9], extend to suitable classes of nonlinear operators. (See also [7].) The first results of this type were the well-known theorems by Riesz (L^p spaces), Marcinkiewicz (L^p and weak L^p) and Salem-Zygmund (H^p and L^p); see [9, Chapter 12] for details and references.

About 1960 the interpolation theory was transformed by the works of Calderón, Gagliardo, Krein, Lions, Peetre, and others. The concept of interpolation methods was introduced, turning theorems of the above type into a definition.

Definitions. A Banach couple (X_0, X_1) is a pair of Banach spaces such that X_0 and X_1 both are continuously included in some Haus- dorff vector space (thus $X_0 + X_1$ is defined). F is an interpola- tion method if, for any Banach couple (X_0, X_1), $F(X_0, X_1)$ is a Banach space such that $X_0 \cap X_1 \subset F(X_0, X_1) \subset X_0 + X_1$, and for any two Banach couples (X_0, X_1) and (Y_0, Y_1), every linear operator that maps X_0 boundedly into Y_0 and X_1 into Y_1 also maps $F(X_0, X_1)$ boundedly into $F(Y_0, Y_1)$.

The search for interpolation theorems of the above type is now broken down into two steps; first an interpolation method has to be constructed; second, the interpolation spaces $F(X_0, X_1)$ and $F(Y_0, Y_1)$ have to be computed.

There exist plenty of interpolation methods, but two suffice for most applications, viz. the complex method $(\ )_\theta$ and the real method $(\ )_{\theta q}$ ($0 < \theta < 1$, $1 \leq q \leq \infty$). See [2] for the definitions. (According to the definition above, these are families of inter- polation methods, one for each choice of θ (θ and q).)

Aronszajn and Gagliardo [1] gave the following two construc- tions of interpolation methods. Given a Banach couple (A_0, A_1) and a Banach space A such that $A_0 \cap A_1 \subset A \subset A_0 + A_1$, we define for any Banach couple (X_0, X_1),

$$G(X_0, X_1) = \left\{ \sum_1^\infty T_n a_n : a_n \in A, \ T_n \in L(A_0, X_0) \cap L(A_1, X_1) \quad \text{and} \right.$$

$$\left. \sum \|a_n\|_A \max(\|T_n\|_{A_0, X_0}, \|T_n\|_{A_1, X_1}) < \infty \right\}$$

and

$$H(X_0, X_1)$$
$$= \{x \in X_0 + X_1 : Tx \in A \text{ for every } T \in L(X_0, A_0) \cap L(X_1, A_1)\}.$$

G and H are interpolation methods and $G(A_0, A_1) \supset A$, $H(A_0, A_1) \subset A$. Furthermore, G is minimal for this property; that is, if F is any interpolation method such that $F(A_0, A_1) \supset A$, then $F(X_0, X_1) \supset G(X_0, X_1)$ for all couples (X_0, X_1). Similarly, H is maximal.

The purpose of this paper is to show that the real and complex methods can be characterized in this way, and that this simplifies many proofs of properties of the methods. This is mainly a summary of [6], where many other results and proofs are given, including duality theorems and applications to other interpolation methods.

THE REAL METHOD

Let $\ell_\theta^p = \{\{a_n\}_{-\infty}^\infty : (\sum_{-\infty}^\infty |2^{-n\theta} a_n|^p)^{1/p} < \infty\}$. Thus $\ell_0^p = \ell^p$.

Theorem 1. The real method $(\ )_{\theta q}$ coincides with the minimal method such that $G(\ell_0^1, \ell_1^1) \supset \ell_\theta^q$, as well as with the maximal method such that $H(\ell_0^\infty, \ell_1^\infty) \subset \ell_\theta^q$. (In fact, equalities hold.)

Sketch of proof. A linear operator $T : \ell_0^1 \to X_0$ and $\ell_1^1 \to X_1$ is given by $T\{a_n\} = \Sigma\ a_n x_n$ for some sequence $\{x_n\}$ with $\sup_n J(2^n, x_n) = \max(\|T\|_{\ell_0^1, X_0}, \|T\|_{\ell_1^1, X_1}) < \infty$. It follows that the construction by Aronszajn and Gagliardo yields the same result as the discrete J-method [2, Chapter 3.2], which is one version of the real method. Similarly, the maximal construction with $\ell_0^\infty, \ell_1^\infty, \ell_\theta^\infty$ yields the discrete K-method.

Theorem 2. Let F be an interpolation method, $0 < \theta < 1$, $1 \le q \le \infty$.

(i) *If $F(\ell_0^1, \ell_1^1) \supset \ell_\theta^q$, then $F(X_0, X_1) \supset (X_0, X_1)_{\theta q}$ for all couples (X_0, X_1).*

(ii) *If $F(\ell_0^\infty, \ell_1^\infty) \subset \ell_\theta^q$, then $F(X_0, X_1) \subset (X_0, X_1)_{\theta q}$ for all couples (X_0, X_1).*

Theorem 3. $(\ )_{\theta q}$ is the only interpolation method F such that $F(\ell_0^1, \ell_1^1) = F(\ell_0^\infty, \ell_1^\infty) = \ell_\theta^q$.

One application of this characterization is the reiteration theorem [2]. For $0 < \theta_0$, θ_1, $\theta < 1$ and $1 \le q_0$, q_1, $q \le \infty$, let $F(X_0, X_1) = ((X_0, X_1)_{\theta_0 q_0}, (X_0, X_1)_{\theta_1 q_1})_{\theta q}$. F is an interpolation method and, if $\theta_0 \ne \theta_1$, $F(\ell_0^1, \ell_1^1) = F(\ell_0^\infty, \ell_1^\infty) = (\ell_{\theta_0}^{q_0}, \ell_{\theta_1}^{q_1})_{\theta q} = \ell_\eta^q$, where $\eta = (1 - \theta)\theta_0 + \theta\theta_1$, whence $F = (\ )_{\eta q}$. Similarly, $F = (\ )_{\theta_0 q}$ if $\theta_0 = \theta_1$ and $\frac{1}{q} = \frac{1 - \theta}{q_0} + \frac{\theta}{q_1}$. The same method proves mixed reiteration theorems combining the real method and another interpolation method, for example,

$$\left((X_0, X_1)_{\theta_0 q_0}, (X_0, X_1)_{\theta_1 q_1} \right)_\theta = (X_0, X_1)_{\eta q}$$

$$\eta = (1 - \theta)\theta_0 + \theta\theta_1,$$

$$\frac{1}{q} = \frac{1 - \theta}{q_0} + \frac{\theta}{q_1} > 0.$$

In general, a mixed reiteration including the real method gives back the real method (cf. [3]).

We can also prove a similar reiteration theorem for interpolation of families of Banach spaces as described in [5], where the corresponding theorem for the complex method is proved. In brief, the definition uses a domain D in the complex plane with nice boundary Γ and a family $\{B(\gamma) \mid \gamma \in \Gamma\}$ of Banach spaces satisfying a measurability condition. The interpolation spaces $B(z)$, $z \in D$, are defined to be the values assumed at z by a certain class of analytic functions having boundary values in $B(\gamma)$ with bounded norms; see [5] for details. Let dP_z be the harmonic measure on Γ with respect to $z \in D$. We will use the fact that if the norms on the spaces $B(\gamma)$ are changed within factors $k(\gamma)$ (retaining the measurability condition) and $\int_\Gamma \log k(\gamma) \, dP_z < \infty$, then the interpolation spaces remain the same.

Theorem 4. Let $0 < \theta(\gamma) < 1$ and $1 \le q(\gamma) \le \infty$ be measurable functions on Γ such that $\int_\Gamma \log \theta(\gamma) \, dP_z(\gamma) > -\infty$, $\int_\Gamma \log(1 - \theta(\gamma)) \, dP_z(\gamma) > -\infty$ and $\int_\Gamma \frac{1}{q(\gamma)} \, dP_z(\gamma) > 0$ for some (and hence all) $z \in D$. Let

X_0, X_1 be a Banach couple and let $B(\gamma) = (X_0, X_1)_{\theta(\gamma), q(\gamma)}$ with the K-method norm. Then the interpolation spaces are $B(z) = (X_0, X_1)_{\theta(z), q(z)}$, where $\theta(z)$ and $1/q(z)$ are the harmonic extensions of $\theta(\gamma)$ and $1/q(\gamma)$ to D.

Proof. It is easy to see that $\{(X_0, X_1)_{\theta(\gamma), q(\gamma)}\}$ satisfy the measurability condition of [5]. (This is the reason for choosing the K-method norm.) Fix $z_0 \in D$ and let $F(X_0, X_1) = B(z_0)$. F is an interpolation method, and we test the couples $(X_0, X_1) = (\ell_0^1, \ell_1^1)$ and $(\ell_0^\infty, \ell_1^\infty)$. In both cases $B(\gamma) = (X_0, X_1)_{\theta(\gamma), q(\gamma)} = \ell_{\theta(\gamma)}^{q(\gamma)}$ and the norms are equivalent within a factor $3/\theta(\gamma)(1 - \theta(\gamma))$. Since $\int \log\big(3/\theta(\gamma)(1 - \theta(\gamma))\big) \, dP_z < \infty$, we may use the $\ell_{\theta(\gamma)}^{q(\gamma)}$ norm on $B(\gamma)$.

Let $\alpha(z) = \theta(z) + i\tilde{\theta}(z)$ and $\beta(z) = 1/q(z) + i(1/q(z))^{\sim}$ be analytic in D and real at z_0.

If $\{a_n\}_{-N}^{N}$ is a finite sequence and $\| \{a_n\} \|_{\ell_{\theta(z_0)}^{q(z_0)}} = 1$, define $F(z) = \{2^{n\alpha(z)} |2^{-n\theta(z_0)} a_n|^{q(z_0)\beta(z)} \text{ sign } a_n\}$. F is analytic in D, $F(z_0) = \{a_n\}$, and $\| F(\gamma) \|_{\ell_{\theta(\gamma)}^{q(\gamma)}} = 1$. Consequently, $\| \{a_n\} \|_{B(z_0)} \leq 1$ and thus $\ell_{\theta(z_0)}^{q(z_0)} \subset B(z_0)$.

Conversely, assume that $F(z) = \{f_n(z)\}$ is analytic and that $\| F(\gamma) \|_{B(\gamma)} \leq 1$. If $\{b_n\}_{-N}^{N}$ is any finite sequence such that $\| \{b_n\} \|_{\ell_{-\theta(z_0)}^{q(z_0)'}} = 1$, define

$$G(z) = \{g_n(z)\} = \{2^{-n\alpha(z)} |2^{n\theta(z_0)} b_n|^{q(z_0)'(1-\beta(z))} \text{ sign } b_n\}.$$

$\Sigma f_n(z) g_n(z)$ is analytic and

$$\left| \Sigma f_n(\gamma) g_n(\gamma) \right| \leq \| F(\gamma) \|_{\ell_{\theta(\gamma)}^{q(\gamma)}} \| G(\gamma) \|_{\ell_{-\theta(\gamma)}^{q(\gamma)'}} = \| F(\gamma) \|_{B(\gamma)} \leq 1, \quad \gamma \in \Gamma.$$

By the maximum principle, $\left| \Sigma f_n(z_0) g_n(z_0) \right| \leq 1$. Since $\{g_n(z_0)\} = \{b_n\}$ is arbitrary, this implies that $\| F(z_0) \|_{\ell_{\theta(z_0)}^{q(z_0)}} \leq 1$. Consequently, $B(z_0) \subset \ell_{\theta(z_0)}^{q(z_0)}$.

We have proved that $F(\ell_0^1, \ell_1^1) = F(\ell_0^\infty, \ell_1^\infty) = \ell_{\theta(z_0)}^{q(z_0)}$. By Theorem 3, $F(X_0, X_1) = (X_0, X_1)_{q(z_0), \theta(z_0)}$ for any X_0, X_1.

Remark. If $q(\gamma) = \infty$ a.e. on Γ, it follows similarly that $B(z)$ equals the closure of $X_0 \cap X_1$ in $(X_0, X_1)_{\theta(z)\infty}$ (except when $\theta(\gamma)$ is constant a.e.).

Corollary. Let $0 \leq 1/p(\gamma)$, $1/q(\gamma) \leq 1$ be measurable functions on Γ and let $1/p(z)$, $1/q(z)$ be their harmonic extensions to D. Let $B(\gamma)$ be the Lorentz space $L^{p(\gamma), q(\gamma)}$. If $\int_\Gamma |\log(p(\gamma) - 1)| dP_z < \infty$ and $q(z) < \infty$, then $B(z) = L^{p(z), q(z)}$, $z \in D$.

Proof. Take $X_0 = L^\infty$, $X_1 = L^1$ and $\theta(\gamma) = 1/p(\gamma)$ in Theorem 4.

THE COMPLEX METHOD

Let $\hat{L}_\theta^1 = \{\{a_n\} : \{2^{-n\theta} a_n\} \in \hat{L}^1\} = \{\{a_n\} : \sum_{-\infty}^\infty 2^{-n\theta} a_n e^{int} \in L^1(0, 2\pi)\}$ and define $\hat{M}_\theta$, $\hat{L}_\theta^\infty$ similarly.

Theorem 5. *The complex method $(\ )_\theta$ coincides with the minimal interpolation method G such that $G(\hat{L}_0^1, \hat{L}_1^1) \supset \hat{L}_\theta^1$. $(\ )^\theta$ coincides with the minimal method G such that $G(\hat{L}_0^1, \hat{L}_1^1) \supset \hat{M}_\theta$, $0 < \theta < 1$.*

Let $^\theta(\ )$ be the maximal method H such that $H(\hat{L}_0^\infty, \hat{L}_1^\infty) = \hat{L}_\theta^\infty$. Then $^\theta(X_0, X_1) \supset (X_0, X_1)^\theta$, but we do not know whether strict inclusion may occur. In any case, $(X_0, X_1)_\theta$ is a closed subspace of $^\theta(X_0, X_1)$; viz. the closure of $X_0 \cap X_1$.

This yields:

Theorem 6. *If F is an interpolation method such that $F(\hat{L}_0^1, \hat{L}_1^1) \supset \hat{L}_\theta^1$, $F(\hat{L}_0^\infty, \hat{L}_1^\infty) \subset \hat{L}_\theta^\infty$, and $X_0 \cap X_1$ is dense in $F(X_0, X_1)$, then $F(X_0, X_1) = (X_0, X_1)_\theta$.*

One application of this is that $(\tilde{X}_0, \tilde{X}_1)_\theta = (X_0, X_1)_\theta$ for any Banach couple, where $\tilde{X}_0$ and $\tilde{X}_1$ are the Gagliardo completions [2] of

X_0 and X_1 in $X_0 + X_1$. This leads (see [6 and 8]) to the following abstract version of the Stein interpolation theorem (cf. [4 and 5]). Let $\bar{S}$ be the strip $\{z : 0 \leq \operatorname{Re} z \leq 1\}$.

Theorem 7. Let T_z, $z \in \bar{S}$, be a family of linear operators from $X_0 \cap X_1$ into $Y_0 + Y_1$ such that for every $x \in X_0 \cap X_1$ and $u \in (Y_0 + Y_1)^*$, $\langle u, T_z x \rangle$ is continuous on $\bar{S}$ and analytic in the interior and $\sup(\|T_{it}\|_{X_0, Y_0}, \|T_{1+it}\|_{X_1, Y_1}) < \infty$. Assume further that Y_0 (or Y_1) is separable or that $t \to \langle u, T_{it} x \rangle$ is continuous for every $x \in X_0 \cap X_1$ and $u \in Y_0^*$ (or the corresponding condition for Y_1).

Then there is a unique extension of T_θ mapping $(X_0, X_1)_\theta$ continuously into $(Y_0, Y_1)_\theta$.

Note that the extra condition involves only one of the spaces Y_0, Y_1. It is automatically satisfied if $Y_0 \supset Y_1$ (or $Y_0 \subset Y_1$). (It is not known whether this condition is needed at all.)

[Note added in proof: Some condition on Y_0 or Y_1 is necessary in Theorem 7. See Cwikel, M., and Janson, S., Interpolation of analytic families of operators (to appear).]

REFERENCES

1. Aronszajn, N., and Gagliardo, E. Interpolation spaces and interpolation methods. *Ann. Math. Pure Appl.* 68 (1965): 51-118.

2. Bergh, J., and Löfström, J. *Interpolation Spaces*. New York: Springer, 1976.

3. Brudnyi, Yu, A., and Krugljak, N. Ja. Real interpolation functors, *Dokl. Akad. Nauk SSSR* 256 (1981): 14-17. (Russian.)

4. Calderón, A. P., and Torchinsky, A. Parabolic maximal functions associated with a distribution, II. *Adv. Math.* 24 (1977): 101-171.

5. Coifman, R.; Cwikel, M.; Rochberg, R.; Sagher, Y.; and Weiss, G. A theory of complex interpolation for families of Banach spaces. *Adv. Math.* 43 (1982): 203-229.

6. Janson, S. Minimal and maximal methods of interpolation. *J. Functional Anal.* 44 (1981): 50-73.

7. Janson, S. On the interpolation of sublinear operators. *Studia Math.* 75 (1982): 51-53.

8. Janson, S., and Jones, P. Interpolation of H^p-spaces: the complex method. To appear.

9. Zygmund, A. *Trigonometric Series*. 2d ed. New York: Cambridge Univ. Press, 1959.

A THEOREM OF KHRUSHCHEV AND PELLER ON RESTRICTIONS OF ANALYTIC FUNCTIONS HAVING FINITE DIRICHLET INTEGRAL TO CLOSED SUBSETS OF THE UNIT CIRCUMFERENCE

Paul Koosis

University of California, Los Angeles

The following exposition is based on some recent work of Khrushchev and Peller [1, Sec. 3.6].

A well-known theorem of Rudin says that every continuous function defined on a closed subset E of the unit circumference coincides thereon with a function analytic in the unit disc and continuous up to its boundary if and only if E has Lebesgue measure zero. S. Khrushchev and V. Peller have asked what additional requirements must be imposed on E in order that some of the above-mentioned analytic extensions of any continuous function defined on it have *finite Dirichlet integral over the unit disc*. Their answer, namely, that this is the case *if and only if E has logarithmic capacity zero,* is a direct consequence of their theorem, given in Sec. 1 below, together with a known result of H. Wallin discussed in Sec. 2.

Our discussion makes use of only the bare rudiments of potential theory; everything needed can be found in Chapter 3 of [3]. We will use the following *notation*:

T is the interval $[-\pi, \pi]$ with the endpoints identified.

This work was partially supported by NSF Grant NO. MC580-02955.

If E is a closed subset of T, $\mathscr{C}(E)$ is the Banach space of continuous functions on E, equipped with the usual sup norm $\| \ \|_E$.

If $f \in L_1(T)$, we write $f(\theta) \sim \sum_{-\infty}^{\infty} \hat{f}(n) e^{in\theta}$ to indicate that the expression on the right is the Fourier series of $f(\theta)$.

The Dirichlet norm, $\| f \|_{\mathscr{D}}$, of an $f \in L_1(T)$ is given by

$$\| f \|_{\mathscr{D}}^2 = \pi \sum_{-\infty}^{\infty} |n| |\hat{f}(n)|^2 .$$

$\mathscr{D}$ denotes the set of f with $\| f \|_{\mathscr{D}} < \infty$.

$\mathscr{D}_A$ denotes the set of f in $\mathscr{D}$ with $\hat{f}(n) = 0$ for all $n < 0$; these are the "analytic" f in $\mathscr{D}$.

$\mathscr{M}(E)$ denotes the set of complex-valued finite Borel measures carried on the closed set E; the total variation norm for such measures is denoted by $\| \ \|$.

We write $\hat{\mu}(n) = \frac{1}{2\pi} \int_{-\pi}^{\pi} e^{-in\theta} \, d\mu(\theta)$ for $\mu \in \mathscr{M}(T)$.

If $\mu \in \mathscr{M}(T)$, the *energy*, $\mathscr{E}(\mu)$, of μ is defined by

$$\mathscr{E}(\mu) = \pi \sum_{n \neq 0} \frac{|\hat{\mu}(n)|^2}{|n|} .$$

If $\mu \in \mathscr{M}(T)$, its *logarithmic potential* is

$$U_\mu(\theta) = \frac{1}{\pi} \int_{-\pi}^{\pi} \log \left\{ \operatorname{cosec} \left| \frac{\theta - t}{2} \right| \right\} d\mu(t) .$$

The logarithmic potential of a measure always belongs to $L_1(T)$, and we always have

$$\| U_\mu \|_{\mathscr{D}}^2 = \mathscr{E}(\mu) = \frac{1}{2} \int_{-\pi}^{\pi} U_\mu(\theta) \, \overline{d\mu(\theta)} .$$

1..

Theorem of Khrushchev and Peller. *If $E \subseteq T$ is closed and of logarithmic capacity zero, the restrictions to E of the functions in $\mathscr{D}_A \cap \mathscr{C}(T)$ fill out $\mathscr{C}(E)$.*

Khrushchev and Peller prove this result by using an ingenious duality argument, which will be given shortly. Two lemmas are needed.

Lemma 1.   *Let* $\mu \in \mathcal{M}(\mathsf{T})$ *and* $\mathcal{E}(\mu) < \infty$. *If a closed* $F \subseteq \mathsf{T}$ *has logarithmic capacity zero,* $\mu(F) = 0$.

Lemma 2.   *Let* $\mu \in \mathcal{M}(\mathsf{T})$. *If* $\sum_1^\infty \frac{|\hat{\mu}(n)|^2}{n} < \infty$, *then* $\sum_{-\infty}^{-1} \frac{|\hat{\mu}(n)|^2}{|n|} < \infty$.

Assuming, for the time being, the two lemmas, let us prove the theorem. The idea of Khrushchev and Peller is to use the product space $\mathcal{D}_A \times \mathcal{C}(\mathsf{T})$ with its natural norm $\| (f, \, g) \|_\times = \| f \|_{\mathcal{D}} + \| g \|_{\mathsf{T}}$ and to examine the special subspace $\mathcal{L}$ of it consisting of the pairs $(f, \, f)$ with $f \in \mathcal{D}_A \cap \mathcal{C}(\mathsf{T})$. Making every such pair $(f, \, f)$ correspond to the restriction of f to E gives us an obvious continuous map of $\mathcal{L}$ into $\mathcal{C}(E)$; we are to prove that this map is *onto*. Let Σ be the unit sphere in $\mathcal{L}$; it is *more* than enough to show that the restriction map just described takes Σ onto a *dense subset* of the unit sphere in $\mathcal{C}(E)$. Indeed, then the map's image of Σ will contain any $\varphi \in \mathcal{C}(E)$ with $\| \varphi \|_E < 1$, as is shown by a well-known approximation argument.

Reasoning by contradiction, let us assume that the restrictions of f to E with $(f, \, f) \in \Sigma$ are *not* dense in the unit sphere of $\mathcal{C}(E)$. The Hahn-Banach theorem then gives us a $\nu \in \mathcal{M}(E)$, $\| \nu \| = 1$, with $\left| \int_{\mathsf{T}} \bar{f} d\nu \right| <$ some $\lambda < 1$ for all $(f, \, f) \in \Sigma$. To ν corresponds a *functional* $(0, \, \nu)$ on $\mathcal{L}$, and since Σ is the unit sphere of $\mathcal{L}$, the inequality just written means that the *norm* $\| (0, \, \nu) \|_{\mathcal{L}}^*$ *of this functional on* $\mathcal{L}$ *is* $< \lambda$.

Let us describe the dual $\mathcal{L}^*$ of $\mathcal{L}$. The dual of $\mathcal{C}(\mathsf{T})$ is of course $\mathcal{M}(\mathsf{T})$. The dual $\mathcal{D}_A^*$ of $\mathcal{D}_A$ consists of all distributions $S(\theta) \sim \sum_1^\infty \hat{S}(n) e^{in\theta}$ with finite norm $\| S \|_{\mathcal{D}}^* = \sqrt{\frac{1}{\pi} \sum_1^\infty \frac{|\hat{S}(n)|^2}{n}}$; action on $\mathcal{D}_A$ is through the inner product $\langle S, \, f \rangle = \sum_1^\infty \overline{\hat{S}(n)} \hat{f}(n)$. The dual of $\mathcal{D}_A \times \mathcal{C}(\mathsf{T})$ is thus $\mathcal{D}_A^* \times \mathcal{M}(\mathsf{T})$, and, finally, the dual of $\mathcal{L}$ is the quotient space $(\mathcal{D}_A^* \times \mathcal{M}(\mathsf{T})) / \mathcal{L}^\perp$, where $\mathcal{L}^\perp$ is the sub-

space of $\mathscr{D}_A^* \times \mathscr{M}(\mathsf{T})$ consisting of the (S, μ) with $\langle S, f \rangle +$
$\int_{\mathsf{T}} \overline{fd\mu} = 0$ for all $(f, f) \in \mathscr{L}$.

Returning to our $\nu \in \mathscr{M}(E)$, let us compute $\| (0, \nu) \|_{\mathscr{L}}^*$. It is
the infimum of $\| (0, \nu) - (S, \mu) \|_\times^*$ with (S, μ) ranging over $\mathscr{L}^\perp$.
So, since $\| (0, \nu) \|_{\mathscr{L}}^* < \lambda$, there is an $(S, \mu) \in \mathscr{L}^\perp$ with
$\sup\{ \| S \|_{\mathscr{D}}^* , \| \nu - \mu \| \} < \lambda$. This, however, is impossible. The pairs
$(e^{in\theta}, e^{in\theta})$ with $n = 0, 1, 2, \ldots$ all belong to $\mathscr{L}$, so, from
$(S, \mu) \in \mathscr{L}^\perp$ we get $2\pi\overline{\hat{\mu}(n)} + \overline{\hat{S}(n)} = 0$ for such n. Since $S \in \mathscr{D}_A^*$,
this makes $\sum_1^\infty \frac{|\hat{\mu}(n)|^2}{n} < \infty$. From Lemma 2, we now obtain $\mathscr{E}(\mu) < \infty$.
However, E has logarithmic capacity zero by hypothesis. So Lemma 1
implies that $\mu(F) = 0$ for any closed subset F of E. Since $\nu \in \mathscr{M}(E)$,
we see from this that $\| \nu - \mu \| \geq \| \nu \| = 1$, contradicting the in-
equality $\sup\{ \| S \|_{\mathscr{D}}^* , \| \nu - \mu \| \} < \lambda$.

The theorem is thus established; we have in fact proved *more*.
If $f \in \mathscr{D}_A \cap \mathscr{C}(\mathsf{T})$, let φ be the restriction of f to E. Clearly
$\| \varphi \|_E \leq \| f \|_{\mathsf{T}}$. Our argument has shown that for *given* $\varphi \in \mathscr{C}(E)$, we
can *find* such f with $\| f \|_{\mathscr{D}} + \| f \|_{\mathsf{T}}$ *arbitrarily close to* $\| \varphi \|_E$; this
means that the *Dirichlet norms* $\| f \|_{\mathscr{D}}$ *of the* $f \in \mathscr{D}_A \cap \mathscr{C}(\mathsf{T})$ *equal to*
φ *on* E *can be taken arbitrarily small*.

We return now to the lemmas .

The *first* one is routine and follows almost from the definition of capacity. For F closed and of capacity zero, take any open
neighborhood $\mathscr{O}$ of F on T and any $\varepsilon > 0$. It is enough to construct
a $\varphi \in \mathscr{D} \cap \mathscr{C}(\mathsf{T})$ with $0 \leq \varphi \leq 1$, $\varphi \equiv 1$ on F and $\varphi < \varepsilon$ outside $\mathscr{O}$, such
that $|\hat{\varphi}(0)|^2 + \| \varphi \|_{\mathscr{D}}^2 < \varepsilon^2$, for then $|\int_{\mathsf{T}} \overline{\varphi}d\mu|$, which differs but
little from $|\mu(F)|$, will be $\leq 2\varepsilon\sqrt{\pi^2|\hat{\mu}(0)|^2 + \mathscr{E}(\mu)}$ by Schwarz's
inequality.

Let $2\delta = \text{dis}(\sim\mathscr{O}, F)$. Since F has logarithmic capacity zero,
we can find a neighborhood $\mathscr{N}$ of F with $\text{dis}(\sim\mathscr{N}, F) < \delta$ such that
the conductor potential of $\bar{\mathscr{N}}$ is a U_σ with a positive $\sigma \in \mathscr{M}(\bar{\mathscr{N}})$
having $\| \sigma \|$ *as small as we please*. We need only put $\varphi(\theta) = U_\sigma(\theta)$,
for $U_\sigma(\theta) \equiv 1$ on $\bar{\mathscr{N}}$ and $0 \leq U_\sigma \leq 1$ everywhere on T. ($\mathscr{N}$ can be
taken as a finite union of intervals and then has no irregular
points.) Also, for $\theta \notin \mathscr{O}$, $U_\sigma(\theta) \leq \frac{1}{\pi} \| \sigma \| \log \text{cosec} \frac{2}{\delta}$, and this is

$< \varepsilon$ if $\| \sigma \|$ is small. Finally, $|\hat{\varphi}(0)| = |\hat{\sigma}(0)| \log 4$ and $\| \varphi \|_{\mathcal{D}}^2 = \frac{1}{2} \int_{\mathbf{T}} U_\sigma d\sigma \leq \frac{1}{2} \| \sigma \|$, showing that $|\hat{\varphi}(0)|^2 + \| \varphi \|_{\mathcal{D}}^2 < \varepsilon$ if $\| \sigma \|$ is small enough. Lemma 1 is thus proved.

There remains Lemma 2. Khrushchev and Peller remark that it can be obtained by duality from a known result in the theory of Gaussian processes (see [2, Chapter 4, Sec. 4]). They also have a quite complicated proof of their own; it uses BMO and the theory of Hankel and Toeplitz operators. This argument is a good deal longer than the one used to prove the theorem.

While walking on the UCLA campus, S. Pichorides and I came upon the following very easy proof (my real motivation for preparing this exposition!); it is based on the idea used to prove Hardy's theorem about the Fourier coefficients of H_1 functions.

It is enough to establish the inequality

$$\left| \sum_1^\infty \frac{|\hat{\mu}(n)|^2}{n} - \sum_{-\infty}^{-1} \frac{|\hat{\mu}(n)|^2}{|n|} \right| \leq \frac{\|\mu\|^2}{4\pi}$$

for *absolutely continuous measures* μ. Let $d\mu(\theta) = f(\theta) d\theta$ with $f \in L_1(\mathbf{T})$; put $f^*(\theta) = \overline{f(-\theta)}$, and let $G = f * f^*$, the convolution of f and f^*. We have $\hat{G}(n) = 2\pi |\hat{f}(n)|^2 = 2\pi |\hat{\mu}(n)|^2$. Take $G(\theta)$, multiply it by

$$\frac{\pi - |\theta|}{2} \operatorname{sgn} \theta \sim \sum_1^\infty \frac{\sin n\theta}{n},$$

and integrate from $-\pi$ to π!. We get $2\pi^2 i \sum_1^\infty \frac{1}{n} \left(|\hat{\mu}(n)|^2 - |\hat{\mu}(-n)|^2 \right)$. This must, however, have absolute value $\leq \frac{\pi}{2} \| G \|_1 \leq \frac{\pi}{2} \| \mu \|^2$. That's all there is to it.

2.

To complete this discussion, it is necessary to show that if E *has logarithmic capacity* > 0, *then the restrictions of functions in* $\mathcal{D}_A \cap \mathcal{C}(\mathbf{T})$ *to* E *do not fill out* $\mathcal{C}(E)$. Wallin proved, however, in

1963 that *in these circumstances not even the restrictions of functions in $\mathscr{D} \cap \mathscr{C}(\mathsf{T})$ to E can exhaust $\mathscr{C}(E)$*. Because Wallin's proof [4] is rather long, I give here a different, easier one, based on duality.

Use the norm $\| f \|_{\mathscr{D}} + \| f \|_{\mathsf{T}}$ for functions $f \in \mathscr{D} \cap \mathscr{C}(\mathsf{T})$. For this norm, restriction of f to E is a *continuous* mapping from $\mathscr{D} \cap \mathscr{C}(\mathsf{T})$ into $\mathscr{C}(E)$. It is claimed that this mapping cannot be *onto*.

Suppose that it *were* onto. Then, by the Baire category theorem, for some N, the restrictions to E of the f with $\| f \|_{\mathscr{D}} + \| f \|_{\mathsf{T}}$ $\leq N$ would be *dense* in some nonempty open *sphere* of $\mathscr{C}(E)$; by taking N larger, we may suppose that sphere to have its center at 0. This means that there is a constant K such that, if $\varphi \in \mathscr{C}(E)$ and $\| \varphi \|_{E} \leq 1$, we have $f \in \mathscr{D} \cap \mathscr{C}(\mathsf{T})$ with $\| f \|_{\mathscr{D}} + \| f \|_{\mathsf{T}} \leq K$, whose restrictions to E are *arbitrarily close* to φ in norm $\| \ \|_{E}$.

We proceed to dualize this last statement. Take any $\nu \in \mathscr{M}(E)$; we can choose a $\varphi \in \mathscr{C}(E)$ with $\| \varphi \|_{E} = 1$ and $\int_{\mathsf{T}} \bar{\varphi} d\nu$ as close as we like to $\| \nu \|$; by approximating φ with restrictions of f to E in the manner just described, we see that $\| \nu \|$ is a limit of some expressions of the form $2\pi \sum_{-\infty}^{\infty} \overline{\hat{f}(n)} \hat{\nu}(n)$ with $f \in \mathscr{D} \cap \mathscr{C}(\mathsf{T})$ and $\| f \|_{\mathscr{D}} + \| f \|_{\mathsf{T}} \leq K$. Suppose now that $\mathscr{E}(\nu) < \infty$. Then Schwarz's inequality shows that

$$\pi | \sum_{-\infty}^{\infty} \overline{\hat{f}(n)} \hat{\nu}(n) | \leq \sqrt{ (|\hat{f}(0)|^2 + \| f \|_{\mathscr{D}}^2)(\pi^2 |\hat{\nu}(0)|^2 + \mathscr{E}(\nu))},$$

yielding

$$\| \nu \| \leq 2K(\pi^2 |\hat{\nu}(0)|^2 + \mathscr{E}(\nu))^{1/2} \qquad (*)$$

for $\nu \in \mathscr{M}(E)$ with $\mathscr{E}(\nu) < \infty$.

We will therefore be done if we show that a relation like (*) *cannot hold*; to do this we use the fact that E has nonzero logarithmic capacity. By definition, there *is* some positive $\mu \in \mathscr{M}(E)$ with $\mathscr{E}(\mu) < \infty$; there is no loss of generality in supposing $\| \mu \| = 1$. Assume, for the moment, that we can construct μ-measurable functions φ_n on T with $|\varphi_n(\theta)| \equiv 1$ such that the measures ν_n

given by $d\nu_n(\theta) = \varphi_n(\theta)\, d\mu(\theta)$ tend to zero weak* in $\mathcal{M}(E)$. Then (*) cannot hold for the ν_n; indeed, $\|\nu_n\| = \|\mu\| = 1$, but as we now show, $\mathcal{E}(\nu_n) \xrightarrow[n]{} 0$.

We have

$$\mathcal{E}(\mu) = \frac{1}{2} \int_{-\pi}^{\pi} \int_{-\pi}^{\pi} \log\left\{\operatorname{cosec}\left|\frac{s-t}{2}\right|\right\} d\mu(s)\, d\mu(t).$$

Write $K(s, t) = \frac{1}{2}\log\left\{\operatorname{cosec}\left|\frac{s-t}{2}\right|\right\}$; because μ is positive and $\mathcal{E}(\mu) < \infty$, we can, for any given $\varepsilon > 0$, get an M such that $\iint_{\Delta_M} K(s, t)\, d\mu(s)\, d\mu(t) < \varepsilon$, where Δ_M is the closed subset of T^2 on which $K(s, t) \geq M$. Put $K_M(s, t) = \min[K(s, t), M]$; then $K_M(s, t)$ is continuous on T^2, so, since the product measures $d\nu_n(s)\overline{d\nu_n(t)}$, converge weak* to zero on T^2, $\iint_{\mathsf{T}^2} K_M(s, t)\, d\nu_n(s)\overline{d\nu_n(t)} \xrightarrow[n]{} 0$. Here, the difference $K(s, t) - K_M(s, t)$ is ≥ 0 on T^2 and different from zero only on Δ_M, where it is, however, $\leq K(s, t)$. So

$$\left|\iint_{\mathsf{T}^2}(K(s, t) - K_M(s, t))\, d\nu_n(s)\overline{d\nu_n(t)}\right| \leq \iint_{\Delta_M} K(s, t)\, d\mu(s)\, d\mu(t) < \varepsilon,$$

since $\left|d\nu_n(t)\right| = d\mu(t)$. This shows that $\mathcal{E}(\nu_n) = \iint_{\mathsf{T}^2} K(s, t)\, d\nu_n(s)\overline{d\nu_n(t)}$ is $< \varepsilon$ for all sufficiently large n, invalidating (*) and thereby establishing our result.

It remains to construct the φ_n. Take any n, and look at the intervals $J_k = \left(-\pi + \frac{2\pi}{n}(k-1), -\pi + \frac{2\pi}{n}k\right]$, $k = 1, 2, \ldots, n$. If $\mu(J_k) = 0$, put $\varphi_n \equiv 1$ on J_k. If $\mu(J_k) > 0$, there must be some point $m_k \in J_k$ for which $\mu(J_k^-) = \mu(J_k^+) = \frac{1}{2}\mu(J_k)$, where J_k^- consists of the $t \in J_k$ that are $\leq m_k$ and J_k^+ of those that are $> m_k$. That's because the positive measure μ can contain no point mass, since $\mathcal{E}(\mu) = \frac{1}{2}\int_{\mathsf{T}} U_\mu\, d\mu < \infty$. For such a J_k, take $\varphi_n(t) \equiv 1$ on J_k^- and $\varphi_n(t) \equiv -1$ on J_k^+. The function φ_n is now defined on all the J_k, and it is clear that $\varphi_n(\theta)\, d\mu(\theta) \xrightarrow[n]{} 0$ weak*. We are done.

3. A CONCLUDING REMARK

It seems like a good idea to try and give an analogous characterization of the closed sets $E \subseteq T$ having α-capacity equal to zero,
replacing the Dirichlet norm for our analytic functions by one
corresponding to the Riesz potentials of order α. In this situation, the reasoning of Sec. 2 goes through as is; one need only
make the necessary changes in the formulas. Lemma 1 of Sec. 1 also
holds; however, *the proof of Lemma 2 breaks down, because*
$\sum_{1}^{\infty} \dfrac{\sin n\theta}{n^{1-\alpha}}$ *is no longer the Fourier series of a bounded function if*
$\alpha > 0$. It is therefore not clear that an analogue of the Khrush-
chev-Peller theorem, referring to *analytic functions*, will hold.

The converse of the result in Sec. 2 *is*, however, true in
the present situation, namely, if E has α-capacity zero, the re-
strictions to E of *all* continuous f having $\sum |n|^{1-\alpha}|\hat{f}(n)|^{2} < \infty$
(*not* just the *analytic* ones) do fill out $\mathscr{C}(E)$. This result is also
due to Wallin [4]; it can be obtained by running the duality argument of Sec. 2 backwards. Such reasoning resembles that followed
in proving the theorem of Sec. 1; it is, however, simpler, using
only an analogue of Lemma 1.

It should be mentioned, finally, that by using their own
methods, Khrushchev and Peller obtain [1, Sec. 3.6] a result analogous to the one in Sec. 1 about restrictions to E of continuous
functions f analytic in the unit disc with

$$D_p(f) = \sqrt[p]{\int_{-\pi}^{\pi} \int_{-\pi}^{\pi} \frac{|f(e^{i(x+t)}) - f(e^{ix})|^p}{|t|^2} dx \, dt} < \infty.$$

They find that these restrictions fill out $\mathscr{C}(E)$ provided that E
has capacity zero with respect to the norm D_p. For $p = 2$, this
reduces to the theorem of Sec. 1.

REFERENCES

1. Khrushchev, S. V., and Peller, V. V. Hankel operators, best approximations and stationary Gaussian processes, II. LOMI preprint E-5-81, Leningrad Department, Steklov Mathematical Institute, Leningrad, 1981. In Russian, this material is now published in *Uspekhi Matem. Nauk* 37 (1982): 53-124.

2. Ibragimov, I. A., and Rosanov, Y. A. *Gaussian Random Processes.* New York: Springer, 1978. English translation of *Gaussovskie sluchainye protsessy,* Nauka, Moscow, 1970.

3. Kahane, J. P., and Salem, R. *Ensembles parfaits et séries trigonométriques.* Paris: Hermann, 1963.

4. Wallin, H. Continuous functions and potential theory. *Arkiv för Matem.* 5 (1963): 55-84.

SEMIEMBEDDINGS IN LINEAR TOPOLOGICAL SPACES

Heinrich P. Lotz
N.T. Peck
Horacio Porta
University of Illinois

INTRODUCTION

The starting point of this paper is an observation about metrics
on the space of measurable functions. If L_0 denotes the space of
(equivalence classes of) Lebesgue measurable functions on $[0, 1]$,
two metrics on L_0 that give the topology of convergence in measure
are $d_\sigma(f, g) = \sigma(f - g)$ and $d_\delta(f, g) = \delta(f - g)$, where $\sigma(f) =$
$\inf\{\varepsilon + m(|f| > \varepsilon); \varepsilon > 0\}$, and $\delta(f) = \int_0^1 |f(t)|(1 + |f(t)|)^{-1} dt$
(here and in the following, $m(E)$ denotes the measure of a set E).
The observation is that σ and δ are badly nonhomogeneous for small
scalars. In other words, the quotients $\sigma(rf)/r$ and $\delta(rf)/r$ considered as functions of r are far from being constant. In fact, it
is easy to see that for $f(t) = 1/t$ $(0 < t \leq 1)$ and $r > 0$ near 0,
they behave, respectively, like $r^{-1/2}$ and $-\ln r$. In particular,
they are unbounded.

It turns out that we can identify rather simply the functions
f for which these quotients are bounded (see below, Proposition 3):
$\sigma(rf)/r$ (resp., $\delta(rf)/r$) is bounded if and only if $f \in L_\infty$ (resp.,
$f \in L_1$), and in that case, $\|f\|_\infty = \lim_{r \to 0^+} \sigma(rf)/r$ (resp., $\|f\|_1 =$
$\lim_{r \to 0^+} \delta(rf)/r$).

Quite naturally, this raises the question: which Banach spaces of measurable functions can be recovered from suitable metrics on L_0 that induce the convergence in measure? We consider this question below (see Sec. 2).

At this point, two lines of generalization suggest themselves:

(i)   identify the Banach spaces that are isomorphic to the Banach spaces of measurable functions recovered from L_0;

(ii) consider the corresponding situation with L_0 replaced by an arbitrary linear metric space.

Concerning (ii), we show in Theorem 1 that a normed space E can be obtained by this procedure from a metric linear space L if and only if E can be semiembedded in L.

In Sec. 3 we consider semiembeddings of Banach lattices in L_p, $0 \leq p \leq 1$, and in Sec. 4 we make some remarks about semiembeddings of spaces of continuous functions in L_0.

1. NORMED SPACES ASSOCIATED TO METRIC LINEAR SPACES

Let L be a real vector space and ρ a *pseudonorm* on L. This simply means that ρ is a function from L into the nonnegative reals such that

$$\rho(x) = 0 \quad \text{if and only if} \quad x = 0; \tag{1}$$

$$\rho(x + y) \leq \rho(x) + \rho(y); \tag{2}$$

$$\rho(-x) = \rho(x); \tag{3}$$

$$\rho(tx) \text{ is a continuous function of} \tag{4}$$
$$t \text{ for each fixed } x \in L.$$

We will say that L provided with ρ is a *metric linear space*. This is justified by the fact that $d(x, y) = \rho(x - y)$ is an invariant metric on L making the operations continuous.

A characteristic property of metric linear spaces is that balls of different sizes also have different shapes. To compare

them, we have to magnify the balls to the same scale with a procedure that can be described as follows. First, for each $r > 0$,
define

$$\rho_r(x) = \rho(rx)/r. \tag{5}$$

We have:

Proposition 1. Let ρ be a pseudonorm on the linear space L. Then:

(i) *for each $r > 0$, ρ_r is a pseudonorm on L defining the same metric topology as ρ;*

(ii) *for each $x \in L$, the limit $\|x\|_\rho = \lim_{r \to 0+} \rho_r(x)$ exists in $[0, +\infty]$, and $\rho_r(x) \leq \|x\|_\rho$ for all $r > 0$;*

(iii) *the set L_ρ of all $x \in L$ with $\|x\|_\rho < +\infty$ is a linear subspace of L and $\|\ \|_\rho$ is a norm on L;*

(iv) *L_ρ is the largest subspace supporting a norm $\|\ \|$ satisfying $\|x\| \geq \rho(x)$, and $\|\ \|_\rho$ is the smallest such norm;*

(v) *let $B_r = \{x \in L; \rho(x) \leq r\}$; then $(1/r)B_r$ is the unit ball of ρ_r and the unit ball of $\|\ \|_\rho$ is the intersection $\bigcap_{r>0} (1/r)B_r$.*

Proof. We leave the proof to the reader except for the technical
point in (ii), which can be obtained as a consequence of Theorem
7.11.1 in [2] or directly in the following way:

Let $a = \lim \inf_{r \to 0+} g(r)$, where $g(r) = \rho_r(x)$, and $x \in L$ is
fixed. Let $\varepsilon > 0$, and pick $r_n > 0$, $r_n \to 0$ such that $g(r_n) \leq a + \varepsilon$.
Since $g(kr) \leq g(r)$ for k a positive integer (use (2)), we conclude
that also $g(kr_n) \leq a + \varepsilon$, for all $k, n = 1, 2, \ldots$. Finally, since
$\{kr_n\}$ is dense in $[0, +\infty]$, it follows from continuity that $g(r) \leq$
$a + \varepsilon$ for all $r > 0$, and so $\sup_{r>0} g(r) \leq a$. Thus, $\lim \sup_{r \to 0+}$
$g(r) \leq \sup_{r>0} g(r) \leq \lim \inf_{r \to 0+} g(r)$, and (ii) of Proposition 1
follows. Observe that $\|x\|_\rho$ coincides also with $\sup_{r>0} \rho_r(x)$, al-
though $\rho_r(x)$ is not a decreasing function of r in general.

Corollary 1. All closed balls of L_ρ are closed in L (for the ρ-topology). This is clear from (v) of Proposition 1.

Proposition 2. If L is complete under ρ, then $(L_\rho, \|\ \|_\rho)$ is a Banach space.

Proof. From $\rho = \rho_1 \leq \|\ \|_\rho$ (see (ii) of Proposition 1) it follows that a $\|\ \|_\rho$-Cauchy sequence $\{x_n\}$ is also ρ-Cauchy, and therefore ρ-convergent to, say, $\tilde{x}$. Now let $\varepsilon > 0$ and pick N such that $\|x_n - x_m\|_\rho \leq \varepsilon$ for $n, m \geq N$. Assume that $n \geq N$ is fixed. Since the ball $B = \{x \in L_\rho; \|x_n - x\|_\rho \leq \varepsilon\}$ is closed for the ρ-convergence (Corollary 1) and $x_m \in B$ when $m \geq N$, it follows that $\tilde{x} \in B$, or $\|x_n - \tilde{x}\|_\rho \leq \varepsilon$, which proves that $\|x_n - x\|_\rho \to 0$ as $n \to \infty$.

Suppose now that L is a vector space provided with a metric linear topology. For each pseudonorm ρ defining the topology of L, a normed space $(L_\rho, \|\ \|_\rho)$ is determined by ρ. We will say that the normed space $(E, \|\ \|)$ is a *normed space associated to the metric linear space* L if $(E, \|\ \|)$ is isometric to some $(L_\rho, \|\ \|_\rho)$. Recall that a continuous linear operator $T : E \to L$ from a normed space E into a topological linear space L is a *semiembedding* if T is one-to-one and the image under T of the unit ball of E (equivalently, any closed ball of E) is a closed subset of L.

The main result of this section is the following:

Theorem 1. Let L be a metric linear space and B a normed space. The following conditions are equivalent:

(i)   B is a normed space associated to L;

(ii) there is a semiembedding from B into L.

Proof. Corollary (1) shows that (i) implies (ii). Assume now that $T : B \to L$ is a semiembedding of B (whose norm we denote by $\|\ \|$) into L, and let ρ be any pseudonorm defining the topology of L. Set, for $x \in L$:

$$\rho'(x) = \inf_{b \in B} \{\|b\| + \rho(x - Tb)^{1/2}\}.$$

Routine calculations show that ρ' is a pseudonorm, and taking $b = 0$, we also get $\rho'(x) \leq \rho(x)^{1/2}$, so that ρ-convergent sequences are also ρ'-convergent. Assume next that $\rho'(x_n) \to 0$. Then there exist elements $b_n \in B$ such that also $\|b_n\| + \rho(x_n - Tb_n)^{1/2} \to 0$ and, in particular, $b_n \to 0$ and $\rho(x_n - Tb_n) \to 0$. But then also $\rho(Tb_n) \to 0$ by continuity of T and therefore $\rho(x_n) \leq \rho(x_n - Tb_n) + \rho(Tb_n)$ implies that $\rho(x_n) \to 0$. Hence ρ' defines the same topology as ρ.

Assume now that $0 < \|x\|_{\rho'} < +\infty$. We claim that $x = Tb$ for some b and that $\|b\| = \|x\|_{\rho'}$. In fact, let $\varepsilon > 0$. Picking $r = 1/n$ in (ii) of Proposition 1, we conclude that for n large, there are elements $b_n \in B$ such that

$$n[\|b_n\| + \rho(x/n - Tb_n)^{1/2}] \leq \|x\|_{\rho'} + \varepsilon.$$

It follows that $\|nb_n\| \leq \|x\|_{\rho'} + \varepsilon$ and also $n\rho(x/n - Tb_n)^{1/2} \leq \|x\|_{\rho'} + \varepsilon$. Hence,

$$\rho(x - nTb_n)^{1/2} = \rho\left(n\left[\frac{x}{n} - Tb_n\right]\right)^{1/2} \leq n^{1/2}\rho\left(\frac{x}{n} - Tb_n\right)^{1/2}$$

$$= n^{-1/2}\left[n\rho\left(\frac{x}{n} - Tb_n\right)^{1/2}\right] \leq n^{-1/2}[\|x\|_{\rho'} + \varepsilon] \to 0.$$

This shows that x is the limit (for the pseudonorm ρ) of the sequence $\{T(nb_n)\}$, which is the image under T of the bounded sequence $\{nb_n\}$. Then, the image under T of a closed ball being closed, we conclude that x is of the form $x = Tb$ with $\|b\| \leq \|x\|_{\rho'} + \varepsilon$. But T being one-to-one, the element b can not depend on ε, and so $\|b\| \leq \|x\|_{\rho'}$.

The reverse inequality follows from

$$\rho'(rx)/r = \rho'(rTb)/r \leq [\|rb\| + \rho(rTb - Trb)^{1/2}]/r$$
$$= \|rb\|/r = \|b\|.$$

and the theorem follows.

2. BANACH SPACES OF MEASURABLE FUNCTIONS

Throughout this section we consider the metric linear space L_0 of (classes of) measurable real functions defined on $[0, 1]$ under the convergence in measure. Since L_0 is complete, all normed spaces associated to L_0 are also complete, that is, Banach spaces.

We begin by proving our claim in the Introduction about the spaces L_1 and L_∞ (the pseudonorms δ and σ were defined in the Introduction).

Proposition 3. *The Banach spaces L_1 and L_∞ are associated to L_0 (by means of δ and σ, respectively).*

Proof. Consider the case of σ. We want to prove that

$$\operatorname*{ess\ sup}_{0 \le t \le 1} |f(t)| = \lim_{r \to 0^+} \sigma(rf)/r \qquad (6)$$

for each measurable function f defined on $[0, 1]$. Set $M = \operatorname{ess\ sup} |f(t)|$, $0 \le t \le 1$ (which is either finite or $+\infty$), and $N = \|f\|_\sigma$ (and again, N could be $+\infty$).

By definition, $\sigma(rf) = \inf\{\varepsilon + m(|f| > \varepsilon/r); \varepsilon > 0\}$ so that picking $\varepsilon = rM$ when $M < +\infty$, we conclude that $\sigma(rf) \le rM$. Thus $N \le M$.

Assume now that $N < M$, and pick $r_n \to 0$, $a > 0$ and $\varepsilon_n > 0$ so that $N + a < M$ and

$$[\varepsilon_n + m(|f| > \varepsilon_n/r_n)]/r_n \le N + a$$

for each $n = 1, 2, \ldots$. In particular, $m(|f| > \varepsilon_n/r_n) \le r_n(N + a)$ and $\varepsilon_n \le r_n(N + a)$. But then

$$m(|f| > N + a) \le m(|f| > \varepsilon_n/r_n) \le r_n(N + a) \to 0$$

whence $m(|f| > N + a) = 0$ and so $M \le N + a$, a contradiction. Thus $\| \ \|_\sigma = \| \ \|_\infty$. For the case of δ, we observe that Beppo Levi's theorem implies that

$$\int_0^1 |f(t)|\ dt = \lim_{r \to 0^+} \int_0^1 |f(t)|(1 + r|f(t)|)^{-1} dt$$

and then the equality $\|\ \|_\delta = \|\ \|_1$ follows easily.

The next theorem gives a list of classical Banach spaces that can be recovered from L_0 and the convergence in measure by this procedure.

Theorem 2.   The following Banach spaces are associated to L_0:

(i) L_1;

(ii) L_∞;

(iii) *all closed subspaces of L_p, $1 < p < \infty$;*

(iv)   *all reflexive subspaces of L_1;*

(v) H_1;

(vi) $Lip(\alpha)$, $0 < \alpha < 1$.

Proof.   In all cases we will simply prove that the natural inclusion $B \subset L_0$, where B is one of the above, is a semiembedding, so that the theorem will follow from Theorem 1. Typically, then, we consider a sequence $\{f_n\}$ in the unit ball U of B converging in measure to $f \in L_0$. The desired conclusion is that $f \in U$.

Proofs of (i) and (ii) are given above.

Proof of (iii): Suppose, taking a subsequence if necessary, that f_n converges weakly in L_p to g. For each $\varepsilon > 0$, Egoroff's theorem implies that $f = g$ on a set E_ε with $m([0, 1] - E_\varepsilon) < \varepsilon$. But then $f = g$ almost everywhere, and $f \in U$ as claimed.

Proof of (iv): The argument just used applies verbatim.

Proof of (v): Here we consider, for simplicity, the interval $[0, 2\pi]$ rather than $[0, 1]$. Using (i), we conclude that $f \in L_1$ and that $\|f\|_1 \leq 1$. Egoroff's theorem again and the Cauchy-Schwarz inequality imply that

$$\lim_n \int_0^{2\pi} |f_n(t) - f(t)|^{1/2} dt = 0$$

so that, by passing to a subsequence, we may assume that

$$\int_0^{2\pi} |f_n(t) - f(t)|^{1/2} dt \leq 2^{-n}. \tag{7}$$

Observing that $\|f_n\|_1 \leq 1$ implies that the functions $f_n(z)$ are uniformly bounded on compact sets of the unit disc $|z| \leq 1$, we also assume (normal families) that there is an analytic function g defined on $|z| < 1$ such that the sequence $\{f_n(z)\}$ converges uniformly on compact subsets of the unit disc to g.

We shall now show that $g \in H_{1/2}$ and that $f = g$ a.e. on the circle $|z| = 1$. This, in conjunction with Theorem 2.11 in [1], will imply that $f \in H_1$ as desired.

A special case of the Hardy-Littlewood maximal theorem (see 1.9 in [1], for example) is that there is a constant K such that for all $h \in H_{1/2}$,

$$\int_0^{2\pi} |Mh(t)|^{1/2} dt \leq K \int_0^{2\pi} |h(t)|^{1/2} dt, \tag{8}$$

where Mh is the maximal function $Mh(t) = \sup_{0<r<1} |h(re^{it})|$. Combining (7) and (8), we get

$$\int_0^{2\pi} |M(g - f_n)(t)|^{1/2} dt$$

$$\leq \int_0^{2\pi} \sup_{0<r<1} \left\{ \sum_{k=n}^{\infty} |f_k(re^{it}) - f_{k+1}(re^{it})|^{1/2} \right\} dt$$

$$\leq K \int_0^{2\pi} \sum_{k=n}^{\infty} |f_k(t) - f_{k+1}(t)|^{1/2} dt \leq K2 \sum_{k=n}^{\infty} 2^{-k} = 2^{-n+2} K.$$

In particular,

$$\int_0^{2\pi} |g(re^{it}) - f_n(re^{it})|^{1/2} dt \leq 2^{-n+2} K$$

for all $r < 1$, and therefore $g = (g - f_n) + f_n$ is in $H_{1/2}$. Taking now limits for $r \to 1$ in the last inequality, we get $g = f$ a.e., and the proof of (v) is complete.

Proof of (vi): Recall that $f \in \text{Lip}(\alpha)$ if and only if $|f(x) - f(y)| \le k|x - y|^\alpha$ for some k and all x, $y \in [0, 1]$, with the norm in $\text{Lip}(\alpha)$ defined by $\|f\| = \|f\|_\infty + k_0$, where $k_0 = \inf\{k\}$ and $\|f\|_\infty$ is the supnorm. In particular, $|f(x) - f(y)| \le \|f\| \cdot \|x - y\|^\alpha$.

Taking a subsequence if necessary, we assume now that $f_n(x) \to f(x)$ for all $x \in E$ where $[0, 1] - E$ has measure zero. Then

$$|f(x) - f(y)| \le |x - y|^\alpha \tag{9}$$

holds for all x, $y \in E$. Since E is dense, and the above condition implies uniform continuity of f on E, we can find a (unique) continuous function $F : [0, 1] \to \mathbf{R}$ satisfying $F|_E = f$. Next we observe that $\{F, f_1, f_2, \ldots\}$ is an equicontinuous family such that $f_n \to F$ on a dense set ($=E$). Thus f_n converges uniformly on $[0, 1]$ to F, and so F also satisfies a Lipschitz condition. Since $F = f$ a.e., we conclude that $f \in U$ as desired.

3. BANACH LATTICES

A Banach lattice E is called σ-1-*boundedly order complete* if for every increasing sequence $\{x_n\}$ of positive elements in E with $\|x_n\| \le 1$, the supremum $x = \sup_n x_n$ exists and $\|x\| \le 1$. A positive element u in a Banach lattice E is a *weak order unit* if $u \wedge x = 0$ implies $x = 0$. A positive element x' in the dual E^* of E is a *strictly positive* linear form on E if $x'(|x|) = 0$ for some x implies $x = 0$; also, an x' in E^* is *order continuous* if for every decreasing net $\{x_a\}$ in E with $\inf_a x_a = 0$ we have $\lim_a x'(x_a) = 0$.

Proposition 4. Let E be a Banach lattice. Then there exist a finite measure space (Ω, Σ, μ) and a semiembedding T from E in $L_1(\mu)$ that is a lattice homomorphism onto an ideal if and only if the following conditions hold:

(i)   *E is σ-1-boundedly order complete;*

(ii)   *E has a weak order unit;*

(iii) *there exist strictly positive order continuous
linear forms on E.*

Proof. We begin with the necessity implication: let T be a lattice isomorphism from E onto an ideal of $L_1(\mu)$. Then $x \longmapsto \int Tx\, d\mu$ is a strictly positive order continuous linear form on E and (iii) holds. Since every maximal family of positive pairwise disjoint elements in $L_1(\mu)$ is at most countable, this is also true in E and therefore E has a weak order unit, which means that (ii) holds. Suppose now that $\{x_n\}$ is a sequence of positive elements in E with $x_n \leq x_{n+1}$ and $\|x_n\| \leq 1$ for each n. Then $\{Tx_n\}$ is a Cauchy sequence in $L_1(\mu)$, and T being a semiembedding, $f = \lim Tx_n = Tx$ for some $x \in E$ with $\|x\| \leq 1$. But since T is an order continuous lattice isomorphism, $x = \sup x_n$, that is, E is σ-1-boundedly order complete.

Sufficiency: Let $w' \in E^*$ be a strictly positive order continuous linear form on E. Then we may consider E to be an ideal in the AL - space (E, w'), since w' is strictly positive and order continuous (see [6, p. 113 and IV.9.3]). Let u be a weak order unit in E; u is also an order unit in (E, w'). Hence, by Kakutani's theorem [6, II.8.5], we may identify (E, w') with a space $L_1(\mu)$, μ finite. Hence we have a lattice isomorphism T from E onto an ideal of $L_1(\mu)$ with $Tu = 1$. We shall show that this T is a semiembedding. Let $\{x_n\}$ be a sequence in the unit ball of E such that $Tx_n = f_n$ converges in measure to $f \in L_0(\mu)$. Then $T(|x_n| \wedge ku) = |f_n| \wedge k$ converges to $|f| \wedge k$ for every $k > 0$. Hence there exist $z_k \in [0, ku]$ with $Tz_k = |f| \wedge k$. Passing to a subsequence we may assume that $|f_n| \wedge k$ converges μ-almost everywhere to $|f| \wedge k$, that is,

$$\bigvee_{m=1}^{i} \bigwedge_{n=m} (|f_n| \wedge k) = |f| \wedge k,$$

which implies that

$$\bigvee_{m=1} \bigwedge_{n=m} (|x_n| \wedge ku) = z_k.$$

Since E is σ-1-boundedly complete, $\|z_k\| \leq 1$. Moreover, since $\{z_k\}$ is an increasing sequence, $\bar{z} = \sup_k z_k$ exists, $\|\bar{z}\| \leq 1$, and $T\bar{z} = |f|$. Hence there exists $z \in E$ with $|z| \leq |\bar{z}|$ and such that $Tz = f$. Clearly $\|z\| \leq 1$, which completes the proof.

Corollary 2. Let E be a separable Banach lattice. Then there exists a semiembedding in L_1 if and only if E is weakly sequentially complete.

Proof. If there exists a semiembedding from E into L_1, then no closed subspace of E is isomorphic to c_0 by Corollary 12 in [4]. Hence E is weakly sequentially complete [3, 2.4]. Now suppose that E is a separable weakly sequentially complete Banach lattice. Then E has a weak order unit and also E^* has a weak order unit w' (see [6, II.6.2, II.6.5, and II.6.6]). It follows easily from (2.1) and (2.4) in [3] that w' is strictly positive and order continuous and that E is σ-1-boundedly complete. Hence there exists a semiembedding from E into a space $L_1(\mu)$ by Proposition 4 above. Since we may assume that $L_1(\mu)$ is separable, $L_1(\mu)$ is lattice isomorphic to a closed sublattice of L_1, which completes the proof.

Observe that in the proof of Proposition 4, the sequence f_n was assumed to converge *in measure* (rather than in L_1). This means that the proof applies to all L_p, $0 \leq p \leq 1$. More precisely:

Corollary 3. Let E be a Banach lattice that satisfies (i), (ii), and (iii) of Proposition 4. Then:

(i) *for each p, $0 \leq p \leq 1$, there exists a measure space with measure μ and a lattice isomorphism from E onto an ideal of $L_p(\mu)$ that is a semiembedding;*

(ii) *for each $0 \leq p < 1$, there is a semiembedding from E into L_p.*

Corollary 4. If E is a separable Banach lattice that does not contain c_0 as a sublattice, then there is a semiembedding of E in L_p for each $0 \leq p \leq 1$.

4. SPACES OF CONTINUOUS FUNCTIONS

We now take up the semiembeddings of spaces $C(X)$ (continuous real functions on a compact space X) in the space $L_0 = L_0[0, 1]$. Recall that X is *scattered* if it contains no (nonempty) perfect sets. The main result of this section is the following.

Theorem 3. Let X be a scattered space. If $C(X)$ has infinite dimension, then there are no semiembeddings of $C(X)$ in L_0.

We indicate some results that are needed for the proof of Theorem 3:

> If $T : c_0 \to L_0$ is linear and continuous for the norm   (10)
> in c_0 (and the convergence in measure in L_0), then T
> is also continuous for the weak topology on any ball
> in c_0 (and again the convergence in measure in L_0).

> All continuous linear maps from c_0 into L_0 are com-   (11)
> pact.

With (10) and (11), we can prove the following proposition of independent interest (cf. Prop. 4 in [4]):

Proposition 5. Let E be a Banach space and $U : E \to L_0$ a semiembedding. Then any continuous linear map $S : c_0 \to E$ has an extension $\tilde{S} : \ell^\infty \to E$ with the same norm.

Proof of Proposition 5. Write $T = US : c_0 \to L_0$. For $x'' \in \ell^\infty$, let $\{x_n\}$ be a sequence in c_0 converging weak* to x'' and with $\|x_n\| \leq \|x''\|$. By (11), there is a subsequence $\{y_n\}$ such that $Ty_n = USy_n$ is convergent in L_0, and, U being a semiembedding, there is a $w \in E$ such that $Uw = \lim Ty_n$. We claim that w depends only on x''. In fact, if

y'_n is a subsequence of a different x'_n also converging weak* to x'', then $y_n - y'_n \to 0$ weakly in c_0, and by (10), $T(y_n - y'_n) \to 0$ so that $Uw = \lim Ty_n$ is well defined. But U is one-to-one, and therefore w is also well defined, as claimed. Observe further that using again the fact that U is a semiembedding, we have $\|w\| \leq \sup \|Sx_n\| \leq \|S\| \|x''\|$, and therefore $\tilde{S}x'' = w$ is an extension of S satisfying $\|\tilde{S}\| = \|S\|$.

Proof of Theorem 3.   Suppose that $U : C(X) \to L_0$ is a semiembedding. Using the remark on page 238 of [4], we see that there is a subspace $V \subset C(X)$ isomorphic to c_0 and such that V is complemented in $C(X)$. According to Proposition 5, the injection $S : V \to C(X)$ has an extension $S : V^{**} \to C(X)$; but then the composition of S with the projection of $C(X)$ onto V gives a projection of V^{**} onto V. Since there are no such projections, we obtain Theorem 3.

For the sake of completeness, we give a proof of (10). Suppose that $\{x_n\}$ is a sequence in c_0 converging to 0 weakly. If $\|x_n\| \to 0$ also, then there is nothing to prove. Assume then (passing to a subsequence) that $0 < a \leq \|x_n\| \leq b$. The usual argument allows us to pass to a second subsequence (also denoted by x_n) equivalent to the standard basis in c_0. This means that for any sequence of coefficients $u_n \to 0$, the series $\sum_{n=1}^{\infty} u_n x_n$ is subseries convergent in c_0. Let $f_n = Tx_n$. Then $\sum_{n=1}^{\infty} u_n f_n$ is also subseries convergent in L_0. Using the Hilfsatz on page 29 of [5], we conclude that $\sum_{n=1}^{\infty} u_n^2 f_n^2 < +\infty$ almost everywhere if $u_n \to 0$. If a subsequence of the f_n converges to 0 in measure, the result follows. Suppose then that there are numbers $r, s > 0$ such that $m(f_n^2 \geq r) \geq s$ for all n. Pick now $F_1, F_2, \ldots$ such that:

$$0 \leq F_n \leq f_n^2; \tag{12}$$

$$F_n = r \quad \text{on a set} \quad E_n \quad \text{of measure} \quad s; \tag{13}$$

$$F_n = 0 \quad \text{off} \quad E_n. \tag{14}$$

Then the functions $g_N = \sum\limits_{n=1}^{N} u_n^2 F_n$ satisfy:

$$\int g_N = rs \sum_{n=1}^{N} u_n^2; \tag{15}$$

$$0 \leq g_1 \leq g_2 \leq \cdots \leq \sum_{n=1}^{\infty} u_n^2 F_n \leq \sum_{n=1}^{\infty} u_n^2 f_n^2. \tag{16}$$

However, if g is a measurable function satisfying $\int g = B$ and $0 \leq g \leq A$, then for $0 \leq C \leq B$, the set $\{g \geq C\}$ has measure at least $(B - C)/(A - C)$. This applied to g_N with the values $A = r \sum\limits_{1 \leq n \leq N} u_n^2$, $B = rs \sum\limits_{1 \leq n \leq N} u_n^2$, and $C = B/2$ yields

$$m\left(g_N \geq \frac{1}{2} rs \sum_1^N u_n^2\right) \geq \frac{s}{2 - s},$$

and therefore, using (16),

$$m\left(\sum_{n=1}^{\infty} u_n^2 f_n^2 \geq \frac{1}{2} rs \sum_{n=1}^{N} u_n^2\right) \geq \frac{s}{2 - s}.$$

Choosing now $u_n = n^{-1/2}$, we get that $\sum\limits_{n=1}^{\infty} n^{-1} f_n^2$ cannot be finite almost everywhere, a contradiction.

Finally, (11) follows from (10), since every bounded sequence in c_0 has a subsequence that is weak Cauchy.

REFERENCES

1. Duren, P. *The Theory of H^p-spaces*. New York: Academic Press, 1970.

2. Hille, E., and Phillips, R. *Functional Analysis and Semigroups*. Providence, R.I.: A.M.S. Colloquium Publications, 1957.

3. Lotz, H. Minimal and reflexive Banach lattices. *Math. Ann.* 209 (1974): 117-126.

4. Lotz, H. P.; Peck, N. T.; and Porta, H. Semi-embeddings of Banach spaces. *Proc. Edinburgh Math. Soc.* 22 (1979): 233-240.

5. Orlicz, W. Über die Divergenz von allgemeinen Orthogonalreihen. *Studia Math.* 4 (1932): 27-32.

6. Schaefer, H. H. *Banach Lattices and Positive Operators.* New York: Springer, 1974.

DIFFUSION ON THE LOOPS

Paul Malliavin
Institut Henri Poincaré, Paris

Given a Riemannian compact simply connected manifold M, the study of the space of closed curves $L(M)$ with a base point had been introduced by Marston Morse [10]. On some smoothness condition $L(M)$ becomes a Hilbert Riemann manifold [3]. Therefore, it has a natural Laplace operator and a diffusion process can be constructed on $L(M)$.

Starting at $t = 0$ from the zero loop, this diffusion makes possible to construct for each $t \in \mathbf{R}^+$ a natural measure γ_t on $L(M)$. Given a vector fiber bundle F of rank r on M, its holonomy defines a natural map h of $L(M)$ in the orthogonal group of dimension r; it is natural to ask what is the asymptotic, for t large, of the image measure $h_\star \gamma_t$.

A construction of γ_t had been given by B. Gaveau and E. Mazet [4] using the structure of infinite dimensional manifold of $L(M)$. In our approach we shall not use functional analysis; a path on $L(M)$ will be a function of two variables with values in M. This means that we will use more the resources of stochastic calculus for two-parameter processes than the theory of infinite dimensional manifolds.

Our starting point will be the loop space on $\mathbf{R}^n$, that is, the study of some random Fourier series of Wiener's type. We shall relate the *loop space on* $\mathbf{R}^n$ to the *Brownian sheet*.

The development on $L(M)$ in the *smooth* case can be reduced, by the moving frame, to the solution of a *nonlinear* hyperbolic system. We will have to solve this nonlinear system when the driving force is a *distribution* belonging to a Sobolev space of order -1. The existence of this solution will be proved using the double stochastic integral of Cairolli and Walsh [2] and an iterative procedure.

In the case of one-dimensional time processes, the Itô's integral makes possible to solve nonlinear systems for one-dimensional time processes.

1. TIME EVOLUTION OF WIENER-FOURIER SERIES

The Brownian Loop

It has been defined by Wiener by the Fourier series

$$u_\omega(\tau) = 2^{1/2} \sum_{k \geq 1} \frac{\sin k\pi\tau}{k} G_k, \tag{1}$$

where G_k are independent normal variables. We denote by ω the random. The probability space Ω corresponds to $\mathbf{R}^{\mathbf{N}}$ with the product measure of the normal law on each component.

The covariance is defined by $\varphi(\tau, \tau') = E\big(u_\omega(\tau)u_\omega(\tau')\big)$.

Lemma 1. $\varphi(\tau, \tau') = \tau'(1 - \tau)$ *for* $0 < \tau' < \tau < 1$.

Regularization of u_ω

It will be convenient to use the convolution by a smooth periodic function as mollifier of u_ω. In fact, we shall use the Poisson kernel and denote, for $0 \leq \rho \leq 1$,

$$u_{\omega,\rho}(\tau) = 2^{1/2} \sum_{k=1}^{+\infty} \frac{\rho^k}{k} G_k \sin k\pi\tau.$$

We remark that 1, $\{2^{1/2} \sin k\pi\tau\}$ form an orthonormal basis of $L^2([0, 1])$. Define, for $f \in L^2$,

$$\hat{f}(0) = \int f(\tau)\, d\tau, \quad \hat{f}(k) = 2^{1/2} \int f(\tau) \sin k\pi\tau\, d\tau.$$

For $f \in L^2$ and $\rho < 1$, we have then

$$I(\omega, f, \rho) = \int_0^{2\pi} f(\tau) u'_{\omega,\rho}(\tau)\, d\tau = \sum_{k \geq 1} \hat{f}(k) G_k \rho^k.$$

Then $I(\omega, f, \rho)$ is a Gaussian variable with variance

$$\sigma_\rho(f) = E\left((I(f, \rho))^2\right) = \sum_{k=1}^{+\infty} (\hat{f}_1(k))^2 \rho^{2k},$$

then, when $\rho \to 1$, we have

$$\lim \sigma_\rho(f) = \int_0^1 \left| f(\tau) - \hat{f}(0) \right|^2 d\tau.$$

White Noise

Given an abstract real Hilbert Space $\mathcal{H}$, we shall call the white noise $W_{\mathcal{H}}$ the canonical cylindrical measure on $\mathcal{H}$. For every finite set $h_i \in \mathcal{H}$, $i \in [1, n]$, the map of $\mathcal{H}$ into $\mathbf{R}^n$ defined by

$$h \to (h|h_i) \qquad i \in [1, n]$$

sent $W_{\mathcal{H}}$ into a Gaussian variable on $\mathbf{R}^n$ having as covariance matrix $(h_i|h_j)$.

Proposition 1. Denote by $\mathcal{H}$ the Hilbert space that is the orthogonal in $L^2([0, 1])$ to the constants. Let $P_{\mathcal{H}}$ be the orthogonal projection of L^2 on $\mathcal{H}$. We define the stochastic integral

$$\int f\, du_\omega = (h|W_{\mathcal{H}}) \quad \text{where} \quad h = P_{\mathcal{H}} f.$$

Then

$$\lim_{r \to 1} \int f \, du_{r,\omega} = \int f \, du_{\omega}.$$

Conditioning White Noise

We can consider the Hilbert space $\mathcal{H}_1 = L^2([0, 1])$. Then the white noise $W_{\mathcal{H}}$ considered can be interpreted as the white noise $W_{\mathcal{H}_1}$, *under the conditioning* $W_{\mathcal{H}}(1) = 0$.

More generally, given an abstract Hilbert space $\mathcal{H}_2$ and the canonical white noise defined on $\mathcal{H}_2$, if we give a finite system of linear form $h_{\alpha_1} \ldots h_{\alpha_s}$ on $\mathcal{H}_2$, we can interpret the white noise $W_{\mathcal{H}_2}$ under the conditioning

$$(W_{\mathcal{H}_2} | h_{\alpha_\ell}) = \beta_\ell, \quad 1 \leq \ell \leq s,$$

where β_ℓ are given constants. We denote by $W_{\mathcal{H}_2}^{h_\alpha}$ this conditioned white noise. Given h_i, $1 \leq i \leq n$, we consider the map

$$\mathcal{H}_2 \to \mathbf{R}^{s+n} \qquad \text{defined by}$$

$$h \to (h|h_\alpha), \ (h|h_i).$$

The image of $W_{\mathcal{H}_2}$ under this map is a Gaussian measure; we shall consider its conditioned law on the affine linear variety $\xi_j = \beta_j | 1 \leq j \leq s|$; we will get a Gaussian measure.

On the other hand, we could introduce the orthogonal $\mathcal{H}_3$ to the linear subspace K generated by the h_α. Then

$$W_{\mathcal{H}_2}^{h_\alpha} = W_{\mathcal{H}_3}. \qquad (2)$$

In this form the conditional white noise does not depend on the values of the conditioning. Denote by W_K the white noise associated to the finite dimensional subspace K (it can be identified to the normal Gaussian variable on K); then we have

$$W_{\mathcal{H}_2} = W_{\mathcal{H}_3} + W_K \qquad\qquad (3)$$

this sum being a sum of independent variables.

Markov Property for the Brownian Loop

*Proposition 2. Let γ be a subinterval of $[0, 1]$, γ^c its comple-
ment in $[0, 1]$. Consider the conditioning:*

*$u_\omega(\tau)$ is given at the two boundary points of γ; we shall de-
note this conditioning by*

$$u_\omega\big|_{\partial\gamma}.$$

*Then the two processes $u_\omega\big|_\gamma$ and $u_\omega\big|_{\gamma^c}$ are independent under the
conditioning $u_\omega\big|_{\partial\gamma}$.*

Proof. The differential du_ω, in the sense of distribution, is the
white noise W on $L^2([0, 1])$, under the conditioning

$$W(1) = 0.$$

As $1_\gamma + 1_{\gamma^c} = 1$, this conditioning can be written as

$$W(1_\gamma) = -W(1_{\gamma^c}).$$

Let P_1, P_2 be the orthogonal projections on $L^2([0, 1])$ defined
by $P_2 f = 1_\gamma f$, $P_3 f = 1_{\gamma^c} f$. Let $\mathcal{H}_i$ be the range of P_i, W_i the white
noise associated to $\mathcal{H}_i$; then the decomposition $L^2 = \mathcal{H}^1 \overset{\perp}{\oplus} \mathcal{H}^2$ implies:

$$W = W_2 + W_3 \quad \text{(sum of independent variables)}.$$

Let $\mathcal{H}$ be the orthogonal of 1 in $L^2([0, 1])$. Define a map of
$\mathcal{H}_2 \oplus \mathcal{H}_3 \to \mathbf{R}^2$ by $(h_2, h_3) \to ((h_2|1), (h_3|1))$.

Note that $(h_2|1) = (h_2|P_2 1) = (h_2|1_\gamma)$.

Denote by K the two-dimensional subspace of $L^2([0, 1])$ generated by 1_γ, 1_{γ^c}:

$$K^{\perp} \cap \mathcal{H}^2 = \{h_2 \in \mathcal{H}_2; \ (1_{\gamma}|h) = 0\}.$$

As $(K^{\perp} \cap \mathcal{H}_2)$ and $(K^{\perp} \cap \mathcal{H}_3)$ are orthogonal, the associated white noise are independent; we shall denote them by W_4 and W_5. The conditioning by 1_{γ}, 1_{γ^c} is stronger than the conditioning by $1_{\gamma} + 1_{\gamma^c}$, and therefore

$$W^{1=0} = \int_{\mathbb{R}} (W_4^{1_{\gamma}} + W_5^{1_{\gamma^c}}) \, d\sigma,$$

where $d\sigma$ is the law of $W^{1=0}(1_{\gamma^c})$.

We have then shown the proposition at the level of stochastic differential. As, by the continuity a.s. of $u_{\omega}(.)$ we have, if $\gamma = [\alpha, \beta]$, $u_{\omega}(\alpha) + \int_{\alpha}^{\tau} du_{\omega} = u_{\omega}(\tau)$, for $\tau \in [\alpha, \beta]$. We have therefore proved the proposition.

Conditional Laws

Proposition 3. Let $u_{\omega}(\tau)$ *be the Brownian loop. Let* $0 < \xi_1 < \xi_3 < \xi_2 < 1$, $\xi_3 = \alpha\xi_1 + \beta\xi_2$, *where* $\alpha + \beta = 1$. *The variable* $u_{\omega}(\tau_3)$, *conditioned by* $u_{\omega}(\tau_1)$ *and* $u_{\omega}(\tau_2)$, *can be written*

$$u_{\omega}(\tau_3) = \alpha u_{\omega}(\tau_1) + \beta u_{\omega}(\tau_2) + \gamma G,$$

where G is a normal variable, independent of the conditioning, and where

$$\gamma = (\xi_2 - \xi_1)^{1/2}.$$

Proof. The covariance computed in Lemma 1.

The Brownian Loop and the Brownian Path

Proposition 4. Let $b_{\omega_1}(\tau)$ *be the Brownian path starting at zero. Then there exists a mapping* $\omega \to \omega_1$ *preserving the probability such that*

$$b_{\omega_1}(\tau) - \tau b_{\omega_1}(1) = u_{\omega}(\tau).$$

Proof. We can use the Wiener's series for the Brownian motion:

$$b_{\omega_1}(\tau) = \tau G_0 + 2^{1/2} \sum_{k>1} G_k \frac{\sin k\pi\tau}{k}$$

(where the G_k are independent normal variables).

Another way to see the same fact will be to use the differential db_{ω_1}, which is the unconditioned white noise on $[0, 1]$.

Stochastic Differential Equation Driven by the Brownian Loop

Let $u_\omega^1, \ldots, u_\omega^n$ n independent samples of the Brownian loop, $n^2 + n$ smooth functions $a_j^i(x)$, $c^i(x)$ on $\mathbf{R}^n$ been given, we want to solve the S.D.E. system defined by

$$dx_\omega^i(\tau) = a_j^i(x_\omega(\tau)) \, du_\omega^j(\tau) + c^i(x_\omega(\tau)) \, d\tau.$$

Using Proposition 4, this system can be written

$$dx_{\omega_1}^i(\tau) = a_j^i(x_{\omega_1}(\tau)) \, db_{\omega_1}^j(\tau) + [c^i(x_{\omega_1}(\tau)) - a_j^i(x_{\omega_1}(\tau)b_{\omega_1}^j(1)] \, d\tau.$$

This system can be solved by iteration of the corresponding integral equation (see J. M. Bismut [1]).

Time Evolution of the Brownian Loop

Our objective is to consider the parameter $\tau \in T$ as "a geometric variable" and to introduce a "time parameter," taking its values in R^+. The increasing in time will correspond to the augmentation of the variance of the Brownian loop. At time $t = 0$, we will start with the function identically to zero. This point of view had been used in [6]. Define the function

$$q_\xi(t, \tau) = \sum_{k\geq 1} \frac{1}{k} G_k(t) \sin k\pi\tau, \tag{4}$$

where $G_k(t)$ are independent samples of the Brownian motion on $\mathbf{R}$, starting at time $t = 0$ from zero. We shall denote by Ξ the infinite product of the probability space of the Brownian on **R** starting from zero. Then $\xi \in \Xi$ denotes the generic point of Ξ.

Theorem 1. *Denote* $\mathscr{C}^\beta$ *the class of* β-*Hölderian function of* τ *defined on* $[0, 1]$, *then a.s. in* ξ *the mapping*

$$t \to q_\xi(t, \cdot)$$

is a continuous map from R^+ *to* $\mathscr{C}^\beta$ *for* $\beta < \dfrac{1}{2}$.

Proof. Cf [7], Proposition 5.1, p. 246].

Definition. We shall call the map

$$(t, \tau) \to q_\xi(t, \tau)$$

the *loop-sheet*.

The Brownian Sheet

We define the Brownian sheet by

$$p_{\xi_1}(t, \tau) = \tau G_0(t) + 2^{1/2} \sum_{k \geq 1} \frac{\sin k\pi\tau}{k\pi} G_k(t).$$

Then the random parameter $\xi_1 \in \Xi \times \Omega$, where the last factor of this product of measure spaces is the probability space of the Brownian on **R**. We have by Proposition 4,

$$q_\xi(t, \tau) = p_{\xi_1}(t, \tau) - \tau p_{\xi_1}(t, 1).$$

The R^n-Valued Sheets

We shall define the *Brownian sheet* on $\mathbf{R}^n$ taking $p_{\xi_1}^1, \ldots, p_{\xi_1}^n$ n independent components of the p_{ξ_1}. In the same way, the loop-sheet on $\mathbf{R}^n$ is defined by taking n independent components $q_\xi^1, \ldots, q_\xi^n$.

2. RIEMANNIAN DEVELOPMENTS

Let M be a compact Riemannian manifold of dimension n. Let $m_o \in M$ be fixed. We shall consider $P_{m_o}(M)$ the path space on M starting from m_o that is the continuous maps u of $[0, 1] \to M$, such that $u(0) = m_o$. We shall denote by

$$L_{m_o}(M) = \{ u \in P_{m_o}(M) ; \ u(1) = m_o \}$$

the loop space with base point m_o.

The development problem is to take some sample path of the $p_\xi(t, \tau)$ in the Euclidian tangent space $T_{m_o}(M)$ and to *develop* it on the manifold, solving some stochastic system, obtaining then a map $\tilde{p}_\xi(t, \tau)$ with values in M.

In [8], such development is realized using the variation of the stochastic differential system defining the parallelism transport. The development that we shall construct will be different; its machinery is based on the natural Riemannian metric on the space of path (or curves) of Sobolev class 1.

The Riemann Hilbert Structure on $P^1(M)$

We shall denote

$$P^1_{m_o}(M) = \left\{ u \in P_{m_o}(M) ; \ \int_0^1 \left\| \frac{du}{d\tau} \right\|^2 d\tau < +\infty \right\}$$

$$L^1_{m_o}(M) = L_{m_o} \cap P^1_{m_o}.$$

Finally, we denote P^∞ the space of smooth path.

Let φ_t be a smooth curve on $P^\infty_{m_o}$. Then φ_t can be seen as a function of two parameters $\varphi(t, \tau)$. We want to compute its length on L^1

$$\int_{t_0}^{t_1} \| d\varphi_t \|_{L^1} .$$

We take by definition

$$\| d\varphi_t \|^2_{L^1} = \int_0^1 \| \frac{\partial \varphi}{\partial t} \|^2 d\tau + \int_0^1 \| \nabla (\frac{\partial \varphi}{\partial t}) \|^2 d\tau, \tag{5}$$

where ∇ denotes the covariant derivative along the curve $\tau \to \varphi_t(\tau)$.

Then it can be proved [3] that this metric defines a Riemann-Hilbert structure on $P_m^1(M)$.

The Normal Charts on $P^1(M)$.

Given $\varphi \in P_{m_0}(M)$, let $\Gamma_\varphi = \varphi_*(T(M))$ be the pull back bundle of the tangent bundle $T(M)$.

Then, given a continuous section s of Γ_φ, with $s(0) = 0$, we can define

$$\exp_\varphi(s) = \psi \in P_{m_0}(M) \quad \text{by the formula}$$

$$\psi(\tau) = \exp_{\varphi(\tau)}(s(\tau)),$$

where $\exp_m$ denotes the normal chart relative to M with center m.

Development of $\Gamma\varphi$

If $\varphi \in L^\infty$, then the Riemannian connection makes possible to identi-fy a section of Γ_φ with a function *defined* on $[0, 1]$ with values in $P_{m_0}(M)$.

The space of sections $\mathcal{H}_1$ are the sections for which the fol-lowing norm is finite:

$$\| s \|^2_{\mathcal{H}} = \int_0^1 \| s(\tau) \|^2 d\tau + \int_0^1 \| \nabla s \|^2 d\tau,$$

where ∇ is the covariant derivative along the smooth path φ. De-note

$$0_{\varphi, \varepsilon} = \{ \varphi ; \| \varphi \|_{\mathcal{H}_1} < \varepsilon \}.$$

Then

$$s \to \exp_\varphi(s)$$

defines, for ε small enough, a normal chart of the Riemann-Hilbert structure on $P^1_{m_0}(M)$ (cf. [3]).

The Development on M

Let p be a C^∞-map of $(R^+)^2$ into $T_{m_0}(M)$, equal to zero on the boundary. We shall define the development $\tilde{p}$ on M by

$$\tilde{p}(0,\ \tau) = \tilde{p}(t,\ 0) = m_0$$

$$\frac{\partial\tilde{p}}{\partial t} = \frac{\partial p}{\partial t}\ , \qquad\qquad (6)$$

equation (6) being understood under the development of Γ_φ of the first member, which is a section of Γ_φ (where $\varphi(\tau) = \tilde{p}(t,\ \tau)$), to $T_{m_0}(M)$-valued function. We shall write equation (6) more explicitly using a lifted curve to the frame bundle.

The Development on $0(M)$

Denote by $0(M)$ the $0(n)$-principal bundle of orthonormal frames on M. Denote by **p** the projection of $0(M)$ over M. Recall that a frame at $m \in M$ is a Euclidian isomorphism r of R^n into $T_m(M)$. Then the orthogonal group $0(n)$ operates on the right by

$$r \to r \circ g, \quad g \in 0(n), \quad r \in 0(M).$$

We can consider the canonic horizontal vector fields on $0(M)$, $A_1,\ \ldots,\ A_n$ defined by the Riemannian connection.

Given $r_1,\ r_2 \in p^{-1}(m)$, we have two bijection $\varphi_1,\ \varphi_2$ of $0(n)$ onto $p^{-1}(m)$ defined by $\varphi_i(g) = r_i \circ g$. As $r_2 = r_1 \circ \gamma$, we deduce that $\varphi_2(g) = \varphi_1(\gamma g)$, which means that $p^{-1}(m)$ can be identified to $0(n)$, in a canonic way, up to left translation. We can therefore identify vertical tangent vectors with the Lie algebra of *left invariant* vector field on $0(n)$. Denote by $\underline{0}(n)$ this Lie algebra. We shall denote by π the differential form with values in $\mathbf{R}^n \times \underline{0}(n)$ defining the parallelism on $0(M)$.

The component of π will be denoted by π^i, π^i_j. We shall also denote by $\dot{\pi}$ the component on $\mathbf{R}^n$, $\ddot{\pi}$ the component on $\underline{O}(n)$.

The horizontal lift of $\tau \to \tilde{p}(t, \tau)$ is defined by the equation

$$\begin{cases} < \frac{\partial r}{\partial \tau}, \dot{\pi} > = r^{-1}\left(\frac{\partial \tilde{p}}{\partial \tau}\right) \\ < \frac{\partial r}{\partial \tau}, \ddot{\pi} > = 0. \end{cases} \tag{7}$$

Then the parallel transport along $\tau \to \tilde{p}(t, \tau)$ is defined by

$$r(t, \tau) \circ r_o^{-1}.$$

Therefore, equation (6) can be written

$$< \frac{\partial r}{\partial t}, \pi > = r_o\left(\frac{\partial p}{\partial t}\right). \tag{8}$$

The two equations (7) and (8) will define the development on $O(M)$. We shall write these equations using the pull back form by r of the parallelism form:

$$(r)^\star \pi = \Theta.$$

Then writing the component of Θ on $\mathbf{R}^n$ and $\underline{O}(n)$ and using (7) and (8), we obtain the following Darboux equation:

$$\begin{cases} \Theta = (\dot{a} + \ddot{a})\, dt + (\dot{c} + \ddot{c})\, d\tau \\ \ddot{c} = 0 \\ \dot{a} = r_o\left(\frac{\partial p}{\partial t}\right) \end{cases} \tag{9}$$

with the boundary condition

$$\begin{cases} \dot{c}(0, \tau) = \ddot{c}(0, \tau) = 0 \\ \dot{a}(t, 0) = \ddot{a}(t, 0). \end{cases} \tag{10}$$

The Structure Equation on $O(M)$

The structure equation can be written $R^j_{i,k,\ell}$ denoting the curvature tensor,

$$\langle d\pi, \, z_1 \wedge z_2 \rangle + [\ddot{\pi}(z_1)\dot{\pi}(z_2) - \ddot{\pi}(z_2)\dot{\pi}(z_1)] = 0$$

$$\langle d\ddot{\pi}, \, z_1 \wedge z_2 \rangle + [\ddot{\pi}(z_1)\ddot{\pi}(z_2) - \ddot{\pi}(z_2)\ddot{\pi}(z_1)] + R_{k,\ell}\overset{\bullet k}{\pi}(z_1)\overset{\bullet \ell}{\pi}(z_2) = 0.$$

3. THE SMOOTH DEVELOPMENT

The system (9) gives *two equations* when we have *three components* to determine: $\dot{c}$, $\dot{a}$, $\ddot{a}$. To complete our computation, we shall use the structure equations

$$r^\star \, d\pi = d(r^\star \pi) = d\Theta,$$

which give, taking in account the fact that $\ddot{c} = 0$,

$$\left(\frac{\partial \dot{a}}{\partial \tau} - \frac{\partial \dot{c}}{\partial t}\right) d\tau \wedge dt = -\ddot{a}\dot{c} \, d\tau \wedge dt$$

$$\left(\frac{\partial \ddot{a}}{\partial \tau}\right) d\tau \wedge dt = (R_{k,\ell}\overset{\bullet k}{a}\overset{\bullet \ell}{c}) \, d\tau \wedge dt.$$

We obtain finally the following system:

$$\dot{a} = P_t \tag{11}$$

$$\frac{\partial \dot{c}}{\partial t} = -\ddot{a}\dot{c} + P_{t\tau}, \quad c(0, \, \tau) = 0 \tag{12}$$

$$\frac{\partial \ddot{a}}{\partial \tau} = R_{k,\ell}\overset{\bullet k}{a}\overset{\bullet \ell}{c}, \quad a(t, \, 0) = 0 \tag{13}$$

4. CONSTRUCTION OF THE DEVELOPMENT

We shall introduce the matrix $X(t)$ solution of the ordinary dif-ferential matrix equation

$$\begin{cases} \dfrac{dX}{dt} = -\ddot{a}X \\[2mm] X(0, \, \tau) = \text{Identity}. \end{cases} \tag{14}$$

Then $X(t, \, \tau) \in 0(n)$ and we can write the solution of the linear equation (12) on the integral form

$$c(t, \tau) = X(t, \tau) \int_0^t \tilde{X}(t' \tau) p_{t,\tau}(t', \tau) \, dt', \tag{15}$$

where $\tilde{X}$ denotes the adjoint of X.

Introduce a matrix $B(t, \tau) \in O(n)$ by the following P.D.E.:

$$\frac{\partial B}{\partial t \partial \tau} = R_{k,\ell} p_t^k(t, \tau) \left[x_{\underline{j}}^k(t, \tau) \int_0^t x_{\underline{j}}^s(t', \tau) p_{t,\tau}^s(t', \tau) \, dt' \right],$$

$$B(0, \tau) = B(t, 0) = 0.$$

(The underlined indices correspond to a summation not included in Einstein convention.) In this formula we consider that the second member of the equation is known, that is finally, *B is defined by the following integral*:

$$B(t_0, \tau_0) \tag{16}$$

$$= \int_0^{t_0} \int_0^{\tau_0} R_{k,\ell} [r(t, \tau)] p_t \, dt \, d\tau \left[\int_0^t x_{\underline{j}}^\ell(t, \tau) x_{\underline{j}}^s(t', \tau) p_{t,\tau}^s(t', \tau) \, dt' \right].$$

Our purpose is to express everything in terms of $p_{t,\tau} = u$. For instance, we have

$$p_t^k = \int_0^\tau u^k(t, \tau) \, dt.$$

On $(R^+)^2$ we introduce the order relation

$$(t', \tau') < (t, \tau) \iff t' \le t \quad \text{and} \quad \tau' \le \tau.$$

We denote by $T_1 = (t, \tau)$ a point of $(R^+)^2$.

We define for this order relation

$$T_1 \vee T_2 = \sup(T_1, T_2)$$

$$Q_T = \{(t', \tau'); \ (t', \tau') < T\}$$

Define on $(R^+)^2 \times (R^+)^2$ the following function h by

$$h(T_1, T_2) = 0 \quad \text{if} \quad T_1 < T_2 \quad \text{or} \quad T_2 < T_1 \quad \text{or} \quad t_1 < t_2$$

$$h(T_1, T_2) = 1 \quad \text{otherwise.}$$

Lemma 2.   We have

$$B(t, \tau)$$

$$= \iiiint_{Q_T^2} R_{k,\ell}\left(r(T_1 \vee T_2)\right)x_j^\ell(T_1 \vee T_2)x_j^{\underline{s}}(T_2)h(T_1, T_2)u^k(dT_1)u^{\underline{s}}(dT_2), \tag{17}$$

where
$$u^s(dT) = p_{t,\tau}^s(t, \tau) \, dt \, d\tau.$$

Proof.   We can fix $T_1 \vee T_2 = T_0$ and integrate in the remaining variables. We will then have to compute a double integral. The introduction of h means that we have only to integrate on the set

$$T_1 = (t_0, \tau) \quad \text{and} \quad T_2 = (t, \tau_0), \qquad t < t_0, \ \tau < \tau_0. \tag{18}$$

Denote $\varphi : (T_1, T_2) \to T_1 \vee T_2$, the relation (18) means that, for the set on which we integrate, we have

$$\varphi(T_1, T_2) = (t_2, \tau_1).$$

This reduces to conditioning by φ to the usual Fubini:

$$\iiint_{Q_T^2} = \iint_{(t_2,\tau_1)<T} dt_1 \, d\tau_2 R_{k,\ell}\left(r(t_1, \tau_2)\right)x_j^k(t_1, \tau_2)$$

$$\left[\int_0^{t_1} x_j^{\underline{s}}(t_2, \tau_2)p_{t,\tau}^s(t_2, \tau_2) \, dt_2\right]\left[\int_0^{\tau_2} p_{t,\tau}^k(t_1, \tau_1) \, d\tau_1\right].$$

Performing the last integration, we recognize (16).

Computation of X

We use the matrix differential equation

$$\frac{\partial X}{\partial t} = \frac{\partial B}{\partial t} X, \qquad X(0, \tau) = \text{Identity}. \tag{19}$$

Computation of r

We will have two different ways to compute $r(t, \tau)$: either to choose a differential equation in τ or in t. For the development of the loop-sheet, the second point of view will be better. This means that r will be defined by the solution of the differential equation

$$\begin{cases} \langle \frac{\partial r}{\partial t}, \dot{\pi} \rangle = p_t \\ \langle \frac{\partial r}{\partial t}, \ddot{\pi} \rangle = \frac{\partial B}{\partial t} \, . \\ r(0, \tau) = r_o \end{cases} \qquad (20)$$

The equations (17), (19), and (20) form an integro-differential system of the development of smooth sheet.

Geometric Signification of X

Suppose that we are developing a smooth loop-sheet that is

$$p(t, 0) = p(t, 1) = m_o \quad \text{for all } t; \text{ then solve}$$

$$\frac{dZ}{dt} = Z\ddot{a}, \ Z(t, 0) = I.$$

We have then that $r(t, 1) = r_o \circ Z(t, 1)$, that is, Z is the *stochastic holonomy*. We have obviously that Z is the transposed of X.

Integration of the System

We shall use an iteration method of perturbative nature. The dominant part of the system will be (17).

We shall take

$$B_1(t, \tau) = R_{k,\ell}(r_o) p^k(t, \tau) p^\ell(t, \tau).$$

Suppose that we have defined $B_{n-1}(t, \tau)$; define X_n by the matrix differential equation

$$\frac{\partial X_n}{\partial t} = \frac{\partial B_{n-1}}{\partial t} X_n.$$

Using $\frac{\partial B_{n-1}}{\partial t}$ instead of $\frac{\partial B}{\partial t}$ in (20), we construct in the same way r_n. Then B_{n+1} is defined by

$$B_{n+1}(t, \tau) = \iiiint\limits_{Q_T^2} R_{k,\ell}(r_n(T_1 \vee T_2))x_{j,n}^{\ell}(T_1 \vee T_2)x_{\underline{j}}^{s}(T_2)\mathbf{h}(T_1, T_2)$$

$$\cdot\ p_{t,\tau}^{k}(T_1)p_{t,\tau}^{s}(T_2)\ dT_1\,dT_2\ .$$

Then if $p_{t,\tau}$ is a continuous function of (t, τ), and if $R_{k,\ell}$ is smooth, then $B_n(t, \tau)$ will converge to the solution of the development.

5. THE REGULARIZATION PROCEDURE AND THE STOCHASTIC SHEET

Denote by W the white noise constructed on the Hilbert space $L^2(R^2)$. Then it is well known that the Brownian sheet on R^n can be written

$$p^k(t, \tau) = w^k([0, t] \times [0, \tau]), \quad k \in [1, n],$$

where the w^k are independent white noise.

Then equation (17) can be written formally

$$\tilde{B}(t, \tau)$$

$$= \iiiint\limits_{Q_T^2} R_{k,\ell}(r(T_1 \vee T_2))x_{\underline{j}}^{\ell}(T_1 \vee T_2)x_{\underline{j}}^{s}(T_2)h(T_1, T_2)w^k(dT_1)w^s(dT_2),$$

which is the form of a Cairolli-Walsh double stochastic integral [2]. We have to use a regularization procedure as in [7] to derive from the ordinary integral equation the stochastic integral equation. We shall denote by $v(\lambda)$ a smooth function with support $[-1, +1]$ and satisfying $\int v(\lambda)\,d\lambda = 1$, $\int v(\lambda)\lambda\,d\lambda = 1$.

We denote
$$v_\varepsilon(\tau) = \frac{1}{\varepsilon}v\left(\frac{t}{\varepsilon}\right),\quad v_\eta(\tau) = \frac{1}{\eta}v\left(\frac{\tau}{\eta}\right)$$

$$p_{\varepsilon,\eta}(t, \tau) = \iint v_\varepsilon(t + t')v_\eta(\tau + \tau')W(dT').$$

Then $p_{\varepsilon,\eta}$ is a C^∞-function of (t, τ) and we can write the corresponding system defining the development on M. We shall first fix ε and let $\eta \to 0$. We shall after let $\eta \to 0$ and then get the equation

of the stochastic development. To obtain the stochastic equation
in one variable, we shall use a transfer principle to compute a
limit of O.D.E. as the solution of an S.D.E. [7]. To compute the
contracted term, we will have to introduce others S.D.E. of the
form of (19) and (20). The contracted expression of B will involve
those solutions depending themselves on B. The iterative scheme
indicated in "Integration of the System" will prove the existence
of the Brownian sheet. The Brownian loop can be obtained using
the point of view of the section "Stochastic Differential Equation
Driven by the Brownian Loop." To finish, we shall compare the development discussed here with the construction of [8]. In this
last paper, the smooth equation was

$$\begin{cases} \langle \dot{\pi}, \dfrac{\partial r}{\partial \tau} \rangle = \dfrac{\partial p}{\partial \tau} \\ \langle \ddot{\pi}, \dfrac{\partial r}{\partial \tau} \rangle = 0 \end{cases}$$

and the use of a structure equation was not necessary for the definition of the sheet; it was appearing only for the computation of
the stochastic differential of the sheet map.

REFERENCES

1. Bismut, J. M. Mécanique aléatoire. Lecture Note, 866,
    Springer-Verlag.

2. Cairolli, R., and Walsh, J. B. Stochastic integrals in the
    plane. *Acta Math*. 134 (1975): 111–183.

3. Flaschel, P., and Klingenberg, W. Riemannsche Hilbert mannig-
    faltigkeiten. Lecture Note, 282, Springer-Verlag.

4. Gaveau, B., and Mazet, E. Diffusions sur des variétés de
    chemins. *Comptes Rendus Acad. Sciences Paris* 289 (1979):
    643–645.

5. Malliavin, P. *Géométrie Différentielle Stochastique*. Presses
    de l'Université de Montréal, 1978.

6. Malliavin, P. Cacul des variation stochastiques subordonies. *Comptes Rendus Acad. des Sciences* 295 (1982).

7. Malliavin, P. Stochastic calculus of variation. *Proc. of the International Conf. on S.D.E., Kyoto,* 1976, Kinokuniya et John Wiley, 1978.

8. Malliavin, P. Stochastic Jacobi field. *Proc. International Conf. on P.D.E. on Manifold,* Salt Lake City, 1978.

9. Michel, D. Formule de Stockes stochastique. *Bull. Sciences Math.* 103 (1979): 193-240.

10. Morse, M. Calculus of variation in the Large. *Colloquium Amer. Math. Soc.* (1932).

WEIGHTED ANALOGUES OF
NIKOLSKII-TYPE INEQUALITIES AND THEIR APPLICATIONS

H.N. Mhaskar
California State University, Los Angeles

1. INTRODUCTION

Nikolskii inequalities relate different L^p-metrics of a function in a specified class. Thus, for example, in the treatise of Timan [15, pp. 229-231] one finds the following inequality concerning trigonometric polynomials.

Theorem 1. Let $0 < p \leq q \leq \infty$. Then, for every trigonometric polynomial T_n of order at most n, we have

$$\left\| T_n \right\|_{L^q[0,2\Pi]} \leq 2n^{\frac{1}{p}-\frac{1}{q}} \left\| T_N \right\|_{L^p[0,2\Pi]} , \quad n \geq 1.$$

This theorem was proved by S. M. Nikolskii for $p \geq 1$. It was also extended for trigonometric polynomials in several variables. Nikolskii has also extended these results to the case of entire functions of finite exponential type [12].

In this paper, we introduce similar type of inequalities for algebraic polynomials and for suitably weighted norms. Of particular interest is the case when the weight is supported on the whole real line. The methods used to prove the inequalities above use a direct estimation of the norms in question. For a general weight function, these methods are not quite suitable to apply. Instead, we shall make use of some estimates on the Christoffel functions and de la Vallee Poussin means associated with polynomials orthogonal with respect to the weight function in question.

Our inequalities play an important role in investigations concerning the convergence of orthogonal polynomial expansions, characterization of entire functions of finite exponential type in terms of the constructive properties of their restrictions to the real line [9], and so on. In this paper, we shall discuss an application to Fourier transforms.

It is well known that the Fourier transform of a function f in $L^p(R)$, $p > 2$, need not exist as a function. We shall give some conditions on the constructive properties of f, under which the Fourier transform $\hat{f}$ of f will be a function in $L^r(R)$ for certain values of r that we shall specify. We shall also establish a relationship between the smoothness of f and that of $\hat{f}$. This smoothness will be measured by a certain modification of Besov space norms, which we shall introduce. As an application of our inequalities, we shall also prove embedding theorems involving these modified Besov spaces.

2. MAIN RESULTS

Let w be a weight function such that $wx^n \in L^1(R) \cap L^\infty(R)$, $n = 0, 1, 2, \ldots$. Let $\{p_k\}_{k=1}^\infty$ be the sequence of orthonormalized polynomials with respect to w^2;

$$p_k(x) = \gamma_k x^k + \cdots, \quad \gamma_k > 0 \tag{1a}$$

$$\int_R p_k(x) p_j(x) w^2(x)\, dx = \delta_{jk} \tag{1b}$$

$$\Gamma_n = \sup_{1 \leq m \leq n} \frac{\gamma_{m-1}}{\gamma_m} \tag{1c}$$

$$M_n = \left\| w^2 \sum_{k=0}^{2n} p_k^2 \right\|_\infty \tag{1d}$$

$$N_n = \frac{\Gamma_{2n}}{2n} M_n . \qquad (1e)$$

Here and elsewhere in the sequel, $\| \cdot \|_p$ will denote the norm of the $L^p(R)$ space, unless otherwise specified. We shall denote the class of all polynomials of degree at most n by π_n. Also, by $A \ll B$, we mean that $A \leq cB$ for some positive constant c depending on w and the metrics involved but on nothing else.

The most general form of our analogue of the Nikolskii-type inequalities is given in the following theorem.

Theorem 2.   Let $P \varepsilon \pi_n$.

(a) *If $1 \leq p \leq r \leq 2$,*

$$\| wP \|_r \ll M_n^{\frac{1}{p} - \frac{1}{r}} N_n^{\frac{1}{r} - \frac{1}{2}} \| wP \|_p . \qquad (2)$$

(b) *If $1 \leq p \leq 2 \leq r$,*

$$\| wP \|_r \ll M_n^{\frac{1}{p} - \frac{1}{r}} \| wP \|_p . \qquad (3)$$

(c) *If $2 \leq p \leq r \leq \infty$,*

$$\| wP \|_r \ll M_n^{\frac{1}{p} - \frac{1}{r}} N_n^{\frac{1}{2} - \frac{1}{p}} \| wP \|_p . \qquad (4)$$

The most interesting case occurs when the sequence $\{N_n\}$ is bounded. Weights that satisfy this condition have already risen to importance in the investigations concerning weighted polynomial approximation, convergence of orthogonal polynomial expansion, zeros of orthogonal polynomials, and so on, recently carried out by Freud [5]. Examples of such weight functions are $w_\alpha(x) = \exp(-|x|^\alpha)$, $\alpha \geq 2$. For such weights, we have the following.

Corollary 1.   If N_n is a bounded sequence and $P \varepsilon \pi_n$, $1 \leq p \leq r \leq \infty$,

$$\| wP \|_r \ll M_n^{\frac{1}{p}-\frac{1}{r}} \| wP \|_p \ll \left(\frac{2n}{\Gamma_{2n}}\right)^{\frac{1}{p}-\frac{1}{r}} \| wP \|_p. \tag{5}$$

In the remaining theorems of this paper, we shall restrict our attention to a special class of weight functions, also including all the weights $\exp(-|x|^\alpha)$, $\alpha \geq 2$, for which the theory of weighted polynomial approximation has been more fruitfully investigated. Thus, we do not yet know Bernstein's inequality and converse theorems in weighted polynomial approximation for weights satisfying only the hypothesis of Corollary 1; but these are known for the more special class of weights that we now introduce.

We say that a weight function $w_Q(x) = \exp(-Q(x))$ is *super-regular* if:

(i) Q is even, convex, twice continuously differentiable on $(0, \infty)$ and $Q'(x) \to \infty$ as $x \to \infty$.

(ii) Q'' is increasing.

(iii) $t\,\dfrac{Q''(t)}{Q'(t)} \leq K$, for all $t \geq a > 0$,   (6)

where K and a are constants.

(iv)   Let

$$q_n Q'(q_n) = n. \tag{*}$$

Then there exist constants c and d such that

$$q_n < cq_n \leq q_{2n} \leq dq_n < 2q_n \tag{7}$$

Corollary 2. Let w_Q be *superregular. Let* $P \in \pi_n$, $1 \leq p \leq r \leq \infty$, and q_n be as in (*). Then

$$\| w_Q P \|_r \ll \left(\frac{n}{q_n}\right)^{\frac{1}{p}-\frac{1}{r}} \| w_Q P \|_p. \tag{8}$$

For superregular weights, we also have the following in-equality.

Theorem 3. Let w_Q be superregular, $P \varepsilon \pi_n$, $1 \leq p \leq r \leq \infty$, and q_n be as in (*). Then

$$\| w_Q P \|_p \ll q_n^{\frac{1}{p} - \frac{1}{r}} \| w_Q P \|_r \ll \left(\frac{n}{q_n} \right)^{\frac{1}{p} - \frac{1}{r}} \| w_Q P \|_r. \qquad (9)$$

We now turn our attention to the definition of Besov spaces modified to incorporate the weight function (Definition 1).

If w_Q is superregular and $w_Q f \varepsilon L^p(R)$, we define the modulus of smoothness of f as follows (equation (10)):

$$\omega_2 (L^p, w_Q; f, \delta)$$

$$= \sup_{|t| \leq \delta} \| \Delta_t^2 (w_Q f) \|_p + \delta \sup_{|t| \leq \delta} \| Q_\delta' \Delta_t (w_Q f) \|_p + \delta^2 \| Q_\delta'^2 w_Q f \|_p,$$

where

$$\Delta_t f(x) = f(x + t) - f(x),$$

$$\Delta_t^2 f = \Delta_t (\Delta_t f), \quad Q_\delta'(x) = \min\{\tfrac{1}{\delta}, (1 + Q'^2(x))^{\frac{1}{2}}\}$$

$$\Omega_2 (f, \delta) \equiv \Omega_2 (L^p, w_Q; f, \delta)$$

$$= \inf\{\omega_2 (L^p, w_Q; f - a - bx, \delta) : a, b \varepsilon R\}. \qquad (10)$$

Let

$$\varepsilon_n (L^p, w_Q, f) = \inf\{ \| w_Q (f - P) \|_p : P \varepsilon \pi_n \}. \qquad (11)$$

In [7], we proved direct and converse theorems for weighted polynomial approximation where Ω_2 plays the same role that the ordinary second-order modulus of continuity plays in direct and converse theorems for trigonometric approximation. In particular, we proved the following equivalence.

Proposition 1. Let w_Q be superregular, $1 \leq p \leq \infty$, $w_Q f \varepsilon L^p(R)$, $0 < \alpha \leq 1$. The following are equivalent:

(a) f is r-times differentiable, $w_Q f^{(r)} \in L^p(R)$, and

$$\sup_{0<\delta\leq 1}\left\{\delta^{-\alpha}\Omega_2(L^p, w_Q; f^{(r)}, \delta)\right\} < \infty.$$

(b) $$\sup_{n\geq 1}\left\{\left(\frac{n}{q_n}\right)^{r+\alpha}\varepsilon_n(L^p, w_Q, f)\right\} < \infty.$$

The Besov spaces consist of functions satisfying smoothness conditions that are somewhat more refined than the Lipschitz-type conditions above.

Definition 1. A function f is said to be in the Besov space $B(w_Q, p, \rho, \theta)$ if $w_Q f \in L^p(R)$, f is r-times differentiable where $\rho = r + \alpha$, $(0 < \alpha \leq 1)$, $w_Q f^{(r)} \in L^p(R)$, and

$$\|f\|_{b,Q,p,\rho,\theta} = \left\{\int_0^1\left[\frac{\Omega_2(L^p, w_Q, f^{(r)}, \delta)}{\delta^\alpha}\right]^\theta \frac{d\delta}{\delta}\right\}^{\frac{1}{\theta}}, \quad 1\leq\theta<\infty \quad (12)$$

or

$$\|f\|_{b,Q,p,\rho,\infty} = \|f\|_{h,Q,p,\rho}$$

$$= \sup_{0<\delta<1}\left\{\delta^{-\alpha}\Omega_2(L^p, w_Q; f^{(r)}, \delta)\right\} < \infty. \quad (13)$$

Put

$$\|f\|_{B,Q,p,\rho,\theta} = \|w_Q f\|_p + \|f\|_{b,Q,p,\rho,\theta}, \quad 1 \leq \theta \leq \infty \quad (14)$$

$$\|f\|_{H,Q,p,\rho} = \|f\|_{B,Q,p,\rho,\infty}. \quad (15)$$

A discussion of the nonweighted Besov spaces, especially the embedding theorems can be found in [12]. See also [13, Chapter 5]. The embeddings for the weighted analogues are expressed in our Theorem 4 below.

Theorem 4.

(a) *Suppose* w_{Q_1}, w_{Q_2} *are superregular weights and* $Q_1(x) \leq Q_2(x)$ *if* $|x| > K$ *for some* K. *Then*

$$\| f \|_{B,Q_2,p,\rho,\theta} \ll \| f \|_{B,Q_1,p,\rho,\theta'} \tag{16}$$

$$1 \leq p \leq \infty, \ 1 \leq \theta \leq \infty, \ \rho > 0.$$

(b) If $1 \leq \theta_1 \leq \theta_2 \leq \infty$ and $\varepsilon > 0$, and w_Q is superregular, then

$$\| f \|_{B,Q,p,\rho,\infty} \ll \| f \|_{B,Q,p,\rho,\theta_2}$$
$$\ll \| f \|_{B,Q,p,\rho,\theta_1} \ll \| f \|_{B,Q,p,\rho+\varepsilon,\infty}. \tag{17}$$

(c) (*Embeddings of different metrics*): If $1 \leq p, \ q, \ \theta \leq \infty$, then

$$\| f \|_{B,Q,q,\rho-\left|\frac{1}{p}-\frac{1}{q}\right|,\theta} \ll \| f \|_{B,Q,p,\rho,\theta'} \tag{18}$$

provided $\rho > \left|\frac{1}{p} - \frac{1}{q}\right|$ and w_Q is superregular.

We are now in a position to state our result about the Fourier transform.

Let, for $f \in L^p(R)$, $f^*(x) = \exp\left(\frac{x^2}{2}\right) f(x)$.

Theorem 5. Suppose $f \in L^2(R)$ and g is its Fourier transform (or inverse Fourier transform) defined as a principal value. We have

$$\| g^* \|_{B,\frac{x^2}{2},2,\rho,\theta} \ll \| f^* \|_{B,\frac{x^2}{2},2,\rho,\theta} \ll \| g^* \|_{B,\frac{x^2}{2},2,\rho,\theta}. \tag{19}$$

Remark. With the help of Theorem 4, we can now find the Besov spaces where g^* will lie, given that f^* is in a certain Besov space. Thus, for example, suppose $f \in L^p(R)$, $p > 2$ and satisfies the modified Lipschitz condition

$$\Omega_2\left(L^p, \ \exp\left(-\frac{x^2}{2}\right), \ f^*, \ \delta\right) = 0(\delta^\alpha)$$

for some α where $\frac{1}{2} - \frac{1}{p} < \alpha \leq 1 - \frac{1}{p}$. Then the Fourier transform g of f will exist as a function in $L^r(R)$ for all r with $\frac{p}{1 + \alpha p} < r < \frac{p}{p - \alpha p - 1}$. Further, for all such r, g will satisfy the modified Lipschitz condition

$$\Omega_2\left(L^r, \exp\left(-\frac{x^2}{2}\right), g^*, \delta\right) = 0(\delta^\rho)$$

where $\rho = \alpha + \frac{1}{p} - \frac{1}{r}$ if $r \leq 2$ and $\alpha + \frac{1}{p} + \frac{1}{r} - 1$ if $r \geq 2$.

Continuing the same chain of thought, we also note the following corollary.

Corollary 3. Let $1 < p < \infty$; $\frac{1}{p} + \frac{1}{p'} = 1$, $\alpha = \max\left(\frac{1}{p}, \frac{1}{p'}\right)$;

$\beta = \min\left(\frac{1}{p}, \frac{1}{p'}\right)$. Suppose $f \in L^p(R)$ and f^* satisfies

$$\Omega_2\left(L^p, \exp\left(-\frac{x^2}{2}\right); f^*, \delta\right) = 0(\delta^\alpha).$$

Then the Fourier transform g of f exists as a function in $L^r(R)$ for all r with $1 < r < \infty$, and, in particular, g^* will satisfy in $L^{p'}$ the condition

$$\Omega_2\left(L^{p'}, \exp\left(-\frac{x^2}{2}\right), g^*, \delta\right) = 0(\delta^\beta).$$

When $1 \leq p \leq 2$ and thus we know the existence of the Fourier transform in $L^{p'}(R)$, where $\frac{1}{p} + \frac{1}{p'} = 1$, a similar relationship between the smoothness of f in $L^p(R)$ and the smoothness of g in $L^{p'}(R)$ was given by *Ky* [8].

Our results neither include his theorem nor are included in it. However, when the hypotheses of both our theorem and Ky's theorem are satisfied, our estimates are sharper than those of Ky.

3. PROOFS

Let $wf \in L^p R$, $p \geq 1$. We then have the orthogonal polynomial expansion, $f \sim \Sigma c_k(f) p_k(w^2, x)$, where

$$c_k(f) = \int_R w^2(x) f(x) p_k(w^2, x)\, dx. \tag{20}$$

Denote

$$s_n(w^2, f) = \sum_{k=0}^{n-1} c_k(f) p_k(w^2) \in \pi_{n-1} \tag{21a}$$

$$v_n(w^2, f) = \frac{1}{n} \sum_{k=n+1}^{2n} s_k(f) \ \varepsilon \ \pi_{2n-1}. \tag{21b}$$

For $g \ \varepsilon \ L^p(R)$, let

$$T_n(g) = wv_n(w^2, w^{-1}g). \tag{22}$$

Proposition 2.

(a) *If* $P \ \varepsilon \ \pi_n$, $v_n(P) = P$.

(b) *The operator* T_n *is self-adjoint in the sense that if*

$g \ \varepsilon \ L^p(R)$, $h \ \varepsilon \ L^{p'}(R)$, $\frac{1}{p} + \frac{1}{p'} = 1$, *then*

$$\int_R T_n(g)h \ dx = \int_R gT_n(h) \ dx. \tag{23}$$

In particular,

$$\left\| T_n \right\|_{L^p(R) \to L^r(R)} = \left\| T_n \right\|_{L^{r'}(R) \to L^{p'}(R)}. \tag{24}$$

(c) $\left\| T_n \right\|_{L^p(R) \to L^p(R)} \ll N_n^{\left|\frac{1}{p} - \frac{1}{2}\right|}$, $p \geq 1$, $n \geq 1$.

(Here, $\ll$ is, in fact, independent of p.)

Proof. Part (a) is obvious. The first part of (b) follows easily by substituting from (22), (21), (20) in the expression on the left of (23) and interchanging the order of integration, which is permissible, since $T_n(g) \ \varepsilon \ L^p(R)$ and $h \ \varepsilon \ L^{p'}(R)$. The equation (24) follows from (23) and the formula

$$\left\| f \right\|_p = \sup\left\{\int_R fg \ dx : \left\| g \right\|_{p'} = 1\right\}. \tag{25}$$

Part (c) was proved by Freud in [4], although it was not stated there in this form. We reproduce a sketch of the proof. Let $g \ \varepsilon \ L^\infty(R)$, $f = w^{-1}g$, $x \ \varepsilon \ R$. Put $f = f_1 + f_2$, where

$$f_1(t) = \begin{cases} f(t) & \text{if} \quad |x - t| \leq \dfrac{\Gamma_{2n}}{2n} \\ 0 & \text{otherwise.} \end{cases}$$

Then

$$s_k(f) = s_k(f_1) + s_k(f_2) \tag{26a}$$

$$v_k(f) = v_k(f_1) + v_k(f_2). \tag{26b}$$

From (21),

$$w(x) \left| s_k(f_1, x) \right| \leq M_n^{\frac{1}{2}} \left\| wf_1 \right\|_2 \ll N_n^{\frac{1}{2}} \left\| wf \right\|_\infty, \quad 1 \leq k \leq 2n.$$

Hence,

$$\left| w(x) v_n(f_1, x) \right| \ll N_n^{\frac{1}{2}} \left\| wf \right\|_\infty \tag{27}$$

Let $f_2{}^* = \dfrac{f_2(t)}{x - t}$. We have [4]

$$w(x) s_k(f_2, x) = w(x) \frac{\gamma_{k-1}}{\gamma_k} \left[c_{k-1}(f_2^*) p_k(x) - c_k(f_2^*) p_{k-1}(x) \right].$$

Hence,

$$\sum_{k=n}^{2n} w(x) \left| s_k(f_2, x) \right| \ll \Gamma_{2n} M_n^{\frac{1}{2}} \left\| wf_2^* \right\|_2 \ll N_n^{\frac{1}{2}} n \left\| wf \right\|_\infty.$$

Thus,

$$\left| v_n(f_2, x) \right| \left| w(x) \right| \ll N_n^{\frac{1}{2}} \left\| wf \right\|_\infty. \tag{28}$$

From (26), (27), (28),

$$\left\| T_n(g) \right\|_\infty = \left\| wv_n(w^{-1}g) \right\|_\infty = \left\| wv_n(f) \right\|_\infty \ll N_n^{\frac{1}{2}} \left\| g \right\|_\infty. \tag{29}$$

Clearly,

$$\left\| T_n(g) \right\|_2 = \left\| wv_n(f) \right\|_2 \ll \left\| wf \right\|_2 = \left\| g \right\|_2. \tag{30}$$

Part (c) now follows for $p \geq 2$ by the Riesz–Thorin interpolation theorem and for $1 \leq p \leq$ by part (b).

Proof of Theorem 2. It is convenient to prove parts (c), (a), and (b) of the theorem in that order.

(c) Let $2 \leq p \leq r \leq \infty$. Observe that if $P \varepsilon \pi_n$, we have

$$P(x) = \sum_{k=0}^{n} c_k(P) p_k(x), \quad \text{where} \quad p_k(x) \equiv p_k(w^2, x).$$

So

$$w(x) |P(x)| \leq \left(\sum_{k=0}^{n} c_k(P)^2 \right)^{\frac{1}{2}} \cdot \left(\sum_{k=0}^{n} w^2(x) p_k^2(x) \right)^{\frac{1}{2}}; \quad x \varepsilon R,$$

that is,

$$\| wP \|_{\infty} \leq \left\| w^2 \sum_{k=0}^{n} p_k^2 \right\|_{\infty}^{\frac{1}{2}} \| wP \|_2.$$

Hence, for $g \varepsilon L^2(R)$,

$$\| T_n(g) \|_{\infty} \leq M_n^{\frac{1}{2}} \| T_n(g) \|_2 \ll M_n^{\frac{1}{2}} \| g \|_2. \tag{31}$$

Also, for $g \varepsilon L^{\infty}(R)$, we have by Proposition 2(c),

$$\| T_n(g) \|_{\infty} \ll N_n^{\frac{1}{2}} \| g \|_{\infty}. \tag{32}$$

Then by the Riesz-Thorin interpolation theorem, for all $g \varepsilon L^p(R)$,

$$\| T_n(g) \|_{\infty} \ll N_n^{\frac{1}{2} - \frac{1}{p}} M_n^{\frac{1}{p}} \| g \|_p. \tag{33}$$

But, by Proposition 2(c), for all $g \varepsilon L^p(R)$, we also have

$$\| T_n(g) \|_p \ll N_n^{\frac{1}{2} - \frac{1}{p}} \| g \|_p. \tag{34}$$

Again, by the Riesz-Thorin interpolation theorem, we get, for all $g \varepsilon L^p(R)$,

$$\| T_n(g) \|_r \ll N_n^{\frac{1}{2} - \frac{1}{p}} M_n^{\frac{1}{p} - \frac{1}{r}} \| g \|_p. \tag{35}$$

In view of Proposition 2(a), this completes the proof of part (c).

(a) Let $1 \leq p \leq r \leq 2$. Let

$$\frac{1}{p} + \frac{1}{p'} = \frac{1}{r} + \frac{1}{r'} = 1, \quad \text{so that} \quad 2 \leq r' \leq p' \leq \infty.$$

Hence, by part (c),

$$\| T_n(g) \|_{p'} \ll N_n^{\frac{1}{2} - \frac{1}{r'}} M_n^{\frac{1}{r'} - \frac{1}{p'}} \| g \|_{r'},$$

for all $g \in L^{r'}(R)$; that is,

$$\| T_n \|_{L^{r'}(R) \to L^{p'}(R)} = \| T_n \|_{L^p(R) \to L^r(R)}$$

$$\ll N_n^{\frac{1}{r} - \frac{1}{2}} M_n^{\frac{1}{p} - \frac{1}{r}}.$$

(The first equation above uses Proposition 2(b).) Thus, for all $g \in L^p(R)$,

$$\| T_n(g) \|_r \ll N_n^{\frac{1}{r} - \frac{1}{2}} M_n^{\frac{1}{p} - \frac{1}{r}} \| g \|_p. \tag{36}$$

In view of Proposition 2(a), this proves part (a) of the theorem.

(b) If $1 \leq p \leq 2 \leq r \leq \infty$ and $P \in \pi_n$, we have by parts (c) and (a) of this theorem,

$$\| wP \|_r \ll M_n^{\frac{1}{2} - \frac{1}{r}} \| wP \|_2 \ll M_n^{\frac{1}{2} - \frac{1}{r}} M_n^{\frac{1}{p} - \frac{1}{2}} \| wP \|_p$$

$$= M_n^{\frac{1}{p} - \frac{1}{r}} \| wP \|_p.$$

This completes the proof.

Corollary 1 is obvious. Corollary 2 becomes obvious if we note the inequality

$$M_n \ll \frac{n}{q_n} \tag{37}$$

proved by Freud for the weights in question [3].

Proof of Theorem 3. If w_Q is superregular, then the sequence N_n is bounded. Hence, for all $n \geq 1$,

$$\| T_n \|_{L^p(R) \to L^p(R)} \leq K, \tag{38}$$

where K is a constant depending on Q alone.

If Q is even, positive, differentiable and $Q''(t) > 0$ for all $t > 0$ and $Q'(t) \to \infty$ as $t \to \infty$; in particular, when w_Q is superregular, we have for every $p \in \pi_n$,

$$\int_R P^2(x) w_Q^2(x) \ dx \ll \int_{-cq_{2n}}^{cq_{2n}} P^2(w) w_Q^2(x) \ dx, \tag{39}$$

where c is a constant depending on Q alone [4, p. 157]. From (39),

$$\| w_Q P \|_2 \ll q_{2n}^{\frac{1}{2}} \| w_Q P \|_\infty. \tag{40}$$

Since w_Q is superregular, equations (7), (38), and (40) imply that

$$\| T_n(g) \|_2 \ll q_n^{\frac{1}{2}} \| g \|_\infty, \quad g \in L^\infty(R). \tag{41}$$

The remainder of the proof is similar to that of Theorem 2:

$$\| T_n(g) \|_2 \ll \| g \|_2, \quad g \in L^2(R). \tag{42}$$

If $2 \leq r \leq \infty$, (41) and (42) give, for all $g \in L^r(R)$,

$$\| T_n(g) \|_2 \ll q_n^{\frac{1}{2} - \frac{1}{r}} \| g \|_r \tag{43}$$

$$\| T_n(g) \|_r \ll \| g \|_r \tag{44}$$

If $2 \leq p \leq r \leq \infty$, (43) and (44) give for all $g \in L^r(R)$,

$$\| T_n(g) \|_p \ll q_n^{\frac{1}{p} - \frac{1}{r}} \| g \|_r. \tag{45}$$

By the duality relation in Proposition 2(b), (45) holds even when $1 \leq p \leq r \leq 2$. Then by Proposition 2(a), for all $P \in \pi_n$,

$$\| wP \|_p \ll q_n^{\frac{1}{p} - \frac{1}{r}} \| wP \|_r$$

when p and r are on the same side of 2. If $1 \leq p \leq 2 \leq r \leq \infty$, we get

$$\| wP \|_p \ll q_n^{\frac{1}{p} - \frac{1}{2}} \| wP \|_2 \ll q_n^{\frac{1}{p} - \frac{1}{r}} \| wP \|_r.$$

This proves the first inequality in (9) for all p, r, $1 \leq p \leq r \leq \infty$. Finally, observe that since

$$\frac{xQ''(x)}{Q'(x)} \leq K,$$

$$\frac{q_n^2 Q''(q_n)}{q_n Q'(q_n)} = \frac{q_n^2 Q''(q_n)}{n} \leq K.$$

Since Q'' is increasing,

$$q_n \ll \frac{n}{q_n Q''(q_n)} \ll \frac{n}{q_n}.$$

This completes the proof of Theorem 3.

To prove Theorem 4, we need a lemma, which we proved in [10], and which gives some equivalent formulas for the Besov space norms. Since these norms are refinements of usual Lipschitz-space norms, the lemma also embodies a refinement of Proposition 1.

Lemma 1.   *Let w_Q be superregular and $w_Q f \in L^p(R)$. Define*

$$\| f \|_{b,1,Q,p,\rho,\theta} = \left\{ \int_0^1 \left[\frac{\Omega_2(f^{(r)},\, \delta)}{\delta^\alpha} \right]^\theta \frac{d\delta}{\delta} \right\}^{\frac{1}{\theta}}$$

$$\| f \|_{b,2,Q,p,\rho,\theta} = \left\{ \sum_{n=0}^\infty \left[\left(\frac{2^n}{q_{2^n}} \right)^\alpha \Omega_2 \left(f^{(r)},\, \frac{q_{2^n}}{2^n} \right) \right]^\theta \right\}^{\frac{1}{\theta}}$$

$$\| f \|_{b,3,Q,p,\rho,\theta} = \left\{ \sum_{n=0}^\infty \left[\left(\frac{2^n}{q_{2^n}} \right)^\rho \varepsilon_{2^n}(L^p,\, w_Q,\, f) \right]^\theta \right\}^{\frac{1}{\theta}}$$

$$\| f \|_{b,4,Q,p,\rho,\theta} = \inf \left\{ \sum_{n=0}^\infty \left[\left(\frac{2^n}{q_{2^n}} \right)^\rho \| w_Q P_{2^n} \|_p \right]^\theta \right\}^{\frac{1}{\theta}} ,$$

where the inf is taken over all sequences of polynomials $\{P_{2^n}\}$ such that

$$P_{2^n} \in \pi_{2^n} \quad \text{and} \quad \left\| w_Q \left(\sum_{n=0}^N P_{2^n} - f \right) \right\| \to 0 \quad \text{as} \quad N \to \infty$$

$$\| f \|_{B,k,Q,p,\rho,\theta} = \| w_Q f \|_p + \| f \|_{b,k,Q,p,\rho,\theta} , \quad k = 1,\, 2,\, 3,\, 4.$$

In all of these definitions, $1 \leq \theta \leq \infty$ and when $\theta = \infty$, the usual convention of taking sup is followed.

Then, for $j,\, k = 1,\, 2,\, 3,\, 4$,

$$\| f \|_{B,k,Q,p,\rho,\theta} \sim \| f \|_{B,j,Q,p,\rho,\theta}.$$

Here and in the sequel, $A \sim B$ will mean $A \ll B$ and $B \ll A$.

Proof of Theorem 4.

(a) We notice the following equivalence proved in [6]:

$$\Omega_2(L^p,\, w_Q,\, f,\, \theta) \sim \inf\{ \| w_Q f_1 \|_p + \delta^2 \| w_Q f_2'' \|_p \}, \qquad (46)$$

where the inf is over all $f_1,\, f_2$ such that $f = f_1 + f_2$, $w_Q f_1 \in L^p(R)$ and f_2 is twice differentiable with

$w_Q f_2'' \in L^p(R)$. Part (a) now follows from the version $\|f\|_{B,1,Q,p,\rho,\theta}$ of the Besov space norms.

(b) This part is clear from the version $\|f\|_{B,3,Q,p,\rho,\theta}$ of the Besov space norms if we observe the inequality

$$\left\{\sum_{k=0}^{\infty} |a_k|^{\theta_2}\right\}^{\frac{1}{\theta_2}} \leq \left\{\sum_{k=0}^{\infty} |a_k|^{\theta_1}\right\}^{\frac{1}{\theta_1}} \tag{47}$$

for $1 \leq \theta_1 \leq \theta_2 \leq \infty$ and for all sequences $\{a_k\}$.

(c) Note that in view of Corollary 2 and Theorem 3, if w_Q is superregular and $P \in \pi_n$, then

$$\|w_Q P\|_p \ll \left(\frac{n}{q_n}\right)^{\left|\frac{1}{p}-\frac{1}{q}\right|} \|w_Q P\|_q, \quad 1 \leq p, q \leq \infty.$$

Now let $f \in B(w_Q, p, \rho, \theta)$ and $\rho > \left|\frac{1}{p} - \frac{1}{q}\right|$.

Choose a sequence $P_{2^k} \in \pi_{2^k}$ such that

$$\lim_{n \to \infty} \left\| w_Q\left(f - \sum_{k=0}^{n} P_{2^k}\right) \right\|_p = 0 \tag{48}$$

$$\left\{\sum_{k=0}^{\infty} \left[\left(\frac{2^k}{q_{2^k}}\right)^{\rho} \|w_Q P_{2^k}\|_p\right]^{\theta}\right\}^{\frac{1}{\theta}} \leq 2\|f\|_{b,4,Q,p,\rho,\theta}. \tag{49}$$

We have

$$\left\{\sum_{k=0}^{\infty} \left[\left(\frac{2^k}{q_{2^k}}\right)^{\rho-\left|\frac{1}{p}-\frac{1}{q}\right|} \|w_Q P_{2^k}\|_q\right]^{\theta}\right\}^{\frac{1}{\theta}}$$

$$\ll \left\{\sum_{k=0}^{\infty} \left[\left(\frac{2^k}{q_{2^k}}\right)^{\rho} \|w_Q P_{2^k}\|_p\right]^{\theta}\right\}^{\frac{1}{\theta}} \leq 2\|f\|_{b,4,Q,p,\rho,\theta}. \tag{50}$$

Since $\rho - \left|\frac{1}{p} - \frac{1}{q}\right| > 0$, we also conclude that

$\lim_{n \to \infty} \left\| w_Q\left(f - \sum_{k=0}^{n} P_{2^k}\right) \right\|_q$ is 0; in particular, $w_Q f \in L^q(R)$.

Further, if $\rho - \left|\frac{1}{p} - \frac{1}{q}\right| = \beta$, then

$$\| w_Q f \|_q \leq \sum_{k=0}^{\infty} \| w_Q P_{2^k} \|_q$$

$$\leq \left\{ \sum_{k=0}^{\infty} \left[\left(\frac{2^k}{q_{2^k}} \right)^{\beta} \| w_Q P_{2^k} \|_q \right]^{\theta} \right\}^{\frac{1}{\theta}} \left\{ \sum_{k=0}^{\infty} \left(\frac{q_{2^k}}{2^k} \right)^{\beta \theta'} \right\}^{\frac{1}{\theta'}}$$

$$\left(\frac{1}{\theta} + \frac{1}{\theta'} = 1 \right)$$

$$\ll \| f \|_{B,4,Q,p,\rho,\theta}. \tag{51}$$

Part (c). follows from (50) and (51). This completes the proof of Theorem 4.

If h_k is orthonormal Hermite polynomial of degree k, and $\psi_k(x) = \exp\left(-\frac{x^2}{2}\right) h_k(x)$, it is well known that

$$\hat{\psi}_k = \lambda_k \psi_k, \tag{52}$$

where $\hat{\psi}_k$ is the Fourier transform of ψ_k and $|\lambda_k| = 1$ [16]. Further if $f \in L^2(R)$, by the completeness of the Hermite polynomial system, we have

$$f(x) = \sum_{k=0}^{\infty} c_k \psi_k(x), \quad \text{where} \quad c_k = \int_R f(x) \psi_k(x)\, dx. \tag{53}$$

The series in (53) is convergent in $L^2(R)$. Then by Plancherel's theorem, if $\hat{f} = g$ is the Fourier transform of f, we get, from (52),

$$g(x) = \sum_{k=0}^{\infty} \lambda_k c_k \psi_k(x) \quad \text{and} \quad \lambda_k c_k = \int_R g \psi_k\, dx. \tag{54}$$

From (53), (52), and (54),

$$\varepsilon_n\left(L^2, \exp\left(-\frac{x^2}{2}\right), g^*\right) = \varepsilon_n\left(L^2, \exp\left(-\frac{x^2}{2}\right), f^*\right), \tag{55}$$

where $f^*(x) = \exp\left(\frac{x^2}{2}\right) f(x)$ and $g^*(x) = \exp\left(\frac{x^2}{2}\right) g(x)$.

Theorem 5 now follows from the version $\| f \|_{B,3,\exp\left(-\frac{x^2}{2}\right),2,\rho,\theta}$ of the Besov space norms.

REFERENCES

1. Freud, G. Extension of the Dirichlet-Jordan criterion to a general class of orthogonal polynomial expansions. *Acta Math. Acad. Sci. Hung.* 25 (1974): 109-122.

2. Freud, G. Markov-Bernstein type inequalities in $L_p(-\infty, \infty)$. In *Approximation Theory II* (Lorentz, Chui, Shumaker, eds.). New York, Academic Press, 1976, pp. 369-377.

3. Freud, G. On Markov-Bernstein type inequalities and their applications. *J. of Approx. Theo.* 19 (1977): 22-37.

4. Freud, G. On polynomial approximation with respect to general weights. In *Lecture Notes* 399 (Garnir, Unni, Williamson, eds.). New York: Springer, 1974, pp. 149-179.

5. Freud, G. On the extension of the Fejer-Lebesgue theorem to orthogonal polynomial series. *Iliev Festschrift* (1975), Sofia, 257-265.

6. Freud, G., and Mhaskar, H. N. K-functionals and moduli of continuity in weighted polynomial approximation. To appear in *Arkiv för Matematik*.

7. Freud, G., and Mhaskar, H. N. Weighted polynomial approximation in rearrangement invariant Banach function spaces on the whole real line. To appear in *Ind. J. of Math.*

8. Ky, N. X. Fourier-transzformaltak modositott folytonossagi modulusarol. *Matem. Lapok* 23 (1972): 99-102.

9. Mhaskar, H. N. Weighted polynomial approximation of entire functions I, II. To appear in *J. of Approx. Theo.*

10. Mhaskar, H. N. Weighted polynomial approximation on the whole real line and related topics. Ph.D. dissertation, The Ohio State Univ. 1980.

11. Nevai, G. P. Orthogonal Polynomials; Memoirs of the American Mathematical Society, 213, 1979.

12. Nikolskii, S. M. *Approximation of Functions of Several Variables and Imbedding Theorems*. New York: Springer, 1975.

13. Stein, E. M. *Singular Integrals and Differentiability Properties of Functions*. Princeton, N.J.: Princeton Univ. Press, 1970.

14. Szegö, G. *Orthogonal Polynomials*. American Mathematical
     Society Colloquium Publications, 23, 4th ed., 1975.

15. Timan, A. F. *Theory of Approximation of Functions of a Real
     Variable*. New York: Pergamon Press and Macmillan Company,
     1963.

16. Wiener, N. *The Fourier Integral and Certain of Its Applica-
     tions*. New York: Cambridge Univ. Press, 1933.

17. Zygmund, A. *Trigonometric Series*, vol. 1. New York: Cam-
     bridge Univ. Press, 1959.

AN APPLICATION OF THE
APPROXIMATION FUNCTIONAL
IN INTERPOLATION THEORY

Yoram Sagher
University of Illinois at Chicago Circle

Real interpolation theory has been dominated by the J and K functionals, while the approximation functional, defined

$$E(t, a; A_0, A_1) = \inf\{|a - a_0|_{A_1} / |a_0|_{A_0} \leq t\}, \tag{1}$$

has taken a secondary role. This is understandable, since J is a quasinorm on $A_0 \cap A_1$, K is a quasinorm on $A_0 + A_1$, while E lacks any such interpretation. Moreover, the statements of some important theorems, most notably the reiteration theorem, in terms of the E functional alone, are quite cumbersome.

We shall see, however, that the approximation functional is useful in the incorporation of W (see [1]) in the $L(p, q)$ interpolation scale (see [4]), as well as in the incorporation of the analogues of W in other interpolation scales. W consist of all functions on a σ-finite measurable space (X, μ), satisfying

$$f^{**}(t) - f^*(t) \leq C, \tag{2}$$

where f^* is the nonincreasing rearrangement of f, and $f^{**}(t) = \frac{1}{t} \int_0^t f^*(u)\, du$. Using the K functional and

$$\int_0^t f^*(u)\, du = K(t, f; L_1, L_\infty), \tag{3}$$

the following theorem is proved in [4]:

Theorem 1.   If
$$T : A_0 \to L(p_0, \infty)$$
$$T : A_1 \to W$$

is quasilinear: $\left|T(a_0 + a_1)\right|(x) \leq k\left(\left|Ta_0\right|(x) + \left|Ta_1\right|(x)\right)$ *a.e., then,*

for $0 < \theta < 1,\ 0 < q \leq \infty,\ \dfrac{1}{p} = \dfrac{1 - \theta}{p_0}$, *we have*

$$T : (A_0, A_1)_{\theta, q; K} \to L(p, q) .$$

An annoying feature of the proof in [4] is that since (3) involves L_1, one must deal separately with the cases $p_0 < 1$, $p_0 = 1$, $1 < p_0$. Using the approximation functional, E, we will get the theorem for all three cases at once. There is another reason to use the E functional. The condition replacing (2) in the general theorem is

$$\frac{1}{t} K(t, b; B_0, B_1) - K'(t, b; B_0, B_1) \leq C. \tag{4}$$

The derivative, contrary to the first impression, presents no great difficulty. As is true almost everywhere in interpolation, the formulas containing the continuous parameter t are in fact just shorthand notation for the series definitions where all problems of measurability and differentiability vanish. The aspect of (4) that presents a practical problem is the subtraction of K', for it dictates a precise determination of K.

This is where the approximation functional proves useful. In some cases it can be determined precisely, almost by observation. We recall two results presented in [3]:

$$E(t, f; L_0, L_\infty) = f^*(t) \tag{5}$$

$$E(t, f; \Lambda_\omega, L_p) = \left(\int_{\{\omega(x) > t\}} |f(x)|^p \, dx\right)^{1/p} . \tag{6}$$

Motivated by these considerations, we present an approach to the inclusion of W and its analogues in interpolation theory, based on the E functional. Since we cannot hope to make a self-contained presentation, we refer the reader to [3] for the definitions and theorems used here. Our notation, except for some minor details, is consistent with that of [3].

Definition. A_j, B_j are quasinormed Abelian groups. $T : A_0 + A_1 \to B_0 + B_1$ is E-quasilinear iff there exist $0 < c, d$ so that for all $a_j \in A_j$ and all $0 < t_0$, t_1, we have

$$E(t_0 + t_1, T(a_0 + a_1); B_0, B_1) \leq c[E(dt_0, Ta_0; B_0, B_1)$$
$$+ E(dt_1, Ta_1; B_0, B_1)]. \tag{7}$$

Theorem 2. If $T : A_0 + A_1 \to Lp_0 + Lp_1$, $0 \leq p_0 < p_1 \leq \infty$, satisfies

$$|T(a_0 + a_1)|(x) \leq k(|Ta_0|(x) + |Ta_1|(x)) \quad \text{a.e.,} \tag{8}$$

then T is E-quasilinear.

Proof. The following facts are easily verified from the definitions:

$$E(t_0 + t_1, f_0 + f_1; Lp_0, Lp_1) \leq C_1 E(t_0/C_0, f_0; Lp_0, Lp_1)$$
$$+ C_1 E(t_1/C_0, f_1; Lp_0, Lp_1), \tag{9}$$

C_0, C_1 depending on p_0 and p_1.

$$E(t, kf; Lp_0, Lp_1) = kE(t/k, f; Lp_0, Lp_1); \quad 0 < k \tag{10}$$

if $0 < p_0$. If $p_0 = 0$, we have:

$$E(t, kf; Lp_0, Lp_1) = kE(t, f; Lp_0, Lp_1). \tag{11}$$

$|f| \leq |g|$ a.e. implies $E(t, f; Lp_0, Lp_1) \leq E(t, g; Lp_0, Lp_1)$.
$$\tag{12}$$

Using (9)-(12), the theorem follows easily:

$$E(t_0 + t_1, T(a_0 + a_1); Lp_0, Lp_1) = E(t_0 + t_1, |T(a_0 + a_1)|; Lp_0, Lp_1)$$
$$\leq E(t_0 + t_1, k|Ta_0| + k|Ta_1|; Lp_0, Lp_1)$$
$$\leq C_1 kE(t_0/kC_0, |Ta_0|; Lp_0, Lp_1) + C_1 kE(t_1/kC_0, |Ta_1|; Lp_0, Lp_1)$$
$$= CE(t_0 d, Ta_0; Lp_0, Lp_1) + CE(t_1 d, Ta_1; Lp_0, Lp_1).$$

Theorem 3. $T : A_0 + A_2 \rightarrow \Lambda\omega + Lp$ *satisfies*

$$|T(a_0 + a_1)|(x) \leq k(|Ta_0|(x) + |Ta_1|(x))   \quad a.e.$$

Then T is E-quasilinear.

Proof.   Given a function g, denote by χ_g the characteristic function of its support. Recall:

$$|g|_{\Lambda\omega} = |\omega(x)\chi_g(x)|_{L_\infty}. \tag{13}$$

One verifies that (9)-(12) hold when Lp_0 is replaced by $\Lambda\omega$, and Lp_1 by Lp, and the proof concludes as that of Theorem 2.

Theorem 4. (A_0, A_1), (B_0, B_1) *are quasinormed Abelian couples.*
$T : A_0 + A_1 \rightarrow B_0 + B_1$ *is E-quasilinear, $1 \leq p \leq \infty$, and*

$$\left(\frac{1}{t}\int_0^t E^p(s, Ta; B_0, B_1)\, ds\right)^{1/p} - E(t, Ta; B_0, B_1) \leq c_1|a|_{A_1} \tag{14}$$

$$E(t, Ta; B_0, B_1) \leq c_0 t^{-\beta}|a|_{A_0}. \tag{15}$$

Then:

$$|Ta|_{\beta(1-\theta),q;E} \leq C(\beta, \theta, q, c_0, c_1, c, d)|a|_{\theta,q;K} \tag{16}$$

where $0 < \theta < 1$ and $0 < q \leq \infty$.

Proof.   Since $\frac{1}{t}\int_0^t E(s, Ta; B_0, B_1)\, ds - E(t, Ta; B_0, B_1) \leq$
$\left(\frac{1}{t}\int_0^t E^p(s, Ta; B_0, B_1)\right)^{1/p} - E(t, Ta; B_0, B_1)$, we can assume in (14) $p = 1$. (The statement with $1 < p$ will be useful later.) Denote $g(t) = \frac{1}{t}\int_0^t E(s, Ta; B_0, B_1)\, ds$. Clearly, $0 \leq -g'(t) \leq \frac{c_1}{t}|a|_{A_1}$.

For $t < s$, we have

$$0 \leq g(t) - g(s) = -\int_t^s g'(u)\, du \leq c_1|a|_{A_1} \log s/t,$$

and so:

$$g(t) \leqq C_1 |a|_{A_1} \log s/t + g(s) \leqq C_1 |a|_{A_1} \log s/t + C_1 |a|_{A_1}$$

$$+ E(s, \, Ta; \, B_0, \, B_1) \leqq C_0 s^{-\beta} |a|_{A_0} + C_1 |a|_{A_1} (1 + \log s/t).$$

Therefore,

$$\int_0^s t^\alpha E(t, \, Ta; \, B_0, \, B_1) \frac{dt}{t} \leqq \int_0^s t^\alpha g(t) \frac{dt}{t} \leqq C \int_0^s t^\alpha (s^{-\beta} |a|_{A_0}$$

$$+ (1 + \log s/t) |a|_{A_1}) \frac{dt}{t}$$

$$\leqq C(C_0, \, C_1, \, \alpha, \, \beta) \, [s^{\alpha - \beta} |a|_{A_0} + s^\alpha |a|_{A_1}].$$

For $s < t$, use (15):

$$\int_s^\infty t^\alpha E(t, \, Ta; \, B_0, \, B_1) \frac{dt}{t} \leqq C_0 |a|_{A_0} \int_s^\infty t^{-(\beta - \alpha)} \frac{dt}{t}$$

$$= C(\alpha, \, \beta, \, C_0) |a|_{A_0} s^{\alpha - \beta}.$$

Taking: $s = \left(\dfrac{|a|_{A_0}}{|a|_{A_1}} \right)^{1/\beta}$, we have

$$|Ta|_{\alpha, 1; E} \leqq C(\alpha, \, \beta, \, C_0, \, C_1) |a|_{A_0}^{\alpha/\beta} |a|_{A_1}^{1 - \alpha/\beta}. \tag{17}$$

$a \to |Ta|_{\alpha, 1; E}$ is a semi-quasinorm on $A_0 \cap A_1$, and by the reiteration theorem, using (17) with different values of α,

$$T : (A_0, \, A_1)_{\theta, q; k} \to (B_{\alpha_0, 1; E}, \, B_{\alpha_1, 1; E})_{\lambda, q; k} \tag{18}$$

with $\theta = (1 - \lambda)(1 - \alpha_0/\beta) + \lambda(1 - \alpha_1/\beta)$.

Further,

$$(B_{\alpha_0, 1; E}, \, B_{\alpha_1, 1; E})_{\lambda, q; K} = B_{\alpha, q; E} \tag{19}$$

with $\alpha = (1 - \lambda)\alpha_0 + \lambda\alpha_1$, so that $\alpha = \beta(1 - \theta)$, and we have

$$T : (A_0, \, A_1)_{\theta, q; K} \to (B_0, \, B_1)_{\beta(1-\theta), q; E}. \tag{20}$$

The theorem is proved.

The first corollary is Theorem 1, which follows simply from
$E(t, f; L_0, L_\infty) = f^*(t)$.

We next consider interpolation of weighted Lp spaces, with
$1 \leq p \leq \infty$. We want first to characterize $(\Lambda\omega, Lp)_{\alpha,q;E}$. For this
we shall need a theorem of J. E. Gilbert [2]:

Theorem 5. *X is a Banach function space.* $\omega(x) > 0$ *a.e. Denote by*
G the class of piecewise continuous functions $\sigma : R^+ \to R^+$, *so that*
both $\sigma(t) \in L_\infty(R^+)$ *and* $t\sigma(t) \in L_\infty(R^+)$. *Denote also* $|f|_{X_\omega} = |\omega f|_X$
and $X_\omega = \{|f/|f|_{X_\omega} < \infty\}$. *Then, for* $0 < \eta < 1$, $0 < q \leq \infty$ *and any*
$\sigma \in G$:

$$|f|_{(X,X_\omega)_{\eta,q;K}} \sim \left(\int_0^\infty t^{(1-\eta)q} \left| \omega(x) \sigma(t\omega(x)) f(x) \right|_X^q \frac{dt}{t} \right)^{1/q} . \qquad (21)$$

Theorem 6. *Let* $0 < \alpha < \infty$, $0 < \eta < 1$, $0 < q \leq \infty$, *and* $\sigma \in G$. *Then*

$$|f|_{(\Lambda_\omega,Lp)_{\eta\alpha,q;E}}$$

$$\sim \left(\int_0^\infty t^{(1-\eta)q} \left(\int |\omega^\alpha(x) \sigma(t\omega^\alpha(x)) f(x)|^p dx \right)^{q/p} \frac{dt}{t} \right)^{1/q} . \qquad (22)$$

Proof. From (6) we have for all γ:

$$(\Lambda_\omega, Lp)_{\gamma,p;E} = Lp, \omega^\gamma .$$ Take any $\alpha_0 < \eta\alpha < \alpha$, and let $0 < \lambda$
< 1 be defined by $\eta\alpha = (1 - \lambda)\alpha_0 + \lambda\alpha$. Using reiteration we have

$$(\Lambda_\omega, Lp)_{\eta\alpha,q;E} = \left((\Lambda_\omega, Lp)_{\alpha_0,p;E}, (\Lambda_\omega, Lp)_{\alpha,p;E} \right)_{\lambda,q;K}$$

$$= (Lp, \omega^{\alpha_0}, Lp, \omega^\alpha)_{\lambda,q;K} .$$

From the case $p = q$ of Theorem 5, we have

$$Lp, \omega^{\alpha_0} = (Lp, Lp, \omega^\alpha)_{\frac{\alpha_0}{\alpha},p;K}$$

Using this, and reiterating again, we get

$$(\Lambda_\omega, Lp)_{\eta\alpha,q;E} = (Lp, Lp,\omega^\alpha)_{\eta,q;K}. \tag{23}$$

Equation (22) is the corresponding norm equivalence.

Theorem 7. *If* $T : A_0 + A_1 \rightarrow \Lambda_\omega + Lp$ *is E-quasilinear.* $1 \leq p \leq \infty$,
$0 < \theta < 1$, $0 < q \leq \infty$, $0 < \beta$, *and if*

$$\left(\frac{1}{t} \int_{\omega(x)\leq t} \omega(x) |Ta|^p(x) \, dx\right)^{1/p} \leq c_1 |a|_{A_1} \tag{24}$$

$$\left(\int_{t<\omega(x)} |Ta|^p(x) \, dx\right)^{1/p} \leq c_0 t^{-\beta} |a|_{A_0}, \tag{25}$$

then

$$\left(\int_0^\infty t^{\theta q}\left(\int |\omega^\beta(x) \sigma(t\omega^\beta(x)) (Ta)(x)|^p \, dx\right)^{q/p} \frac{dt}{t}\right)^{1/q}$$
$$\leq c |a|_{(A_0,A_1)_{\theta,q;K}} \tag{26}$$

Proof. We want to apply Theorem 4:

$$\left(\frac{1}{t} \int_0^t E^p(s, Ta; \Lambda_\omega, Lp) \, ds\right)^{1/p} - E(t, Ta; \Lambda_\omega, Lp)$$

$$= \left(\frac{1}{t} \int_0^t \left(\int_{s<\omega(x)} |Ta|^p(x) \, dx\right) ds\right)^{1/p} - \left(\int_{t<\omega(x)} |Ta|^p(x) \, dx\right)^{1/p}$$

$$= \left(\frac{1}{t} \int |Ta|^p(x) \left(\int_0^{\omega(x)\wedge t} ds\right) dx\right)^{1/p} - \left(\int_{t<\omega(x)} |Ta|^p(x) \, dx\right)^{1/p}$$

$$= \left(\frac{1}{t} \int_{\omega(x)\leq t} |Ta|^p(x) \omega(x) \, dx + \int_{t<\omega(x)} |Ta|^p(x) \, dx\right)^{1/p}$$

$$- \left(\int_{t<\omega(x)} |Ta|^p(x) \, dx\right)^{1/p} \leq \left(\frac{1}{t} \int_{\omega(x)\leq t} |Ta|^p(x) \omega(x) \, dx\right)^{1/p}.$$

Condition (24) therefore implies (14).

Condition (25) is the same as (15) for our interpolation couple, so that we have

$$T : (A_0, A_1)_{\theta,q;K} \rightarrow (\Lambda_\omega, Lp)_{\beta(1-\theta),q;E}.$$

(26) is the corresponding norm inequality, by Theorem 6.

Finally, let us compare Theorem 7 with Gilbert's theorem. (24) is clearly a weaker hypothesis than

$$T : A_1 \rightarrow Lp. \tag{27}$$

Since $(\Lambda_\omega, Lp)_{\beta,p;E} \subset (\Lambda_\omega, Lp)_{\beta,\infty;E}$, condition (25), which is $T : A_0 \to (\Lambda_\omega, Lp)_{\beta,\infty;E}$, is weaker than

$$T : A_0 \to Lp,\omega^\beta. \tag{28}$$

Conditions (27) and (28) yield

$$T : (A_0, A_1)_{\theta,q;K} \to (Lp, Lp,\omega^\beta)_{1-\theta,q;K} = (\Lambda_\omega, Lp)_{\beta(1-\theta),q;E}. \tag{29}$$

The equality of the interpolation spaces above is (23).

We therefore have, from weaker hypotheses, the conclusion of Gilbert's theorem. The situation is analogous to the relation between the Marcinkiewicz and the M. Riesz interpolation theorems.

REFERENCES

1. Bennett, C.; DeVore, R. A.; and Sharpley, R. Weak L^∞ and BMO. *Ann. Math.* 113 (1981): 601-611.

2. Gilbert, J. E. Interpolation between weighted *Lp* spaces. *Arkiv för Mat.* 10 (1972): 234-249.

3. Peetre, J., and Sparr, G. Interpolation of normed Abelian groups. *Annali di Matematica pura ed Applicata (IV)* 92 (1972): 217-262.

4. Sagher, Y. A new interpolation theorem. *Proc. of Conf. in Harmonic Analysis, Minneapolis,* 1981. Springer-Verlag lecture notes in *Mathematics* 908: 189-198.

AN ANALOGUE OF THE
FEJER-RIESZ THEOREM FOR THE DIRICHLET SPACE

Allen L. Shields
University of Michigan

L. Fejér and F. Riesz proved that if $f \in H^2$, then

$$\int_0^1 |f(x)|^2 \, dx \leq \pi \, \|f\|_2^2 \, . \tag{1}$$

(See [5] or [4, Theorem 3.13].) Here H^2 denotes the Hardy space of square-summable power series. We shall prove an analogue for the Dirichlet space, D. This is the space of analytic functions $f(z) = \Sigma \, a_n z^n$ in the open unit disc Δ, with $f(0) = 0$, such that

$$\|f\|^2 = \sum_1^\infty n |a_n|^2 = \frac{1}{\pi} \iint_\Delta |f'(re^{i\theta})|^2 \, r dr \, d\theta < \infty. \tag{2}$$

We prove the following result, which answers a question of A. J. Nagel.

Theorem 1. If $f \in D$, then

$$\int_0^1 |f(x)|^2 \frac{x^2}{1-x} \left(\log \frac{1}{1-x}\right)^{-2} dx \leq c \, \|f\|^2. \tag{3}$$

Here c is a constant independent of f.

This research was supported in part by the National Science Foundation.

We note that the exponent -2 on the logarithm cannot be replaced by $-\alpha$ for any $\alpha < 2$, since for $\beta < 1/2$, the function $(\log(1 - z))^{\beta}$ is in D. On the other hand, the result is trivially true if -2 is replaced by $-\alpha$ for any $\alpha > 2$. This is a consequence of the following growth estimate for functions in the Dirichlet space. Applying the Cauchy (-Buniakowsky-Schwarz) inequality to the power series for f, evaluated at x, one has:

$$|f(x)|^2 = |\sum_1^{\infty} (n^{1/2} a_n) (x^n/n^{1/2})|^2$$

$$\leq \|f\|^2 \log \frac{1}{1 - x^2}, \quad 0 \leq x < 1. \tag{4}$$

For any fixed x, this result cannot be improved (that is, there is a nonzero $f \in D$ for which equality is obtained). However, this result does not imply (3). (Similarly, in the Fejér-Riesz theorem, if $f \in H^2$, then $|f(x)|^2 \leq 1/(1 - x^2)$, which is not integrable.) Theorem 1 is a consequence of the following result.

Theorem 2. Let μ be a positive finite Borel measure on the half-open interval $0 \leq x < 1$. Then there is a constant c_1 such that

$$\int |f(x)|^2 d\mu(x) \leq c_1 \|f\|^2 \quad (f \in D), \tag{5}$$

if and only if there is a constant c_2 such that

$$\mu([x, 1)) \leq c_2 [-\log(1 - x)]^{-1}, \quad (x < 1). \tag{6}$$

Note. If μ is a finite positive Borel measure on the open unit disc Δ such that $\int |f|^2 d\mu < \infty$ for all $f \in D$, then there is a constant c such that $\int |f|^2 d\mu \leq c \|f\|^2$; this follows from the closed graph theorem. These measures have been characterized in a very nice paper by David Stegenga [10, Theorem 2.3]; the characterization involves comparing the logarithmic capacity of a finite union of open intervals on $\partial\Delta$ with the μ-measure of a certain associated set in Δ. In an important paper by Nagel, Rudin, and Shapiro [8, Theorem 4.5], a simpler sufficient condition is given (in a more general context). When their result is specialized to our situation, it says that (6) implies (5) (which is the difficult half of Theorem 2). Since our proof is simpler than either of the above, it seems worth presenting it; we use the theory of quadratic forms. Afterward we point out the connection of Theorem 1 to the theory of multipliers.

Warning: The same symbol c will be used to denote various constants, not necessarily the same at each occurrence.

We require two lemmas; μ will denote a fixed positive finite Borel measure on $[0, 1)$.

Lemma 1. If μ satisfies (6), then

$$\int_0^1 x^n \, d\mu(x) \le c[\log n]^{-1} \qquad (n \ge 2).$$

Proof. Let $\psi(x) = \mu([x, 1))$. Integrating by parts, we have

$$\int_0^1 x^n \, d\mu(x) = n \int_0^1 x^{n-1} \psi(x) \, dx = n\left(\int_0^{t_n} + \int_{t_n}^1\right) x^{n-1} \psi(x) \, dx,$$

where $t_n = 1 - n^{-1}$. Then

$$n \int_{t_n}^1 x^{n-1} \psi(x) \, dx \le n\psi(t_n)(1 - t_n) \le c[\log n]^{-1}.$$

Next, one verifies that the function $\varphi(x) = (1 - x)[-x^{-1} \log(1 - x)]$ is decreasing on $[0, 1)$. Hence

$$n \int_0^{t_n} x^{n-1} \psi(x) \, dx \le cn \int_0^{t_n} x^{n-2}(1 - x)[\varphi(x)]^{-1} \, dx$$

$$\le cn[\varphi(t_n)]^{-1} \int_0^1 x^{n-2}(1 - x) \, dx$$

$$= cn[\log n]^{-1}(n - 1)^{-1}.$$

This completes the proof.

Lemma 2.   There is a constant c such that

$$\sum_{n=1}^{\infty} \frac{1}{n[\log(n+1)]^{1/2} \log(n+m)} \le c/[\log(m+1)]^{1/2},$$

$$m = 1, 2, \ldots .$$

Proof. Let m be given. For simplicity, the summand on the left side of (9) will be omitted. Then

$$\sum_{n=1}^{m} \le \frac{1}{\log(m+1)} \sum_{1}^{m} \frac{1}{n[\log(n+1)]^{1/2}} \le c/[\log(m+1)]^{1/2},$$

as one sees by comparing the sum to an integral. Also,

$$\sum_{n=m+1}^{\infty} \le \sum_{m+1}^{\infty} \frac{1}{n[\log(n+1)]^{3/2}} \le c/[\log(m+1)]^{1/2},$$

again by comparison with an integral. This completes the proof of the lemma.

We also require the following sufficient condition for a matrix operator to be bounded.

Proposition 1 (Schur Test).   Let $a_{nm} \ge 0$ ($n, m \ge 1$) be given. If there is a function $p(n) > 0$ ($n \ge 1$) such that

(i) $\sum_{n} a_{nm} p(n) \le c_1 p(m)$ *(for all   m)*,

(ii) $\sum_{m} a_{nm} p(m) \le c_2 p(n)$ *(for all   n)*,

then

$$\left| \sum_{n,m} a_{nm} x_n \bar{y}_m \right| \le (c_1 c_2)^{1/2} (\sum |x_n|^2)^{1/2} (\sum |y_m|^2)^{1/2}$$

for all square-summable sequences $\{x_n\}$ and $\{y_m\}$.

For a proof, see, for example, [2, p. 126] (where the designation Schur test is used). See [7, Problems 37, 38, pp. 22, 23] for an application of the test to the Hilbert matrix $a_{nm} = (n + m + 1)^{-1}$ ($n, m \ge 0$). This is the matrix arising from (1); in this case the test gives the exact bound, π, and thus furnishes

another proof of the Fejér-Riesz theorem (the usual proof uses
Cauchy's integral theorem). Schur himself considered the case $p(n)$
$\equiv 1$ (see [9, I, p. 6]). This case can be regarded as an early example of the interpolation of linear operators, since now the
conditions (i) and (ii) state that the matrix $A = (a_{nm})$ is a
bounded operator on the spaces ℓ^1 and ℓ^∞, respectively, with
bounds c_1 and c_2. Schur's conclusion, then, is that A is bounded
on ℓ^2 with bound $(c_1 c_2)^{1/2}$. I do not know when the general function $p(n)$ was first introduced, but Carleman [3, Chapter 3, p.
112] establishes the test in this form (for integral transforms
with real and symmetric kernel). There is a still more general
form with two functions $p(n)$ and $q(m)$, where $q(m)$ replaces $p(m)$ on
the right side of (i) and on the left side of (ii), and the other
sides are unchanged. See [1, Theorem 6.3, especially (6.8),
p. 240], where the condition (suitably modified) is shown to imply
boundedness from L^p to L^p. In this form the test is necessary as
well as sufficient (see E. Gagliardo [6]).

Proof of Theorem 2. We first show that (6) implies (5). So let μ
be given and assume that (6) is satisfied. For $f(z) = \Sigma\, a_n z^n \in D$,
we have, by Lemma 1,

$$\int |f|^2 \, d\mu \leq \sum_{n,m=1}^{\infty} |a_n a_m| \int x^{n+m} \, d\mu(x) \leq c \sum \alpha_{nm} b_n b_m \,,$$

where $b_n = n^{1/2} a_n$, and $\alpha_{nm} = (nm)^{-1/2} [\log(n + m)]^{-1}$. We apply the
Schur test with $p(n) = [n \log(n + 1)]^{-1/2}$, $n = 1, 2, \ldots$. By Lemma
2 we have $\Sigma\, \alpha_{nm} p(n) \leq cp(m)$, $m = 1, 2, \ldots$; since $\alpha_{nm} = \alpha_{mn}$, we
have the same inequality when n and m are interchanged. This com-
pletes the proof of (5).

    For the converse, assume that μ satisfies condition (5). For
$w \in \Delta$, let $k_w(z) = -\log(1 - \bar{w}z)$. This function is the "reproducing
kernel" for the space D; that is, in terms of the inner product induced by (2), we have

$$f(w) = (f, k_w) \qquad (f \in D).$$

Hence, $\| k_w \|^2 = (k_w, k_w) = k_w(w) = -\log(1 - |w|^2)$. Now choose $t \in [0, 1)$, and let $I = [t, 1)$. Since $\| k_t \|^4 \leq |k_t(x)|^2$ for $x \in I$, we have, by (5),

$$\mu([t, 1)) \| k_t \|^4 \leq \int_I |k_t(x)|^2 \, d\mu(x) \leq \int_0^1 |k_t|^2 \, d\mu \leq c \| k_t \|^2.$$

Hence $\mu([t, 1)) \leq c[-\log(1 - t^2)]^{-1}$ and (6) follows from this. (There is a slight problem, since we have t^2 on the right side, unlike (6). And $\log(1 - t)/\log(1 - t^2) \to \infty$ as $t \to 0^+$. However, there is no problem away from $t = 0$, and (6) is automatically satisfied near 0 for any finite measure μ. This completes the proof of the theorem.

Next we discuss Theorem 1 from the viewpoint of multipliers. A function φ, analytic in Δ, is said to multiply D into H^2 if $\varphi D \subset H^2$. These multipliers (and many others) are classified by Stegenga in [10] (see Theorems 1.1(b) and 2.3). Namely, he showed that φ is such a multiplier if and only if the measure $d\mu(z) = |\varphi'(z)|^2(1 - |z|) \, dx \, dy$ satisfies the condition: $\int |f|^2 \, d\mu < \infty$ for all f in D. And as pointed out earlier, in the same paper he gave a characterization of all such measures. We now show that Theorem 1 is equivalent to a multiplier result. The factor z in the function φ below is inserted to cancel the pole at the origin; this could be omitted, since $f(0) = 0$ for $f \in D$.

Proposition 2. *The following two statements are equivalent:*

(a) *The function* $\phi(z) = z(1 - z)^{-1/2}[-\log(1 - z)]^{-1}$ *is a multiplier from D to H^2.*

(b) *Inequality (3) is valid.*

Proof. In view of the Fejér-Riesz inequality (1), we see that the first condition above implies the second.

For the reverse implication, we use the fact that an analytic function g in the unit disc is in H^2 if and only if

$\iint |g'|^2 (1 - r) \, r \, dr \, d\theta < \infty$ (this may be seen, for example, by ex-pressing the integral in terms of Taylor coefficients). Thus, an analytic function ψ is a multiplier from D to H^2 if and only if $g = \psi f$ satisfies the above condition. Since the function f is in D, its derivative is square-integrable with respect to area measure (see (2)). Thus a sufficient condition for $|\psi f'|^2 (1 - r)$ to have a finite area integral is that $(1 - |z|) |\psi(z)|^2$ be bounded. Actually, the stronger condition

$$|\psi(z)|^2 \le \frac{c}{(1 - |z|^2)(-\log(1 - |z|^2))} , \qquad |z| < 1, \qquad (7)$$

is necessary for ψ to multiply D into H^2. This follows from Theorem 1 of [10] if one specializes to the particular spaces we are con-cerned with. (The logarithmic factor comes from inequality (4) for D.) And of course the function ϕ of the present proposition satis-fies an even stronger condition. Thus a general function ψ satis-fying (7) will multiply D into H^2 if and only if

$$\iint |f\psi'|^2 (1 - r) \, r \, dr \, d\theta < \infty \qquad (f \in D). \qquad (8)$$

To prove this for ϕ, we need the following result.

Lemma 3. If $[\log(1 - z)]^2 = \sum_2^\infty a_n z^n$, then $a_2 > a_3 > \cdots \to 0$.

Proof. We have

$$a_n = \sum_1^{n-1} \frac{1}{k(n - k)} = \frac{1}{n} \sum_1^{n-1} \left(\frac{1}{k} + \frac{1}{n - k} \right) = \frac{2}{n} \sum_1^{n-1} \frac{1}{k} .$$

Clearly, $a_n \to 0$. Thus it remains to show that

$$(n + 1) \sum_1^{n-1} \frac{1}{k} > n \sum_1^n \frac{1}{k} = 1 + n \sum_1^{n-1} \frac{1}{k} ,$$

and this is obvious.

Corollary. If $g(z) = (1 - z)[\log(1 - z)]^2$, then $|g(z)| \ge g(|z|)$ for $|z| < 1$.

Proof.   With the notations of the lemma, we have

$$g(z) = a_2 z^2 - \sum_3^\infty (a_{n-1} - a_n) z^n.$$

Hence, $|g(z)| \geq a_2 |z|^2 - \Sigma (a_{n-1} - a_n) |z|^n = g(|z|)$.

To complete the proof of the proposition, we must show that (8) holds with ψ replaced by ϕ. Let $f = \Sigma b_n z^n$ be in D and let $h = \Sigma |b_n| z^n$. Then h is in D and has the same norm as f. Also, $|f(z)| \leq h(|z|)$. We use the function g of the above corollary. Also,

$$\phi'(z) = (1 - z)^{-3/2} (-\log(1 - z))^{-1} [z/2 - z(-\log(1 - z))^{-1} + 1 - z].$$

By the above corollary, and since the function in square brackets is bounded, we have (with $r = |z|$): $|\phi'(z)|^2 \leq cg(r)^{-1}|1 - z|^2.$

Thus

$$\iint |f\phi'|^2 (1 - r) \leq c \int_0^1 h(r)^2 g(r)^{-1} (1 - r) \int_0^{2\pi} |1 - z|^{-2} \delta\theta \; r dr$$

$$\leq c \int_0^1 h(r)^2 g(r)^{-1} \; dr \leq c \| f \|^2$$

by (3). This completes the proof.

Finally, we would like to say something about the extent to which Theorem 1 is a "best possible" result. Theorem 2 really gives the whole story and we shall just make a few comments. First, if $\psi(x)$ is a positive measurable function on the interval $0 \leq x < 1$ such that $\int |f\psi|^2 \; dx < \infty$ for all f in D, then

$$\lim_{x\uparrow 1} \left(\frac{\psi(x)}{\varphi(x)} \right) < \infty, \tag{9}$$

where φ is the function from the preceding proposition. This is an easy consequence of condition (6), applied to the measure $d\mu = \psi^2 \; dx$.

For the Fejér-Riesz theorem, one has a better result: if ψ is a positive increasing function on the half-open interval $[0, 1)$, and if $\int |f\psi|^2 < \infty$ for all $f \in H^2$, then ψ is bounded. This follows, since $\psi^2 \; dx$ is a Carleson measure (see [4, Theorem 9.3, p. 157]).

It turns out that the analogue of this result is false for the Dirichlet space.

Proposition 3.   There is an increasing function $\psi(x)$, $0 \le x < 1$, such that

(i) $\psi(x) \ge \phi(x)$, $0 \le x < 1$, *where ϕ is the function defined in Theorem 2;*

(ii) $\overline{\lim} \, \psi(x)/\phi(x) = \infty$, $x \to 1$;

(iii) $\int_0^1 |(\psi f)(x)|^2 \, dx < \infty$, $f \in D$.

Proof.   We shall choose a sequence of numbers $\{\varepsilon_n\}$ such that $\varepsilon_1 < 1$ and $n\varepsilon_n < \varepsilon_{n-1}$ $(n > 1)$. Hence $\varepsilon_n < 1/n!$ There will be an additional condition to be specified later. Let $x_n = 1 - \varepsilon_n$, $x'_n = 1 - \varepsilon_n/n$. Then $0 < x_1 \le x'_1 < x_2 < x'_2 < \cdots \to 1$. We define $\psi(x)$ as follows:

$$\psi(x) = \phi(x'_n), \quad x_n \le x \le x'_n,$$

$$\psi(x) = \phi(x), \quad x'_n \le x < x_{n+1},$$

$n = 1, 2, \ldots,$ and $\psi(x) = \phi(x)$ for $0 \le x \le x_1$. Then clearly ψ is increasing and $\psi(x) \ge \phi(x)$ for all x. (If desired, ψ could be modified slightly to be continuous, without affecting the proof.) Also,

$$\frac{\psi(x_n)}{\phi(x_n)} = \frac{\phi(x'_n)}{\phi(x_n)} = n^{1/2} \frac{x'_n}{x_n} \frac{\log(1/\varepsilon_n)}{\log(n/\varepsilon_n)} \to \infty,$$

since the factor multiplying $n^{1/2}$ tends to 1.

Thus, to complete the proof it only remains to establish (iii). In view of inequality (3), to prove (iii) we only need to consider those intervals where ψ is larger than ϕ. Let $f \in D$ with $\|f\| \le 1$. Then for $x \in [x_n, x'_n]$, we have, by (4),

$$|f(x)|^2 \le -\log(1 - x'^2_n) < -\log(1 - x'_n).$$

Hence

$$\sum_{n=2}^{\infty} \int_{x_n}^{x'_n} |(\psi f)(x)|^2 dx < \sum_{n>2} [\phi(x'_n)]^2 \left(\log \frac{1}{1-x'_n}\right)(x'_n - x_n)$$

$$< \sum_{n>2} \frac{n}{\varepsilon_n} \left(\log \frac{n}{\varepsilon_n}\right)^{-1} \varepsilon_n < \Sigma n \left(\log \frac{1}{\varepsilon_n}\right)^{-1}.$$

This series will converge if ε_n approaches 0 fast enough, for example, if $\varepsilon_n = \exp(-n^3)$. This completes the proof.

REFERENCES

1. Aronszajn, N.; Mulla, F.; and Szeptycki, P. On spaces of potentials connected with L^p spaces. *Ann. Inst. Fourier (Grenoble)*, 13, Fasc. 2 (1963): 211-306.

2. Brown, Arlen; Halmos, P. R.; and Shields, A. L. Cesàro operators. *Acta Scient. Math. (Szeged)* 26 (1965): 125-137.

3. Carleman, T. *Sur les Équations Intégrales Singulières a Noyau Rèel et Symétrique.* Uppsala: Almqvist and Wiksells, 1923.

4. Duren, Peter L. *Theory of H^p Spaces.* New York: Academic Press, 1970.

5. Fejér, L., and Riesz, F. Über einige funktionentheoretische Ungleichungen. *Math. Zeitschrift* 11 (1921): 305-314.

6. Gagliardo, Emilio. On integral transformations with positive kernel. *Proc. Amer. Math. Soc.* 16 (1965): 429-434.

7. Halmos, P. R. *A Hilbert Space Problem Book.* New York: Van Nostrand, 1967.

8. Nagel, Alexander; Rudin, Walter; and Shapiro, Joel H. Tangential boundary behaviour of functions in Dirichlet-type spaces. To appear in *Ann. of Math.*

9. Schur, I. Bemerkungen zur theorie der beschränkten Bilinearformen mit unendlichvielen Veränderlichen. *J. reine angew. Math.* 140 (1911): 1-28.

10.   Stegenga, D. A. Multipliers of the Dirichlet space. *Ill. J. Math.* 24 (1980): 113-139.

11.   Taylor, G. D. Multipliers on D_α. *Trans. Amer. Math. Soc.* 123 (1966): 229-240.

POTENTIAL THEORY AND
DIFFUSION ON RIEMANNIAN MANIFOLDS

N.T. Varopoulos
Université Paris

INTRODUCTION

Let M be a complete connected Riemannian manifold and let us denote by Δ = divgrad its invarient Laplacian. I shall follow [1] in
recalling the definitions and the basic properties of the Green's
function and the harmonic measure on M.

Let $p \in M$ be fixed and let V_p be the family of functions that
are defined and subharmonic on $M \setminus p$ and identically zero outside
some compact subset of M and such that for all $v \in V_p$ we can find
some $E = E_p(m)$, $m \in \Omega$, where Ω is an open Nhd of p, that satisfies

$$\frac{1}{2} \Delta E = -\delta_p$$

in the distribution sense in Ω (that is, E is an elementary solution of Δ in Ω) and such that

$$\overline{\lim_{m \to p}} \left(v(m) - E(m) \right) < +\infty.$$

V_p is clearly a Perron family of functions (cf. [1, Chapter 9],
for a definition), and therefore the function

$$g(p, .) = \sup\{v; \ v \in V_p\},$$

when not identically $+\infty$, is harmonic on $M \setminus p$. When $g \neq +\infty$, we say that Green's function exists on M (at p); g is then the Green's function. Here are some of the basic properties of the Green's function (cf. [1, Chapter 9] for the proofs).

If g exists at $p \in M$, it also exists at every other point $p_1 \in M$. When g exists, it satisfies

$$\frac{1}{2} \Delta g(p, .) = -\delta p$$

in the distribution sense. In fact, g is then the minimal positive solution of the above equation, in the sense that if $0 \leq G \in L^1_{loc}$ (M) is any other (distributional) solution of $\frac{1}{2} \Delta G = -\delta_p$, we then have $g(p, m) \leq G(m)$ $(m \in M \setminus p)$. Indeed, by the local behavior of G (near p) and the definition of V, we see that

$$(1 + \varepsilon) G(m) - u(m) \geq 0, \quad \varepsilon > 0,$$

when $u \in V$ and $m \neq p$ lies in some Nhd of p. The above inequality also holds in some Nhd of ∞ of M and therefore everywhere. Letting $\varepsilon \to 0$, we get that

$$G(m) \geq u(m).$$

Now let K be a compact subset of M with nonempty interior $(K^\circ \neq \emptyset)$. Let also U_K be the Perron family of functions that are defined and subharmonic on $M \setminus K$ and such that each $u \in U_K$ satisfies

$$u \leq 1 \quad \text{on} \quad M \setminus K; \quad \overline{\lim_{m \to \infty}} u(m) \leq 0.$$

$u_K = \sup\{u; u \in U_K\}$ is then a harmonic function on $M \setminus K$, and when $u_K \neq 1$, we call it the harmonic measure of K (observe that $0 \leq u_K \leq 1$ and $0 \neq u_K$ by $K^\circ \neq \emptyset$) and say that harmonic measure exists (for K). It is then a well-known fact (cf. [1, Chapter 9]) that if harmonic measure exists for K, it exists for every other compact subset K_1 $(K_1^\circ \neq \emptyset)$ and that on any Riemannian manifold harmonic measure exists if and only if the Green's function exists.

Now let $\{z(t); \ 0 \leq t \leq e\}$ be the canonical diffusion on M with $\frac{1}{2}\Delta$ as an infinitesimal generator, and let

$$p_t(x, \ y) \ dV = \mathbf{P}_x[z(t) \in dV, \ t < e]$$

be the heat diffusion Kernel (dV denotes the Riemannian volume element). The above diffusion, which is sometimes called the Brownian motion of M, has been constructed by Ito (cf. [8, Chapter 4]). e is a random time called explosion time and we have $z(t) \xrightarrow[t \to e]{} \infty_M$, and if $f(x) = \mathbf{P}_x[e = +\infty] = 1$ for some $x \in M$, then $f(y) = \mathbf{P}_y[e = +\infty] = 1$ for all $y \in M$. This is because $f(x)$ is a harmonic function on M (cf. [8, 4.4]). We then say that there is no explosion. The nonexplosion of Brownian motion on M is equivalent to

$$\int_M p_t(x, \ y) \ dV(y) = 1, \quad \forall \, x \in M, \quad t \geq 0, \tag{1}$$

and in fact, everything that will be said about Brownian motion here has an interpretation in terms of the heat diffusion semigroup and the language of P.D.E. and semigroups. (1) has been proved under the condition $\mathrm{Ric}(M) \geq -K$ (for some $K \geq 0$) by Yau [11].

We have:

Theorem 1.   Let $C(r)(r \geq 0)$ be a smooth $(C^5$, say) positive nondecreasing function such that

$$\int_1^\infty \frac{dr}{C^{\frac{1}{2}}(r)} = +\infty. \tag{2}$$

Let also M be a complete connected Riemannian manifold that satisfies

$$\lambda(x) \geq -C(r), \quad r = d(x, \ m), \quad x \in M, \tag{3}$$

where m is some fixed point on M, d denotes Riemannian distance, and λ denotes the lowest eigenvalue of $\mathrm{Ric}(M)$. Under the above conditions, Brownian motion does not explode.

We shall also see that if we replace the Ricci curvature by the sectional curvature, then the condition (2) becomes optimal.

Explosion or not on M, let us define for a fixed $x \in M$

$$0 \leq R_\lambda(x,\ y) = \int_0^\infty \bar{e}^{\lambda t} p_t(x,\ y)\ dt \in L^1_{loc}(M;\ dV(y))$$

the resolvant operator. By the general theory (cf. [8, Sec. 1.5, and 13, IX 4]) we have

$$\frac{1}{2} \Delta_y R_\lambda(x,\ y) = \lambda R_\lambda(x,\ y) - \delta_x. \tag{4}$$

It follows that if we make the assumption that

$$G(x,\ y) = \int_0^\infty p_t(x,\ y)\ dt \in L^1_{loc}(M;\ dV(y))$$

$(x \in M$ fixed), then $\left(\text{let } \lambda \to 0 \text{ in } (4)\right)$ G satisfies $\frac{1}{2} \Delta G = -\delta$. The conclusion is that the Green's function g exists on M and that $g(x,\ y) \leq (G\ x,\ y)$.

Conversely, if the Green's function g exists on M, then $G(x,\ y) \in L^1_{loc}(M;\ dV(y))$ (for $\forall\ x \in M$ fixed) and we have, in fact, $G(x,\ y) = g(x,\ y)$. To see this, we can argue as follows.

Let $x \in \Omega \subset M$ be a relatively compact subset of M with smooth boundary and let

$$p_t^\Omega(x,\ y)\ dV(y) = \mathbf{P}_x[z(t) \in dV(y);\ z(s) \in \Omega \quad 0 \leq s \leq t]$$

be the relative heat kernel. By the general theory of boundary value problems in P.D.E.'s, we know then that (the relative Green's function)

$$G_\Omega(x,\ y) = \int_0^\infty p_t^\Omega(x,\ y)\ dt \in L^1(\Omega;\ dV(y))$$

satisfies

$$\frac{1}{2} \Delta_y G_\Omega(x,\ y) = -\delta_x$$

in Ω. It follows from the definition of g that

$$G_\Omega(x,\ y) \leq g(x,\ y).$$

It is clear, on the other hand, that for $\Omega_1 \subset \Omega_2 \subset \cdots \subset \Omega_n \subset \cdots M = U\Omega_n$ relatively compact subsets as above, we have

$$P_t^{\Omega_n}(x, y) \xrightarrow[n \to \infty]{} P_t(x, y)$$

and that the p^{Ω_n}'s are an increasing sequence. Our assertion follows.

Let $\Omega \subset M$ be a relatively compact subset with smooth boundary and let

$$\widetilde{u}_\Omega(x) = \mathbf{P}_x [\exists \, t \geq 0 \quad z(t) \in \bar{\Omega}].$$

It is a standard fact (cf. [8, Sec. 3]) that $\widetilde{u}_\Omega(x)$ is a harmonic function on $M \setminus \bar{\Omega}$ that extends continuously to $\partial\Omega$ with $\widetilde{u}_\Omega|_{\partial\Omega} \equiv 1$.

We say that the Brownian motion is *recurrent* if $\widetilde{u}_\Omega(x) = 1 \; \forall \; x$ and $\forall \; \Omega$ as above. We say that the Brownian motion is *transient* if $\widetilde{u}_\Omega(x) < 1 \; \forall \; \Omega$ and $\forall \; x \notin \bar{\Omega}$.

From what has been said, it follows that if for some x and Ω as above with $x \notin \bar{\Omega}$ we have $\widetilde{u}_\Omega(x) < 1$, then harmonic measure exists on M. Conversely, if for some $x \notin \bar{\Omega}$ as above we have $\widetilde{u}_\Omega(x) = 1$, then harmonic measure does not exist. Brownian motion is therefore transient (resp. recurrent) if and only if harmonic measure exits (resp. does not exist) on M. To see that last point, we can argue as follows.

Let $\Omega = \Omega_0 \subset \Omega_1 \subset \cdots \subset M = U\Omega_j$ be an increasing family of relatively compact open subsets with smooth boundary and let $u_j \in C(\bar{\Omega}_j \setminus \Omega_0)$ be the solution of the Dirichlet problem with $u_j|_{\partial\Omega_0} = 1$ $u_j|_{\partial\Omega_j} = 0$. We have then (cf. [8, Sec. 3])

$$u_j(m) = \mathbf{P}_m [z(t) \text{ hits } \partial\Omega_0 \text{ before it hits } \partial\Omega_j].$$

Now Brownian motion cannot stay in a compact set; if it did, it would do so with probability one (by the zero one law) and this is absurd. It follows, therefore, that $u_j(x) \xrightarrow[j \to \infty]{} \widetilde{u}_\Omega$ (on $M \setminus \Omega$)

and so by our hypothesis, $u_j(x) \xrightarrow[j \to \infty]{} 1$ and therefore we have $u_k(x) = 1$ (by definition of u_k). This proves our assertion.

Now let M be a complete Riemannian manifold as before, and let

$$B_t(m) = \{x \in M;\ d(x, m) \leq t\}$$

be the Riemannian ball ($t \geq 0$; $m \in M$). We shall denote by $V_m(t) =$ $\mathrm{Vol}\bigl(B_t(m)\bigr)$ the Riemannian volume of that ball. We have then:

Theorem 2. Let M be as above. Then a necessary condition for the existence of Green's function on M is that we have

$$\int_1^\infty \frac{t\,dt}{V_m(t)} < +\infty \tag{5}$$

for all $m \in M$. Conversely, if we suppose in addition that $\mathrm{Ric}(M) \geq 0$ and if (5) is verified for some $m \in M$, the Green's function exists on M.

The first half of the above theorem (that is, the necessity of the condition) for the two-dimensional case (that is, Riemann surfaces) is a classical theorem of Ahlfors [2].

By the following example (which is a simplified version of an example given to me by R. Greene), we see that (5) is not in general sufficient to ensure the existence of Green's function.

Example. Let $M = \mathbf{C}$ be the complex plane and let $ds^2 = \varphi(y)|dz|^2$ ($z = x + iy \in \mathbf{C}$), where $0 \leq \varphi \in C^\infty(M)$, $\varphi(y) = 1/y$ ($y \geq 1/10$), and $\varphi(y) = 1$ ($y < 0$). Green's function clearly does not exist (since the conformal structural is that of $\mathbf{C}$). An easy computation in hyperbolic geometry, on the other hand, shows that (5) is very generously verified $\bigl(V(t)$ increases exponentially!$\bigr)$.

If we assume that (5) holds on M and that M has nonnegative *sectional* curvature everywhere, then we can even obtain a nice estimate on the Green's function $g(m, x)$. Indeed, I shall show elsewhere that under these conditions for all $m \in M$, we can find $C = C(m) > 0$ such that:

$$c^{-1} \leq g(x, m) \left\{ \int_{d(x,m)}^{\infty} \frac{tdt}{V_m(t)} \right\}^{-1} \leq c.$$

PROOF OF THEOREM 1

Let M be a smooth complete connected Riemannian manifold. Theorem 1 is a consequence of the following two propositions.

Proposition 1. Let M be as above and let us suppose that there exists $U > 0$ a locally Lipschitz function on M and two contents $A, B > 0$ such that $U(x) \to \infty$ $(x \to \infty)$ and such that

$$\Delta U \leq AU + B \tag{6}$$

in the distribution sense; that is, we suppose that

$$\int_M \Delta U \cdot \varphi \, dV \leq \int_M (AU + B)\varphi \, dV$$

for all $0 \leq \varphi \in C_0^{\infty}(M)$. Under the above conditions, there is no explosion on M; that is, we have $\mathbf{P}_m[e = +\infty] = 1$ $(m \in M)$.

Proposition 2. Let $0 < C(r) \in C^5([0, +\infty))$ be a nondecreasing function that satisfies condition (2), and let M be a complete Riemannian manifold that satisfies

$$\mathrm{Ric}_x(X, X) \geq -C(r)|X|^2 \tag{7}$$

$\forall x \in M$, $X \in T_x M$, where $r = d(x, m)$ denotes the distance from some fixed point m. Then a function U exists on M that satisfies the conditions of Proposition 1.

Proof of Proposition 1. The proof is based on two distinct technical components. The first regularizes $U(x)$ locally, that is, on compact subsets, and the second deals with the singularity at infinity.

Lemma 1. Let U, A, B be as in Proposition 1 and let $K \subset M$ be a compact subset of M. There exists then $0 \leq \varphi_n \in C_0^{\infty}(M)$ such that

$$\Delta \varphi_n \leq A\varphi_n + B + 1, \quad on \ K \quad (n \geq 1)$$

$$\varphi_n \xrightarrow[n \to \infty]{} U \quad uniformly \ on \ K.$$

This lemma is essentially contained in [6], and its proof is based on the following idea of P. Malliavin.

Let us first modify U and consider $U_K \in C_0(M)$, which is Lipschitz and coincides with U on some Nhd of K. Let $\{P_t; \ t \geq 0\}$ denote the heat diffusion semigroup on M and let us set

$$\varphi_\varepsilon = (P_\varepsilon U_K)\Psi, \quad 0 < \varepsilon < 1,$$

with $\Psi \in C_0^\infty(M)$, $0 \leq \Psi \leq 1$, and $\Psi \equiv 1$ on some Nhd of K. If we then set $\varphi_n = \varphi_{1/n}$, we have our sequence (at least for n large enough). The reason for this is that the heat diffusion semigroup commutes with the Laplacian, that is, $(\Delta P_t - P_t \Delta)\varphi = 0$, $\forall \ \varphi \in C_0^\infty(M)$, and that both Δ and P_t are self-adjoint.

If the above proof frightens you, you can look at [10, Lemma 2.1], where I prove a weaker lemma with more conventional methods (the Friedrich mollifier), which is also sufficient for our purpose.

To cope with the singularity at ∞, it is very advantageous to use probability theory (rather than try to cope directly with the formulation (1)).

Let $\{z(t); \ 0 < t < e\}$ be the Brownian motion on M and let

$$T_a = \inf\{t/d(m, \ z(t)) \geq a\}$$

(for some fixed point $m \in M$) be the exit time from the ball $B_t(m)$.

Lemma 2.   Let U be as in Proposition 1. We then have

$$\mathbf{E}_m U\{z(t \wedge T_a)\} \leq \left(U(m) + 2 + \frac{B+1}{A}\right)e^{At} = K(m)e^{At},$$

where m is an arbitrary fixed point of M.

Proof.   Let m and $a > 0$ be fixed and let $\varphi_n \in C_0^\infty(M) \ (n \geq 1)$ be a sequence that satisfies the conditions of Lemma 1 with $K =$

$\{x \in M/d(x, m) \leq 2a\}$. Dynkin's formula (cf. [8, Sec. 2, n.6]) applies then, since $z(t)$ has the strong Markov property (the strong Markov property is already contained in Ito's construction (cf. [8, Chapter 4]) and we get

$$\mathbf{E}_m \varphi_n\big(z(t \wedge T_a)\big) - \varphi_n(m) = \mathbf{E}_m \int_0^{t \wedge Ta} \Delta \varphi_n\big(z(s)\big)\, ds$$

$$\leq \mathbf{E}_m \int_0^{t \wedge Ta} \big(A\varphi_n\big(z(s)\big) + B + 1\big)\, ds$$

$$\leq \mathbf{E}_m \int_0^t \big[A\varphi_n\big(z(s \wedge T_a)\big) + B + 1\big]\, ds.$$

This implies that the expectation

$$\Psi_n(t) = \mathbf{E}_m \varphi_n [z(t \wedge T_a)], \ t \geq 0$$

satisfies the differential inequality.

$$\Psi_n(t) \leq U(m) + 1 + \int_0^t \big(A\Psi_n(s) + B + 1\big)\, ds$$

for n large enough.

Let θ be the unique solution of the equation:

$$\theta(t) = U(m) + 2 + \int_0^t \big(A\theta(s) + B + 1\big)\, ds, \ t \geq 0.$$

It follows that $\Psi_n - \theta \leq 0$ for $t \geq 0$ and n large enough.

If we let $n \to \infty$, we obtain that $\mathbf{E}_m U[z(t \wedge T_a)] \leq \theta(t)$ $(t \geq 0)$. If we substitute the explicit value of $\theta(t)$, we obtain our lemma.

We are now in a position to give the:

End of the Proof of Proposition 1. By Lemma 2, it follows that for all $t \geq 0$ and $a \geq 0$ (and m fixed), we have

$$\inf\{U(z); z \in M; d(m, z) = a\} \cdot \mathbf{P}_m[T_a \leq t] \leq K(m) e^{At}.$$

By our hypothesis on U, it follows therefore that

$$\mathbf{P}_m[T_a \leq t] \xrightarrow[a \to \infty]{} 0.$$

On the other hand, we clearly have $[e < t] = \bigcap_{a \geq 0} [T_a < t]$, which is a monotone intersection. Our proposition follows.

Now let $F(r)$ $(r \geq 0)$ be a smooth function on $[0, +\infty)$ (say, c^{10}) such that $F(0) = 0$ $F'(0) = 1$. The quadratic differential

$$ds^2 = dr^2 + F(r)^2 \, d^2 \circleddash$$

on $\mathbf{R}^n \setminus 0$, where $r^2 = \sum_{r=1}^{n} x_r^2$ and $d^2 \circleddash$ is the standard Euclidean metric on the unit (Euclidean) sphere, determines then a smooth metric on $\mathbf{R}^n \setminus 0$ that extends to a continuous (but not necessarily smooth) metric on $\mathbf{R}^n$. The Jacobi equations give us then

$$F''(r) = -K(r)F(r),$$

where $K(r)$ is the curvature (the *radial* curvature) at the point $x \neq 0$ with $r = d(0, x)$ for any section that contains the ray $0x$. (We see, therefore, that the metric will certainly not be smooth at 0 unless $F''(r) \xrightarrow[r \to 0]{} 0$.) We have also $\mathrm{Ric}\left(\frac{\partial}{\partial r}, \frac{\partial}{\partial r}\right) = (n - 1)K(r)$. The above Riemannian space is called an (n-dimensional) model.

For $u(r)$ a smooth radial function on $\mathbf{R}^n \setminus 0$, the Laplacian with respect to the above metric takes the form

$$\Delta u = \frac{d^2 u}{dr^2} + (n - 1) \frac{F'(r)}{F(r)} \frac{du}{dr}, \quad r > 0.$$

We shall base the proof of Proposition 2 on the following classical theorem (cf. [7, Sec. 2]).

Comparison Theorem. *Let M be a smooth manifold and M_0 a (smooth at 0) model of the same dimension, and let us suppose that $\lambda(x)$ the lowest eigenvalue of Ric(M) at the point $x \in M$ satisfies*

$$\lambda(x) \geq \mathrm{Ric}_0 \left[\frac{\partial}{\partial r}, \frac{\partial}{\partial r}\right] = (n - 1)K(r), \tag{8}$$

where $r = d(m, x)$ with m some fixed point on M, and Ric_0 is the Ricci curvature of M_0. Let us furthermore assume that $u(r)$ $(r > 0)$ is a nondecreasing smooth radial function on $M_0 \setminus 0$ such that $\Delta_0 u \leq Au + B$ on $M_0 \setminus 0$ for two positive constants A and B. Under the above conditions, the function $U(x) = u\bigl(d(m, x)\bigr)$, $x \in M \setminus m$, satisfies $\Delta U \leq AU + B$ in the distribution sense on $M \setminus m$.

In fact, for $M \ni x \neq m$ and $x \notin C m =$ the cut locus of m, we have the pointwise estimate $\Delta U(x) \leq \Delta u(d(m, x))$ (cf. [7, Sec. 2]). To obtain the distributional inequality on the whole of $M \setminus m$, we can use the general principle formulated by Yau at the very end of [12] (and which is essentially due to Cheeger and Gromoll [4]) and which says just this: namely, that we can extend an inequality, like the one we have, across the cut locus of m provided that we interpret that inequality in the distribution sense. We are now in a position to give the:

Proof of Proposition 2. Let $0 < \varphi(t) \in C^{10} \left([0, +\infty) \right)$ be fixed and let

$$\Phi(t) = \frac{1}{t} + \varphi(t), \ t > 0; \ F(r) = A \exp \left[\int_1^r \Phi(t) \, dt \right], \ r > 0,$$

where we can extend the definition of F to $F(0) = 0$ and also choose A such that $F'(0) = \lim_{r \to 0} F'(r) = 1$ so as to be able to define a model M_0 by $ds^2 = dr^2 + F^2(r) \, d^2\oplus$. The radial curvature of M_0 is then given by

$$-K(r) = \frac{F''}{F} = \Phi' + \Phi^2 = \varphi' + \frac{2\varphi}{t} + \varphi^2.$$

If we set $\varphi(t) = \sqrt{C(t)}$, where $C(t)$ is as in the prosposition for $t \geq t_0 > 0$ and $\varphi(t) = \alpha t$ for $t < t_0/2$, then for t_0 small enough and $\alpha > 0$ large enough, we can complete the definition of $\varphi(f)$ in $[t_0/2, t_0]$ so that M_0 becomes smooth at 0 and satisfies (cf. [7], pp. 58-61)

$$\text{Ric}_0 \left(\frac{\partial}{\partial r}, \frac{\partial}{\partial r} \right) = (n - 1) K(r) \leq -C(r) \tag{9}$$

for that model.

If we set

$$u(r) = \exp \left(\int_1^r \frac{dt}{\Phi(t)} \right), \ r > 0,$$

we have for the above choice $\varphi = \sqrt{C(r)}$

$$\Delta u = u'' + (n - 1)\Phi u' = \left[-\frac{\Phi'(r)}{\Phi^2(r)} + \frac{1}{\Phi^2(r)} + n - 1 \right], \ u \le Au \quad (10)$$

for some fixed $A > 0$ (by conditions on $C(t)$). We also have $u(r) \xrightarrow[r \to \infty]{} \infty$, again by the conditions on $C(t)$.

Now let M satisfy condition (7) of Proposition 2. By (9) it then satisfies condition (8). The comparison theorem and (10) tell us, therefore, that $U(x) = u(d(m, x))$ satisfies (6) on $M \setminus m$. We can therefore smooth U on some small Nhd of m on M and obtain condition (6) everywhere on M. This completes the proof of the proposition.

To show that Theorem 1 is optimal, we shall prove the following.

Theorem 3. Let M be a complete connected Riemannian manifold and let $m \in M$ be a fixed point of M such that the sectional curvature $K(V)$ of M satisfies

$$K(V) \le -C(r), \ r \ge 0,$$

for every two-dimensional section $V \subset T_x(M)$ with $d(m, x) = r \ge 0$, where now $C(r)$ is assumed to be a positive C^1-function on $r \ge 0$ that satisfies

$$\int_1^\infty \frac{dr}{\sqrt{C(r)}} < +\infty$$

and

$$C'(r) \le aC^{3/2}(r), \ r \ge 0 \quad (11)$$

for some positive a. Under the above conditions, the explosion time e satisfies

$$\mathbf{P}[e < +\infty] = 1. \quad (12)$$

Proof. Observe first that by renormalizing the metric, we can suppose that a in (11) is as small as we like. Let then

$$\varphi(t) = \lambda t, \ t \in [0, t_0]; \ \varphi(t) = u\sqrt{C(t)}, \ t \in [t_0, +\infty],$$

where t_0 is the first point if anywhere $y = \lambda t$ intersects $y = \mu\sqrt{C(t)}$ and where $\lambda, \mu > 0$. By choosing λ and μ appropriately and then a appropriately small and smoothing the definition of $\varphi(t)$ out at the point t_0, we see that $\varphi(t)$ then verifies

$$\varphi'(t) + \frac{2\varphi(t)}{t} + \varphi^2(t) \leq C(t)$$

$$\left(\frac{1}{t^2} - \varphi'(t) + 1\right) + \left[\frac{1}{t} + \varphi(t)\right]^{-2} + n - 1 \geq \frac{1}{2} > 0.$$

A model M_0 can be constructed then as in the proof of Proposition 2 by $F(r) = A \exp\left[\int_1^r \left(\frac{1}{t} + \varphi(t)\right) dt\right]$. With the use of that model and the Hessian comparison theorem of [7, p. 19] (this is where the full strength of the sectional curvature is used), we can then construct, just as before, $u(x) \in C^2(M) \cap L^\infty(M)$ that satisfies

$$\Delta u(x) \geq \alpha u(x) \geq 0, \ \forall \ x \in M; \ \lim_{m \to \infty} u(m) = \|u\|_\infty \tag{13}$$

for some fixed positive $\alpha > 0$.

To terminate the proof, we use the following.

Lemma 3.   Let M complete connected manifold be such that $u \in C^2 \cap L^\infty(M)$ exists that satisfies (13). Then (12) holds.

Proof of the Lemma.   Fix $m \in M$ and set

$$T = \inf\{t/d(z(t), m) \geq r\}$$
$$X(t) = \exp(-\alpha(T \wedge t)) \, u(z(t \wedge T))$$

$(r \geq 0$ and $z(t)$ is the Brownian motion on M). The process $X(t)$ is a submartingale (compute dX by ITO calculus), by letting $r \to \infty$ in the inequality

$$u(m) \leq \mathbf{E}_m[\exp(-\alpha T) u(z(T))];$$

we then obtain

$$u(m) \leq \|u\|_\infty \mathbf{E}_m(\exp(-\alpha\Theta)) \leq \|U\|_\infty \mathbf{P}_m[e < +\infty].$$

The lemma follows (observe that $v(m) = \mathbf{P}_m[e < +\infty]$ is a harmonic function on M).

PROOF OF THEOREM 2

Let M be an arbitrary complete Riemannian manifold, let $m \in M$ be a fixed point on M, and let us suppose that

$$L(R) = \int_1^R \frac{t\,dt}{\text{Vol } B_t(m)} \xrightarrow[R\to\infty]{} +\infty.$$

I shall proceed to show that harmonic measure does not exist on M.
Let

$$F(r) = F_R(r) = 1 - \frac{L(r)}{L(R)}, \quad (r > 0),$$

and

$$\Phi(x) = \Phi_R(x) = F[d(m, x)] \quad (x \in M \quad x \neq m).$$

Clearly, then, Φ is a locally Lipschitz function on $M \setminus m$ and C^1-smooth on the complement of $C(m)$ the cut locus of m. For $x \neq m$, $x \notin C(m)$, we have

$$d\Phi(x) = -\frac{1}{L(R)} \frac{r}{\text{Vol } B_r(m)}dr, \quad r = d(x, m).$$

We have, therefore, if we denote by $|\ |$ the Riemannian norm on $T^{\star}M$,

$$I_R = \int_M \left[x \notin C(m); \tfrac{1}{2} \leq d(x, m) \leq R \right] |d\Phi|^2 \, d\text{Vol}(x)$$

$$\leq \frac{1}{L^2(R)} \int_{1/2 \leq d(x,m) \leq R} \left[\frac{r}{\text{Vol } B_r(m)}\right]^2 d\text{Vol}(x)$$

$$= \frac{1}{L^2(R)} \int_{1/2}^R r^2 \frac{1}{(\text{Vol } B_r(m))^2} \frac{d}{dr}(\text{Vol } B_r(m))\, dr$$

$$= -\frac{1}{L^2(R)} \left[\frac{r^2}{\text{Vol } B_r(m)}\right]_{1/2}^R + 2\frac{L(R) - L(1/2)}{L^2(R)}$$

$$\leq \frac{1}{4L^2(R)\text{Vol } B_{1/2}(M)} + 2\frac{L(R) - L(1/2)}{L^2(R)}.$$

The last step is obtained by integration by parts. The upshot, by our hypothesis, is that

$$I_R \xrightarrow[R\to\infty]{} 0.$$

(14)

Using the fact that Φ is locally Lipschitz and that $C(m)$ is of measure zero on M, we see that we can find $\varphi_\varepsilon \in C_0^\infty(M \setminus m)$ such that $\varphi_\varepsilon \xrightarrow[\varepsilon\to 0]{} \Phi$ uniformly on the compact of $M \setminus m$ and such that

$$\int_K |d\varphi_\varepsilon|^2 \, d\,\text{Vol} \xrightarrow[\varepsilon\to 0]{} \int_{K \setminus C(m)} |d\Phi|^2 \, d\,\text{Vol}$$

for every compact subset $K \subset M \setminus m$. This is done by the classical Friedrich mollifier (cf. [10, Lemma 2.1]). This together with (14) shows that for every $\varepsilon > 0$, we can find some $R > 1$ and some $\varphi \in C_0^\infty(M)$ such that

$$1 < \varphi(x) < 1 + \varepsilon; \quad x \in M, \ d(m, x) = 1$$
$$0 < \varphi(x) < \varepsilon; \quad x \in M, \ d(m, x) = R$$
$$\int_{1 \le d(x,m) \le R} |d\varphi|^2 \, dv \le \varepsilon.$$

If we denote then by u_R $(R > 0)$ the solution of the obvious Dirichlet problem in $C_R = \{x \in M; \ 1 \le d(x, m) \le R\}$, it follows from the above and the minimizing energy property of u_R (for example, ch. [3]) that

$$\int_{C_{R_j}} |du_{R_j}|^2 \, d\,\text{Vol} \xrightarrow[j\to\infty]{} 0$$

for some sequence $R_j \xrightarrow[j\to\infty]{} \infty$. The conclusion is of course that harmonic measure does not exist on M. This proves the first half of this theorem. The second and (by far) harder half has been proved in [10] and I shall not repeat myself here. I shall finish up with a few remarks, allowing myself to become sketchy and incomplete in my presentation.

Proposition 3. Let $a_j > 0$ $(j = 1, 2, \ldots)$. *Then we have*

$$\left(\frac{1}{a_1} + \frac{1}{a_2} + \cdots + \frac{1}{a_n} \right) \ge \frac{1}{100} \left(\sum_{j=1}^{n} \frac{j}{a_1 + \cdots + a_j} \right), \quad n \ge 1.$$

The above inequality is clear when $a_1 \leq a_2 \leq \cdots$. The proof consists in reducing the general case to the special nondecreasing one. Indeed, if for some $1 \leq j < n$ we have $a_j > a_{j+1}$, without changing any other a_r we can replace a_j by $a_j^\star$ and a_{j+1} by $a_{j+1}^\star$ so that

$$a_j > a_j^\star > a_{j+1}^\star > a_{j+1} \, , \quad \frac{1}{a_j} + \frac{1}{a_{j+1}} = \frac{1}{a_j^\star} + \frac{1}{a_{j+1}^\star} \, .$$

This operation does not change the left-hand side of the above inequality while it increases the right-hand side. This completes the proof.

The same inequality holds, therefore, for a locally integrable $f \geq 0$:

$$\int_a^b \frac{dt}{f(t)} \geq \frac{1}{100} \int_a^b \frac{t \, dt}{\int_0^t f(s) \, ds} \tag{15}$$

(at least this is certainly so if f is Riemann integrable).

Let us now return to our manifold M and let us denote by $\sigma_m(t)$ for some fixed $m \in M$ the $(n - 1)$-dimensional Hausdorff measure (induced by the Riemannian metric (cf. [5, 3.2.46]) of the "sphere of radius t around m") $\{x \in M;\ d(x,\ m) = t\}$. We then have

$$\sigma(t) = \frac{d}{dt} \operatorname{Vol} B_t(m), \qquad \text{a.a.}\ t \geq 0,$$

and we can even use this as a definition for $\sigma(t)$. It follows from (15) that

$$\Lambda(R) = \int_1^R \frac{dt}{\sigma(t)} \geq \frac{1}{100} \int_1^R \frac{t}{\operatorname{Vol} B_t(m)} \, dt. \tag{16}$$

We can then repeat the proof of Theorem 2 with $F(r) = 1 - \frac{\Lambda(r)}{\Lambda(R)}$ (this is technically more difficult, since $F(r)$ is no longer locally Lipschitz) and deduce that when

$$\int^\infty \frac{dt}{\sigma(t)} = +\infty,$$

there is no harmonic measure on M. This in view of (16) is a
sharper result and it is in this form that L. Ahlfors proved it for
$n = 2$. (His proof is different; it uses the conformal structure of
a Riemann surface.)

REFERENCES

1. Ahlfors, L. V. Conformal invariants. In *Topics in Geometric Function Theory*. New York: McGraw Hill.

2. Ahlfors, L. V. Sur le type d'une surface de Riemann. *C.R.A.S.* (1935).

3. Brelot, M. Elements de la theorie classique du potentiel. Cours de la Sorbonne, 1959.

4. Cheeger, J., and Gromoll, D. The splitting theorem for manifolds of nonnegative Ricci curvature. *J. Diff. Geometry* 6 (1971): 119–128.

5. Federer, H. *Geometric Measure Theory*. New York: Springer-Verlag, 1969.

6. Greene, R. E., and Wu, H. C^{∞} approximation of convex subharmonic and plurisabharmonic functions. *Ann. Scient. Ec. Norm. Sup.*, 4ème série t. 12 (1979): 47–84.

7. Greene, R. E., and Wu, H. Function theory on manifolds which possess a pole. *Lecture Notes in Math.*, 699.

8. Ito, K. *Lectures on Stochastic Processes*. Tata Institute of Fundamental Research, Bombay.

9. McKean, H. P., Jr. *Stochastic Integrals*. New York: Academic Press, 1969.

10. Varopoulos, N. T. The Poisson kernel on positively curved manifolds. To appear in *J. Func. Anal*.

11. Yau, S. T. On the heat kernel of a complete Riemannian manifold. *J. Math. Pure et Appl.* 57 (1978): 191–201.

12. Yau, S. T. Some function-theoretic properties of complete Riemannian manifolds and their applications to geometry. *Ind. Univ. Math. J.* 25 (1976): 659–670.

13. Yosida, K. *Functional Analysis*. New York: Springer-Verlag, 1978.

T - #0013 - 071024 - C0 - 234/156/27 [29] - CB - 9780534980412 - Gloss Lamination